Nanostructured
Thin Films and Coatings

Handbook of
Nanostructured Thin Films and Coatings

Nanostructured Thin Films and Coatings

Functional Properties

Edited by
Sam Zhang

CRC Press is an imprint of the
Taylor & Francis Group, an **informa** business

CRC Press
Taylor & Francis Group
6000 Broken Sound Parkway NW, Suite 300
Boca Raton, FL 33487-2742

First issued in paperback 2017

ISBN 13: 978-1-138-11475-3 (pbk)
ISBN 13: 978-1-4200-9395-7 (hbk)

Visit the Taylor & Francis Web site at
http://www.taylorandfrancis.com

and the CRC Press Web site at
http://www.crcpress.com

Contents

Preface

The twenty-first century is said to be the century of nanotechnologies. In a way, it is. The development of science and technology has come to a stage where "microscopic" is no longer enough to properly describe or depict a scientific phenomenon or a technological process. With the advance of nanoscience and nanotechnology, the world technological landscape changes not only affect the way scientists do research, technologists carry out development, and engineers manufacture products, but also the way ordinary people go about their daily life, through, for instance, nanomedicine, cell phones, controlled drug delivery, no-pain operations, solar cell–powered gadgets, etc. Thin films and coatings play a very important and indispensable role in all of these. This three-volume book set aims to capture the development in the films and coatings area in relation to nanoscience and nanotechnology so as to provide a timely handbook series for researchers to refer to and for newcomers to learn from, and thus contribute to the advancement of the technology.

The three-volume book set, *Handbook of Nanostructured Thin Films and Coatings*, has 25 chapters where 11 chapters in volume 1 concentrate on the mechanical properties (hardness, toughness, adhesion, etc.) of thin films and coatings, including processing, properties, and performance, as well as a detailed analysis of theories and size effect, etc., as listed here: Chapter 1, The Fundamentals of Hard and Superhard Nanocomposites and Heterostructures; Chapter 2, Determination of Hardness and Modulus of Thin Films; Chapter 3, Fracture Toughness and Interfacial Adhesion Strength of Thin Films: Indentation and Scratch Experiments and Analysis; Chapter 4, Toughness and Toughening of Hard Nanocomposite Coatings; Chapter 5, Processing and Mechanical Properties of Hybrid Sol-Gel- Derived Nanocomposite Coatings; Chapter 6, Using Nanomechanics to Optimize Coatings for Cutting Tools; Chapter 7, Electrolytic Deposition of Nanocomposite Coatings: Processing, Properties, and Applications; Chapter 8, Diamond Coatings: The Industrial Perspective; Chapter 9, Amorphous Carbon Coatings; Chapter 10, Transition Metal Nitride–Based Nanolayered Multilayer Coatings and Nanocomposite Coatings as Novel Superhard Materials; and Chapter 11, Plasma Polymer Films: From Nanoscale Synthesis to Macroscale Functionality.

Volume 2 contains eight chapters focusing on functional properties, i.e., optical, electronic, and electrical properties, and the related devices and applications: Chapter 1, Large-Scale Fabrication of Functional Thin Films with Nanoarchitecture via Chemical Routes; Chapter 2, Fabrication and Characterization of SiC Nanostructured/Nanocomposite Films; Chapter 3, Low-Dimensional Nanocomposite Fabrication and its Applications; Chapter 4, Optical and Optoelectronic Properties of Silicon Nanocrystals Embedded in SiO_2 Matrix; Chapter 5, Electrical Properties of Silicon Nanocrystals Embedded in Amorphous SiO_2 Films; Chapter 6, Properties and Applications of Sol-Gel-Derived Nanostructured Thin Films: Optical Aspects; Chapter 7, Controllably Micro/Nanostructured Films and Devices; and Chapter 8, Thin Film Shape Memory Alloy for Microsystem Applications.

Volume 3 focuses on organic nanostructured thin-film devices and coatings for clean energy with six chapters discussing the processing and properties of organic thin films, devices, and coatings for clean energy applications: Chapter 1, Thin Film Solar Cells Based on the Use of Polycrystalline Thin Film Materials; Chapter 2, Anodized Titania Nanotube Array and its Application in Dye-Sensitized Solar Cells; Chapter 3, Progress and Challenges of Photovoltaic Applications of Silicon Nanocrystalline Materials; Chapter 4, Semiconductive Nanocomposite Films for Clean Environment; Chapter 5, Thin Coating Technologies and Applications in High-Temperature Solid Oxide Fuel Cells; and Chapter 6, Nanoscale Organic Molecular Thin Films for Information Memory Applications.

A striking feature of these books is that both novice and experts have been considered while they were written: the chapters are written in such a way that for newcomers in the relevant field, the handbooks would serve as an introduction and a stepping stone to enter the field with least confusion, while for the experts, the handbooks would provide up-to-date information through the figures, tables, and images that could assist their research. I sincerely hope this aim is achieved.

The chapter authors come from all over the globe: Belgium, China, the Czech Republic, Egypt, Germany, India, Korea, Singapore, Taiwan, the Netherlands, the United Kingdom, and the United States. Being top researchers at the forefront of their relevant research fields, naturally, all the contributors are very busy. As editor, I am very grateful that they all made special efforts to ensure timely response and progress of their respective chapters. I am extremely indebted to many people who accepted my request and acted as reviewers for all the chapters—as the nature of the writing is to cater to both novice and experts, the chapters are inevitably lengthy. To ensure the highest quality of the chapters, more than 50 reviewers (at least two per chapter) painstakingly went through all the chapters and came out with sincere and frank criticism and suggestions that helped make the chapters complete. Though I am not able to list all the names, I would like to take this opportunity to say a big thank you to all of them. Last but not least, I would like to convey my gratitude to many CRC Press staff, especially Allison Shatkin and Jennifer Ahringer at Taylor & Francis Group, for their invaluable assistance rendered to me throughout the entire endeavor that made the smooth publication of the handbook set a reality.

Sam Zhang
Singapore

Editor

Sam Zhang Shanyong, better known as Sam Zhang, received his BEng in materials in 1982 from Northeastern University (Shenyang, China), his MEng in materials in 1984 from the Central Iron and Steel Research Institute (Beijing, China), and his PhD in ceramics in 1991 from the University of Wisconsin-Madison (Madison, Wisconsin). Since 2006, he has been a full professor at the School of Mechanical and Aerospace Engineering, Nanyang Technological University (Singapore).

Professor Zhang serves as editor in chief for *Nanoscience and Nanotechnology Letters* (United States) and as principal editor for the *Journal of Materials Research* (United States), among other editorial commitments for international journals. He has been involved in the fields of processing and characterization of thin films and coatings for the past 20 years, his interests ranging from hard coatings to biological coatings and from electronic thin films to energy films and coatings. He has authored/coauthored more than 200 peer-reviewed international journal articles, 14 book chapters, and guest-edited 9 journal volumes in *Surface and Coatings Technology* and *Thin Solid Films*. Including this handbook, he has authored and/or edited 6 books so far: *CRC Handbook of Nanocomposite Films and Coatings*: Vol. 1, *Nanocomposite Films and Coatings*: *Mechanical Properties*; Vol. 2, *Nanocomposite Films and Coatings*: *Functional Properties*; Vol. 3, *Organic Nanostructured Film Devices and Coatings for Clean Energy*, and *Materials Characterization Techniques* (Sam Zhang, Lin Li, Ashok Kumar, published by CRC Press/Taylor & Francis Group, 2008); *Nanocomposite Films and Coatings—Processing, Properties and Performance* (edited by Sam Zhang and Nasar Ali, Published by Imperial College Press, U.K., 2007), and *CRC Handbook of Biological and Biomedical Coatings* (scheduled for a 2010 publication by CRC Press/Taylor & Francis Group).

Professor Zhang is a fellow at the Institute of Materials, Minerals and Mining (U.K.), an honorary professor at the Institute of Solid State Physics, Chinese Academy of Sciences, and a guest professor at Zhejiang University and at Harbin Institute of Technology. He was featured in the first edition of *Who's Who in Engineering Singapore* (2007), and featured in the 26th and 27th editions of *Who's Who in the World* (2009 and 2010). Since 1998, he has been frequently invited to present plenary keynote lectures at international conferences including in Japan, the United States, France, Spain, Germany, China, Portugal, New Zealand, and Russia. He is also frequently invited by industries and universities to conduct short courses and workshops in Singapore, Malaysia, Portugal, the United States, and China.

Professor Zhang has been actively involved in organizing international conferences: 10 conferences as chairman, 12 conferences as member of the organizing committee, and 6 conferences as member of the scientific committee. The Thin Films conference series (The International Conference on Technological Advances of Thin Films & Surface Coatings), initiated and, since, chaired by Professor Zhang, has grown from 70 members in 2002 at the time of its inauguration to 800 in 2008. It has now become a biannual feature at Singapore.

Professor Zhang served as a consultant to a city government in China and to industrial organizations in China and Singapore. He also served in numerous research evaluation/advisory panels in Singapore, Israel, Estonia, China, Brunei, and Japan. Details of Professor Zhang's research and publications are easily accessible at his personal Web site: http://www.ntu.edu.sg/home/msyzhang.

Contributors

Weiping Cai
Key Laboratory of Materials Physics
Institute of Solid State Physics
Chinese Academy of Sciences
Hefei, China

G.P. Carman
Department of Mechanical and Aerospace Engineering
University of California
Los Angeles, California

T.P. Chen
Division of Microelectronics
School of Electrical and Electronic Engineering
Nanyang Technological University
Singapore, Singapore

Pierre Chevalier
Dow Corning Europe S.A.
Surface and Interface Solutions Center
Seneffe, Belgium

Chen-Kuei Chung
Department of Mechanical Engineering
Center for Micro/Nano Science and Technology
National Cheng Kung University
Tainan, Taiwan

L. Ding
Division of Microelectronics
School of Electrical and Electronic Engineering
Nanyang Technological University
Singapore, Singapore

Guotao Duan
Key Laboratory of Materials Physics
Institute of Solid State Physics
Chinese Academy of Sciences
Hefei, China

Lichao Jia
Key Laboratory of Materials Physics
Institute of Solid State Physics
Chinese Academy of Sciences
Hefei, China

Yue Li
Key Laboratory of Materials Physics
Institute of Solid State Physics
Chinese Academy of Sciences
Hefei, China

Fei Liu
State Key Laboratory of Fine Chemicals
Department of Materials Science and Chemical Engineering
School of Chemical Engineering
Dalian University of Technology
Dalian, China

Jun Liu
State Key Laboratory of Fine Chemicals
Department of Materials Science and Chemical Engineering
School of Chemical Engineering
Dalian University of Technology
Dalian, China

Zhen Liu
School of Electrical and Electronic Engineering
Nanyang Technological University
Singapore, Singapore

Ana C. Marques
Dow Corning Europe S.A.
Surface and Interface Solutions Center
Seneffe, Belgium

K.P. Mohanchandra
Department of Mechanical and Aerospace Engineering
University of California
Los Angeles, California

Georgia C. Papaefthymiou
Department of Physics
Villanova University
Villanova, Pennsylvania

and

Institute of Materials Science
National Center of Scientific Research "Demokritos"
Athens, Greece

Bo-Hsiung Wu
Department of Mechanical Engineering
Center for Micro/Nano Science and Technology
National Cheng Kung University
Tainan, Taiwan

Junshu Wu
State Key Laboratory of Fine Chemicals
Department of Materials Science and Chemical Engineering
School of Chemical Engineering
Dalian University of Technology
Dalian, China

Dongfeng Xue
State Key Laboratory of Fine Chemicals
Department of Materials Science and Chemical Engineering
School of Chemical Engineering
Dalian University of Technology
Dalian, China

Dong Kee Yi
Division of BioNanotechnology
College of BioNanotechnology
Gachon Bionano Research Institute
Kyungwon University
Sung Nam City, Republic of Korea

Sam Zhang
School of Mechanical and Aerospace Engineering
Nanyang Technological University
Singapore, Singapore

Wali Zhang
School of Mechanical and Aerospace Engineering
Nanyang Technological University
Singapore, Singapore

1 Large-Scale Fabrication of Functional Thin Films with Nanoarchitecture via Chemical Routes

Fei Liu, Junshu Wu, Jun Liu, and Dongfeng Xue

CONTENTS

1.1 INTRODUCTION

1.1.1 General Instructions

Thin films are thin material layers with thickness ranging from a nanometer to several micrometers. Electronic semiconductor devices and optical coatings are the main applications benefiting from thin film construction [1–3]; ferromagnetic thin films are recently used for high-density data storage [4,5]; it is also being applied to biology and iatrology, via thin film drug delivery [6] and biocompatible coating [7]; ceramic thin films are in wide use, the relatively high hardness and inertness of ceramic materials make this type of thin coating of interest for protection of substrate materials from corrosion, oxidation, and wear [8–10].

Nanoscience and nanotechnology involve studying and working with matter on an ultrasmall scale, which have reached within the last decade the status of a leading science with fundamental and applied research prospects, stretching across the whole spectrum of science, touching medicine [11], physics [12–14], engineering [15–17], and chemistry [18–21]. An important feature of nanoscience and nanotechnology is their ability to bridge the crucial dimensional gap between the atomic and molecular scale of fundamental sciences and the microstructural scale of engineering and manufacturing. Whereas bulk inorganic crystals possess their innate characteristics and properties, such as color and melting point, at the nanometer scale, these properties can be tailored as desired. For example, it is found that the wavelength (color) of strong fluorescent light emitted by semiconductor quantum dots under ultraviolet (UV) light illumination depends sensitively on their size [22,23]. Although there are enough atoms in these particles to effectively validate the concepts of solid-state physics, which include electron bands, forbidden energy bandgaps, and electron- and hole-effective masses, still, these particles are small enough to be called an "artificial atom," characterized by discrete sharp electron energy states, and discrete sharp absorption and emission wavelengths for photons. Novel phenomena like this that appear in nanoscale materials prefigure out their totally different applications in a vast number of areas.

Functional thin films with nanoarchitecture have attracted much attention in recent years because they benefit from both thin film constructions and nanounits functionalities [24–29]. A complex nanostructure to be useful may need multiple connections to its outer environment. An example is the array of wires attached to the edges of a computer chip in a computer [30,31]. If the premise is that the engineered nanounit is to perform a function useful to the human-scale world, a large number of connections scaling up from the nanoscale to the centimeter scale may be needed. Arranging nanoscale building block into functional films or the fabrication of thin films with nanoarchitecture is an excellent scheme to fulfill the requirement of connecting functional nanounits to the outside world. And it should not be a surprise that these nanounits working together may show unique properties that individual units do not have. Good examples are superhydrophilic and superhydrophobic films constructed by nano-components [32,33]. These bio-inspired nanostructured thin films have been fabricated with different materials, including polymers and carbon nanotubes [33]; while individual components do not possess superhydrophilic or superhydrophobic ability, the collective effect makes these thin films with significant application values.

The ability to produce nanostructured thin films with a higher level of design and complexity, combining for instance hollow and ordered nano- and microstructures with controlled and tunable porosity is yet an important challenge to address to promote their performance and develop new applications in devices, such as dye-sensitized solar cells [34], Li-ion batteries [35,36], electrochemical supercapacitors [37,38], hydrogen storage devices [39,40], as well as chemical, gas, and biosensors [41–44]. The basic goals have been to develop synthesis techniques to fabricate thin films with controllable nanoarchitecture. Although, the control of the size has theoretically and experimentally been achieved [45,46], the strict control of the particle size over the orders of magnitude has not yet been clearly demonstrated or quantitatively established. The control of the shape of nanocrystallites also represents an essential task to fulfill. The control of the orientation of anisotropic units on a substrate is another important challenge. Several classes of materials, including insulators, semiconductors, metallic, and magnetic oxides have been synthesized with nanostructured films using photolithography, replication, and template techniques [47–50]. However, such techniques showed strong limitations when upscaling in practical applications. Indeed, to unfold the full potential of applications and the prospects of nanotechnology, the most serious challenge scientists and engineers have to face is the development of sustainable large-scale manufacturing techniques for the time and cost-effective productions of functional films with applicable size and high quality. Although the material development over the last decade has been outstanding due to a plethora of novel and ingenious synthetic techniques for the fabrication of smart and functionalized materials, no large-scale techniques are currently available to meet the economic, environmental, and technological requirements of the mass production of nanostructured thin films.

Chemical processes are based on the manipulation of atoms and molecules, and indeed have a very long history in the synthesis of materials. The demand of a synthesis method for nanostructured thin films not only revitalizes the use of many traditional chemical methods, but also motivates many new ones to be continually developed. The key point of chemical strategies for the fabrication of nanostructures lies on the flexible use of various chemical reactions. Chemical reactions for material synthesis can be carried out in solid [51–53], liquid [54–77], gaseous [78–82], and heterogeneous states [83]. In basic chemistry terms, the starting materials of a chemical reaction are called the reactants, and the products are the material to which the reactants are converted. The reactants can be solids, liquids, or gases in any combination, in the form of a single element or multicomponent compounds. The chemical synthesis approach is an especially powerful tool for the convenient and reproducible shape-controlled synthesis of functional thin films with nanoarchitecture, because this method allows the resulting products to be precisely tuned in terms of their size, shape, and composition at the nanometer scale.

With the intention to take part in the fascinating nanoworld, and to contribute to its ambitious challenges, this chapter will give a background on the general thermodynamic concept of nucleation, growth, and aging control for nanocrystals and focus on chemical thin film processing methods, which are dedicated to the rational low-cost fabrication of nanostructured thin films and multidimensional arrays. The focus of this chapter is on inorganic materials, which represents the most common, the most diverse, and probably the richest class of materials in terms of optical, optoelectronic, magnetic, electrical, thermal, electrochemical, photoelectrochemical, mechanical, and catalytic properties [84–87]. The combinations of such a variety of properties and applications with the unique effects of nanomaterials of low dimensionality make the studies of novel inorganic nanostructures a very important issue of research and development from both the fundamental and the industrial standpoints. For instance, developing the ability and competence to hierarchically order, connect, and integrate 0D quantum dots and 1D nanowires and nanotubes, in thin films with functional networks, as well as multidimensional arrays is of significant importance to generate and operate practical nanodevices.

Such future devices will certainly revolutionize materials science and engineering due to their unique design, architecture, and physical properties, as well as their capability to connect the nanoworld to the outside world.

1.1.2 Basic Concepts of Nanoarchitecture Control by Chemical Routes

The concepts mentioned here are aimed to develop a strategy to reach the rational design of nanomaterials with the desired and controlled particle size, morphology, orientation, and crystal structure for harvesting their full potential applications.

1.1.2.1 Nucleation and Particle Growth

Two processes are inevitably involved in the generation of solids, the formation of stable nuclei, and the subsequent particle growth (Figure 1.1), if carried out in the solution phase, this series of events can be described as "precipitation" [88]. Nucleation plays an important role whenever condensation (precipitation) occurs. The solubility of a solute is constant for a particular solvent at a certain temperature, and the existence of any excess solute will result in precipitation. Therefore, for material fabrications, in order to achieve precipitation, the solution must be supersaturated either by directly concentrating the solution or by adding necessary reactants to produce a supersaturation during the reaction. The supersaturation makes the whole system unstable in a thermodynamic viewpoint. Transformation from the solution phase to solid does happen when the supersaturation passes over a threshold, since free energy of the solid phase is lower than that of the solution phase. After product nucleation, the thermodynamic equilibrium state of the system is restored.

Generally, two kinds of nucleation are involved in the precipitation process: homogeneous nucleation and heterogeneous nucleation [88]. Homogeneous nucleation takes place in the condition that no foreign species nucleating aids are involved. Heterogeneous nucleation involves foreign substances as surfaces on which the nucleation occurs. Nucleation generally occurs with much more difficulties by the process of homogeneous nucleation. The creation of nuclei is an energy-consuming process, because it brings a new phase, and generates an interface at the boundary of new phase.

For the fabrication of thin films, heterogeneous nucleation processes are mostly involved, crystal seeds are precoated to the substrate surface, or nucleation directly happens at the interface of the substrate and the saturated solution [89]. Heterogeneous nucleation can take place at significantly lower supersaturation as homogeneous nucleation, because some energy is released by the partial destruction of the previous interface [89]. For example, when a gas bubble forms on the interface of the water and the bottle, the energy inherent in the water–bottle interface is released wherever a layer of gas intervenes, and this energy goes toward the formation of the bubble–water and the bubble–bottle interfaces. A similar effect can cause particles to precipitate

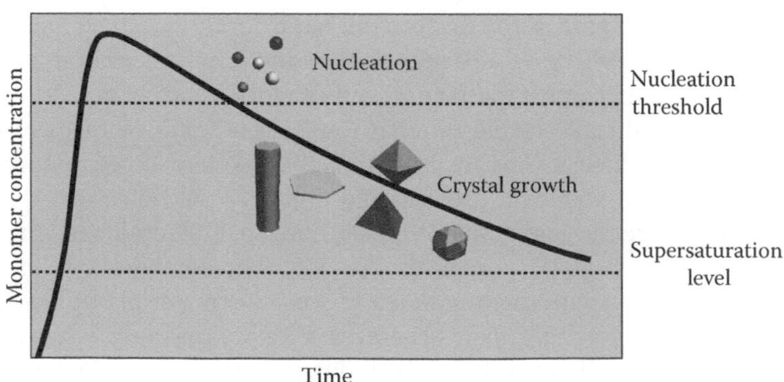

FIGURE 1.1 Crystal nucleation and growth diagram.

at the interface of the saturated solution and the existing solid. This can affect the precipitation strengthening, which relies on homogeneous nucleation to produce a uniform distribution of precipitate particles.

After a short-time nucleation burst, the solute concentration drops below the critical level, nucleation stops and the particles continue to grow by molecular addition until the equilibrium concentration of the product is reached, as illustrated in Figure 1.1. Nanounits with a relatively narrow size distribution are usually required for practical applications, so an insight into the particle growth and size control process is extremely important from both a scientific and a technological point of view.

1.1.2.2 Size Control

Crystal size control is undoubtedly with great importance because it directly influences the thickness and texture of nanostructured thin films. The main idea of size control is to tune the thermodynamics and the kinetics of nucleation, growth, and aging of a system by controlling its interfacial energy experimentally. (The solid–liquid interfacial energy is defined thermodynamically as the energy required to create a new soli–liquid interface. It can also be defined as the difference of the free energy between the interface and the bulk of the solid.)

Figure 1.2 illustrates the dependence of growth rate on the nanocrystal radius. Crystals having radii smaller than the critical size are unstable owing to their large fraction of active surface atoms, leading to a high interfacial energy, therefore possessing negative growth rate [88]. In contrast, larger crystals with smaller surface-to-volume ratios are stable in energy, possessing positive growth rate. Crystals just with the critical size would neither grow nor shrink. The critical size depends on the monomer concentration; hence, the low monomer concentration favors a larger critical size. Monodispersed particles can be obtained if the monomer concentration is sufficiently high, by which small crystals grow more rapidly than larger ones. On the other hand, when the monomer concentration is depleted owing to crystal growth, the critical size becomes larger, and the opposite phenomenon occurs, where the larger particles continue to grow, and the smaller ones get smaller and finally dissolve [46].

The interfacial energy of the system can also be altered by the chemical composition and the charge density of the interface [89–91]. For example, a high ionic strength will contribute to an increase of the surface charge density. In such case, the surface charge may therefore reach its maximum charge density, which ultimately depends on the composition of the interface. At maximum charge density, the interfacial energy of the system will reach a minimum. Adsorption phenomenon at the interface is another route to decrease the interfacial energy. Thermodynamic colloidal

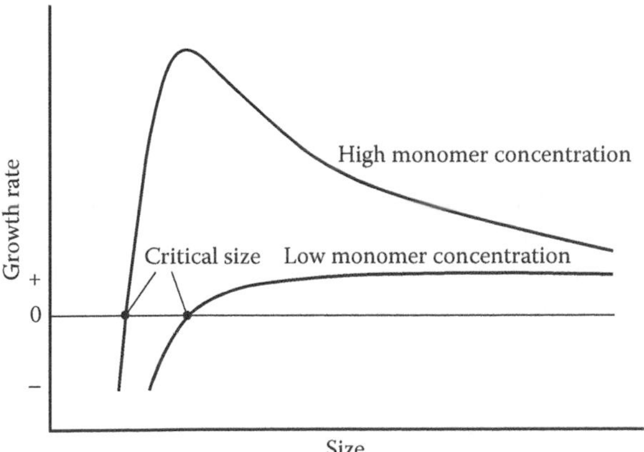

FIGURE 1.2 Relation of growth rate and nanocrystal radius.

stability may thus be reached resulting in a considerable lowering of the secondary aging processes, henceforth preventing the nanoparticle size (and distribution size) to increase as well as avoiding crystal phase (and morphological) transformations.

1.1.2.3 Shape Control

When the thermodynamic stabilization is achieved, not only the size is tailored but also the shape as well as the crystallographic structure may be controlled. The particle growth in solution is interface controlled when the particle is small, so nanoparticles develop a spherical morphology to minimize their surface energy because sphere represents the smallest surface-to-volume ratio, thus obtaining the lowest interfacial tension. After reaching a critical size, the particle growth becomes diffusion controlled, the shape of nanoparticles does not necessarily require being spherical; kinetic factors such as the reaction and transport rates of reactants, colloidal stability, recrystallization, and the aging can influence the particle growth. These factors compete with the symmetry of the crystal structure (ideal crystal shape), leading to the final morphology.

Since the kinetic factors can be affected by tuning the growth conditions such as the concentration of reactants, temperature, and pH value, for understanding the shape development process and achieving the desired architectures, a theoretical prediction of the ideal crystal shape under thermodynamic equilibrium conditions is extraordinarily important. Xue and coworkers [92–97] have proven that the chemical bonding theory can be successfully used to calculate the normal growth rate of every selected plane of single crystals, thus scale the area ratio and the emergence possibility of every selected plane, which provides a convenient and reliable method to obtain the ideal shape. Experimental evidence from the growth of potassium dihydrogen phosphate (KDP), ammonium dihydrogen phosphate (ADP), and magnesium hydroxide sulfate hydrate (512MHSH) have verified the feasibility and effectiveness of this method [92–98]. Based on the theoretical prediction of crystal shape under thermodynamic equilibrium conditions [99], suitable routes can be designed to fabricate desired architectures [100].

The chemical bonding theory can be described as the following formula:

$$R = K\frac{N \cdot S}{D \cdot A} \tag{1.1}$$

where

R is the vertical growth rate of the selected plane
N is the bond number
S is the bond strength
D is the interplanar distance
A is the area of selected plane
K is a constant for different planes in an equilibrium system

According to Equation 1.1, the vertical growth rate of selected planes of any single crystal can be quantitatively calculated. Combining the relative growth rates of each face with its inherent symmetry leads to the prediction of the overall crystal shape of a single crystal grown in ideal conditions. Table 1.1 lists the ideal morphologies of some selected inorganic crystals predicted by this method.

1.1.2.4 Crystal Structure Control

The same material with different crystal structures may possess totally different properties; so in addition to particle size and shape control, crystal structure control is also an important issue for the fabrication of functional materials. Crystal phase transitions in solution are generally conducted

TABLE 1.1
Predicted Growth Morphologies of Some Crystals by Chemical Bonding Theory

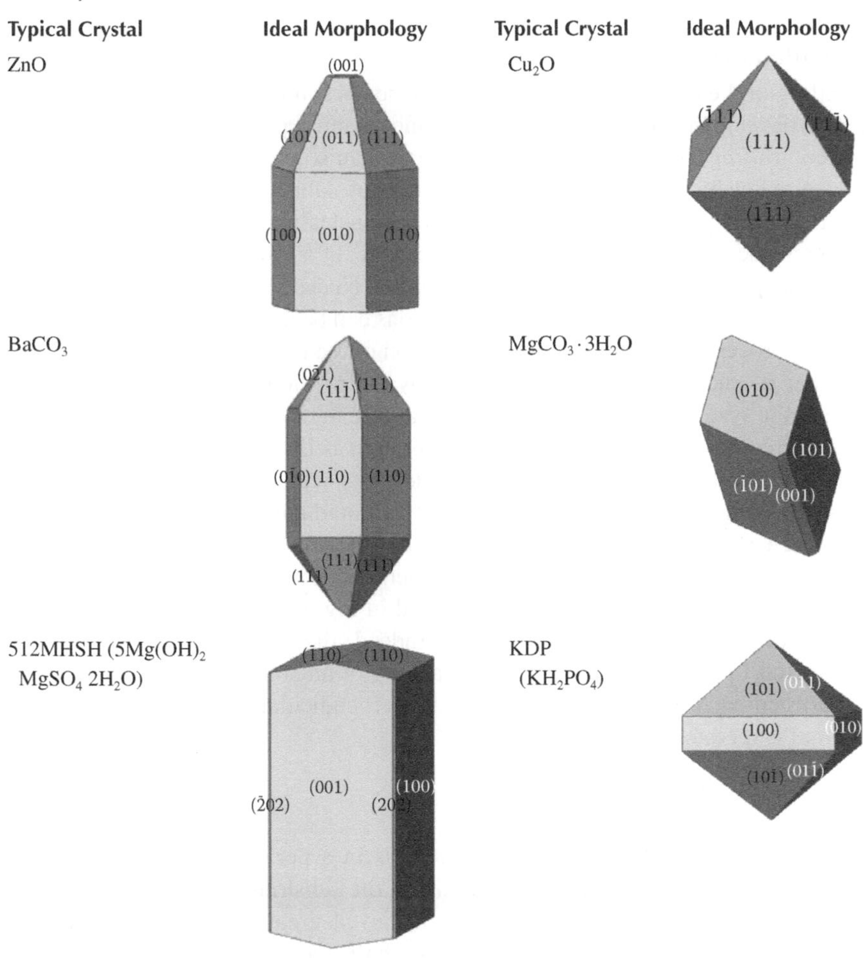

Typical Crystal	Ideal Morphology	Typical Crystal	Ideal Morphology
ZnO		Cu_2O	
$BaCO_3$		$MgCO_3 \cdot 3H_2O$	
512MHSH ($5Mg(OH)_2$ $MgSO_4$ $2H_2O$)		KDP (KH_2PO_4)	

by a dissolution–recrystallization process, to meet the surface energy minimization requirement of the system [89]. Indeed, when a solid has several allotropic phases and polymorphs, the one that precipitates first typically possess the highest solubility, and consequently the lowest stability (i.e., the crystallographic metastable phase). This is understood by considering the nucleation kinetics in solution. At a given supersaturation ratio, lower interfacial tension leads to the smaller nuclei size (higher nucleation rate). Thus, considering that the solubility is inversely proportional to the interfacial tension, the precipitation of the most soluble phase is therefore kinetically favored. The precipitation of nanoparticles at low interfacial tension allows the stabilization of some metastable crystal structures, but in most cases, this metastable phase is more sensitive to secondary growth and aging, which leads to thermodynamically more stable phases.

Practically, reaction and aging conditions can be chosen depending on the crystal phase wanted. For fabricating metastable phases, careful consideration of the precipitation conditions is needed, when nanosystems are precipitated at low interfacial tension, phase transition to more stable polymorphs may be avoided. The case is much easier if a thermodynamically more stable phase is desired;

elevating the system temperature or prolonging the reaction time mostly leads to phase transitions. This case is typically for aluminum hydroxide (boehmite→bayerite or gibbsite), titanium dioxide (anatase→rutile) [101], and more attractively niobium oxide (pseudo-hexagonal→monoclinic) [102], which can be used to drive voids generation and will be discussed in detail in Section 1.3.3.1.

1.1.2.5 Orientation Control

An understanding of the differences between homogeneous and heterogeneous nucleation phenomena is needed to develop the capability of growing nano- to microparticulate thin films from aqueous solution, and controlling the orientation of large arrays of anisotropic nanoparticles onto a substrate. As discussed previously, the homogeneous nucleation of solid phases from solutions requires a higher activation energy barrier and, therefore, heteronucleation is energetically more favorable if foreign species nucleating aids are available. Therefore, at a lower saturation ratio, nucleation is more likely to take place onto a substrate than in solution. Nuclei will grow by heteronucleation onto the substrate, and various morphologies and orientations can be obtained by experimental control of the seed-coating process and the chemical composition of the precipitation medium [89].

If the number of nuclei is exceedingly limited by the precipitation conditions or by the seed-coating process, the system will promote multiangular orientation growth of anisotropic units along their easy axis, irregular oriented array may be generated, as illustrated in Figure 1.3a.

If the number of nuclei is moderate, anisotropic units with a perpendicular orientation onto substrates may be obtained. The slow appearance of a limited number of nuclei will allow the slow growth along the easy direction of crystallization. As a result, a condensed phase of anisotropic nanorods parallel to the substrate normal will be generated as illustrated in Figure 1.3b.

If the nucleation site is overabundant as illustrated in Figure 1.3c, anisotropic units with a parallel orientation onto substrates can be obtained. Indeed, the fast appearance of a large number of nuclei will result in a rapid 2D growth. The stacking of anisotropic nanoparticles with random orientation between each other but with an overall perpendicular orientation with respect to the substrate normal is therefore promoted.

1.1.2.6 Patterning

The patterning of semiconductor and other thin films in a periodic fashion is of great interest for industrial applications. Patterning of nanounits on the substrate using lithographic techniques (top-down) is easy to be understood, and it has achieved great success in the field of silicon-based microtechnology. But it is still an important challenge using bottom-up strategies to pattern, assemble, and integrate nanounits as functional networks on various substrates. The term

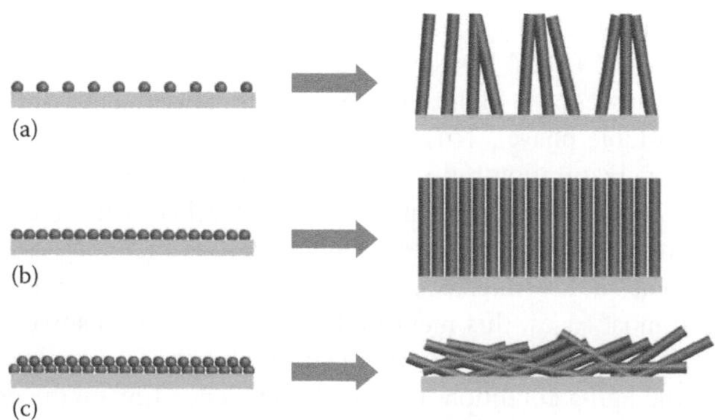

FIGURE 1.3 Illustration of crystal orientation control with (a) limited nuclei number; (b) moderate nuclei number; (c) overabundance nuclei number.

"nanopatterning" refers to approaches that provide periodic arrangements with feature sizes below 100 nm. The interest in the ongoing search for new nanopatterning techniques is based on the desire for inexpensive methods to precisely pattern areas on square-centimeter scale, and beyond. These methods should also be easier to realize than conventional lithography. Most of the attempts are still at the laboratory stage; however, the possibility of parallel sequenced writing based on such nanopatterning methods has already been predicted. The realization of patterned metal arrays as catalysts for the growth of semiconductor nanowires is a new application of the nanopatterning techniques [103].

The self-organization of submicrometer spheres into a monolayer with a hexagonal close-packed structure has been used as a mask to pattern the depositing site on the substrate, as shown in Figure 1.4. Typical spheres used are silica and polystyrene, which are commercially available with narrow size distributions. The deposition of a single layer of monodisperse spheres on a substrate can be used as a lithography shadow mask and selectively expose the substrate surface for deposition. After the spheres are dissolved, a periodical pattern of triangular metal dots is obtained. The size and distance of the metal dots are determined by the size of the spheres used. These periodical metal dots can be then used as a catalyst to produce nanowires via the vapor–liquid–solid (VLS) process, leading to a nanowire array with a defined pattern. Thin films obtained by this approach can reach an area of more than 1 cm^2. Furthermore, other patterns can also be realized if a multi-nanosphere layer or a hybrid layer in which spheres with different sizes are employed. This method offers a

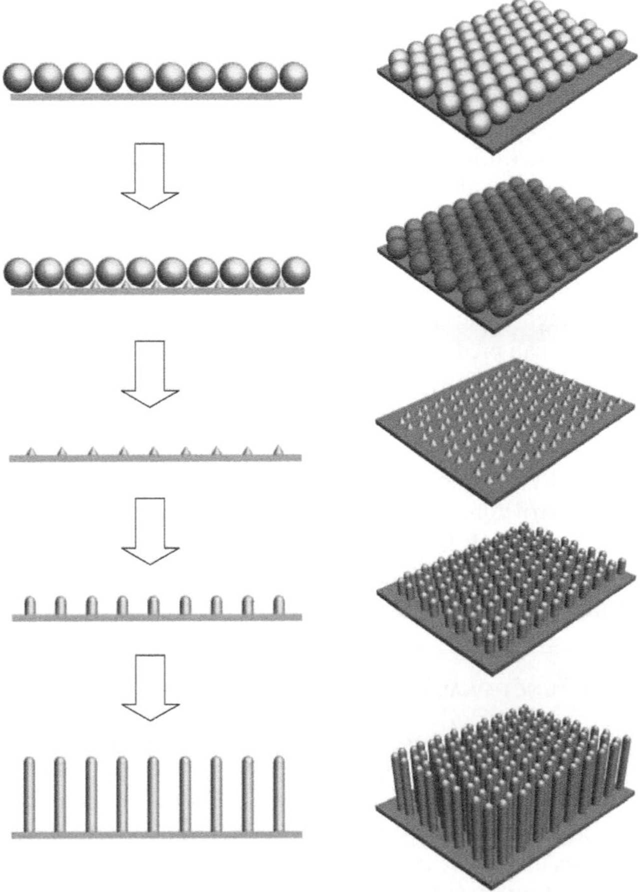

FIGURE 1.4 Position control of nanowires using nanospheres.

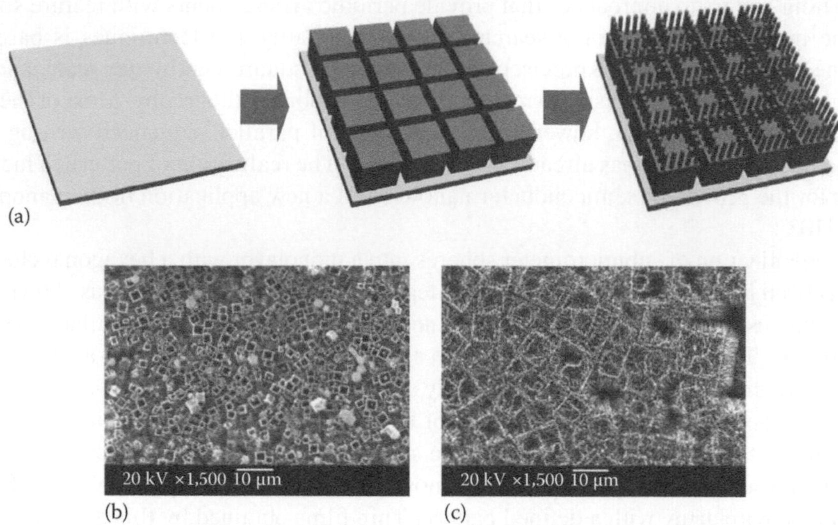

FIGURE 1.5 Using cubic crystals to pattern nanowires; (a) schematical illustration; and (b,c) SEM images of the products.

simple, cost-effective, and high-throughput patterning approach, and has been widely utilized for the fabrication of a wealth of nanostructure arrays.

The fabrication of patterned arrays is also possible via the solution method: a recent method developed in our lab showed that nanowires can be fabricated in a squared pattern on a substrate precoated with cubic crystals [104]. This process is schematically illustrated in Figure 1.5a, and scanning electron microscope (SEM) images in Figure 1.5b and c indicated that these nanowires are preferred to grow on the edge of cubic crystals, thus forming an array with a squared pattern. This phenomenon can be explained by the Berg effect, which has always been employed to explain the dendrite or skeletal crystal growth in nonequilibrium conditions [105]. According to Berg's examination, the density or supersaturation distribution in a solution is not uniform at the interface of a crystal: the density is high at the edges and smallest in the centers. By extending this theory to a heterogeneous system, we can predict that the density of nanowire nutrient is much higher on the edge than the centre area, and the growth rate of nanowire at an interface is proportional to the local supersaturation of the product, then nanowires on edges grow more than other regions, therefore, a well-aligned array can be maintained.

In summary, by rationally designed routes and carefully controlled chemical conditions, the size, the shape, the crystal structure, and the arrangement manner of nanounits of functional thin films may be tuned and optimized. It allows the ability to develop purpose-built nanostructured thin films and probe the influence of such parameters on the physical and chemical properties and performance in practical applications.

1.1.3 Applications of Functional Thin Films with Nanoarchitecture

Many other chapters in this handbook have been discussing the novel applications of nanostructured thin films in a variety of areas, including superhard coatings, sensing, microfluidics, and data storage. In this chapter, our intention is not to focus on just one special property of nanostructured thin films; only brief introductions are given to illustrate the importance of this kind of material, as well as some subject forefront examples are included.

An important discovery for the optic application of thin films constructed by nanowires is the room-temperature ultraviolet laser [106]. It was observed by Yang et al. that under optical excitation, the self-organized, ⟨0001⟩-oriented zinc oxide nanowires grown on sapphire substrates can emit laser with a wavelength of 385 nm, and emission linewidth less than 0.3 nm. The chemical flexibility and the one-dimensionality of the nanowires make this system ideal miniaturized laser light sources. These short-wavelength nanolasers were expected to have applications in optical computing, information storage, and microanalysis.

Another promising use of the same kind of ZnO nanowire array thin film is as "nanogenerator," which was reported by Wang and his colleagues [107]. They have presented a series of work on the fabrication and modification of nanowire nanogenerator based on the piezoelectric zinc oxide. With these nanowire arrays, nanoscale mechanical energy can be converted into electrical energy. In the very first prototype of a nanogenerator, the aligned nanowires are deflected with a conductive atomic force microscope tip in contact mode. The coupling of piezoelectric and semiconducting properties in zinc oxide creates a strain field and charge separation across the nanowire as a result of its bending. The rectifying characteristic of the Schottky barrier formed between the metal tip and the nanowire leads to electrical current generation. This device was then modified to be driven by an ultrasonic wave to produce continuous direct-current output [108]; zinc oxide nanowire arrays were placed beneath a zigzag metal electrode with a small gap. The wave drives the electrode up and down to bend and/or vibrate the nanowires. The most recent development of this scheme is using piezoelectric zinc oxide nanowires grown radially around textile fibers to convert low-frequency vibration/friction energy into electricity [109]. By entangling two fibers and brushing the nanowires rooted on them with respect to each other, mechanical energy is converted into electricity owing to a coupled piezoelectric-semiconductor process. These approaches present an adaptable, mobile, and cost-effective technology for harvesting energy from the environment, and they offer a potential solution for powering nanodevices and nanosystems.

Nature has provided a vast number of high-performance devices that essentially are composed of nanostructured materials. For example, the adhesive strategy of the gecko relies on foot pads composed of specialized keratinous foot-hairs called setae, which are subdivided into terminal spatulae of approximately 200 nm. Inspired by this, Lee and coresearchers reported a hybrid adhesive consisting of an array of nanofabricated polymer pillars coated with a thin layer of a synthetic polymer that mimics the wet adhesive proteins found in mussel holdfasts [110]. Wet adhesion of the nanostructured polymer pillar arrays increased nearly 15-fold when coated with mussel-mimetic polymer. This kind of a thin film maintains its adhesive performance for over a thousand contact cycles in both dry and wet environments. This hybrid adhesive, which combines the salient design elements of both gecko and mussel adhesives, should be useful for reversible attachment to a variety of surfaces in any environment.

1.2 CHEMICAL METHODS FOR FABRICATING THIN FILMS WITH NANOARCHITECTURE

Top-down and bottom-up approaches are the two approaches used for fabricating nanoscale materials and devices. The top-down approach often uses the traditional methods where externally controlled tools are employed to cut, mill, and shape materials into the desired structure. Bottom-up approaches, in contrast, use the chemical properties of single molecules or small particles to automatically arrange themselves into some functional conformations. Such bottom-up approaches should, broadly speaking, be able to produce devices in parallel and much cheaper than top-down methods.

1.2.1 Top-Down Process

There is a range of top-down processes that can be used to produce nanostructured films. The most significant advances of these are physical methods (different lithographic techniques [111]), and the combination of physical and chemical methods (stamp printing [112–115]). Since this chapter focuses on the chemical fabrication of nanostructures, no further introduction will be given herein; readers interested in this area can find more information from the reference [116].

1.2.2 Bottom-Up Processes

Bottom-up processes refer to the producing of matters atom by atom or molecule by molecule. Dominantly, synthesis using this approach relies heavily on chemical processes, either performed in liquid and/or gas phase. For thin film fabrication, the most common techniques are chemical vapor deposition (CVD) [117], sol-gel-coating process [118,119], and chemical precipitation [120,121]. Section 1.2.2.1 includes general introduction and illustration of these techniques.

1.2.2.1 Chemical Vapor Deposition

CVD is a method that can produce high-purity, high-performance solid materials [117]. This process is often used in the semiconductor industry to produce thin films. The typical CVD process is illustrated in Figure 1.6, where the wafer (substrate) is exposed to one or more volatile precursors, and the vaporized precursors adsorb onto a substance held at an elevated temperature. These adsorbed molecules either thermally decompose or react with other gases/vapors to form products. Frequently, volatile by-products are also produced, which are removed by gas flow through the reaction chamber. Nucleation in the gas phase is homogeneous, whereas nucleation on the substrate is heterogeneous.

Many recent successful syntheses of inorganic 1D nanostructure arrays are based on the metal-catalyst-assisted CVD process, which undergoes a VLS mechanism of anisotropic crystal growth. This mechanism was first proposed by Wagner and Ellis in the 1960s [122]. The VLS growth mechanism involves three stages, as shown in Figure 1.7. Initially, the source material from the gas phase is absorpted into a nanosized liquid droplet of the catalyst (usually a molten particle of gold) located on the substrate. After the liquid alloy is supersaturated, the source material condenses and generates a solid precipitate at the solid–liquid interface (the solid phase

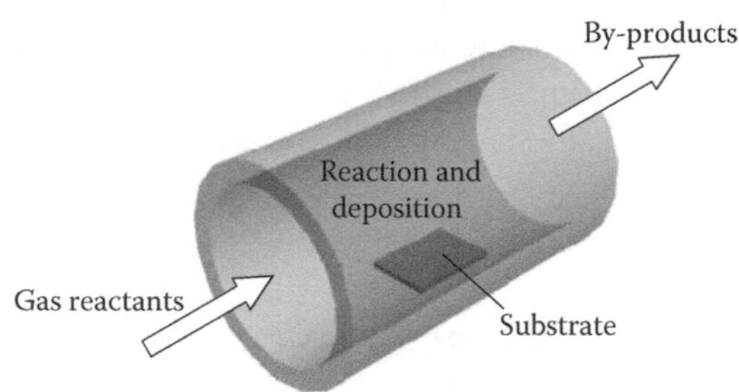

FIGURE 1.6 Illustration of a typical CVD reactor.

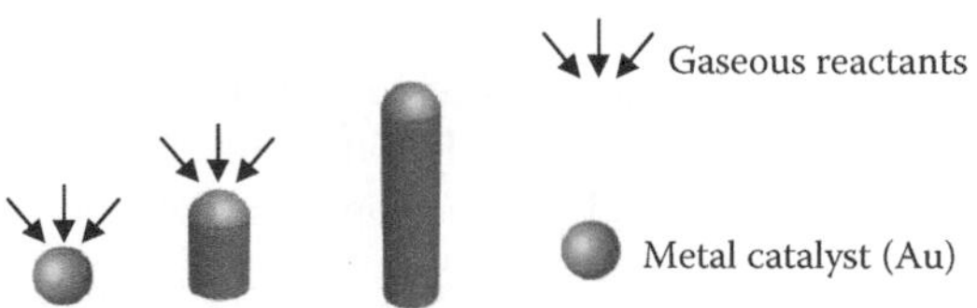

FIGURE 1.7 Schematic drawing of a VLS process to 1D nanostructures.

could be the substrate at the beginning or the generated 1D nanostructure later); The formed solid precipitate serves as a preferred site for further deposition, promoting the elongation of the product into a nanowire or a whisker. Since the liquid droplet catalyzes the incorporation of material from the gas source to the growing crystal, the deposit grows anisotropically as a whisker whose diameter is determined by the diameter of the liquid alloy droplet. In this sense, the diameter of the product can be precisely controlled with carefully prepared catalysts.

1.2.2.2 Sol-Gel Coating Process

The sol-gel process is a wet-chemical technique for the fabrication of film and powder. The starting material can be either a homogeneous chemical solution or a colloidal suspension (sol), condensed to produce an integrated network (gel) [56,60]. Soluble metal alkoxides and metal chlorides are typically employed as precursors, which undergo hydrolysis and polycondensation reactions to produce solid particles with size ranging from 1 nm to 1 μm dispersed in a solvent (sol). The precursor sol can be either deposited on a substrate to form a film or used to synthesize powders. This kind of a colloidal system then condenses to form a gel with an inorganic substance as the continuous phase while containing a discrete liquid phase. A subsequent drying process is usually performed to remove the liquid phase within the pores leading to a porous material named "xerogel." Inorganic xerogel is rarely used as obtained after a simple drying; a thermal treatment is often applied in order to facilitate further polycondensation and crystallization, thus enhancing the functional and mechanical properties [88].

The sol-gel approach has great advantage for thin film fabrication, because it is a cheap and low-temperature technique that allows fine control on the product composition, and doping elements can be easily introduced. An important issue for applying sol-gel technique to thin films is uniformly coating the substrates with precursor sol. Several deliberate and useful coating methods have been developed, among which spin-coating and dip-coating are most frequently used.

Spin coating is a procedure used to apply uniform thin films to flat substrates [123,124]. In short, an excess amount of a solution is placed on the substrate, which is then rotated at high speed in order to spread the fluid by centrifugal force. Figure 1.8 shows a typical machine used for spin coating, called a spin coater.

Rotation is continued while the fluid spins off the edges of the substrate, until the desired thickness of the film is achieved. The applied solvent is usually volatile, and simultaneously evaporates. So, the higher the angular speed of spinning, the thinner the film. The thickness of the film also depends on the concentration of the solution and the solvent. Thin films with thicknesses below 10 nm have been achieved with this technique.

Dip coating is another effective way of creating thin film-coated materials along with the spin-coating procedure [125]. In a dip-coating process (Figure 1.9), a substrate is dipped into a solution containing coating material, and then is withdrawn from the solution at a controlled speed, allowing

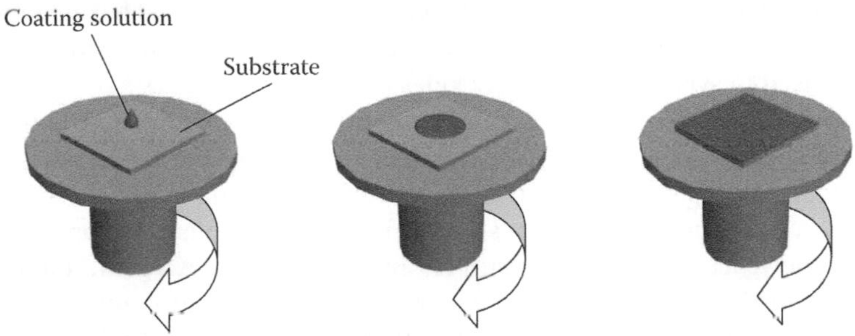

FIGURE 1.8 Schematic illustration of the spin-coating process.

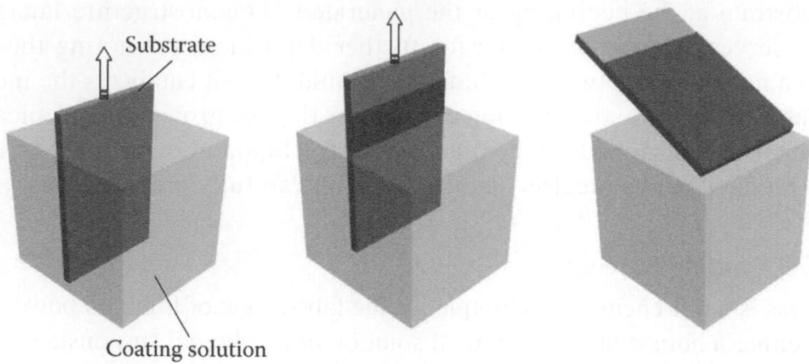

FIGURE 1.9 Schematic illustration of the dip-coating process.

it to drain. Coating thickness generally increases with faster withdrawal speed. The thickness is determined by the balance of forces at the stagnation point on the liquid surface. A faster withdrawal speed pulls more fluid up onto the surface of the substrate before it has time to flow back down into the solution. The thickness is primarily affected by fluid viscosity, fluid density, and surface tension.

While excellent for producing high-quality, uniform films, both spin-coating and dip-coating processes require precise control and a clean environment. The obtained coating film may remain wet for several minutes until the solvent evaporates completely. This process can be accelerated by heated drying. In addition, the coating may be cured by a variety of means including conventional thermal, UV, or IR techniques depending on the chemical composition of the coating solution. Once a layer is cured, another coating/curing process can be applied to cover a new layer on top of the existing one. In this way, a multilayer thin film can be constructed.

1.2.2.3 Chemical Precipitation

Chemical precipitation is a novel and convenient bottom-up approach, and this technique has been developed to produce nanostructured thin films and coatings of functional materials onto various substrates at a low cost and mild temperatures [88,89]. Such a method allows generating advanced nano-, meso-, and microparticulate thin films as well as 2D and 3D arrays. Given that the crystallites are nucleating and growing directly from the substrate at mild temperatures, no specific requirements in substrate activation, thermal stability, or crystallinity is needed, which greatly extends the range of substrates (e.g., amorphous, polymer, etc.) that can be used. Such diversity provides a higher degree of nanomaterials engineering and design.

Chemical precipitation occurs when an insoluble product is formed as a consequence of a chemical reaction [126]. The process involves the mixing of the precipitating agent with metal ions (dissolving salts) in a liquid medium, hydrolysis-condensation of hydrated metal ions and complexes (inorganic polymerization), and their heteronucleation onto substrates. The precipitation can also be started with a homogeneous mixture by using chemicals that can be decomposed to generate precipitating agents *in situ* throughout the liquid phase, for example, urea and hexamethylenetetramine [127]. Co-precipitation can be carried out by using salt solutions containing mixed metal ions; this method is widely used to synthesize complex oxides [128].

Such a process does not require expensive and complicated apparatus. The whole process is entirely recyclable, safe, and environment friendly. In addition, the kinetics of nucleation and particle growth can be adjusted by the controlled release of anions and cations. Careful control of the kinetics of precipitation can result in high-quality films. The precipitation kinetics is influenced by the reactant concentration, pH, and temperature. Organic molecules can also be employed to control the release of reagents and ions in the solution during the precipitation process. By engineering these factors, nanostructured films with desired architectures can be produced.

1.2.2.4 Hydrothermal and Solvothermal Syntheses

Hydrothermal and solvothermal processes are promising chemical routes for the preparation of nanosized powders and films [67–73]. In this case, a suspension or sometimes a homogeneous solution is subjected to elevated temperatures and pressures. The combined effect of pressure and temperature can reduce the free energy of the phases that are not stable under atmospheric conditions. The properties of reactants, including their solubility and reactivity, also change at high temperatures. These changes mentioned above lead to products with special motif and structure, which are difficult to achieve using other methods. Hydrothermal synthesis can be considered as a dissolution–reprecipitation process [129]. The particles involved in the hydrothermal process are partially or gradually dissolved in the solution. Eventually, the solution is supersaturated, which causes nucleation and growth into crystalline particles with a specific size, which are difficult to be obtained at low temperatures. During the synthesis of nanocrystals, parameters such as water pressure, temperature, reaction time, and the respective precursor–product system can be tuned to achieve a high simultaneous nucleation rate and a good size distribution. A variety of nanoparticles have been successfully synthesized in this way [67–73]. The solvent is not limited to water but also includes other polar or nonpolar solvents, such as benzene and ethanol, and these processes are more appropriately called solvothermal syntheses.

Hydrothermal and solvothermal syntheses are performed in an apparatus called the autoclave. These are usually thick-walled steel cylinders with a hermetic seal that must withstand high temperatures and pressures for a prolonged period of time. Furthermore, the autoclave material must be inert with respect to the solvent. To prevent corrosion of the internal cavity of the autoclave, protective inserts are usually used. These inserts can be made of gold, platinum, titanium, glass, quartz, or Teflon, depending on the temperature and solution used.

1.2.2.5 Self-Assembly

Self-assembly refers to the process of a disordered system of preexisting components that form an organized structure or pattern as a consequence of specific, local interactions (noncovalent) among the components themselves, without external directions [130–135]. Self-assembly is the strategy nature chose to develop complex, functional biological structures, which is far advanced than what the modern fabrication technology can achieve. An important aspect of self-assembly is the key role of weak interactions (e.g., van der Waals, capillary, π–π, hydrogen bonds) with respect to more "traditional" covalent, ionic, or metallic bonds, which makes the building blocks not only atoms and molecules, but consist of a wide range of nano- and mesoscopic structures, with different chemical compositions, shapes, and functionalities. Structures produced by self-assembly are mostly thermodynamically stable, and tend to be relatively defect-free and self-healing.

The physical and chemical functional specificity and selectivity of nanoparticles suggest them as ideal building blocks for self-assembly (into forms of powder, thin film, and solid bulk), in which the particles behave as well-defined molecular matter and are arranged with long-range order [136,137]. This approach to functional networks avoids many difficult steps that require very highly developed techniques. Well-defined ordered solids prepared from tailored nanocrystalline building blocks provide new opportunities for optimizing, tuning, and/or enhancing the properties and performance of the materials. Tuning particle size and interparticle forces could effectively tune the manner in which they are arranged, and thus change the properties of the assembled nanounits.

Thin films have been successfully fabricated by self-assembly including metal, semiconductor, dielectric, and many other functional materials [138–140], which possess novel architecture and are with fundamental interests and technological importance. The thin films are formed primarily on a solid substrate via a self-assembling process of nanocrystals. An alternative technique for forming monolayer nanocrystal thin films is at the liquid surface using the Langmuir–Blodgett films [141,142], which has also shown great potential in synthesis and assembling nanocrystals.

Self-assembly extends the scope of chemistry aiming at synthesizing products with order and functionality properties, extending chemical bonds to weak interactions, and encompassing the self-assembly of building blocks on all length scales. Synthesis of new materials using self-assembly is an effective approach that has the potential of producing high quality, large quantity, and chemically and structurally controlled new materials.

1.3 RECENT DEVELOPMENT OF CHEMICAL FABRICATION OF FUNCTIONAL THIN FILMS

Nanostructured thin films with controlled architectures are desirable for many applications in optics, electronics, biology, medicine, and energy conversions. The nanoscale units possess novel physical and chemical properties that differ from isolated atoms or molecules and their corresponding bulk materials. Another advantage of nanomaterials is that their optical, electronic, magnetic, and chemical properties can be significantly altered [143], depending on their size, morphology, and spatial arrangement. Evidently, fabrications of thin films with controlled nanoarchitectures are with significant importance to harvest the full potential application of this kind of materials, and represent a significant challenge in the field of nanoscale science. New chemical strategies for precisely tuning the microstructure of particles are of fundamental importance in both science and technology. In the following part, we use ZnO, ZnS, ZnSe, CuCNS, $(NH4)_2Cu(MoO_4)_2$, Nb_2O_5, $NaNbO_3$, $KNbO_3$, and other semiconductor materials as examples to show how we can finely fabricate functional thin films with nanoarchitecture by selecting suitable chemical routes. These works show that even though some obvious disadvantages exist in the solution chemistry routes, it is still promising to improve these chemical techniques for nanodevices due to their many flexibilities.

1.3.1 Thin Films Constructed by Zn-Based Materials

As a direct wide bandgap (3.37 eV) semiconductor with a large excitation binding energy (60 meV), zinc oxide (ZnO) is one of the most important multifunctional oxide materials possessing a suite of useful properties such as optical absorption and emission, piezoelectricity, high voltage nonlinearity, field emission, energy conversion, sensitivity to gases, and photocatalysis [144–148]. A wide range of ZnO nanostructures such as nanowires, nanosprings, nanotubes, and more others have been synthesized by thermal evaporation, metal organic chemical vapor deposition (MOCVD), pulsed laser deposition (PLD), templating against anodic alumina membrane, and aqueous chemical growth processes [89,144–148]. Especially, the synthesis of arrayed 1D nanostructures has been greatly spurred due to their essential scientific and technological significance [149–151]. There are two prerequisites for the realization of ZnO nanodevices. One is the development of a simple and low-cost method of preparation ZnO nanomaterials for industrial mass production. The other is the ability to control the position and morphology of the ZnO nanostructures. Various low-temperature solution-phase approaches have recently been developed to fabricate ordered ZnO nanoarrays because of their good potential for scale-up production and commercial feasibility. The aqueous chemical growth technique has been applied to produce functionalized thin films and coatings of metal oxide materials on various substrates including single-crystalline, polycrystalline, and flexible ones [89]. Such a method does not require any template, surfactant membrane, or epitaxial substrate to control orientation. Furthermore, catalyst-free and low-cost growth is conducted at moderate temperatures.

1.3.1.1 ZnO Films Fabricated on Zinc Foil by Self-Seeding Method

Thin films and coatings of oriented nanostructures, particularly, free-standing 1D ones growing directly on a current-collecting substrate are desirable for applications in many electronic and

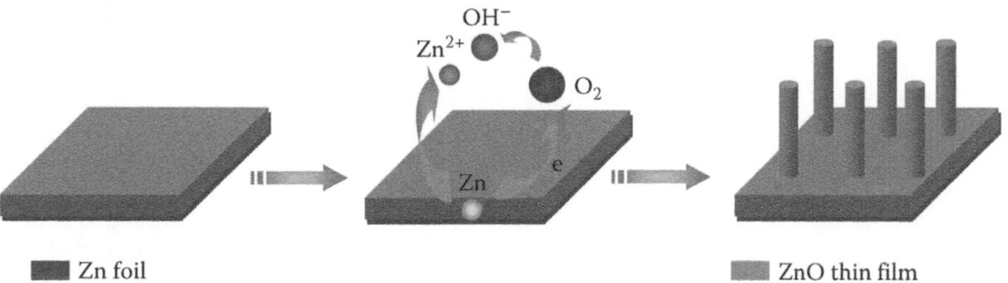

FIGURE 1.10 Schematic illustration of heterogeneous nucleation and the subsequent crystal growth of 1D nanostructures on the zinc metal foil.

photonic device [107]. Notably, the direct growth of oriented ZnO nanorod/nanoneedle arrays on a zinc substrate has been recently realized by the natural surface oxidation of zinc foil in solution at room temperature or under hydrothermal conditions. The process is based on heterogeneous nucleation and subsequent crystal growth of 1D nanostructures on the zinc metal foil, which is shown in Figure 1.10. This direct, one-pot approach allows the well-controlled growth of high-quality ZnO nanoarrays with large-area homogeneity and consisting of nanowires or nanorods with predictable morphologies, such as identical lengths and tunable diameters. In contrast to the traditional CVD method and hard template approaches, this self-seeding approach for ZnO thin film growth on Zn foil can provide an alternative, convenient, low temperature, and environmentally friendly pathway for fabrication of arrayed nanostructures, which is an important issue for their applications.

1.3.1.1.1 NaCl Solution Corrosion-Based Strategy to ZnO Nanorod Arrays

Corrosion is defined as the deterioration of a material through a chemical or electrochemical reaction with its environment. The corrosion of metal is a process that returns metal to its natural form [152]. Steel corrodes actively in oxygen-rich environments due to its large content of iron; the related cost for corrosion damage of infrastructure in the United States alone exceeds billions of dollars each year. Chloride ions found in deicing salt and seawater accelerate the corrosion of steel, and therefore are a concern for reinforced concrete structures. Structures located in areas of harsh winters or near marine environments are very susceptible to corrosion due to the high presence of chloride ions.

But corrosion has shown its good side when it was introduced to the synthesis of nanomaterials. A useful method based on the NaCl solution corrosion strategy has been successfully designed to synthesize ZnO nanorod arrays [153]. Zinc foil was used here, and directly acted as the reactant and the substrate to support the obtained 1D nanorods. The oxidation of zinc metal by naturally dissolved oxygen in water is a very slow process in ambient pressures and temperatures. However, in the presence of NaCl in solution, this spontaneous oxidation reaction can be accelerated drastically due to its strong corrosion ability to Zn metal. The corrosion of zinc foil in NaCl solution is an electrochemical process. Adding NaCl to the reaction solution leads to more charge carriers. NaCl breaks up into Na^+ and Cl^- in solution and allows the electrons to pass through the solution more easily. This corrosion of zinc metal by aqueous electrolytic solution causes the dissolution of zinc. Zn^{2+} ions are released continuously from Zn foil while oxygen is simultaneously reduced. The anodic reaction in the corrosion process is the oxidation of zinc metal atom to its ion. This can be written in the form below:

$$Zn \rightarrow Zn^{2+} + 2e \tag{1.2}$$

The NaCl aqueous solution is nearly neutral, and it contains dissolved oxygen. Thus, the cathodic reaction is

$$O_2 + 2H_2O + 4e \rightarrow 4OH^- \quad (1.3)$$

The overall reaction can be described

$$2Zn + 2H_2O + O_2 \rightarrow 2Zn^{2+} + 4OH^- \rightarrow 2Zn(OH)_2 \quad (1.4)$$

Figure 1.11 shows SEM images of the large arrays of oriented ZnO nanorods, which are vertically grown on a zinc foil substrate. The top view (Figure 1.11a) shows that ZnO nanorods cover the zinc surface uniformly, smoothly, and compactly. Such an alignment can be better perceived from the high-magnification view in Figure 1.11b and the side view in Figure 1.11c. It can be clearly seen that ZnO nanorods are aligned approximately normal to the zinc substrate with a diameter of about 200 nm and a thickness of about 4 μm. Note that these nanorods are better aligned and have good mechanical strength due to the fact that ZnO nanorods are indeed grown from the zinc foil instead of a random deposition of ZnO particles on the zinc foil. Transmission electron microscopy (TEM) image and high-resolution TEM image shown in Figure 1.11d verify the fact that these nanorods are single-crystalline wurtzite-type ZnO grown along the [001] direction.

In this condition, ZnO prefers to grow into 1D nanostructure due to its highly anisotropic bonding in the crystallographic structure. With respect to the structure of wurtzite ZnO, the polar growth of ZnO crystal along the [001] direction proceeds through the adsorption of growth units onto the (001) plane, and the crystal growth habit exhibits a basal polar plane $(00\bar{1})$ and two types of low-index faces, a nonpolar $(\bar{1}00)$ face, and a tetrahedron corner-exposed polar (001) face [99]. Figure 1.12 shows the crystallographic model of ZnO; hexagonal wurtzite structure of ZnO can be simply described as the number of alternating planes composed of tetrahedrally coordinated O^{2-} and Zn^{2+}

FIGURE 1.11 (a–c) SEM images of the large arrays of oriented ZnO nanorods; (d) TEM and HRTEM images taken from an individual nanorod.

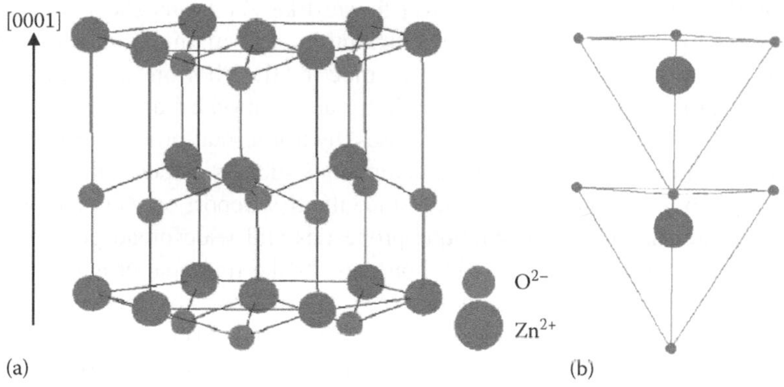

FIGURE 1.12 Crystallographic structure of ZnO.

ions, stacked alternately along the *c*-axis. The relative stacking rate of the constituent tetrahedral in various crystal faces is strongly dependent on the bonding force of atoms in the tetrahedral at the interface. Figure 1.12b shows that each $\{ZnO_4\}$ tetrahedron has a corner in the [001] direction. The atom at the corner of a tetrahedron has the strongest bonding force as compared with the atoms at the other positions. Therefore, ZnO crystal grows fast along the direction in which the tetrahedron corners point (i.e., [001] direction).

The lattice matching between ZnO and Zn crystals facilitates the growth of ZnO nanorods on the surface of zinc metal, and thus ZnO nanorod arrays can readily grow on the zinc foil substrate. The photoluminescence (PL) spectrum shown in Figure 1.13 reveals that thc as-obtained ZnO nanorods films have negligible oxygen vacancies. The size and shape of the as-obtained ZnO nanorods can be effectively tuned by adjusting the concentration of NaCl solution and pH value of the reaction solution.

1.3.1.1.2 Self-Seeding Method to ZnO Films with Other Nanoarchitectures

A novel 2D pattern of flower-like ZnO nanosheets have been prepared by formamide-induced sequential nucleation and growth on zinc foil in the $Zn(OH)_4^{2-}$ solution. The secondary growth on initially formed nuclei was necessary for the growth of complex patterns of ZnO nanocrystals. These flower-like ZnO crystals are composed of 2D nanosheets that were simultaneously grown from pre-agglomerated seeds. The appropriate volume of formamide was critical and the enhanced oxidation of metal zinc led to numerous nuclei that could be quickly formed. Secondary growth produced flower-like ZnO nanosheets from those already formed nuclei in

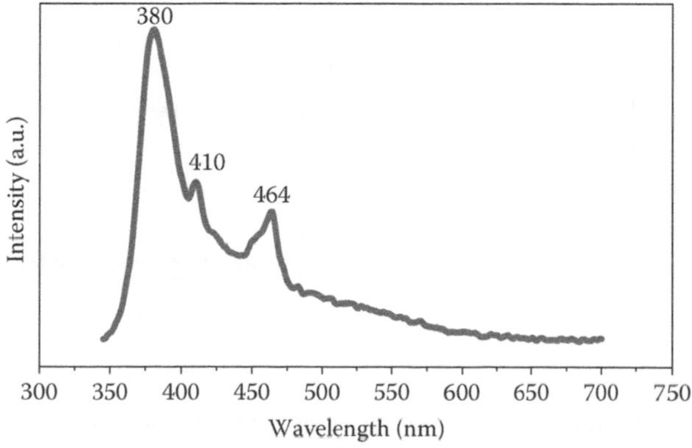

FIGURE 1.13 Photoluminescence (PL) spectrum of the as-obtained ZnO nanorod arrays.

the aqueous solution. The preparation of novel flower-like ZnO nanosheet thin films is rarely reported, which is suggested to be due to the contribution of formamide molecules in the present reaction system. The present technique might be nice for the "bottom-up" design of nanoscale photonic and electronic devices that are available for application cheaply and easily.

As one kind of solution-based chemical processes, hydrothermal process has been widely used for the synthesis of a huge range of solid-state compounds such as oxides, sulfides, molecular zeolites, and other microporous phases [154–156]. Since the first report on ZnO nanobelts, they have attracted intensive interests due to their unique properties and widespread potential applications. Recently, ultralong ZnO nanobelt arrays with honeycomb-like micropatterns were grown directly on Zn substrates under hydrothermal conditions [157]. The synthesis of large-scale ZnO nanowires has been realized by the simple surface oxidation of zinc metal surface in an aqueous solution of NaOH and $(NH_4)_2S_2O_8$ at 150°C. The aligned ZnO nanowires grew vertically on the Zn substrate in the solution and then the capillary action forced these long nanowires to bend and bundle together during drying. When the zinc foil was immersed in the alkaline oxidant solution at room temperature for 2 days under ambient conditions, large-area arrayed ZnO nanorods ~30 nm in diameter were grown on the Zn foil. The growth process of 1D ZnO nanostructures is shown in Figure 1.14. The length of the aligned ZnO 1D nanostructures could be tuned via simply changing the reaction temperature. This solution approach to the fabrication of 1D ZnO nanostructures with controlled morphologies is highly technologically important for II–VI semiconductors, and even can be rapidly extended to prepare other transition metal semiconductors.

1.3.1.2 ZnO Films Fabricated on Other Substrates

For various novel applications, it is important that low-temperature and low-cost controllable growth of arrayed ZnO nanorods on properly fitting substrates can be achieved. Using a coated ZnO seed layer, large-scale arrays of highly oriented ZnO nanorods can be formed via a simple low-temperature liquid-phase route on a variety of substrates (including glass, silicon, PET, etc.).

Figure 1.15 is a schematic illustration of the growth of ZnO nanorods on a patterned substrate [158]. Typically, a ZnO-seed-coated substrate was loaded flatly on a sample vial with a formamide aqueous solution, and two Zn foils were immersed vertically in the solution. Zinc precursors were continuously supplied from Zn foil, and there was a concentration gradient of the precursors from Zn foil to ZnO-film-coated substrate in the liquid phase. At the low concentration region around the

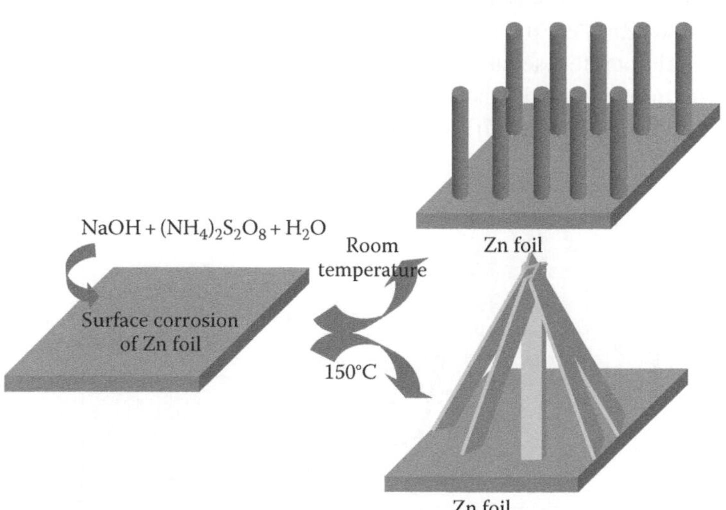

FIGURE 1.14 Schematic illustration of the preparation of ZnO nanorods and nanowires on Zn foils at different temperatures.

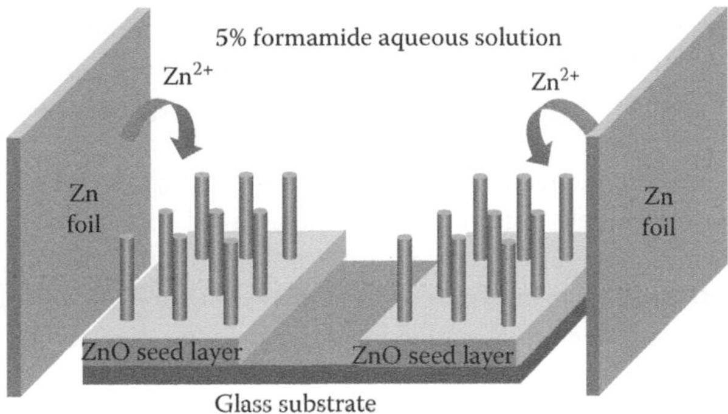

FIGURE 1.15 Schematic illustration of the growth of ZnO nanorods on a patterned substrate.

ZnO-film-coated substrate, solid nanorods grew uniformly throughout the surface of seed particles. At the region near the zinc foils, the zinc concentrations were largely decreased close to the top area of 1D ZnO nanostructures, leading to the preferential growth of ZnO nanotubes. The hollow structures reduced the top (001) metastable areas and enlarged the lateral areas of the most stable low-index nonpolar surfaces.

In aqueous solutions, the conditions such as the concentration, temperature, and pH are tunable. Furthermore, by selecting the adequate quantities and the kinds of additives, such as complexing or chelating agents, extended and oriented nanostructures are available. Recently, a low-temperature, environmentally benign, solution-based method to prepare complex ZnO nanostructures has been developed [159]. The large arrays of oriented ZnO nanorods possess complex film morphologies made of oriented nanocolumns and complex bilayers that are remarkably similar to biomineral structures in red abalone shells. During the preparation process, a seeded growth procedure is used. Organic molecules are known to promote or inhibit crystal growth, and citrate anions are applied in the process to adsorb on ZnO basal planes as the structure-directing reagent.

A substrate with a high density of nucleation sites is crucial for growing large arrays of oriented nanorods. A glass slide contained the ZnO nanoparticle coating, and was placed vertically in a sealed Teflon bottle containing an aqueous solution of $Zn(NO_3)_2$ and hexamethyltetramine (HMT). The reaction was performed for up to 3 days at 60°C. When a substrate was used without ZnO nanoparticle seeds, sparsely populated and randomly oriented large ZnO rods were produced at the same reaction conditions. Time-dependent experiments showed that in the early stages of crystal growth, oriented structures were not formed on the substrate. By prolonging the reaction time, ZnO nanoparticle seeds became hexagonal, and finally the seed layer was [001]-oriented, which induced the growth of ZnO nanorod array.

Citrate anions would preferably bind on the (001) facet of ZnO to prohibit the nucleation and crystal growth along the [001] direction, and therefore providing an effective approach to control the aspect ratio of the ZnO rods. With the addition of citrate, the ZnO rods became short and fat. Finally, higher citrate concentrations led to the appearance of ZnO hexagonal plates. It was found that the growth of fat hexagonal crystals can proceed in the solution with high citrate content. Although the morphology of the microcrystals remained hexagonal, the crystals could still grow sideways in the form of thin platelets, which were quite uniform in thickness throughout the sample. Interestingly, this kind of plate-like nanofeatures could be healed in the $Zn(NO_3)_2$ and HMT aqueous solution without citrate, that is, the conversion between the rod and the plate is reversible via controlling the addition of citrate. It was also exhibited that oriented ZnO nanorods on substrates could also perform secondary and tertiary growth in the presence of citrate ions. The resultant single ZnO rod was thicker, but not much longer, owing to the binding of citrate ions on the (001) surface. By using a multiple-step seeded-growth way, large arrays of ZnO nanostructures remarkably similar to those from biogenic calcium carbonates in abalone could be fabricated, which showed enhanced photocatalytic property. The simple, "bottom-up" chemical-solution approach might readily be expanded to create many other application-targeted functional materials from nanometer to micrometer scales.

According to the results mentioned above, the solution-based chemical strategy has been developed to grow nanorods, nanobelts, and other nanostructured thin films on various substrates. The method is simple, mild, cost-effective, and does not need templates. Individual nanorods or other building blocks directly assembly into ordered arrays. Problems such as templates/catalyst removal, tedious operation procedures, and poor adhesion and electrical contacts between the nanorods and the underlying substrate can be avoided, thus facilitating device fabrication and characterization.

1.3.1.3 ZnO Porous Films by Selective Core Dissolution

Among many inorganic materials, hollow micro-/nanostructures are of great interest in many current and emerging areas of technology. Thin films constructed by free-standing hollow micro-, meso-, and nanounits have distinct structural and geometrical features, and represent an important class of novel materials due to their potential applications. The synthetic approaches for hollow structures can be divided into four categories: hard templating synthesis, soft templating synthesis, sacrificial templating synthesis, and template-free synthesis. Using these methods mentioned above, many hollow structures have been achieved, which exhibit a wide variety of potential applications including the delivery of drugs, catalysts, chemical storage, microcapsule reactors, and so on [160–164].

Selective core etching of the solid structure is a newly emerged strategy for fabricating hollow structures [165]. The key factor for selective etching is to create an unbalanced environment to selectively dissolve the core while maintaining the integrity of the shell. A ZnO porous thin film constructed by hexagonal taper-tubes has been prepared through taper-tubes in an acidic medium based on a one-step in situ crystallization/etching mechanism. During electroless deposition reaction, the formation process may be rationalized by considering that the etching of ZnO with acetic acid starts from six sites located at the six corners of the top surface of ZnO truncated taper, resulting in a gradual widening of the open holes and tubular ZnO.

The electroless deposition reaction is carried out in an acidic solution containing $CuCl_2 \cdot 2H_2O$, acetic acid, and zinc foil together with the addition of H_2O_2 to produce O_2. Considering the total standard reduction potential of Cu^{2+}/Cu and O_2/OH (0.3803 V vs. standard hydrogen electrode (SHE)) is higher than that of the Zn^{2+}/Zn pair (0.7626 V vs. SHE), Zn foil will lose electrons and dissolute as Zn^{2+} ions; Cu^{2+} ions in the solution then accept the electrons and get deposited as Cu nanoclusters, which are uniformly dispersed on the surface of the zinc foil. These Cu nanoclusters combined with metal Zn could form numerous micrometer-sized cells on the surface of the zinc foil, in which Cu and Zn act as local cathodes and anodes, respectively. Owing to the presence of dissolved oxygen, a modified galvanic replacement reaction can be realized, in which the redox reactions can be described as follows:

At the anode:

$$Zn \rightarrow Zn^{2+} + 2e^- \tag{1.5}$$

At the cathode:

$$Cu^{2+} + 2e^- \rightarrow Cu \tag{1.6}$$

$$O_2 + 2H_2O + 4e^- \rightarrow 4OH^- \tag{1.7}$$

OH^- ions generated at the cathode then migrate toward the anode region, where ZnO can be locally formed on the surface of zinc foil by the reaction of Zn^{2+} and OH^-.

Figure 1.16 shows typical SEM images of the synthesized taper-tubular ZnO arrays at different magnifications. It can be seen from Figure 1.16a that hexagonal ZnO taper-tubes are well aligned, and uniformly grown from a zinc foil substrate. The detailed geometrical morphologies of ZnO taper-tubes are shown in Figure 1.16b, which indicates that the taper-tubular ZnO structures have outer diameters in the range of 1–2 μm and wall thickness in the range of 200–300 nm. The diameter of each tube has some variation from bottom to top, which thus shows a unique taper-tube morphology. The hexagonal open end of the ZnO taper-tubes is well consistent with the symmetry of the wurtzite ZnO. The PL spectrum of the aligned ZnO tubes demonstrates that a very strong band edge ultraviolet peak at ~382 nm can be observed (Figure 1.17). Inspection of the SEM images suggests the selective-etching mechanism for producing taper-tubular ZnO arrays.

The etching process can be divided into four distinctive stages, which is clearly illustrated in Figure 1.18. At the early stage, ZnO tapers were formed by the precipitation of Zn^{2+} in the anode region into ZnO with migrated OH^- from the cathode (Figure 1.18a). The etching of the top surface of the truncated ZnO tapers with acetic acid starts when the precipitation reaction reached a balance, the etching is initiated at the corners of the hexagonal top surface due to their high surface energy, six pits are observed in the corresponding six corners of an individual truncated taper (Figure 1.18b), These newly formed pits on the corners should represent the most active sites for further etching. As expected, six gradual widening holes located at the hexagonal top surface of an individual truncated taper have been generated by prolonging the reaction time. As shown in Figure 1.18c, the opening on the six corners of the top surface of each truncated taper is, indeed,

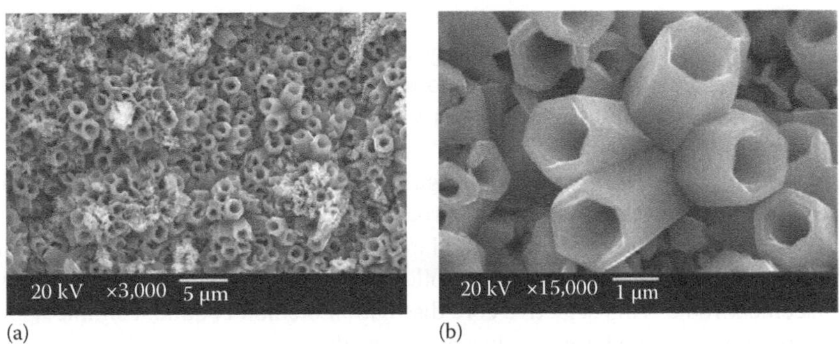

FIGURE 1.16 (a,b) SEM images of the synthesized taper-tubular ZnO arrays.

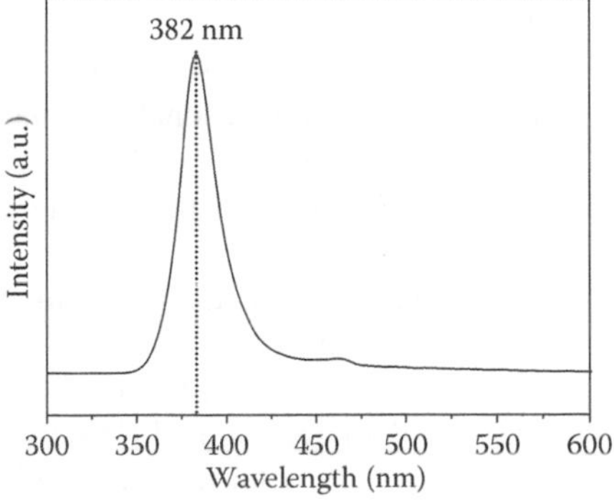

FIGURE 1.17 Photoluminescence spectrum of aligned ZnO taper-tube arrays.

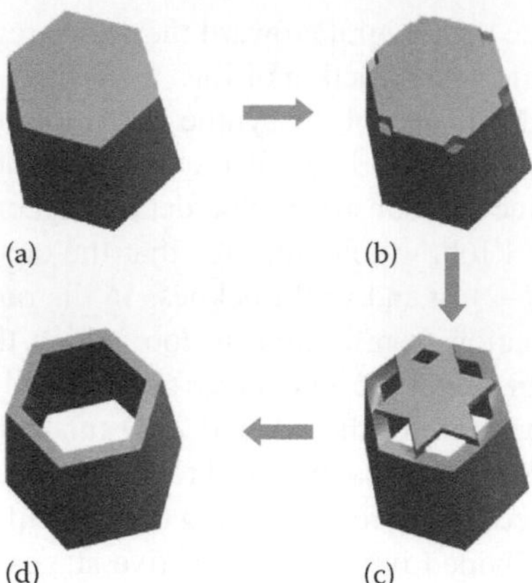

FIGURE 1.18 Illustration of the etching process of ZnO tapers; (a) complete ZnO tapers; (b) etching started in the six corners; (c) opening on the six corners of the top surface; (d) tubular structures.

increased in area as compared to those shown in Figure 1.18b, and these holes continue to serve as active sites for the subsequent etching reaction. Tubular structures (Figure 1.18d) are finally formed after complete etching of the core of the truncated taper. The diameter of the final tubes is slightly larger than those of the truncated taper precursors, which indicates that Ostwald ripening occurs during the whole acetic acid selective-etching processes.

The selective etching of the ZnO {0001} face (top-surface face of truncated tapers) is related to its structure polarity and chemical activities. ZnO is a polar crystal and consists of a positive polar (0001) plane rich in Zn cations; the positive charges would cause structural surface metastability on the (0001) plane, which can be selectively etched under proper conditions. The existence of acetic acid is also essential in this process, for its ability to accelerate the etching process, thus the corresponding hollow structure can be achieved.

This work represents an extension of applying galvanic replacement reaction in fabricating porous thin films from metals to metal oxides. This approach not only enriches metal oxide chemistry, but provides a new strategy to synthesize aligned tube arrays of nonlayered materials, which could be applicable to the synthesis of other inorganic hollow structures through designing an appropriate galvanic replacement route. This selective core dissolution method is quite a simple approach, and furthermore, nanotubes obtained are mostly single crystals.

1.3.1.4 ZnS Porous Films by Sacrificial Template Method

Zinc sulfide (ZnS) is a II–VI group compound semiconductor with direct bandgap energy of 3.6 eV. ZnS thin films have attracted considerable research interest due to its extensive usage in electroluminescent devices, infrared windows, flat panel displays, sensors, and lasers [166–169]. Extensive efforts have been made on the synthesis of low-dimensional ZnS nanostructures, including nanoparticles, nanowires, nanobelts, nanocables, and nanotubes [170]. However, there are relatively few reports on ZnS nanocable and nanotube arrays, although these thin film structures are highly desirable for many applications.

As discussed in the previous part, ZnO nanorod arrays with uniform diameter and length have been prepared on a zinc foil by self-seeding method without any templates or additives [153,155]. The prepared ZnO nanorod arrays could be used as an ideal sacrificial template to prepare ZnS nanocable and nanotube arrays [171,172]. Thioglycolic acid ($HSCH_2COOH$) was used to improve the activity of Zn^{2+}. As illustrated in Figure 1.19, after the reaction between Na_2S and the surface layer of ZnO with the assistance of $HSCH_2COOH$, ZnO/ZnS core/shell nanocable arrays could be gained on the Zn substrate. The typical chemical reactions could be described as follows:

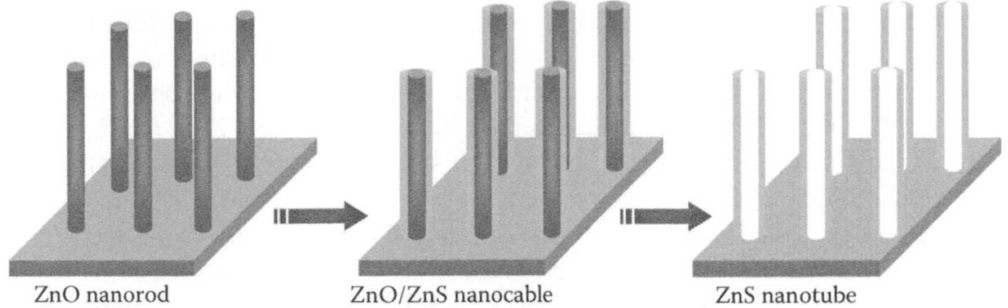

FIGURE 1.19 Illustration of synthetic strategies of ZnO/ZnS nanocables and ZnS nanotubes.

$$HSCH_2COOH + Zn^{2+} \rightarrow ZnHS^{+} + CH_2COOH^{+} \tag{1.8}$$

$$2ZnHS^{+} + S^{2-} \rightarrow 2ZnS + H_2S \tag{1.9}$$

The activity of ZnO was greatly enhanced according to reaction (1.7). The complexes $ZnHS^+$ accelerated the nucleation on the surface of the ZnO rods, leading to the subsequent growth of the ZnS shell. To fabricate ZnS nanotube arrays, ZnO/ZnS nanocable arrays with ZnO as the inner core and ZnS as the outer shell are firstly synthesized via a thioglycolic acid-assisted sulfurization route. Subsequently, removing ZnO core leads to the formation of ZnS nanotubes. Figure 1.20a

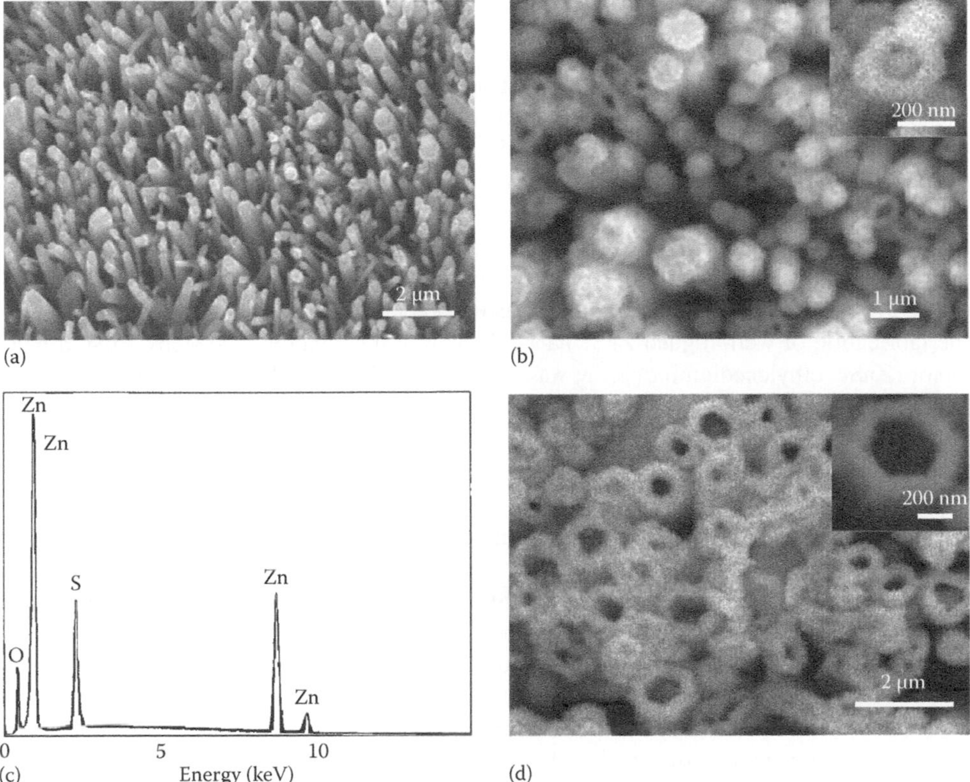

FIGURE 1.20 (a) SEM image of ZnO nanorod arrays on the zinc foil substrate; (b) SEM image of ZnO/ZnS nanocable arrays via a thioglycolic acid-assisted solution method; (c) EDX spectrum taken from the surface of ZnO/ZnS nanocables; (d) SEM image of ZnS nanotubes.

shows the typical SEM image of ZnO nanorod arrays: the nanorods are vertically grown on a zinc foil substrate along the [001] direction, with a diameter of about 200 nm and a length of about 4 μm. By the reaction between ZnO and S^{2-}, ZnS nucleates and grows on consuming ZnO nanorods. Due to the difference of mutual diffusion rates between ZnO and ZnS, nanoscale pores can be generated inside nanocrystals with a mechanism analogous to void formation in the Kirkendall effect, and ZnO/ZnS nanocables can be obtained. Figure 1.20b show the SEM image of ZnO/ZnS nanocable arrays on the zinc foil substrate, with ZnO as the core and ZnS as the shell. The chemical composition of ZnO/ZnS nanocables is determined by using EDX analysis, as shown in Figure 1.20c, which exhibits the presence of Zn, O, and S elements and confirms the composition of the ZnO/ZnS nanocables. Since ZnO has an amphoteric characteristic, treating ZnO/ZnS nanocables in KOH solution leads to the dissolution of the ZnO cores, thus ZnS nanotube arrays can be successfully obtained. From Figure 1.20d, we can see that the ZnS tubes have open ends and are with pore sizes of about 400 nm. The high-magnification SEM image of a single ZnS nanotube in the inset of Figure 1.20d indicates that ZnS nanotubes obtained were in a polycrystalline state, and the cross section of the nanotubes showed a perfect hexagonal shape with a uniform pore size of about 400 nm.

Considering that thin films of metal oxides can be fabricated with a variety of nanoarchitecture and composition, and the method to prepare metal sulfide is relatively lacking, the sacrificial template strategy presented here can be extended to a great number of systems; metal sulfide thin films with porous nanostructure can be easily fabricated based on their corresponding oxide precursors.

1.3.1.5 ZnSe Nanobelt Arrays by Metal-Organic Precursor Method

As one of the Zn-based II–VI compounds, ZnSe is a direct bandgap semiconductor, with a bandgap energy of 2.7 eV, suggesting it to be a promising material for blue-diode lasers and other photoelectronic devices [162,173]. Stimulated by these novel properties, ZnSe are expected be fabricated into nanostructured thin films to offer new opportunities for applications in the emerging field of nanoscience and nanotechnology. Although ZnSe nanostructures with various morphologies have been prepared by diverse methods and different synthesizing systems, the synthesis of ZnSe nanobelt arrays is still a challenge. Recently, large-scale synthesis of well-aligned ZnSe nanobelt array films by a simple solvothermal reaction and a subsequent heat treatment was achieved [104]. The reaction medium played an important role in the morphology of the product. The high quality of the nanostructures shows their potential applications for optoelectronic devices.

The fabrication of well-aligned ZnSe nanobelt array films contains two steps. First, a belt-like precursor (ZnSe·ethylenediamine) array was synthesized by a simple template-free solvothermal route. And subsequently annealed, these inorganic–organic precursor arrays in N_2 atmosphere produce a phase-pure ZnSe thin film of nanobelt arrays.

X-ray diffraction (XRD) characterization shows that the as-synthesized ZnSe thin film has a cubic phase without any impurities. To investigate the morphology and the microstructure of the as-prepared ZnSe film, SEM was adopted. From Figure 1.21a, it can be concluded that the Zn foil is covered with uniform and dense ZnSe nanobelts. High-magnification SEM images exhibit that the products are composed of vertically standing nanobelts (Figure 1.21b), and each of them has a clear rectangle top section.

The ability to grow ZnSe nanobelts with strong PL emission is expected to increase the versatility and power of these building blocks in nanoscale photonic and electronic devices. The PL emission spectrum of ZnSe nanobelt arrays is shown in Figure 1.22. A sharp and stable peak in deep-level emission, which is usually associated with dislocations, stacking faults, and non-stoichiometric defects, was observed at 578 nm. Besides the very strong emission peak, two weak emission peaks in the near-band-edge emission region were also observed at about 420 and 433 nm, respectively. These weak emission bands are usually attributed to the edge emission of ZnSe.

FIGURE 1.21 SEM images of ZnSe nanobelt arrays synthesized at 160°C for 40h and post-annealed for 4h under a pure N_2 atmosphere. (a) Low-magnification image. (b) Local magnified image of (a).

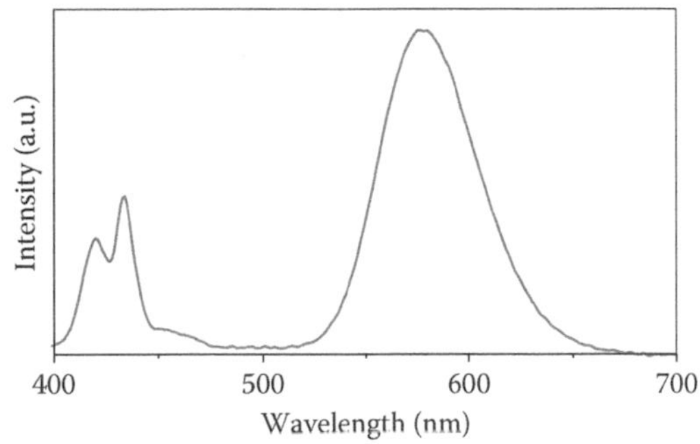

FIGURE 1.22 Room-temperature PL spectrum of ZnSe nanobelt arrays.

1.3.2 Thin Films Constructed by Cu-Based Materials

1.3.2.1 Cu_2O and CuO Porous Films

Porous materials attract great research interest and have made a great impact in many applications, including catalysis, sorption, and separation. As a p-type semiconductor (direct bandgap ~ 2.17 eV) with unique optical and magnetic properties, cuprous oxide (Cu_2O) is a promising material with potential applications in solar energy conversion, micro-/nanoelectronics, magnetic storage devices, catalysis, and biosensing [174,175]. Moreover, Cu_2O crystals have been at the center of research on the Bose–Einstein condensation of excitons [176]. Many recent efforts have been devoted to the shape-controlled synthesis of Cu_2O micro- and nanocrystals. However, it is still a real challenge to prepare uniform porous CuO films. Recently, Xu and coworkers has designed a colloidal crystal template route for fabrication of 3D macroporous (3DOM) Cu_2O thin film [177]. This template route includes three steps. First, well-ordered colloidal crystal templates were assembled on an ITO glass substrate that is submerged vertically into an aqueous dispersion of PS spheres at a low temperature (Figure 1.23a). Then Cu_2O materials were deposited on these PS hard templates through electrochemical reduction of copper lactate in an alkaline solution (Figure 1.23b). Finally, the 3DOM Cu_2O thin film was obtained after PS was removed in tetrahydrofuran (THF) (Figure 1.23c).

Similarly with Cu_2O, cupric oxide (CuO) has been extensively studied because of its close connection to high-T_c superconductors [178–180]. The valence of Cu and its fluctuation play important roles in the superconductivity of various types of cupric compounds. Because of its commercial value, CuO has been widely exploited as a powerful heterogeneous catalyst and in the fabrication of lithium copper oxide electrochemical cells [181,182]. It is reasonable to expect that the ability

FIGURE 1.23 Schematic illustration for fabrication of three-dimensional ordered macroporous (3DOM) Cu_2O thin film; (a) coating of PS spheres; (b) deposition of Cu_2O; (c) removing PS in THF.

to process CuO into nanostructured thin films should enrich our understanding of its fundamental properties and enhance its performance in currently existing applications.

Recently, Liu and Xue have successfully designed a new thermal oxidation strategy for the fabrication of transition-metal-oxide porous thin films [183]. The current thermal oxidation strategy is based on a nonequilibrium interdiffusion (the Kirkendall effect) process. The strategy is schematically illustrated in Figure 1.24: initially, through hydrothermal or solvothermal process, a serious of transition-metal sulfides or selenides micro-/nanoparticles was fabricated; then with these solid chalcogenide particles as building blocks, well-ordered colloidal crystal templates on silicon substrate were achieved through a self-assembly process. Finally, these well-ordered colloidal crystal templates that were composed of chalcogenide colloids were thermal oxidated at 700°C in air atmosphere for about 4 h; owing to the combination of the Kirkendall effect, volume loss, and gas (SO_2) release during this oxidation process, porous transition-metal-oxide thin films were obtained. The thickness of these porous metal oxides films can be controlled via the self-assembly process. More importantly, these obtained porous thin films show high thermal stability as the products are fabricated at 700°C, which is important to some of the applications such as catalysis, sensor, and wastewater treatment.

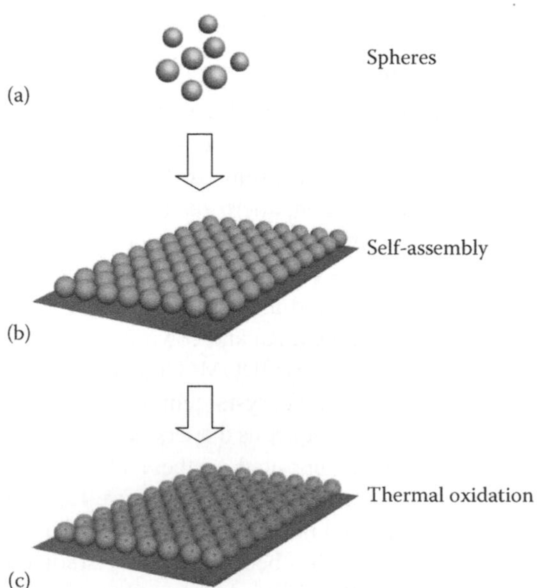

FIGURE 1.24 Schematic illustration of preparing CuO porous film (a) through solvothermal process, Cu_2S spheres were fabricated; (b) with these solid chalcogenide particles as building blocks, well-ordered colloidal crystal templates on a silicon substrate were achieved through a self-assembly process; (c) these well-ordered colloidal crystal templates that were composed of chalcogenide colloids were thermal oxidated at 700°C.

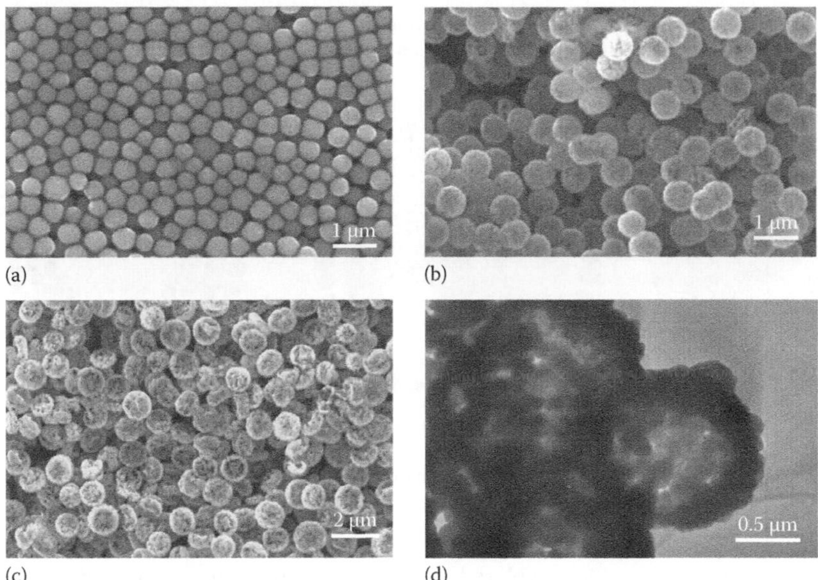

FIGURE 1.25 SEM images of (a) CuS precursor film; (b,c) CuO porous thin films; (d) TEM image of CuO capsules.

The morphology of the precursor CuS film and the final product CuO film is shown in Figure 1.25. The low-magnification SEM image (Figure 1.25b) shows that the obtained CuO thin films are composed of porous particles with uniform size. Particles of CuO products are hollow spheres, which preserve their precursor's (Cu_2S) morphology (Figure 1.25a). Figure 1.25c clearly indicated that these CuO hollow particles have a porous appearance and are composed of many CuO nanoparticles and nanopores. The average diameter of these sphere particles is about 2 μm. TEM observation of a typical porous structure shown in Figure 1.25d indicates that this porous structure is hollow in nature, and the porous surface is also clear to be seen.

In this process, the precursor can be converted into hollow CuO completely, and the formation of hollow structures is believed to be the result of nonequilibrium interdiffusion (the Kirkendall effect), which is introduced into nanomaterials synthesis. At the beginning of thermal oxidation of CuS, a thin intermediate shell was formed on the surface of these particles. This thin layer acts as a barrier and separates the CuS core from the outside atmospheric oxygen. However, the barrier consists of an intermediate shell with lots of vacancies. This kind of shell structures allows the out diffusion of inner sulfides and vacancies. And a net directional flow of precursor/target materials at the template/reactant interface may lead to the formation of hollow structures.

The volume loss and release of internally born SO_2 during the conversion from CuS and Cu_2S to CuO generate some pores in the as-prepared architectures, so the products are not only hollow but also have porous structures composed of nanoparticles and nanopores. As illustrated in Figure 1.26, the newly formed intermediate layer was very densely coated on the perfect CuS microcrystals, which can effectively block the release of internally born SO_2. More SO_2 was gathered gradually in the interior of the structure until the pressure was increased to a certain point to break the structure for the formation of a small pore, and finally, a porous shell formed.

1.3.2.2 $(NH_4)_2Cu(MoO_4)_2$ Thin Film Composed with Nanoflakes

Among copper-based materials, copper-containing molybdates are of intensive research interest due to their promising applications in various fields [184]. The preparation of a number of copper

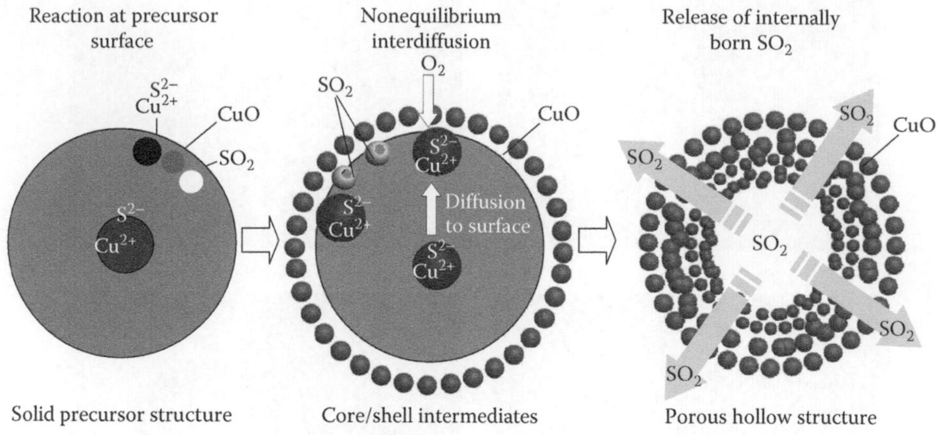

FIGURE 1.26 Illustration of CuO porous hollow spheres formation process.

molybdate phases have been widely explored, while most of these phases were prepared by hydrothermal solid-state reactions of the oxides of copper and molybdenum [185,186]. Recently, a new strategy has been well designed to synthesize thin films of curved ammonium copper molybdate nanoflake and its hierarchical architecture on a copper surface at room temperature [187]. The curved ammonium copper molybdate nanoflakes grow directly on the conducting copper substrate with good electrical contact, which is of crucial importance for the development of electronic or optoelectronic device with good performances. This method includes a liquid–solid reaction between Na_2MoO_4 solution and the copper substrate itself in the assistance of formamide at room temperature.

The morphology of these hierarchical $(NH_4)_2Cu(MoO_4)_2$ crystals is shown in Figure 1.27, from which we can see that the film is constructed by lamellar nanounits over the entire copper surface uniformly and smoothly. These lamellar structures are separated from each other (not compactly packed), as shown in Figure 1.27b. It can also be seen clearly that the lamellar $(NH_4)_2Cu(MoO_4)_2$ crystals are twisty and can form a crescent shape. Some lamellar crystals can bend into two kinds of "S" patterns (left- and right-hand mirror symmetry).

Through investigating the formation processes of $(NH_4)_2Cu(MoO_4)_2$ crystals prepared at different reaction stages, we can clearly understand the formation mechanism of this hierarchical

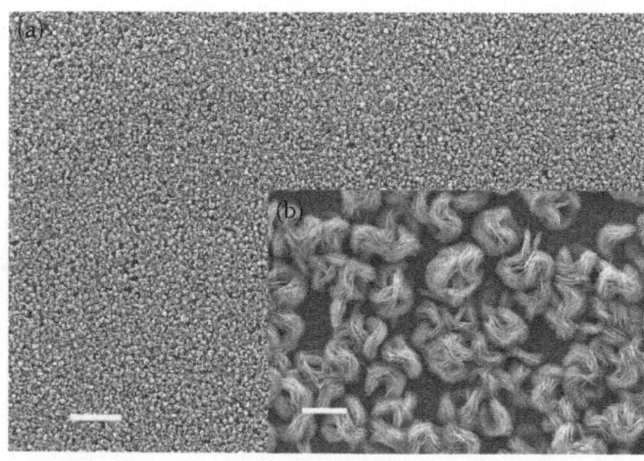

FIGURE 1.27 SEM images of $(NH_4)_2Cu(MoO_4)_2$ crystals on a copper foil substrate. Scale bar: (a) 100 μm; (b) 10 μm.

structure. Due to the surface oxide layer, the natural oxidation of copper metal by naturally dissolved oxygen in water is very slow. However, in the presence of formamide, the spontaneous oxidation reaction can be accelerated drastically. Cu^{2+} ions were released continuously from the copper foil into formamide solution while oxygen was simultaneously reduced, and Cu^{2+} ions can be immediately captured and coordinated by formamide molecules, forming copper-complex ions. When Na_2MoO_4 aqueous solution was added, $[Cu(HCONH_2)_4]^{2+}$-complex ions can react with MoO_4^{2-} to form $(NH_4)_2Cu(MoO_4)_2$ crystals on the copper surface, as shown in Figure 1.27. During the whole process, Cu^{2+} ions were released slowly and gradually from the copper foil, so we can control the morphology of $(NH_4)_2Cu(MoO_4)_2$ crystal facilely.

1.3.2.3 CuCNS Thin Film Composed Taper-Like Structured Arrays

The liquid–solid reaction strategy mentioned above has also been applied for the fabrication of CuCNS thin film composed taper-like structured arrays. As a transparent wide-gap p-type semiconductor, cuprous thiocyanate (CuCNS) possesses excellent chemical and thermal stability, and is expected to have important applications in various areas. For example, CuCNS has the ability to photooxidize water in the presence of sacrificial electron acceptors more readily than most photocatalytic materials. It is also an excellent candidate material for application in dye-sensitized solid-state photovoltaic cells due to photographic cyanine dyes being capable of injecting holes into CuCNS with very high efficiencies [188].

Taper-like structured CuCNS arrays can be successfully fabricated by employing a liquid–solid reaction between an aqueous solution of NH_4CNS and a Cu substrate at room temperature [189]. The copper substrates are used not only as a source of copper but also as a substrate for CuCNS crystals. The controlled reactions allow the large-scale, template-free, cost-effective synthesis of CuCNS with an ordered, uniform, and stable structure.

The morphology of CuCNS film on a copper substrate is shown in Figure 1.28. Figure 1.28a clearly shows a large-area array of CuCNS presenting upended taper-like structure. A higher-magnification image (Figure 1.28b) indicates the diameter of these arrays is about 2–4 μm. Through investigating the morphologies of products at different reaction stages, we have found that the heterogeneous nucleation and growth is dominant during the formation of upended taper-shaped CuCNS crystals.

FIGURE 1.28 SEM images of cuprous thiocyanate crystals on a copper foil substrate (a,b) and on copper mesh (c,d). Scale bar: (a) 50 μm, (b) 2 μm, (c) 100 μm, (d) 20 μm.

The primary factor that determines the surface morphology of products is the crystal structure β-CuCNS.

When copper mesh is used as a copper substrate, CuCNS crystals with a similar morphology can be obtained on the copper mesh substrate, as shown in Figure 1.28c. It is clear that every single crystal also shows an upended taper-like morphology (Figure 1.28d). Every tip end points into the surface of copper substrate and the other big end of the taper exposes out, just like a dart thrusting into the copper substrate.

These current room-temperature soft-solution approaches provide a general route to the synthesis of functional thin films with unique micromorphologies and complex hierarchies. Furthermore, this new strategy may be applicable for making other interesting structures or changing the substrate to another metal, alloy, and any substrate coated with copper.

1.3.3 Thin Films Constructed by Nb-Based Materials

Niobium oxide and alkaline niobates have gained great research interest in the past decades. As an important n-type semiconductor with a wide bandgap of about 3.4 eV, niobium oxide has promising applications in bioelectrochemistry, solar cells, advanced catalysts, and electrochromic devices [190–195]. While alkaline niobates have emerged as novel materials with enormous technological and scientific interest because of their excellent nonlinear optical, ferroelectric, piezoelectric, electricoptic, ionic conductivity, pyroelectric, photorefractive, selective ion exchange, and photocatalytic properties. For instance, potassium/sodium niobates are potential substitutes for the lead zirconium titanate (PZT) as high-performance piezoelectric ceramics because of their piezoelectric characteristics and the feature that they are lead-free. Lithium niobate has been used for holographic data storage, frequency doubling, waveguide structures, solid-state lasers, and so on [196–198].

1.3.3.1 Nb_2O_5 Functional Thin Films and Nanostructure Evolution

Controlled growth of hierarchical architectures has received particular attention recently to support application demands for more complex structures to impart greater control over material and device properties. Although fabrication of structures with the complexity observed in natural materials or biominerals remains a significant challenge, substantial progress has been made in the control of branched hierarchical crystals.

In the following part, we investigate the thin film structural evolution during the treatment of Nb foil in aqueous hydrogen fluoride (HF) solution under hydrothermal conditions [104]. Such a structural evolution provides a marvelous opportunity for selecting the Nb_2O_5 thin films with the desired morphologies and structures through kinetic control of the reaction.

The oxidation process of metal Nb is quite analogous to the NaCl solution corrosion-based fabrication of ZnO nanowires mentioned before. In this case, we believe that F^- ion plays a role as catalyzing the oxidation of Nb. Nb is a very inert metal, and it is hard to form oxides in most cases at low temperatures. But as the most reactive element, F^- ions can release Nb^{5+} from Nb foil to form complexes, and then transform to Nb_2O_5 in certain conditions (Figure 1.29). The whole reaction was based on an oxidation etching process; O_2, F^- ions, and H_2O are all instrumental for the pitting on the surfaces of Nb foil. As documented in the literature, corrosion most commonly occurs when the O_2 in air is dissolved in a thin film of water on the surface of a metal, thereby picking up electrons from the cathode to form hydroxide ions that then migrate toward the anode. Adsorption of F^- ions can greatly increase the rate of surface diffusion and therefore expedite the oxidation process.

$$4Nb + 20F^- + 5O_2 + 6H_2O \rightarrow 4NbOF_5^{2-} + 12OH^- \tag{1.10}$$

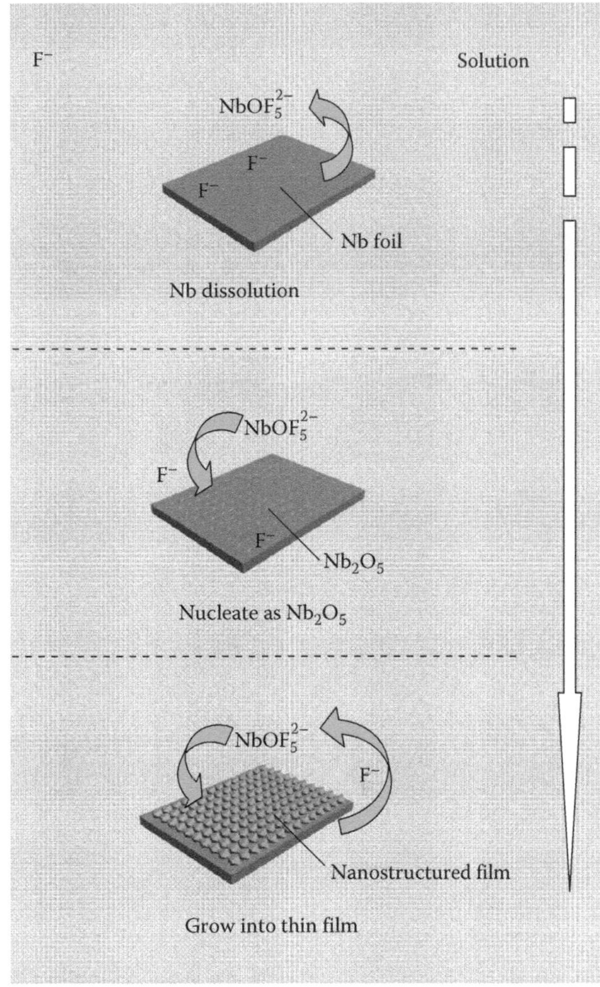

FIGURE 1.29 Illustration of the F^--assisted oxidation process of metal Nb.

$$2NbOF_5^{2-} + 3H_2O \rightarrow Nb_2O_5 + 10F^- + 6H^+ \quad (1.11)$$

After a reaction at elevated temperature for 2 days, a layer of Nb_2O_5 was formed on the surface of Nb foil. It has been found that the nanostructure of this oxide layer was very sensitive to the concentration of HF added in the solution. Figure 1.30 illustrates a structure evolution process of the thin film with different concentrations of HF, from which we can see that nanostructures including ribbons, polyhedrons, plates, and wires can be obtained via selecting different reaction conditions.

The HF can control Nb^{5+} solubility through two important effects on solution chemistry: (1) HF is acid and thus decreases the pH, and (2) the F^- complex with Nb^{5+}, which increases zinc solubility. Because Nb^{5+}, F^-, and H_2O form complexes, and precipitate as Nb_2O_5 and release H^+ via a dehydration reaction at elevated temperature, the pH is important since it can affect the balance of the dehydration reaction.

In this structure evolution process, it is more exciting to find that within a precisely controlled pH value range, branched nanowires can be obtained, and the branching degree can be deliberately adjusted. The most common branched nanocrystals formed in solution are multipod, star-shaped,

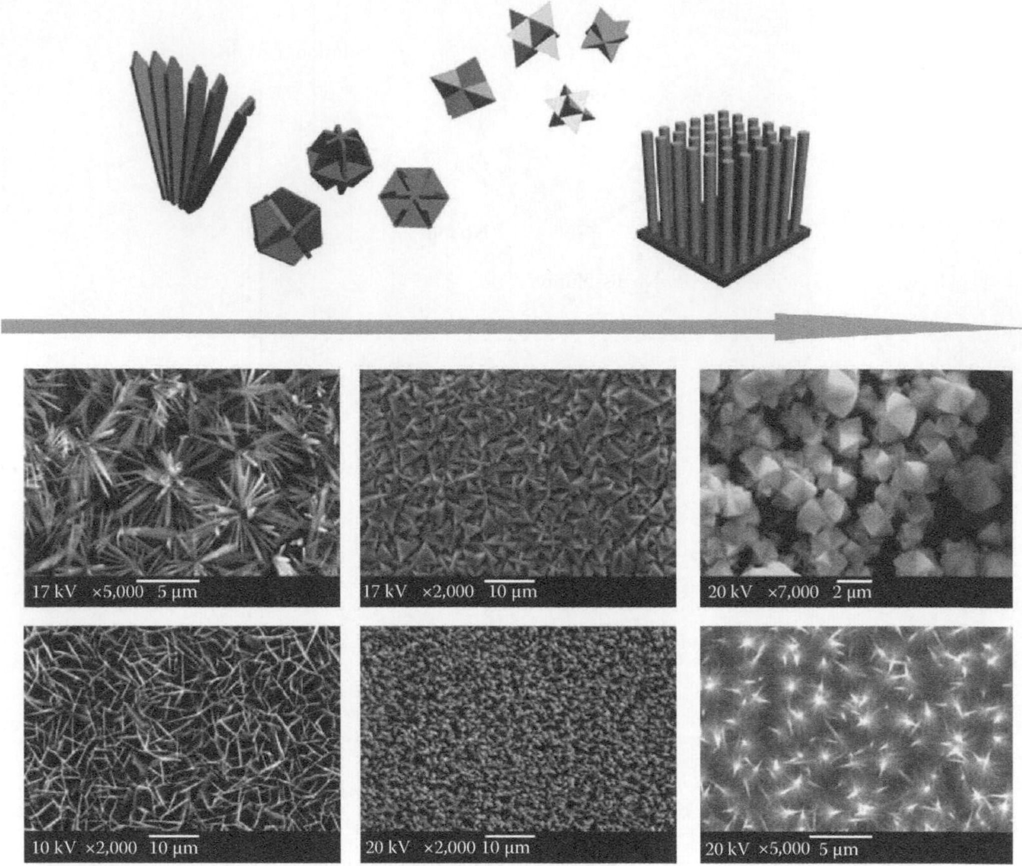

FIGURE 1.30 Illustration and SEM images of Nb_2O_5 thin films with different concentrations of HF.

and nanoflower or nanothorn structures, in which branches grow outward from a central nucleus in a single reaction stage. The underlying mechanism offered for many of these are either anisotropic growth in preferred directions from a twinned nucleus or polymorphic growth through stacking faults on the fastest-growing facets of the nucleus. Hyperbranched crystals with numerous branches might grow from a core of multiple nuclei, and other mechanisms might be inciting further nucleation on existing branched crystals, a phenomenon that is not well understood. Dendritic nanocrystals have also been observed and are often explained by a mechanism of oriented attachment but may also be created by renucleation events.

In the present system, tree-like branched structures were formed by further nucleation and growth of secondary dendrites on the primary stems, which grew perpendicularly on the Nb substrate; this kind of thin film can be produced with several square centimeters. Figure 1.31 shows SEM images of film obtained at different pH values, from which we can see that the branching degree of these nanotrees can be deliberately adjusted. With a reaction under pH 3.7, no branch on the nanorods were observed (Figure 1.31a), but a discrete transition occurs at pH 3.5, at which point very few branches grow (Figure 1.31b) and the branches are monodisperse, aligned needles (Figure 1.31c). Further decreasing the pH value encouraged the growth of the branches, and the density of braches have a significant increase, as well as their diameter, but the size of the central stem has a shrinking tendency. The branch growth follows a controlled, monotonic trend of increasing branch size and number density with decreasing pH value. At the upper limit of the branching pH value range, nanotrees with tertiary branches are observed (Figure 1.31d). The pH value range for branch growth

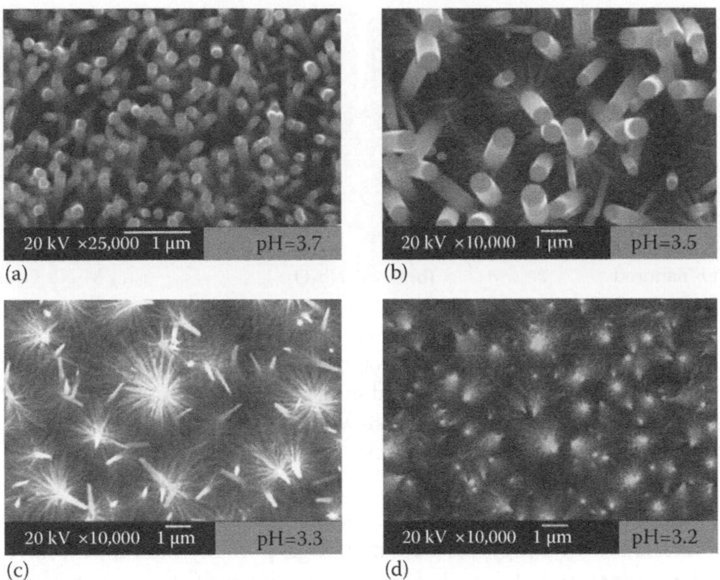

FIGURE 1.31 SEM images of Nb_2O_5 film obtained at different pH values; pH (a) 3.7; (b) 3.5; (c) 3.3; (d) 3.2.

is quite small, spanning less than 0.5. There are two critical pH values that bound the branch growth range: a critical nucleation value, above which no branches are nucleated; and a lowest value, below which leading to the total dissolution of Nb_2O_5, no oxide layer can be obtained on the metal surface. As the critical pH value is reached, branches grow until a concentration is reached at which the level of supersaturation is too low to overcome the free energy of nucleation. Further increase leads to the dissolution of the primary crystals. Thus, branch growth always occurs for HF concentrations that regulate Nb^{5+} just above saturation.

It is a very interesting phenomenon in secondary branch growth that the branches are aligned in a preferred orientation. Liu et al. have observed similar results in the fabrication of branched ZnO nanorods. They suggest a mechanism for this that is analogous to Ostwald ripening [127], whereby the nuclei with the most energetically favorable orientation grow at the expense of the misaligned crystals, which dissolve and recrystallize by epitaxial growth on the oriented crystals. This process is driven by the minimization of interfacial energy between the new branches and the primary stem, which lowers the overall energy of the system after the nucleation burst.

The branch nucleation and growth process includes four steps: Step 1: When the critical pH value is reached, light surface etching induces nucleation of a polycrystalline layer by increasing the surface energy and creating nucleation sites. Step 2: Nuclei that are aligned in the most energetically favorable orientation grow at the expense of the misaligned nuclei. This is energetically favored because it reduces the interfacial stress at the primary stem surface. Step 3: The misaligned nuclei are ultimately eliminated in favor of the oriented branches, which continue to grow larger at a higher rate along the 1D direction. Step 4: Growth stops when Nb in solution is reduced below the solubility limit.

1.3.3.2 Nb_2O_5, $NaNbO_3$, and $KNbO_3$ Porous Thin Films by Secondary Conversion Method

The structural evolution process described above provides a marvelous ability to fabricating Nb_2O_5 thin films with the desired morphologies and structures through kinetic control of the reaction, and also provides an opportunity for employing these finely controlled nanostructures as excellent precursors to fabricate thin films with more advanced architecture.

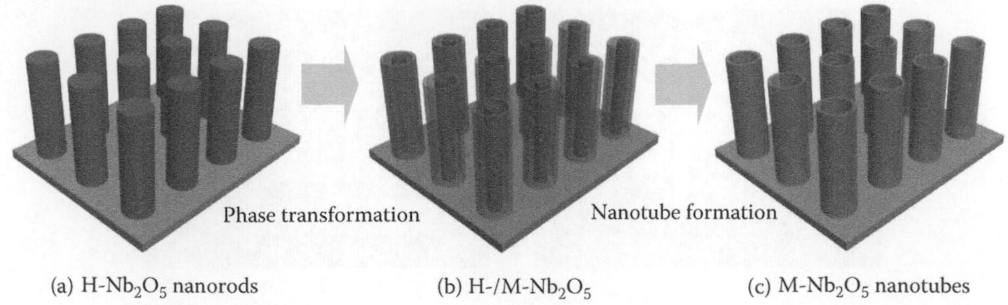

FIGURE 1.32 Schematic illustration of the Nb_2O_5 phase transformation accompanied by void generation as it is applied to the synthesis of monoclinic Nb_2O_5 nanotube arrays: (a) H-Nb_2O_5 nanorod array is used as starting precursor; (b) phase transformation at the nanorod surface leads to the formation of core/shell Nb_2O_5 with H-Nb_2O_5 as the core and M-Nb_2O_5 as the shell; (c) hollow Nb_2O_5 nanotube arrays are formed finally based on phase transformation during the void formation process.

Tubular nanomaterials with a high surface area, high-porosity nanotubes, have emerged as an intriguing class of nanomaterials because of their superior performance in applications such as electronics, optoelectronics, energy storage, sensing, drug delivery, and nanodevices. It has been discussed in Section 1.3.1.4 that thin films of ZnS constructed by nanotubes can be rationally obtained via the secondary conversion of ZnO nanorod arrays. The tube formation process was driven by the difference of the diffusion rate at the interface of ZnO/ZnS, which can be called Kirkendall effect. Here, we demonstrate an important extension of the strategy of fabricating hollow structures based on Kirkendall effect. A thin film of Nb_2O_5 constructed by nanorods was employed as the starting material. The void formation process induced by phase transition was observed in the conversion of Nb_2O_5 nanorods to nanotubes, where voids can be intriguingly generated during phase transformation [102].

Nb_2O_5 is found to exist in different crystal phases, H-Nb_2O_5 (pseudo-hexagonal), O-Nb_2O_5 (orthorhombic), and M-Nb_2O_5 (monoclinic). Among these polymorphic forms, the M-phase is most stable from a thermodynamical viewpoint, while the H-phase is the least stable one and can be readily transformed into the M-phase by heat treatment at a high temperature.

A schematic illustration of the tube formation process is shown in Figure 1.32. Monoclinic Nb_2O_5 nanotube arrays can be successfully synthesized from pseudo-hexagonal Nb_2O_5 nanorod arrays, based on the phase transformation accompanied by void formation. H-Nb_2O_5 nanorod films synthesized via the chemical route described in Section 1.3.3.1 were employed as the starting precursors. To achieve the phase transformation from the H- to M-Nb_2O_5, the H-Nb_2O_5 nanorods are treated in a mixed solution of $NH_2CH_2CH_2NH_2$ and lithium hydroxide under hydrothermal condition. Nonequilibrium interdiffusion is involved during the Nb_2O_5 phase-transformation process, leading to Nb_2O_5 core/shell structure with H-Nb_2O_5 as the core and M-Nb_2O_5 as the shell; with elongated reaction time, hollow M-Nb_2O_5 nanotube arrays can be obtained.

To confirm the phase-transition process, XRD characterization was carried out along with micromorphology analysis. XRD patterns shown in Figure 1.33a identified the precursors and products as H-Nb_2O_5 (pseudo-hexagonal) and M-Nb_2O_5 (monoclinic). SEM images of the H-Nb_2O_5 nanorod precursor and prepared M-Nb_2O_5 nanotube arrays are shown in Figure 1.33b. The voids formation during the phase-transition process was observed in the synthesis of Nb_2O_5 nanotube arrays. As shown in Figure 1.33b, the precursor H-Nb_2O_5 films were composed of well-aligned nanorod arrays that grew perpendicularly to the niobium substrate. Due to the relatively greater thermodynamic stability of M-Nb_2O_5, phase transformation occurred spontaneously under designed solution environment. Nonequilibrium interdiffusion is involved during the

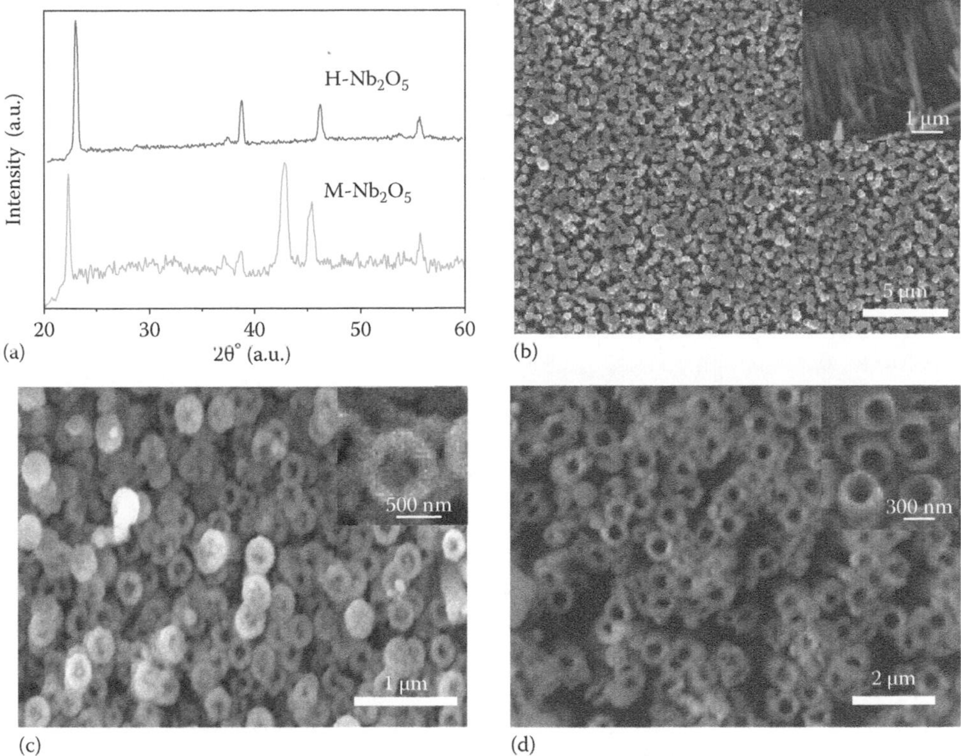

FIGURE 1.33 XRD patterns and SEM images of H-Nb_2O_5 precursor and M-Nb_2O_5 nanotubes, (a) XRD patterns of H-Nb_2O_5 precursor and M-Nb_2O_5 nanotubes; SEM images of (b) H-Nb_2O_5 nanorod arrays; (c) H-/M-Nb_2O_5 nanocables; (d) M-Nb_2O_5 nanotubes.

Nb_2O_5 phase-transformation process, leading to the core/shell structure with H-Nb_2O_5 as core and M-Nb_2O_5 as shell. Figure 1.33c displays the SEM image of products obtained at a reaction period of 6 h; a clear gap can be seen between these two phases. With reaction time prolonged to 48 h, the phase-transformation process was completed, finally resulting in the pure M-Nb_2O_5 nanotube arrays, as shown in Figure 1.33d.

An essential factor for achieving nonequilibrium interdiffusion in this case is the energy difference between H- and M-phase Nb_2O_5, which is the driven force of phase transformation. From a crystallographic viewpoint, H-Nb_2O_5 possesses a pseudo-hexagonal unit cell and forms tetragonal and pentagonal bipyramids with six or seven oxygen atoms coordinated to the Nb atom. Differently, the structure of M-Nb_2O_5 is composed of corner- or edge-sharing NbO_6 octahedral. Therefore, rearrangement of niobium and oxygen atoms is required to accomplish phase transformation from H- to M-Nb_2O_5. The surface atoms are shifted from the seven-coordination state to the six-coordination state. This favorable reorganization of the structure stabilizes the six-coordinate M-Nb_2O_5 structure, so the stable M-Nb_2O_5 nanotubes are preferred during nucleation and growth at the expense of less stable H-Nb_2O_5 nanorods.

The phase transition initially occurred on the surfaces of the H-Nb_2O_5 nanorods, forming a thin layer of M-Nb_2O_5 surrounding the H-Nb_2O_5 core. Further, the shell grows at the expense of the core materials. Therefore, the core and the shell start to get separated by a clear gap as revealed by the difference in contrast visible in the SEM image of Figure 1.33. Small voids can be observed between H-Nb_2O_5 core and M-Nb_2O_5 shell, indicating condensation of vacancies at the boundary. With prolonging growth time, small voids gradually coalesce into bigger ones and the H-Nb_2O_5

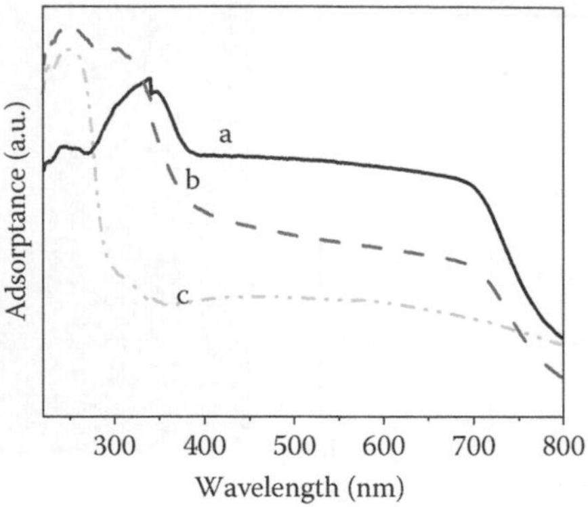

FIGURE 1.34 UV-vis spectra of Nb_2O_5 with different phase compositions, (a) H-Nb_2O_5 nanorod arrays with a pure pseudo-hexagonal phase; (b) core/shelled Nb_2O_5 arrays with a mixed phase of both pseudo-hexagonal and monoclinic phases; (c) M-Nb_2O_5 nanotube arrays with a pure monoclinic phase.

core gradually disappears. Further phase transition and void formation depend on the diffusion or migration of H-Nb_2O_5 (either in the form of niobium and oxygen ions, or in the form of a niobium–oxygen cluster) through the formed M-Nb_2O_5 shell. $NH_2CH_2CH_2NH_2$ molecules are introduced to significantly reduce the activation barrier for phase transition, which can accelerate the outward diffusion of core material through M-Nb_2O_5 shells. The increased interface diffusion of the atoms during the phase-transition process enhances void formation. Therefore, as the reaction time proceeds, more H-Nb_2O_5 core materials diffuse out to the shell and the accompanying transport of vacancies leads to the merging and growth of voids, forming hollow M-Nb_2O_5 nanotubes (Figure 1.33d). The diameter of the final M-Nb_2O_5 nanotubes is larger than that of the H-Nb_2O_5 nanorods, which is due to the hollow interior of M-Nb_2O_5.

This conversion process not only leads to porous thin films, but also makes the manipulation of bandgap in a relatively large regime possible. The UV-vis absorption spectra of films with different phase compositions ranging from H- to M-Nb_2O_5 are shown in Figure 1.34, from which we can see the transformation from H-Nb_2O_5 nanorods to M-Nb_2O_5 nanotubes is accompanied by remarkable changes in the UV-vis absorption properties. In detail, the starting H-Nb_2O_5 nanorods arrays possessed an absorption peak at around 340 nm (Figure 1.34). When the phase transition begins, this peak is gradually blue-shifted. After the formation of M-Nb_2O_5 nanotube arrays, the peak is blue-shifted to below 300 nm (Figure 1.34). Since both bulk H- and M-Nb_2O_5 have the same bandgap, this blue-shift should be attributed to the quantum size effect in the monoclinic Nb_2O_5 nanotube arrays. Although the diameter of tubes is relatively thick, the walls are thin enough to reach the quantum confinement region. The UV-vis peak of the obtained Nb_2O_5 nanostructures could be tuned in a large regime from 340 to 250 nm. This offers huge opportunity for the application in optoelectric devices.

The results shown above suggest that phase transition can also be used in fabricating nanomaterials, since it is a common phenomenon in a wide range of materials. This discovery will motivate theoretical and experimental studies in the fabrication of functional hollow structures.

Thin films that can be fabricated via this secondary conversion strategy are not limited to Nb_2O_5 nanotube arrays. Based on the branched nanorods described in the previous part, arrays of nanotube with branched structure can also be achieved [104]. The branched tube formation is implemented by treating branched Nb_2O_5 nanorod arrays in NaOH or KOH solution. The void generation is driven by the chemical reaction between Nb_2O_5 and NaOH (KOH), which produces $NaNbO_3$ ($KNbO_3$) dielectric thin films [104].

Figure 1.35a and b shows the SEM images of the above hierarchical nanostructures after treating in alkaline solution. The samples conserve the branched morphology of the starting Nb_2O_5 as

FIGURE 1.35 SEM (a,b) and TEM (c) images of $NaNbO_3$ branched nanotubes. (d) SAED pattern.

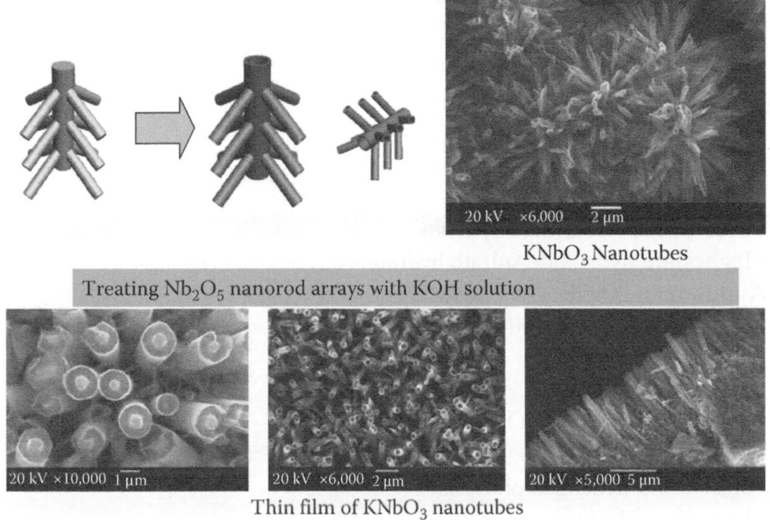

FIGURE 1.36 Illustration and SEM images of $KNbO_3$ branched nanotubes.

demonstrated in Figure 1.35a and b. TEM investigation (Figure 1.35c) reveals that all the nanobranches have been transferred into nanotubes. From the corresponding electron diffraction (ED) pattern in Figure 1.35d, this indicates the single crystal nature of these tubes, and a complete transformation of Nb_2O_5 into $NaNbO_3$. The formation of nanotubes from interfacial liquid–solid reaction of branched nanorods and alkaline solution can be explained by the nanoscale Kirkendall effect, where Nb_2O_5 is identified as the dominant diffusing species. Similar structures of $KNbO_3$ can also be obtained, which is shown in Figure 1.36.

Because of the large surface area and hierarchical hollow nanostructure, these thin films are of particular interests for possible applications in filters, solar cells, catalysts, and controlled release.

1.4 SUMMARY AND OUTLOOK

The rational fabrication of advanced nanostructured thin films was demonstrated by using various chemical strategies such as precipitation, hydrothermal, and thermal oxidation methods. The ability to manipulate atoms and molecules provides a powerful arsenal for modeling, design, and fabrication of purpose-built functional thin films with well-defined textures onto various substrates at low temperature and at a low cost. Such purpose-built films should contribute to the development of novel and optimized devices for high technological applications, as well as biological and medical related ones. This chapter shows that even though some significant challenges are existed in these chemical routes, such as precise film thickness control and particle locating, it is still promising to improve these chemical techniques for nanodevices due to their many attractive advantages. Some recent studies show that programmed DNA can be introduced to control the crystallization of nanomaterials [199–204]. These approaches may push the chemical (bottom-up) fabrication of nanostructured thin films to a new level.

ACKNOWLEDGMENTS

The authors gratefully acknowledge the financial support of the program for New Century Excellent Talents in University (Grant No. NCET-05-0278), the Natural Science Foundation of China (Grant No. 20471012), a Foundation for the Author of National Excellent Doctoral Dissertation of P. R. China (Grant No. 200322), and the Research Fund for the Doctoral Program of Higher Education (Grant No. 20040141004).

ABBREVIATIONS

1D	One-dimensional
2D	Two-dimensional
3DOM	Three-dimensional macroporous
512MHSH	Magnesium hydroxide sulfate hydrate
ADP	Ammonium dihydrogen phosphate
CVD	Chemical vapor deposition
ED	Electron diffraction
HMT	Hexamethyltetramine
H-Nb_2O_5	Pseudo-hexagonal Nb_2O_5
IR	Infrared
KDP	Potassium dihydrogen phosphate
M-Nb_2O_5	Monoclinic Nb_2O_5
O-Nb_2O_5	Orthorhombic Nb_2O_5
PL	Photoluminescence
PZT	Lead zirconium titanate
SEM	Scanning electron microscope
SHE	Standard hydrogen electrode
TEM	Transmission electron microscopy
UV	Ultraviolet
VLS	Vapor–liquid–solid
XRD	X-ray diffraction

SYMBOLS

A Area of selected plane
D Interplanar distance
K Constant for different planes in an equilibrium system
N Bond number
R Vertical growth rate of the selected plane
S Bond strength

REFERENCES

1. Talapin, D. V. and Murray, C. B. 2005. PbSe nanocrystal solids for n- and p-channel thin film field-effect transistors. *Science* 310 (5745): 86–89.
2. Byrd, H., Suponeva, E. P., Bocarsly, A. B., and Thompson, M. E. 1996. Photocurrent generation in metal bisphosphonate multilayer thin films. *Nature* 380 (6575): 610–612.
3. Yang, P. D., Wirnsberger, G., Huang, H. C., Cordero, S. R., McGehee, M. D., Scott, B., Deng, T., Whitesides, G. M., Chmelka, B. F., Buratto, S. K., and Stucky, G. D. 2000. Mirrorless lasing from mesostructured waveguides patterned by soft lithography. *Science* 287 (5452): 465–467.
4. Weisheit, M., Fahler, S., Marty, A., Souche, Y., Poinsignon, C., and Givord, D. 2007. Electric field-induced modification of magnetism in thin-film ferromagnets. *Science* 315 (5810): 349–351.
5. Allwood, D. A., Xiong, G., Faulkner, C. C., Atkinson, D., Petit, D., and Cowburn, R. P. 2005. Magnetic domain-wall logic. *Science* 309 (5741): 1688–1692.
6. Martin, C. R. 1994. Nanomaterials–A membrane-based synthetic approach. *Science* 266 (5193): 1961–1966.
7. Amaratunga, G. A. J., Chhowalla, M., Kiely, C. J., Alexandrou, I., Aharonov, R., and Devenish, R. M. 1996. Hard elastic carbon thin films from linking of carbon nanoparticles. *Nature* 383 (6598): 321–323.
8. Chhowalla, M. and Unalan, H. E. 2005. Thin films of hard cubic Zr_3N_4 stabilized by stress. *Nature Materials* 4 (4): 317–322.
9. Chhowalla, M. and Amaratunga, G. A. J. 2000. Thin films of fullerene-like MoS_2 nanoparticles with ultra-low friction and wear. *Nature* 407 (6801): 164–167.
10. Bhushan, B., Israelachvili, J. N., and Landman, U. 1995. Nanotribology: Friction, wear and lubrication at the atomic-scale. *Nature* 374 (6523): 607–616.
11. Hood, L., Heath, J. R., Phelps, M. E., and Lin, B. Y. 2004. Systems biology and new technologies enable predictive and preventative medicine. *Science* 306 (5296): 640–643.
12. Williams, J. R., DiCarlo, L., and Marcus, C. M. 2007. Quantum hall effect in a gate-controlled p-n junction of graphene. *Science* 317 (5838): 638–641.
13. Service, R. F. 2001. Solid-state physics: Nanotube 'peapods' show electrifying promise. *Science* 292 (5514): 45–45.
14. Doh, Y. J., van Dam, J. A., Roest, A. L., Bakkers, E. P. A. M., Kouwenhoven, L. P., and De Franceschi, S. 2005. Tunable supercurrent through semiconductor nanowires. *Science* 309 (5732): 272–275.
15. Leong, M., Doris, B., Kedzierski, J., Rim, K., and Yang, M. 2004. Silicon device scaling to the sub-10-nm regime. *Science* 306 (5704): 2057–2060.
16. Guo, X. J. and Silva, S. R. P. 2008. Engineering: High-performance transistors by design. *Science* 320 (5876): 618–619.
17. Akahane, Y., Asano, T., Song, B. S., and Noda, S. 2003. High-Q photonic nanocavity in a two-dimensional photonic crystal. *Nature* 425 (6961): 944–947.
18. Whitesides, G. M. and Ismagilov, R. F. 1999. Complexity in chemistry. *Science* 284 (5411): 89–92.
19. Ugarte, D., Chatelain, A., and deHeer, W. A. 1996. Nanocapillarity and chemistry in carbon nanotubes. *Science* 274 (5294): 1897–1899.
20. Service, R. F. 2006. Chemistry–New in nanotech: Self-folding delivery boxes. *Science* 313 (5790): 1032–1033.
21. Pitcher, M. W. 2006. Nanochemistry—A chemical approach to nanomaterials. *Science* 313 (5785): 300.
22. Michalet, X., Pinaud, F. F., Bentolila, L. A., Tsay, J. M., Doose, S., Li, J. J., Sundaresan, G., Wu, A. M., Gambhir, S. S., and Weiss, S. 2005. Quantum dots for live cells, in vivo imaging, and diagnostics. *Science* 307 (5709): 538–544.
23. Klimov, V. I., Ivanov, S. A., Nanda, J., Achermann, M., Bezel, I., McGuire, J. A., and Piryatinski, A. 2007. Single-exciton optical gain in semiconductor nanocrystals. *Nature* 447 (7143): 441–446.

24. Wang, J., Neaton, J. B., Zheng, H., Nagarajan, V., Ogale, S. B., Liu, B., Viehland, D., Vaithyanathan, V., Schlom, D. G., Waghmare, U. V., Spaldin, N. A., Rabe, K. M., Wuttig, M., and Ramesh, R. 2003. Epitaxial $BiFeO_3$ multiferroic thin film heterostructures. *Science* 299 (5613): 1719–1722.
25. Lee, H. N., Hesse, D., Zakharov, N., and Gosele, U. 2002. Ferroelectric $Bi_{3.25}La_{0.75}Ti_3O_{12}$ films of uniform a-axis orientation on silicon substrates. *Science* 296 (5575): 2006–2009.
26. Lange, F. F. 1996. Chemical solution routes to single-crystal thin films. *Science* 273 (5277): 903–909.
27. Yang, H., Kuperman, A., Coombs, N., MamicheAfara, S., and Ozin, G. A. 1996. Synthesis of oriented films of mesoporous silica on mica. *Nature* 379 (6567): 703–705.
28. Shi, H. Q., Tsai, W. B., Garrison, M. D., Ferrari, S., and Ratner, B. D. 1999. Template-imprinted nanostructured surfaces for protein recognition. *Nature* 398 (6728): 593–597.
29. Kawakami, R. K., Rotenberg, E., Choi, H. J., Escorcia-Aparicio, E. J., Bowen, M. O., Wolfe, J. H., Arenholz, E., Zhang, Z. D., Smith, N. V., and Qiu, Z. Q. 1999. Quantum-well states in copper thin films. *Nature* 398 (6723): 132–134.
30. Bachtold, A., Hadley, P., Nakanishi, T., and Dekker, C. 2001. Logic circuits with carbon nanotube transistors. *Science* 294 (5545): 1317–1320.
31. Melosh, N. A., Boukai, A., Diana, F., Gerardot, B., Badolato, A., Petroff, P. M., and Heath, J. R. 2003. Ultrahigh-density nanowire lattices and circuits. *Science* 300 (5616): 112–115.
32. Zhang, X. Y., Zhu, Y. X., and Granick, S. 2002. Hydrophobicity at a Janus interface. *Science* 295 (5555): 663–666.
33. Xia, F. and Jiang, L. 2008. Bio-inspired, smart, multiscale interfacial materials. *Advanced Materials* 20 (15): 2842–2858.
34. Law, M., Greene, L. E., Johnson, J. C., Saykally, R., and Yang, P. D. 2005. Nanowire dye-sensitized solar cells. *Nature Materials* 4 (6): 455–459.
35. Tarascon, J. M. and Armand, M. 2001. Issues and challenges facing rechargeable lithium batteries. *Nature* 414 (6861): 359–367.
36. Wang, Y., Zeng, H. C., and Lee, J. Y. 2006. Highly reversible lithium storage in porous SnO_2 nanotubes with coaxially grown carbon nanotube overlayers. *Advanced Materials* 18 (5): 645.
37. Chmiola, J., Yushin, G., Gogotsi, Y., Portet, C., Simon, P., and Taberna, P. L. 2006. Anomalous increase in carbon capacitance at pore sizes less than 1 nanometer. *Science* 313 (5794): 1760–1763.
38. Dikin, D. A., Stankovich, S., Zimney, E. J., Piner, R. D., Dommett, G. H. B., Evmenenko, G., Nguyen, S. T., and Ruoff, R. S. 2007. Preparation and characterization of graphene oxide paper. *Nature* 448 (7152): 457–460.
39. Heben, M. J. and Dillon, A. C. 2000. Room-temperature hydrogen storage in nanotubes. *Science* 287 (5453): 593.
40. Chen, P., Xiong, Z. T., Luo, J. Z., Lin, J. Y., and Tan, K. L. 2002. Interaction of hydrogen with metal nitrides and imides. *Nature* 420 (6913): 302–304.
41. Xiao, Y., Patolsky, F., Katz, E., Hainfeld, J. F., and Willner, I. 2003. "Plugging into enzymes": Nanowiring of redox enzymes by a gold nanoparticle. *Science* 299 (5614): 1877–1881.
42. Shah, R. R. and Abbott, N. L. 2001. Principles for measurement of chemical exposure based on recognition-driven anchoring transitions in liquid crystals. *Science* 293 (5533): 1296–1299.
43. Kong, J., Franklin, N. R., Zhou, C. W., Chapline, M. G., Peng, S., Cho, K. J., and Dai, H. J. 2000. Nanotube molecular wires as chemical sensors. *Science* 287 (5453): 622–625.
44. Janata, J. and Josowicz, M. 2003. Conducting polymers in electronic chemical sensors. *Nature Materials* 2 (1): 19–24.
45. Puntes, V. F., Krishnan, K. M., and Alivisatos, A. P. 2001. Colloidal nanocrystal shape and size control: The case of cobalt. *Science* 291 (5511): 2115–2117.
46. Yin, Y. and Alivisatos, A. P. 2005. Colloidal nanocrystal synthesis and the organic-inorganic interface. *Nature* 437 (7059): 664–670.
47. Service, R. F. 1999. Nanotechnology–Patterning plastic with plentiful pillars. *Science* 286 (5442): 1067.
48. Park, M., Harrison, C., Chaikin, P. M., Register, R. A., and Adamson, D. H. 1997. Block copolymer lithography: Periodic arrays of similar to 10(11) holes in 1 square centimeter. *Science* 276 (5317): 1401–1404.
49. Jacobs, H. O. and Whitesides, G. M. 2001. Submicrometer patterning of charge in thin-film electrets. *Science* 291 (5509): 1763–1766.
50. Xia, Y. N. and Whitesides, G. M. 1998. Soft lithography. *Angewandte Chemie-International Edition* 37 (5): 551–575.
51. Liu, M. and Xue, D. 2007. Effect of heating rate on the crystal composition of ferroelectric lithium niobate crystallites. *Journal of Alloys and Compounds* 427 (1–2): 256–259.

52. Liu, M., Xue, D., and Luo, C. 2006. Facile synthesis of lithium niobate squares by a combustion route. *Journal of the American Ceramic Society* 89 (5): 1551–1556.
53. Wang, L., Kang, H., Li, K., Xue, D., and Liu, C. 2008. Phase evolution of BaTiO3 nanoparticles: An identification of BaTi2O5 intermediate phase in calcined stearic acid gel. *Journal of Physical Chemistry C* 112 (7): 2382–2388.
54. Liu, F., Sun, C., Yan, C., and Xue, D. 2008. Solution-based chemical strategies to purposely control the microstructure of functional materials. *Journal of Materials Science & Technology* 24 (4): 641–648.
55. Liu, M. and Xue, D. 2008. Amine-assisted route to fabricate $LiNbO_3$ particles with a tunable shape. *Journal of Physical Chemistry C* 112 (16): 6346–6351.
56. Liu, M., Xue, D., and Li, K. 2008. Soft-chemistry synthesis of $LiNbO_3$ crystallites. *Journal of Alloys and Compounds* 449 (1–2): 28–31.
57. Liu, M. and Xue, D. 2005. A solvothermal route to crystalline lithium niobate. *Materials Letters* 59 (23): 2908–2910.
58. Liu, M., Xue, D., Zhang, S., Zhu, H., Wang, J., and Kitamura, K. 2005. Chemical synthesis of stoichiometric lithium niobate powders. *Materials Letters* 59 (8–9): 1095–1097.
59. Luo, C. and Xue, D. 2006. Mild, quasireverse emulsion route to submicrometer lithium niobate hollow spheres. *Langmuir* 22 (24): 9914–9918.
60. Wang, L., Liu, L., Xue, D., Kang, H., and Liu, C. 2007. Wet routes of high purity BaTiO3 nanopowders. *Journal of Alloys and Compounds* 440 (1–2): 78–83.
61. Xu, J. and Xue, D. 2006. Chemical synthesis of $BaCO_3$ with a hexagonal pencil-like morphology. *Journal of Physics and Chemistry of Solids* 67 (7): 1427–1431.
62. Xu, J. and Xue, D. 2005. Fabrication of malachite with a hierarchical sphere-like architecture. *Journal of Physical Chemistry B* 109 (36): 17157–17161.
63. Xu, J. and Xue, D. 2006. Fabrication of copper hydroxyphosphate with complex architectures. *Journal of Physical Chemistry B* 110 (15): 7750–7756.
64. Xu, J. and Xue, D. 2007. Five branching growth patterns in the cubic crystal system: A direct observation of cuprous oxide microcrystals. *Acta Materialia* 55 (7): 2397–2406.
65. Xu, J. and Xue, D. 2007. Hydrothermal synthesis of lindgrenite with a hollow and prickly sphere-like architecture. *Journal of Solid State Chemistry* 180 (1): 119–126.
66. Xu, J., Xue, D., and Yan, C. 2005. Chemical synthesis of $NaTaO_3$ powder at low-temperature. *Materials Letters* 59 (23): 2920–2922.
67. Yan, C., Sun, C., Shi, Y., and Xue, D. 2007. Surface fabrication of oxides via solution chemistry, in *15th International Conference on Crystal Growth*, Salt Lake City, UT, pp. 1708–1712.
68. Yan, C. and Xue, D. 2005. Novel self-assembled MgO nanosheet and its precursors. *Journal of Physical Chemistry B* 109 (25): 12358–12361.
69. Yan, C. and Xue, D. 2006. Morphosynthesis of hierarchical hydrozincite with tunable surface architectures and hollow zinc oxide. *Journal of Physical Chemistry B* 110 (23): 11076–11080.
70. Yan, C. and Xue, D. 2006. Room temperature fabrication of hollow ZnS and ZnO architectures by a sacrificial template route. *Journal of Physical Chemistry B* 110 (14): 7102–7106.
71. Yan, C. and Xue, D. 2006. General, spontaneous ion replacement reaction for the synthesis of micro- and nanostructured metal oxides. *Journal of Physical Chemistry B* 110 (4): 1581–1586.
72. Yan, C. and Xue, D. 2007. Mild solution-based fabrication of high-quality device-dependent ZnO nanoarrays and ZnS nanotube arrays. *Physica Scripta* T129: 288–292.
73. Yan, C. and Xue, D. 2007. Polyhedral construction of hollow ZnO microspheres by CO_2 bubble templates. *Journal of Alloys and Compounds* 431 (1–2): 241–245.
74. Yan, C. and Xue, D. 2008. A modified electroless deposition route to dendritic Cu metal nanostructures. *Crystal Growth & Design* 8 (6): 1849–1854.
75. Yan, C., Xue, D., and Zou, L. 2006. Fabrication of hexagonal MgO and its precursors by a homogeneous precipitation method. *Materials Research Bulletin* 41 (12): 2341–2348.
76. Yan, C., Xue, D., and Zou, L. 2008. A solution-phase approach to the chemical synthesis of ZnO nanostructures via a low-temperature route. *Journal of Alloys and Compounds* 453 (1–2): 87–92.
77. Yan, C., Xue, D., Zou, L., Yan, X., and Wang, W. 2005. Preparation of magnesium hydroxide nanoflowers. *Journal of Crystal Growth* 282 (3–4): 448–454.
78. Zhu, Y., Bando, Y., and Xue, D. 2003. Spontaneous growth and luminescence of zinc sulfide nanobelts. *Applied Physics Letters* 82 (11): 1769–1771.
79. Zhu, Y., Bando, Y., Xue, D., and Golberg, D. 2003. Nanocable-aligned ZnS tetrapod nanocrystals. *Journal of the American Chemical Society* 125 (52): 16196–16197.

80. Zhu, Y., Bando, Y., Xue, D., and Golberg, D. 2004. Oriented assemblies of ZnS one-dimensional nanostructures. *Advanced Materials* 16 (9–10): 831–834.
81. Zhu, Y., Bando, Y., Xue, D., Sekiguchi, T., Golberg, D., Xu, F., and Liu, Q. 2004. New boron nitride whiskers: Showing strong ultraviolet and visible light luminescence. *Journal of Physical Chemistry B* 108 (20): 6193–6196.
82. Zhu, Y., Bando, Y., Xue, D., Xu, F., and Golberg, D. 2003. Insulating tubular BN sheathing on semiconducting nanowires. *Journal of the American Chemical Society* 125 (47): 14226–14227.
83. Wang, X., Zhuang, J., Peng, Q., and Li, Y. D. 2005. A general strategy for nanocrystal synthesis. *Nature* 437 (7055): 121–124.
84. Parilla, P. A., Dillon, A. C., Jones, K. M., Riker, G., Schulz, D. L., Ginley, D. S., and Heben, M. J. 1999. The first true inorganic fullerenes? *Nature* 397 (6715): 114.
85. Tenne, R. 2006. Inorganic nanotubes and fullerene-like nanoparticles. *Nature Nanotechnology* 1 (2): 103–111.
86. Deville, S., Saiz, E., Nalla, R. K., and Tomsia, A. P. 2006. Freezing as a path to build complex composites. *Science* 311 (5760): 515–518.
87. Gur, I. 2005. Air-stable all-inorganic nanocrystal solar cells processed from solution (21 Oct, pg 462, 2005). *Science* 310 (5754): 1618.
88. Burda, C., Chen, X. B., Narayanan, R., and El-Sayed, M. A. 2005. Chemistry and properties of nanocrystals of different shapes. *Chemical Reviews* 105 (4): 1025–1102.
89. Vayssieres, L. 2004. On the design of advanced metal oxide nanomaterials. *International Journal of Nanotechnology* 1 (1): 1–41.
90. Renaud, G., Lazzari, R., Revenant, C., Barbier, A., Noblet, M., Ulrich, O., Leroy, F., Jupille, J., Borensztein, Y., Henry, C. R., Deville, J. P., Scheurer, F., Mane-Mane, J., and Fruchart, O. 2003. Real-time monitoring of growing nanoparticles. *Science* 300 (5624): 1416–1419.
91. Robinson, R. D., Sadtler, B., Demchenko, D. O., Erdonmez, C. K., Wang, L. W., and Alivisatos, A. P. 2007. Spontaneous superlattice formation in nanorods through partial cation exchange. *Science* 317 (5836): 355–358.
92. Xu, D. and Xue, D. 2005. Morphology control of KDP crystallites. *Physica B-Condensed Matter* 370 (1–4): 84–89.
93. Xu, D. and Xue, D. 2006. Chemical bond calculations of crystal growth of KDP and ADP. *Journal of Rare Earths* 24: 144–148.
94. Xu, D. and Xue, D. 2006. Crystallographic analysis of tapering of ADP crystallites. *Journal of Rare Earths* 24: 228–232.
95. Xu, D. and Xue, D. 2006. Chemical bond analysis of the crystal growth of KDP and ADP. *Journal of Crystal Growth* 286 (1): 108–113.
96. Xu, D. and Xue, D. 2008. Computational study of crystal growth habit and cleavage. *Journal of Alloys and Compounds* 449 (1–2): 353–356.
97. Xu, D., Xue, D., and Ratajczak, H. Y. 2005. Morphology and structure studies of KDP and ADP crystallites in the water and ethanol solutions. *Journal of Molecular Structure* 740 (1–3): 37–45.
98. Yan, X., Xu, D., and Xue, D. 2007. SO_4^{2-} ions direct the one-dimensional growth of $5Mg(OH)_2 \cdot MgSO_4 \cdot 2H_2O$. *Acta Materialia* 55 (17): 5747–5757.
99. Sun, C. and Xue, D. 2008. In *Astrophysics and Condensed Matter*, Hardwell, T. G. (Ed.). Hauppauge, NY: Nova Science Publishers.
100. Yan, C., Zou, L., Xue, D., Xu, J., and Liu, M. 2008. Chemical tuning polymorphology of functional materials by hydrothermal and solvothermal reactions. *Journal of Materials Science* 43 (7): 2263–2269.
101. Li, J. and Zeng, H. C. 2007. Hollowing Sn-doped TiO_2 nanospheres via Ostwald ripening. *Journal of the American Chemical Society* 129 (51): 15839–15847.
102. Yan, C. and Xue, D. 2008. Formation of Nb_2O_5 nanotube arrays through phase transformation. *Advanced Materials* 20 (5): 1055–1058.
103. Fan, H. J., Werner, P., and Zacharias, M. 2006. Semiconductor nanowires: From self-organization to patterned growth. *Small* 2 (6): 700–717.
104. Liu, F. and Xue, D. Unpublished work.
105. Byrappa, K. and Ohachi, T. 2003. *Crystal Growth Technology*. New York: William Andrew.
106. Huang, M. H., Mao, S., Feick, H., Yan, H. Q., Wu, Y. Y., Kind, H., Weber, E., Russo, R., and Yang, P. D. 2001. Room-temperature ultraviolet nanowire nanolasers. *Science* 292 (5523): 1897–1899.
107. Wang, X. D., Song, J. H., Liu, J., and Wang, Z. L. 2007. Direct-current nanogenerator driven by ultrasonic waves. *Science* 316 (5821): 102–105.

108. Wang, Z. L. and Song, J. H. 2006. Piezoelectric nanogenerators based on zinc oxide nanowire arrays. *Science* 312 (5771): 242–246.
109. Qin, Y., Wang, X. D., and Wang, Z. L. 2008. Microfibre-nanowire hybrid structure for energy scavenging. *Nature* 451 (7180): 809–813.
110. Lee, H., Lee, B. P., and Messersmith, P. B. 2007. A reversible wet/dry adhesive inspired by mussels and geckos. *Nature* 448 (7151): 338–341.
111. Qin, L. D., Park, S., Huang, L., and Mirkin, C. A. 2005. On-wire lithography. *Science* 309 (5731): 113–115.
112. Geissler, M. and Xia, Y. N. 2004. Patterning: Principles and some new developments. *Advanced Materials* 16 (15): 1249–1269.
113. Xia, Y. N., Qin, D., and Whitesides, G. M. 1996. Microcontact printing with a cylindrical rolling stamp: A practical step toward automatic manufacturing of patterns with submicrometer-sized features. *Advanced Materials* 8 (12): 1015–1017.
114. Geissler, M., McLellan, J. M., and Xia, Y. N. 2005. Edge-spreading lithography: Use of patterned photoresist structures to direct the spreading of alkanethiols on gold. *Nano Letters* 5 (1): 31–36.
115. Meitl, M. A., Zhu, Z. T., Kumar, V., Lee, K. J., Feng, X., Huang, Y. Y., Adesida, I., Nuzzo, R. G., and Rogers, J. A. 2006. Transfer printing by kinetic control of adhesion to an elastomeric stamp. *Nature Materials* 5 (1): 33–38.
116. Wolf, E. L. 2006. *Nanophysics and Nanotechnology*. Weilheim, Germany: WILET-VCH & Co. KGaA.
117. Gregory, P. 1995. Advanced materials and chemical-vapor-deposition. *Advanced Materials* 7 (7): 605–606.
118. Lu, Y. F., Ganguli, R., Drewien, C. A., Anderson, M. T., Brinker, C. J., Gong, W. L., Guo, Y. X., Soyez, H., Dunn, B., Huang, M. H., and Zink, J. I. 1997. Continuous formation of supported cubic and hexagonal mesoporous films by sol gel dip-coating. *Nature* 389 (6649): 364–368.
119. Wu, N. L., Wang, S. Y., and Rusakova, I. A. 1999. Inhibition of crystallite growth in the sol-gel synthesis of nanocrystalline metal oxides. *Science* 285 (5432): 1375–1377.
120. Aldana, J., Lavelle, N., Wang, Y. J., and Peng, X. G. 2005. Size-dependent dissociation pH of thiolate ligands from cadmium chalcogenide nanocrystals. *Journal of the American Chemical Society* 127 (8): 2496–2504.
121. Aldana, J., Wang, Y. A., and Peng, X. G. 2001. Photochemical instability of CdSe nanocrystals coated by hydrophilic thiols. *Journal of the American Chemical Society* 123 (36): 8844–8850.
122. Wagner, R. S. and Ellis, W. C. 1964. Vapor-liquid-solid mechanism of single crystal growth. *Applied Physics Letters* 4: 89.
123. Ozer, N., Chen, D. G., and Lampert, C. M. 1996. Preparation and properties of spin-coated Nb_2O_5 films by the sol-gel process for electrochromic applications. *Thin Solid Films* 277 (1–2): 162–168.
124. Mitzi, D. B., Kosbar, L. L., Murray, C. E., Copel, M., and Afzali, A. 2004. High-mobility ultrathin semiconducting films prepared by spin coating. *Nature* 428 (6980): 299–303.
125. Huang, J. X., Fan, R., Connor, S., and Yang, P. D. 2007. One-step patterning of aligned nanowire arrays by programmed dip coating. *Angewandte Chemie-International Edition* 46 (14): 2414–2417.
126. Nielsen, A. E. 1964. *Kinetics of Precipitation*. London, U.K.: Pergamon Press.
127. Zhang, T. R., Dong, W. J., Keeter-Brewer, M., Konar, S., Njabon, R. N., and Tian, Z. R. 2006. Site-specific nucleation and growth kinetics in hierarchical nanosyntheses of branched ZnO crystallites. *Journal of the American Chemical Society* 128 (33): 10960–10968.
128. Jayakumar, O. D., Salunke, H. G., Kadam, R. M., Mohapatra, M., Yaswant, G., and Kulshreshtha, S. K. 2006. Magnetism in Mn-doped ZnO nanoparticles prepared by a co-precipitation method. *Nanotechnology* 17 (5): 1278–1285.
129. Byrappa, K. and Yoshimura, M. 2001. *Handbook of Hydrothermal Technology*. Norwich, NY: Noyes Publications.
130. Aizenberg, J., Black, A. J., and Whitesides, G. M. 1998. Controlling local disorder in self-assembled monolayers by patterning the topography of their metallic supports. *Nature* 394 (6696): 868–871.
131. Aizenberg, J., Black, A. J., and Whitesides, G. M. 1999. Control of crystal nucleation by patterned self-assembled monolayers. *Nature* 398 (6727): 495–498.
132. Bowden, N., Brittain, S., Evans, A. G., Hutchinson, J. W., and Whitesides, G. M. 1998. Spontaneous formation of ordered structures in thin films of metals supported on an elastomeric polymer. *Nature* 393 (6681): 146–149.
133. Bowden, N., Terfort, A., Carbeck, J., and Whitesides, G. M. 1997. Self-assembly of mesoscale objects into ordered two-dimensional arrays. *Science* 276 (5310): 233–235.

134. Breen, T. L., Tien, J., Oliver, S. R. J., Hadzic, T., and Whitesides, G. M. 1999. Design and self-assembly of open, regular, 3D mesostructures. *Science* 284 (5416): 948–951.
135. Whitesides, G. M. and Grzybowski, B. 2002. Self-assembly at all scales. *Science* 295 (5564): 2418–2421.
136. Lockwood, D. J. 2003. *Self-Assembled Nanostructures*. New York: Kluwer Academic/Plenum Publishers.
137. Rabani, E., Reichman, D. R., Geissler, P. L., and Brus, L. E. 2003. Drying-mediated self-assembly of nanoparticles. *Nature* 426 (6964): 271–274.
138. Xia, Y. N., Gates, B., and Li, Z. Y. 2001. Self-assembly approaches to three-dimensional photonic crystals. *Advanced Materials* 13 (6): 409–413.
139. Bonderer, L. J., Studart, A. R., and Gauckler, L. J. 2008. Bioinspired design and assembly of platelet reinforced polymer films. *Science* 319 (5866): 1069–1073.
140. Walcarius, A., Sibottier, E., Etienne, M., and Ghanbaja, J. 2007. Electrochemically assisted self-assembly of mesoporous silica thin films. *Nature Materials* 6 (8): 602–608.
141. Kim, F., Kwan, S., Akana, J., and Yang, P. D. 2001. Langmuir-Blodgett nanorod assembly. *Journal of the American Chemical Society* 123 (18): 4360–4361.
142. Yu, G. H., Cao, A. Y., and Lieber, C. M. 2007. Large-area blown bubble films of aligned nanowires and carbon nanotubes. *Nature Nanotechnology* 2 (6): 372–377.
143. Alivisatos, A. P. 1996. Semiconductor clusters, nanocrystals, and quantum dots. *Science* 271 (5251): 933–937.
144. He, H., Hsin, C. L., Liu, J., Chen, L. J., and Wang, Z. L. 2007. Piezoelectric gated diode of a single ZnO nanowire. *Advanced Materials* 19 (6): 781–784.
145. Liu, C. H., Zapien, J. A., Yao, Y., Meng, X. M., Lee, C. S., Fan, S. S., Lifshitz, Y., and Lee, S. T. 2003. High-density, ordered ultraviolet light-emitting ZnO nanowire arrays. *Advanced Materials* 15 (10): 838–841.
146. Schmidt-Mende, L. and MacManus-Driscoll, J. L. 2007. ZnO–nanostructures, defects, and devices. *Materials Today* 10 (5): 40–48.
147. Buchine, B. A., Hughes, W. L., Degertekin, F. L., and Wang, Z. L. 2006. Bulk acoustic resonator based on piezoelectric ZnO belts. *Nano Letters* 6 (6): 1155–1159.
148. Johnson, J. C., Knutsen, K. P., Yan, H. Q., Law, M., Zhang, Y. F., Yang, P. D., and Saykally, R. J. 2004. Ultrafast carrier dynamics in single ZnO nanowire and nanoribbon lasers. *Nano Letters* 4 (2): 197–204.
149. Lao, C. S., Liu, J., Gao, P. X., Zhang, L. Y., Davidovic, D., Tummala, R., and Wang, Z. L. 2006. ZnO nanobelt/nanowire Schottky diodes formed by dielectrophoresis alignment across Au electrodes. *Nano Letters* 6 (2): 263–266.
150. Law, M., Sirbuly, D. J., Johnson, J. C., Goldberger, J., Saykally, R. J., and Yang, P. D. 2004. Nanoribbon waveguides for subwavelength photonics integration. *Science* 305 (5688): 1269–1273.
151. Nomura, K., Ohta, H., Ueda, K., Kamiya, T., Hirano, M., and Hosono, H. 2003. Thin-film transistor fabricated in single-crystalline transparent oxide semiconductor. *Science* 300 (5623): 1269–1272.
152. Ryan, M. P., Williams, D. E., Chater, R. J., Hutton, B. M., and McPhail, D. S. 2002. Why stainless steel corrodes. *Nature* 415 (6873): 770–774.
153. Yan, C. and Xue, D. 2007. Solution growth of nano- to microscopic ZnO on Zn, in *15th International Conference on Crystal Growth*, Salt Lake City, UT, pp. 1836–1840.
154. Liu, J. F., Li, Q. H., Wang, T. H., Yu, D. P., and Li, Y. D. 2004. Metastable vanadium dioxide nanobelts: Hydrothermal synthesis, electrical transport, and magnetic properties. *Angewandte Chemie-International Edition* 43 (38): 5048–5052.
155. Liu, B. and Zeng, H. C. 2003. Hydrothermal synthesis of ZnO nanorods in the diameter regime of 50 nm. *Journal of the American Chemical Society* 125 (15): 4430–4431.
156. Song, H., Rioux, R. M., Hoefelmeyer, J. D., Komor, R., Niesz, K., Grass, M., Yang, P. D., and Somorjai, G. A. 2006. Hydrothermal growth of mesoporous SBA-15 silica in the presence of PVP-stabilized Pt nanoparticles: Synthesis, characterization, and catalytic properties. *Journal of the American Chemical Society* 128 (9): 3027–3037.
157. Wang, Z. L. 2003. Nanobelts, nanowires, and nanodiskettes of semiconducting oxides—From materials to nanodevices. *Advanced Materials* 15 (5): 432–436.
158. Yu, H. D., Zhang, Z. P., Han, M. Y., Hao, X. T., and Zhu, F. R. 2005. A general low-temperature route for large-scale fabrication of highly oriented ZnO nanorod/nanotube arrays. *Journal of the American Chemical Society* 127 (8): 2378–2379.
159. Tian, Z. R. R., Voigt, J. A., Liu, J., McKenzie, B., McDermott, M. J., Rodriguez, M. A., Konishi, H., and Xu, H. F. 2003. Complex and oriented ZnO nanostructures. *Nature Materials* 2 (12): 821–826.

160. Chen, Z., Zhan, P., Wang, Z. L., Zhang, J. H., Zhang, W. Y., Ming, N. B., Chan, C. T., and Sheng, P. 2004. Two- and three-dimensional ordered structures of hollow silver spheres prepared by colloidal crystal templating. *Advanced Materials* 16 (5): 417–422.
161. Jiang, X. C., Mayers, B., Herricks, T., and Xia, Y. N. 2003. Direct synthesis of Se@CdSe nanocables and CdSe nanotubes by reacting cadmium salts with Se nanowires. *Advanced Materials* 15 (20): 1740–1743.
162. Peng, Q., Dong, Y. J., and Li, Y. D. 2003. ZnSe semiconductor hollow microspheres. *Angewandte Chemie-International Edition* 42 (26): 3027–3030.
163. Sun, Y. G., Mayers, B. T., and Xia, Y. N. 2002. Template-engaged replacement reaction: A one-step approach to the large-scale synthesis of metal nanostructures with hollow interiors. *Nano Letters* 2 (5): 481–485.
164. Goldberger, J., He, R. R., Zhang, Y. F., Lee, S. W., Yan, H. Q., Choi, H. J., and Yang, P. D. 2003. Single-crystal gallium nitride nanotubes. *Nature* 422 (6932): 599–602.
165. Yan, C. and Xue, D. 2007. Electroless deposition of aligned ZnO taper-tubes in a strong acidic medium. *Electrochemistry Communications* 9 (6): 1247–1251.
166. Ma, C., Moore, D., Li, J., and Wang, Z. L. 2003. Nanobelts, nanocombs, and nanowindmills of wurtzite ZnS. *Advanced Materials* 15 (3): 228–231.
167. Hao, Y. F., Meng, G. W., Wang, Z. L., Ye, C. H., and Zhang, L. D. 2006. Periodically twinned nanowires and polytypic nanobelts of ZnS: The role of mass diffusion in vapor-liquid-solid growth. *Nano Letters* 6 (8): 1650–1655.
168. Labrenz, M., Druschel, G. K., Thomsen-Ebert, T., Gilbert, B., Welch, S. A., Kemner, K. M., Logan, G. A., Summons, R. E., De Stasio, G., Bond, P. L., Lai, B., Kelly, S. D., and Banfield, J. F. 2000. Formation of sphalerite (ZnS) deposits in natural biofilms of sulfate-reducing bacteria. *Science* 290 (5497): 1744–1747.
169. Norris, D. J., Efros, A. L., and Erwin, S. C. 2008. Doped nanocrystals. *Science* 319 (5871): 1776–1779.
170. Wang, X. D., Gao, P. X., Li, J., Summers, C. J., and Wang, Z. L. 2002. Rectangular porous ZnO-ZnS nanocables and ZnS nanotubes. *Advanced Materials* 14 (23): 1732–1735.
171. Yan, C. and Xue, D. 2006. Conversion of ZnO nanorod arrays into ZnO/ZnS nanocable and ZnS nanotube arrays via an in situ chemistry strategy. *Journal of Physical Chemistry B* 110 (51): 25850–25855.
172. Yan, C., Liu, J., Liu, F., Wu, J., Gao, K., and Xue, D. 2008. Tube formation in nanoscale materials. *Nanoscale Research Letters* 3 (12): 473–480.
173. Pradhan, N. and Peng, X. G. 2007. Efficient and color-tunable Mn-doped ZnSe nanocrystal emitters: Control of optical performance via greener synthetic chemistry. *Journal of the American Chemical Society* 129 (11): 3339–3347.
174. Mclaughlin, A. C., Sher, F., and Attfield, J. P. 2005. Negative lattice expansion from the superconductivity-antiferromagnetism crossover in ruthenium copper oxides. *Nature* 436 (7052): 829–832.
175. Lieber, C. M. and Yang, P. D. 1997. High-temperature superconductors. *Science* 277 (5334): 1909–1910.
176. Snoke, D. 2002. Spontaneous Bose coherence of excitons and polaritons. *Science* 298 (5597): 1368–1372.
177. Li, X., Jiang, Y., Shi, Z. W., and Xu, Z. 2007. Two growth modes of metal oxide in the colloidal crystal template leading to the formation of two different macroporous materials. *Chemistry of Materials* 19 (22): 5424–5430.
178. Mostovoy, M. 2008. Transition metal oxides—Multiferroics go high-T_C. *Nature Materials* 7 (4): 269–270.
179. Kimura, T., Sekio, Y., Nakamura, H., Siegrist, T., and Ramirez, A. P. 2008. Cupric oxide as an induced-multiferroic with high-T_C. *Nature Materials* 7 (4): 291–294.
180. Jiang, X. C., Herricks, T., and Xia, Y. N. 2002. CuO nanowires can be synthesized by heating copper substrates in air. *Nano Letters* 2 (12): 1333–1338.
181. Reitz, J. B. and Solomon, E. I. 1998. Propylene oxidation on copper oxide surfaces: Electronic and geometric contributions to reactivity and selectivity. *Journal of the American Chemical Society* 120 (44): 11467–11478.
182. Giardina, M. D., Feduzi, R., Inzaghi, D., Manara, A., Giori, C., Sora, I. N., and Dallacasa, V. 1997. Crystalline phases and electronic structures in superconducting Bi-Sr-Ca-Cu oxides. *Journal of Materials Research* 12 (8): 2009–2013.
183. Liu, J. and Xue, D. 2008. Thermal oxidation strategy towards porous metal oxide hollow architectures. *Advanced Materials* 20 (13): 2622–2627.
184. Cheetham, A. K. 1994. Advanced inorganic materials: An open horizon. *Science* 264: 794–795.

185. Abrahams, S. C., Bernstein, J. L., and Jamieson, P. B. 1968. Crystal structure of the transition-metal molybdates and tungstates. IV. Paramagnetic $CuMoO_4$. *Journal of Chemical Physics* 48: 2619.
186. Moini, A., Peascoe, R., Rudolf, P. R., and Clearfield, A. 1986. Hydrothermal synthesis of copper molybdates. *Inorganic Chemistry* 25: 3782.
187. Xu, J., Xue, D., and Zhu, Y. 2006. Room temperature synthesis of curved ammonium copper molybdate nanoflake and its hierarchical architecture. *Journal of Physical Chemistry B* 110 (35): 17400–17405.
188. Levy-Clement, C., Tena-Zaera, R., Ryan, M. A., Katty, A., and Hodes, G. 2005. CdSe-Sensitized p-CuSCN/nanowire n-ZnO heterojunctions. *Advanced Materials* 17 (12): 1512–1515.
189. Xu, J. and Xue, D. 2006. Fabrication of upended taper-shaped cuprous thiocyanate arrays on a copper surface at room temperature. *Journal of Physical Chemistry B* 110 (23): 11232–11236.
190. Xu, X., Tian, B. Z., Kong, J. L., Zhang, S., Liu, B. H., and Zhao, D. Y. 2003. Ordered mesoporous niobium oxide film: A novel matrix for assembling functional proteins for bioelectrochemical applications. *Advanced Materials* 15 (22): 1932.
191. Cherian, M., Rao, M. S., and Deo, G. 2003. Niobium oxide as support material for the oxidative dehydrogenation of propane. *Catalysis Today* 78 (1–4): 397–409.
192. Tanabe, K. 2003. Catalytic application of niobium compounds. *Catalysis Today* 78 (1–4): 65–77.
193. Braga, V. S., Dias, J. A., Dias, S. C. L., and de Macedo, J. L. 2005. Catalyst materials based on Nb_2O_5 supported on SiO_2-Al_2O_3: Preparation and structural characterization. *Chemistry of Materials* 17 (3): 690–695.
194. Lee, J., Orilall, M. C., Warren, S. C., Kamperman, M., Disalvo, F. J., and Wiesner, U. 2008. Direct access to thermally stable and highly crystalline mesoporous transition-metal oxides with uniform pores. *Nature Materials* 7 (3): 222–228.
195. Ozer, N., Rubin, M. D., and Lampert, C. M. 1996. Optical and electrochemical characteristics of niobium oxide films prepared by sol-gel process and magnetron sputtering: A comparison. *Solar Energy Materials and Solar Cells* 40 (4): 285–296.
196. Zielinska, B., Borowlak-Palen, E., and Kalenzuk, R. J. 2008. Preparation and characterization of lithium niobate as a novel photocatalyst in hydrogen generation. *Journal of Physics and Chemistry of Solids* 69 (1): 236–242.
197. Haugsrud, R. and Norby, T. 2006. Proton conduction in rare-earth ortho-niobates and ortho-tantalates. *Nature Materials* 5 (3): 193–196.
198. Noronha, F. B., Perez, C. A., Schmal, M., and Frety, R. 1999. Determination of cobalt species in niobia supported catalysts. *Physical Chemistry Chemical Physics* 1 (11): 2861–2867.
199. Braun, E., Eichen, Y., Sivan, U., and Ben-Yoseph, G. 1998. DNA-templated assembly and electrode attachment of a conducting silver wire. *Nature* 391 (6669): 775–778.
200. Eckardt, L. H., Naumann, K., Pankau, W. M., Rein, M., Schweitzer, M., Windhab, N., and von Kiedrowski, G. 2002. DNA nanotechnology: Chemical copying of connectivity. *Nature* 420 (6913): 286.
201. Mirkin, C. A., Letsinger, R. L., Mucic, R. C., and Storhoff, J. J. 1996. A DNA-based method for rationally assembling nanoparticles into macroscopic materials. *Nature* 382 (6592): 607–609.
202. Mirkin, C. A. and Taton, T. A. 2000. Materials chemistry: Semiconductors meet biology. *Nature* 405 (6787): 626–627.
203. Nykypanchuk, D., Maye, M. M., van der Lelie, D., and Gang, O. 2008. DNA-guided crystallization of colloidal nanoparticles. *Nature* 451 (7178): 549–552.
204. Park, S. Y., Lytton-Jean, A. K. R., Lee, B., Weigand, S., Schatz, G. C., and Mirkin, C. A. 2008. DNA-programmable nanoparticle crystallization. *Nature* 451 (7178): 553–556.

2 Fabrication and Characterization of SiC Nanostructured/Nanocomposite Films

Chen-Kuei Chung and Bo-Hsiung Wu

CONTENTS

2.1 INTRODUCTION: BRIEF REVIEW OF THE FABRICATION AND PROPERTIES OF SiC AND ITS NANOSTRUCTURES AND NANOCOMPOSITE THIN FILMS

Compared with silicon (Si), silicon carbide (SiC) is a wide-bandgap semiconductor with superior physical and chemical properties and is recognized as one of the most promising materials for applications in high-power and high-temperature semiconductor devices [1,2] and severe, hard biomaterials [1–3]. The superior material properties include high strength, high thermal conductivity, high temperature stability, high refractive index, low thermal expansion, low density, variable wide bandgap, and chemical inertness [3]. In terms of physical properties of semiconductor electronics, crystalline SiC exhibits a larger bandgap (2.3–3.4 eV), a higher breakdown field (30×10^5 V/cm), a higher thermal conductivity (3.2–4.9 W/cm K), and a higher saturation velocity (2×10^7 cm/s) than Si [4]. In terms of mechanical properties, SiC specifically exhibits excellent hardness and wear resistance, among others. The Knoop hardness of SiC is about 2480 kg/mm^2, which is comparable to that of other hard materials such as diamond (7000 kg/mm^2) and Al_2O_3 (2100 kg/mm^2), and is much higher than that of Si (850 kg/mm^2) [5]. The wear resistance value of SiC is 9.15 comparable with that of diamond, which is 10, and Al_2O_3, which is 9.00 [6]. Also, SiC has a high Young's modulus of 700 GPa, about 3.5 times higher than that of Si [5]. In terms of chemical properties, SiC is resistant to most acids, but can be etched by alkaline hydroxide bases at melting temperatures (>600°C). Note that SiC does not melt, but sublimes above 1800°C (Si melts at 1410°C). The SiC surface can be passivated by the formation of a thermal SiO_2 layer, but the oxidation rate is very slow compared to Si [1]. Combining its excellent mechanical properties and high temperature stability, SiC offers new possibilities for developing more challenging applications of MEMS devices than those possible with Si [1,7–9].

From the viewpoint of atomic arrangement, SiC exists in a large number of polytypes, i.e., different crystal structures made from different stacking sequences of C–Si double layers in the Si–C subunit [1,3]. The polytypes of crystal structures lead to a variety of relevant physical properties, which is the reason why SiC is an important compound in materials science [3]. Such polytypes are recognized by the type of the crystal lattice, i.e., hexagonal or cubic, and the number of layers for the periodic pattern. For example, 4H-SiC has a hexagonal lattice with a four-layer repeat structure. More than 100 polytypes are known, but research and development is mainly concentrated on three types: 3C, 6H, and 4H-SiC [1]. The 4H polytype is the most common for electronic devices because of its overall superior material properties. The bandgap of 4H-SiC is 3.23 eV at room temperature (RT) (compared to Si with 1.12 eV) and allows high-temperature operation of SiC electronic devices [10]. The 3C-SiC polytype is more common for MEMS-based sensors, because it may be grown in a polycrystalline form on Si wafers at a lower cost compared with a pure bulk SiC technology [3,11]. In addition, amorphous SiC (a-SiC) has practical importance and is fabricated through low-temperature chemical vapor deposition (CVD) [12] or physical vapor deposition (PVD), ion-beam synthesis [13,14], and neutron bombardment [15], and can replace Si in numerous applications. For example, hydrogen-incorporated SiC, i.e., a-SiC:H, formed by CVD is a potential candidate in single solar-cell material applications [3].

In recent developments in electronics, thin-film and nanofabrication technologies have been able to synthesize new forms of SiC, such as nanotubes, nanowires (NWs), and nanopores and cluster structures together with superlattices for promising applications [3]. The SiC nanostructures of nanotubes and nanowires have been fabricated and considered as promising candidates for hydrogen storage, nanoelectronics, and reinforcements of matrix composites and microelectromechanical systems (MEMS). Mélinon et al. [3] demonstrated the SiC cage-like clusters using the synthesis of carbon fullerenes with a substitutional doping of Si heteroatoms for the controlled altering of electronic and optical properties. Shim and Huang [16] reported that the high-temperature-evaporated SiC NWs embedded in a SiC matrix consisting of microscale fibers can significantly improve the toughness of composites [17]. Self-assembled SiC NWs embedded in a microscale framework may further promote their toughening. Bechelany et al. [18] performed a Raman spectroscopy study of individual SiC NWs for understanding their fundamental physical properties and realizing technological applications based on these materials.

Various synthetic methods have been used to grow SiC NWs. For example, SiC NWs can be fabricated from carbon nanotubes (CNTs) and a Si-rich vapor phase [19,20], or from carbon nanoparticles embedded in a silica matrix [21]. Mechanical activation and ball milling of a mixture of Si and SiO_2 powders that can subsequently be pyrolyzed under a flow of propane can be used to prepare SiC NWs [22]. SiC NWs can also be prepared using a catalytic chemical vapor deposition (CCVD) process and the vapor–liquid–solid (VLS) growth mechanism [23]. The VLS principle can also be used for the synthesis of SiC NWs by laser ablation [24] and an arc-discharge process [25]. Starting from commercial micrometer-sized Si particles, a versatile, noncatalytic, and inexpensive method has been developed for the selective preparation of SiO_2 [26], SiC, and/or Si_3N_4 NWs [27,28]. The main drawbacks of the above-discussed techniques are the limited amount of SiC NWs per batch and the need for additional purification to remove the mixed residual particles. Therefore, a commercially competitive method based on a vapor–solid (VS) growth mechanism is developed for large amounts of SiC-based nanostructures with tunable geometric features [29]. Another two-step process can also be used for the synthesis of SiC NWs by means of a CNT template [19]: hot filament CVD [30] or carbothermal reduction [31]. The fabrication and characterization of SiC nanostructures and nanocomposite thin films by using physical and chemical syntheses methods are described in Sections 2.2 and 2.3.

2.2 LITERATURE SURVEY OF FABRICATION AND CHARACTERIZATION OF SiC NANOSTRUCTURES AND NANOCOMPOSITE THIN FILMS BY USING PHYSICAL AND CHEMICAL SYNTHESES

2.2.1 Fabrication and Characterization of SiC Nanostructures and Nanocomposite Thin Films by Using Physical Synthesis

Physical methods for the synthesis of SiC nanostructures and nanocomposite thin films commonly include thermal/e-beam evaporation, radio frequency (RF) sputtering, laser ablation, and arc discharge, which input energy to solid targets or powers with a Si and C mixture to produce a SiC vapor or gas phase and then condense it on the surface of substrates, or melt the surface and resolidify it to form the nanostructures and/or nanocomposite thin films. Given below are some examples for this statement.

He et al. [32] reported that the sp^3-bonded carbon (sp^3 C)–SiC composite films can be fabricated using e-beam physical vapor deposition (EB-PVD) of SiC materials. The diamond–SiC composite film combines the unique physical, chemical, and mechanical properties of diamond and SiC [33,34]. The SiC film may contain either the diamond or the sp^3 C phase, which tends to uniformly distribute in SiC [35]. The EB-PVD can evaporate many materials for thin-film and coating fabrication and be used to produce gradient structures and composite films [36]. The sp^3 C-doped SiC nanocomposite films were deposited on the thermally oxidized 316 stainless steel substrate by EB-PVD at an electron gun power of 60 kW at different substrate temperatures of 700°C–900°C. An ultrafine 3C-SiC powder with an average particle size of ~0.3 μm was sintered into ingots under 1500°C and 5 MPa pressure in vacuum. The deposition rate was about 1 μm/min and the deposition time was 30 s. The hardness was measured using Hysitron indentation, and it reached 50 GPa, which is significantly higher than the hardness of SiC (28 GPa). The superhardness of the thin films is due to the formation of nanocrystalline SiC/diamond-like carbon (DLC) composites.

Kulikovsky et al. [37] reported that hydrogen-free, amorphous, and nanocrystalline SiC films were prepared by magnetron sputtering from a sintered SiC target in an argon atmosphere (purity 99.999%) at a pressure of 0.2 Pa. The discharge power was typically 150–300 W with a bias of −100 to −200 V. The Si(111) substrates were preheated to 200°C, 400°C, 600°C, and 750°C for deposition. The Si/(Si+C) atomic ratio for all the films ranges within 0.474–0.485. The oxygen content in the films was 0.1% or lesser. Mechanical properties (hardness, elastic modulus, and intrinsic stress) and film structures were investigated on the dependence of the substrate bias and temperature. Both the hardness and the elastic modulus of all a-SiC films prepared at different substrate temperatures and biases are always lower than those for bulk single-crystalline α-SiC films, while the hardness of partially crystalline SiC films is higher and their elastic modulus lower than those for α-SiC films.

The increased hardness of the SiC films compared to bulk α-SiC is attributed to the spontaneous formation of a nanocomposite structure, β-SiC/a-Si/a-C, during deposition.

Zhao et al. [38] reported that the optical absorption and the photocurrent of 6H-SiC can be enhanced by a laser surface–nanostructuring technology. The femtosecond laser was applied to cut the SiC wafer into small pieces and to produce the nanostructured surface up to $4 \times 4\,mm^2$ area. The femtosecond laser scanned the SiC surface at a fluence of up to $12\,J/cm^2$, with scanning speeds ranging from 10 to 3000 μm/s and with the step of 100 μm/s. Ohmic Ti contacts were deposited on the SiC surface by evaporation and rapid isothermal annealing at 450°C in pure nitrogen. A single-beam ultrashort pulsed laser can generate the nanoscale ripple structures on 6H-SiC surface with spatial periods of up to 155 nm. The period of nanoripples is not influenced by the laser fluence and the number of pulses. And about 40% enhancement of optical absorption and a threefold increase of photocurrent are obtained after laser nanostructuring.

Fedorenko et al. [39] reported that nanostructures on the surface of 6H-SiC crystal were generated by pulsed laser radiation. Structural changes in surface and near-surface layers of SiC were studied by atomic force microscopy images, photoluminescence, Raman spectra, and field emission current–voltage characteristics of initial and irradiated surfaces. The cone-shaped nanostructures with typical dimensions of 100–200 nm height and 5–10 nm width at the edge are formed on an SiC surface under nitrogen laser exposure ($\lambda = 337$ nm, $t_p = 7$ ns, and $E_p = 1.5$ mJ). The process of the nanocrystal formation on 6H-SiC(N) surfaces after laser irradiation was concerned with the formation of the liquid phase and the solid–liquid phase transition. The thermal gradient mechanism of mass transfer in the initial stage of the nano-hill formation process on SiC surfaces is important for the formation of a nanostructured SiC. The tunneling nature of the emission current and the increase of emission efficiency occurred from nanostructured SiC surfaces compared to nonirradiated surfaces.

Liu et al. [40] reported large-scale synthesis and photoluminescence properties of SiC/SiO_x nanocables using a simple and low-cost arc-discharge method in deionized water. A DC welding machine was used as the power supply. During the arc-discharge process, the voltage between two graphite electrodes was gathered and sent to a feedback system, which could adjust the feeding rate of the anode to keep the discharge voltage constant, consequently permitting continuous and automatic production. A mixture of graphite, Si, and Fe powders in the weight ratio of 5:5:1 was used as the source for the formation of SiC/SiO_x nanocables. The electrodes together with the quartz tube were submerged into deionized water and arc discharge was activated at a voltage of 22 V and a current of 30 A. The synthesized high-purity SiC/SiO_x nanocables consisted of a uniform cubic β-SiC core and an amorphous silicon oxide shell. The length was about several hundred nanometers to several microns and the smallest average diameter of the β-SiC core was only about 5 nm, which was attributed to the arc discharge in the deionized water approach and to the constraint of the outer sheath. The diameters of the as-grown nanocables can be controlled through adjusting the processing parameters. The SiC/SiO_x nanocables emitted a stable violet-blue light at wavelengths of about 315 and 360–400 nm. Compared to the reported results in β-SiC NWs or nanocables, the photoluminescence of the synthesized nanocables showed a significant blueshift, which resulted from the small diameter of the β-SiC core. The photoluminescence intensity can also be enhanced by annealing the as-prepared SiC/SiO_x nanocables.

2.2.2 Fabrication and Characterization of SiC Nanostructures and Nanocomposite Thin Films by Using Chemical Synthesis

The chemical methods used for the synthesis of SiC crystals, nanostructures, and nanocomposite thin films commonly include CVD, thermal CVD, catalytic CVD, electrochemical etching, and solgel, which bring about the chemical reaction of precursors with Si and C elements to form SiC nanostructures and/or nanocomposite thin films. Given below are some examples for this statement.

From the well-known information [2], CVD is the most popular method for the fabrication of crystalline SiC films. Atmospheric pressure chemical vapor deposition (APCVD) using silicon- and carbon-containing precursors is a popular approach to grow homoepitaxial α-SiC films on α-SiC substrates. Commonly used precursor gases are silane and propane. Hydrogen is used as the carrier gas. Typical dopant gases are trimethyl-aluminum ($Al(CH_3)$)) and diborane (B_2H_6) for p-type films, and N_2, ammonia (NH_3), and phosphine (PH_3) for n-type films. APCVD has also been used to grow 3C-SiC films on 6H-SiC substrates. Growth temperatures range from 1500°C to 1700°C. Unlike all other SiC polytypes, single-crystal 3C-SiC can be heteroepitaxially grown on Si substrates. APCVD is also the most common method, which uses hydrogen as the carrier gas, silane as the Si source gas, and propane as the C source gas [41]. Less common processes use dichlorosilane as a Si source and acetylene as a C source [42]. Single-source gases containing both C and Si, such as hexamethyldisilazane (HMDS), have also been used [43]. Heteroepitaxy has also been demonstrated using LPCVD processes [44]. With regard to polycrystalline 3C-SiC (poly-SiC), it can be deposited on a wide variety of substrate materials, including SiO_2 and Si_3N_4, and used as a structural material for micro-actuators in polysilicon MEMS fabrication. Poly-SiC can also be deposited by plasma-enhanced chemical vapor deposition (PECVD), APCVD, and LPCVD, and even sputtering and electron beam evaporation at substrate temperatures ranging from 200°C to 1000°C [45–48]. In addition to the crystalline SiC film, the synthesis of SiC nanostructures and nanocomposite films has made much progress using different methods for their potentially unique nano-induced properties.

Fan et al. [49,50] reported an optical emission of SiC nanocrystallite films, which clearly shows the quantum confinement effect. Bulk poly-SiC was first electrochemically etched in a hydrofluoric acid (HF)–ethanol solution, and then the fabricated porous silicon carbide was ultrasonically treated in a water or toluene suspension to disperse into colloidal nanoparticles. The SiC wafer was laterally anodized for 60 min when immersed in the HF-ethanol solution (46% HF: 99.8% $C_2H_5OH=2:1$), at a current density of $40\,mA/cm^2$, under a photoassisted illumination with a 15 W halogen lamp from 15 cm above. Transmission electron microscopy images clearly showed that the colloidal nanoparticles have a 3C-SiC lattice structure, with sizes varying from about 6 nm down to below 1 nm. The suspension of 3C-SiC nanocrystallites exhibited ultrabright emission with wavelengths ranging from 400 to 520 nm, when the excitation wavelength varies from 250 to 480 nm, in accordance with the quantum confinement effect. By adding polystyrene to the toluene suspension containing SiC nanoparticles and coating the mixing solution onto a Si wafer, the SiC/polystyrene films exhibited luminescence. The colloidal 3C-SiC nanocrystals would be robust and intense emitters with good chemical stability and biocompatibility. The crystallites with sizes smaller than 8 nm and good surface passivation would be necessary for quantum confinement.

Zhang et al. [23] reported that large quantities of the helical, crystalline SiC/SiO_2 core–shell NWs, tens of micrometers long, could be synthesized by a CVD technique. The composite, helical silicon carbide NWs were synthesized in a flow-tube furnace using iron-catalyzed decomposition of methane at 1100°C. Iron powders were held in an alumina sample boat upstream in the flow tube, and a silicon wafer located downstream was used as the substrate. A reaction time of 5–10 min plus a 1 h post-reaction treatment under the same high-temperature conditions were sufficient to grow a large quantity of nanostructures on the silicon wafer. The SiC core typically had diameters of 10–40 nm with a helical periodicity of 40–80 nm, and was covered by a uniform layer of 30–60 nm thick amorphous SiO_2. A screw-dislocation-driven growth process was proposed for the formation of this structure, based on the detailed characterizations. The helical nanostructures can be used as building blocks in nanomechanical and nanoelectronic devices. This was most likely controlled by the initial nucleation sites, which might be too small to encompass a whole screw dislocation. The formation of the helical NWs could be viewed as an epitaxy growth, following the initial screw dislocation sites on the crystalline SiC nanoparticles. The simultaneous shifts and rotations between successive layers were inherent characteristics of crystal screw dislocations and provided the driving force for the helical structure formation. The screw-dislocation-induced growth mechanism suggested that similar helical nanostructures for a wide range of materials could be synthesized.

Bechelany et al. [18] reported an effective thermal CVD method for the synthesis of SiC NWs. In a typical experiment, an alumina boat containing 6 g of polypropylene (Aldrich, >99.99%) was placed in an alumina tube in a convective furnace. A second alumina boat containing 4 g of a mixture of silicon powder (Aldrich, 99.999%) and silicon oxide powder (Aldrich, >99.99%; 1:1 molar ratio) was placed near the first boat so that the flowing gas first came into contact with the polypropylene. The boat containing the silicon species was then covered with a rectangular piece of graphite (Carbone Lorraine, >99.99%). The system was heated to 1400°C under an argon flow (10 mL/min), held at this temperature for 10 h, and then allowed to cool down to RT. Finally, 250 mg of a cotton-like solid was collected directly from the lower face of the graphite piece. The entire sample was etched in aqueous HF (48%) in order to remove any residual silica coating on the surface of the NWs. The authors studied the Raman spectroscopy behavior of individual SiC NWs for understanding their fundamental physical properties, and promoted for different applications. For example, Raman spectroscopy of a single 40 nm 3C-SiC NW has been achieved at RT with the use of SERS. The nanostructure used to enhance the Raman scattering process is based on a tungsten tip covered by a thin gold layer, with the NW attached at the apex. High-resolution transmission electron microscopy (HRTEM) images showed both atomic arrangements in the SiC NWs, with growth predominantly in the [111] direction, and abundant structural defects. The effort was focused on the correlation between the Raman spectroscopic profiles and the structural deformations. The doping of an individual NW was evaluated. The free carrier concentration between 10^{18} and 10^{19} cm^{-3} was evaluated by the Fano interference features of sharp phonon lines.

Ryu et al. [51] reported that a simple direct growth of SiC–SiO_2 core–shell NWs could be synthesized by heating NiO-catalyzed silicon substrates. A carbothermal reduction of WO_3 provided a reductive environment and carbon source to synthesize crystalline SiC NWs covered with SiO_2 sheaths at a growth temperature of 1000°C–1100°C. The obtained SiC–SiO_2 core–shell NWs were immersed in a 0.25% HF aqueous solution to remove the SiO_2 shell layer from the NWs. Transmission electron microscopy showed that the SiC core was 15–25 nm in diameter and the SiO_2 shell layer was an average of 20 nm in thickness. The thickness of the SiO_2 shell layer could be controlled by using HF etching. Field emission results of SiC–SiO_2 and bare SiC core–shell NWs showed that the SiC NWs coated with an optimum SiO_2 thickness (10 nm) had a higher field emission current than that of the bare SiC NWs. That is, the bare SiC NWs showed that the turn-on field (E_{to}) at the emission current of 10 μA/cm^2 was 4.0 V/μm (bare SiC NWs), 3.3 V/μm (10 nm SiO_2-coated SiC NWs), and 4.5 V/μm (20 nm SiO_2-coated SiC NWs).

Sharma et al. [52] reported that β-SiC nanoparticles along with nanocrystalline SiO_x nanoparticles were grown by using a solgel technique. A homogeneous gel was obtained from the solution of tetraethylorthosilicate, citric acid, and ethylene glycol mixed in fixed a molar ratio and stirred continuously to obtain the gel. The gel obtained was dried at 300°C in air for 24 h to obtain a black powder. This powder was finely ground and was then subjected to heat treatment at 1400°C in a hydrogen atmosphere. A grayish powder was thus obtained. The conversion of silicon dioxide to silicon carbide nanoparticles proceeds via the intermediate formation of SiO in the vapor phase, which readily reacts with carbon at reduced temperature. An HF treatment of the as-prepared powder sample resulted into the dissolution of the silicon oxide phase. The x-ray diffraction data and the calculated lattice parameters were in close match with the standard values known for β-SiC lattices. Microstructure studies indicated that these particles are of nanometer size. Transmission electron microscopy, selected area electron diffraction (SAED), and in situ energy dispersive x-ray (EDX) spectroscopy confirmed the existence of nanocrystalline silicon oxide along with nanocrystalline β-silicon carbide nanoparticles.

Zhu et al. [53] reported that SiC/ZnO nanocomposites were synthesized by a two-stage process. SiC nanoparticles were synthesized by an electrical explosion method, and by using the SiC nanoparticles as precursors, for the synthesis of SiC/ZnO nanocomposites using co-precipitation processes. In these procedures, the explosive cavity was filled with 40 mL hexamethyldisilazane. A strong, pulsed high voltage of 36 kV was applied to the anode, and a dark fluid with nanoparticles was obtained. The nanoparticles were collected and treated with dilute HF to remove SiO_2, and pure SiC nanoparticles were obtained. The treated hydrophilic SiC nanoparticles (0.4 g) were dispersed into a 100 mL ethanol–water solution with a high-intensity ultrasonic irradiation for 1 h. Then, 100 mL of 0.02 mol/L $Zn(NO_3)_2$ and 1.5 mol/L triethanolamine were dropped simultaneously into the suspension under strong stirring for 1 h at 90°C. Finally, the resultant product was centrifuged, washed repeatedly with distilled water, and dried at 80°C. Then the product was separated into five samples and annealed for 2 h in air, respectively, at 200°C, 300°C, 400°C, and 500°C. The SiC/ZnO nanocomposites exhibited energy tuning of band-to-band transitions. In addition to the UV emission, the green and orange emissions were also observed. The UV emission shifted to short wavelengths with decreasing diameter due to the quantum confinement effect. The green emission first decreased and then increased with increasing annealing temperature. This could be explained by a model based on oxygen vacancy, band bending, and impurity. The lower annealing temperature was predominated by the oxygen vacancy while the higher annealing temperature was concerned with band bending and impurity.

2.2.3 Fabrication and Characterization of SiC Nanoparticles by Using Physical and Chemical Syntheses

In this section, we focus on the survey of the fabrication of SiC nanoparticles (np-SiCs) in order to compare it with our novel method described in Section 2.3. With regard to the fabrication of np-SiC on wafers, Xu et al. [54] used RF plasma sputtering and a compound SiC target (Si:C~ 1:1) to deposit SiC nanoparticles on Si(100) substrates with AlN buffer layers at a substrate temperature of 350°C for 1.0 h. Xu et al. [55] applied low-energy mass-selected ion beam deposition (MSIBD) to select C^+ ions at 100 eV for bombarding a 4°-off Si(111) substrate, in order to grow nanosized SiC dots at temperatures of 800°C–950°C. Liu et al. [56] reported that SiC nanocrystals embedded in a SiO_2 matrix on the single-crystal Si(111) substrates were prepared by means of RF co-sputtering with Si, C, and SiO_2 targets, and subsequent high-temperature annealing. The co-sputtering of C, Si, and SiO_2 targets was applied by a 300 W RF power in Ar gas at a working pressure of 2 Pa and a substrate temperature of 500°C. Then the deposited composite film was covered by a 100 nm Si cap for the subsequent annealing in a N_2 atmosphere at 800°C–1100°C for 60 min. Kametani et al. [57] reported that the CVD ferrocene ($Fe(C_5H_5)_2$) precursor was used to form SiC nanodots on Si(111) substrates at a deposition temperature of 600°C, followed by annealing at 600°C–800°C. Flores et al. [58] reported that Si(111) substrates with high C content were used to form SiC nanodots by annealing at 660°C–810°C. Then the formation of SiC nanodots was achieved through the segregation of C atoms from the bulk to the surface and the reaction with Si. Lin et al. [59] reported that the a-SiC nanoparticles of 4–6 nm in size were synthesized by means of the decomposition of a tetramethylsilane (TMS) precursor in a microwave plasma reactor, and appropriate reaction conditions at low temperature and low precursor partial pressure (0.001–0.02 Torr) using argon as the carrier gas (3 Torr). Takagi et al. [60] reported that SiC nanoparticles were formed on the clean Si(111) substrate that was soaked in C_2H_5OH solution for 3 min. The c-SiC nanoparticles with the epitaxial relationship of (111)||(111) were formed by heating the Si(111) substrate at around 1000°C in an ultrahigh vacuum (UHV). The source of carbon was the carbon-bearing molecules desorbed from the sample or the sample holder.

Laser pyrolysis is also one popular method for np-SiC synthesis. Kassiba et al. [61] reported that SiC nanoparticles were synthesized by means of CO_2 laser pyrolysis at a 750 W laser power for a gaseous mixture of silane (SiH_4) and acetylene (C_2H_2) with a proper reactant flux. The atomic ratio of np-SiC was $C/Si = 1.02 \pm 0.02$, close to a stoichiometric one. Kassiba et al. [62] also used the laser pyrolysis process to adjust the SiC nanoparticles stoichiometry by controlling the initial fluxes of the reactants (SiH_4 and C_2H_2) for Si-rich or C-rich compounds. Vicens et al. [63] applied CO_2 laser pyrolysis from a C_2H_2 and SiH_4 mixture to grow SiC nanoparticles of 10 nm diameter at temperatures of 875°C–1100°C. The total gas mixture composed of C_2H_2 (80 cm^3/min), SiH_4

(150 cm³/min), and He (1470 cm³/min). The influences of helium were to ensure the dilution of the reactive species and increase the velocity of the gaseous species in the reaction zone. Clement et al. [64] reported that SiC nanoparticles were produced from the gas phase by pulsed CO_2 laser pyrolysis. The gas reactants composed of SiH_4, C_2H_2, and NH_3 for nitrogen doping. Helium was also used as a buffer gas for cooling the reaction products and thereby causing particle condensation.

2.3 RECENT DEVELOPMENT OF FABRICATION AND CHARACTERIZATION OF NANOSTRUTURED SiC THIN FILMS BY USING THERMAL ANNEALING

2.3.1 Formation of np-SiC Films from Si/C/Si Multilayers by Using Post-Thermal Annealing

Recently, a novel approach to the formation of SiC nanoparticles (np-SiC) composite films from Si/C/Si multilayers has been demonstrated by using post-thermal annealing [65,66]. The particle sizes, the density, and the distribution of np-SiC are influenced by the annealing temperature, the layer thickness, and the number. The Si/C/Si multilayers were initially deposited on the single-crystalline Si(100) wafers by means of UHV ion beam sputtering (IBS). An electron cyclotron resonance (ECR) microwave plasma–based ion beam source was used in the UHV IBS system. Both types of three-layer Si/C/Si films with respective thicknesses of 50/200/50 and 75/150/75 nm were investigated. The single-layer C film (250 nm) and the two-layer C/Si film (200/50 nm) were also prepared for the reaction, in comparison with the three-layer Si/C/Si film. The purity of both C and Si targets is the same, i.e., 99.999%. The base pressure and the working pressure were maintained at about 10^{-7} to 10^{-8} Pa and 4.5×10^{-2} Pa, respectively, to avoid potential oxygen contamination. The substrate was rotated at 10 rpm to keep the deposition uniformity and maintained at RT. The post-high-vacuum annealing was performed at 10^{-4} Pa at temperatures of 500°C–900°C for 0.5, 1.0, and 1.5 h to study the thermally induced formation of np-SiC from the Si/C/Si multilayers. The particle sizes, the density, and the morphology of np-SiC were examined by a high-resolution field emission scanning electron microscope (HR-FESEM). The composition of the particles was analyzed by the EDX spectroscopy during the observation by HR-FESEM. The operation conditions were set at a low voltage of 5 kV and a low probe current of 7 A corresponding to a very small beam diameter of 0.03 nm with a short penetration depth, to increase spatial resolution and minimize the background noise. A Si(Li) atmospheric thin-window EDS analyzer with a good sensitivity and energy resolution less than 133 eV was used. The phase identification and crystallinity of films were characterized by a grazing incidence x-ray diffractometer (GIXRD). The reaction and bonding behavior of C and Si in the Si/C/Si multilayer after the reaction was characterized by Raman spectroscopy with high sensitivity for carbon and its compound, in the variations of the graphite peak (G-peak), the disorder-induced peak (D-peak), the Si–Si peak, and the Si–C peak, at specific wavenumber shifts. The interdiffusion and reaction behaviors between C and Si in the Si/C/Si multilayer were examined by the depth profile of Auger electron spectroscopy (AES).

Figure 2.1a through c shows the FESEM micrographs of the three-layer Si/C/Si multilayers of 75/150/75 nm annealed at 500°C, 700°C, and 900°C for 1.0 h, respectively. No particles show up on the specimen treated at 500°C and few particles are present on the specimen treated at 700°C. Many np-SiCs appear on the surface of Si/C/Si multilayers after 900°C annealing. Many large particles of several hundreds of nanometers (labeled A) were surrounded by small particles of tens of nanometers (labeled B) throughout the wafer. The sizes of the large submicron particles synthesized at 900°C exhibit a diameter ranging from 300 to 600 nm, while small np-SiCs have the diameter of about 10–50 nm. The density of large submicron particles with uniform distribution is about 1.37×10^8 cm^{-2}. In contrast, the sizes of np-SiCs annealed at 700°C are about 100–300 nm with a density of about 8.33×10^6 cm^{-2}, which is about 16 times smaller than that observed on the specimen treated at 900°C.

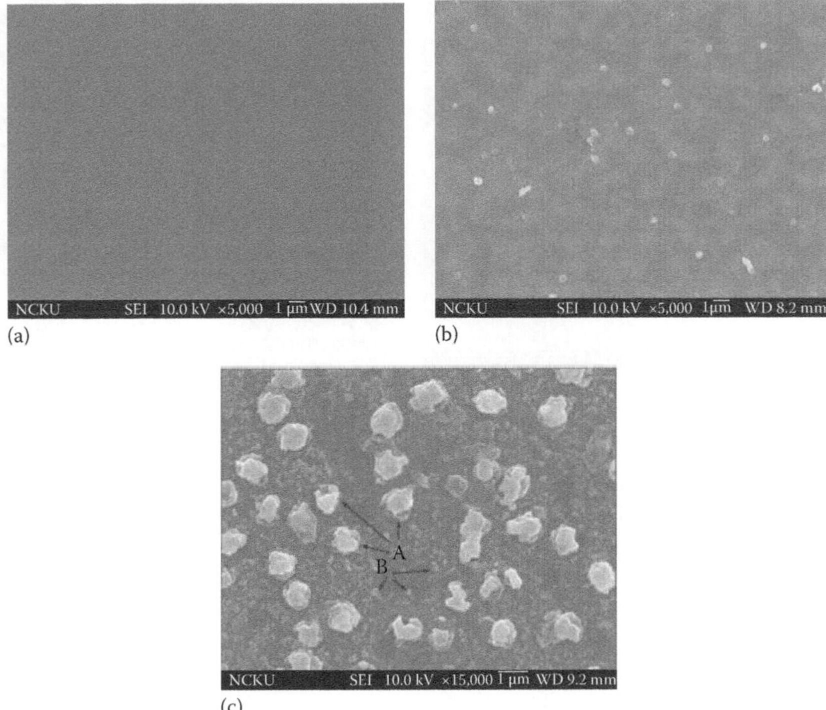

FIGURE 2.1 FESEM micrographs of np-SiC formed on the three-layer Si/C/Si composite films annealed at (a) 500°C, (b) 700°C, and (c) 900°C for 1.0 h, respectively. Many large particles of several hundreds of nanometers (labeled A) surrounded by small particles of tens of nanometers (labeled B) appear on the surface annealed at 900°C. (From Chung, C.K. and Wu, B.H., *Nanotechnology*, 17, 3129, 2006. With permission.)

The higher the annealing temperature, the larger the nanoparticle size became, and consequently the higher the density. The Si/C/Si multilayer structure with interfaces between Si and C layers induces a large driving force due to the increased interfacial energy for reactions at high temperature. Thermal energy is needed for the interdiffusion between Si and C layers for the formation of SiC phase and to reduce the interfacial energy. High temperature gives rise to higher mobility or more rapid diffusion of the atoms together with the larger driving force of superheating for the formation of SiC.

Figure 2.2a and b shows a magnified HR-FESEM micrograph and an EDX spectrum of a large SiC particle formed after 900°C annealing for 1.0 h, respectively. The formation of the shape of particles during growth is not a pure spherical enlargement but seems modified to be a disk-like shape. The atomic composition ratio of Si to C (Si/C) of np-SiC formed at 900°C is detected from the square area within the particle (Figure 2.2a) and analyzed from the peak intensity of EDX spectra (Figure 2.2b) to be about 53/47, i.e., 1.13, close to the stable stoichiometric SiC phase with Si/C atomic ratio of 1.0 in the binary Si and C system. The phase identification and crystallinity of the three-layer Si/C/Si structure were characterized by a GIXRD. Curves A through D in Figure 2.3 show the GIXRD patterns of the as-deposited three-layer Si/C/Si films and those annealed at 500°C, 700°C, and 900°C for 1.0 h, respectively. The broad peaks centered at about 28° and 52° indicate that the as-deposited Si film is amorphous (a-Si). It is stable up to 500°C annealing and becomes polycrystalline with obvious diffraction peaks of Si(111), (220), and (311), as annealed at 700°C and 900°C. Another distinct SiC(111) diffraction peak appears at 900°C annealing. It indicates that the SiC phase is crystalline and primarily grows along the [111] direction, the normal direction of the close-packed (111) plane. It is consistent with the HR-FESEM result. No identifiable SiC diffraction peak at 700°C annealing is attributed to the small amount of SiC particles on the surface (Figure 2.1b).

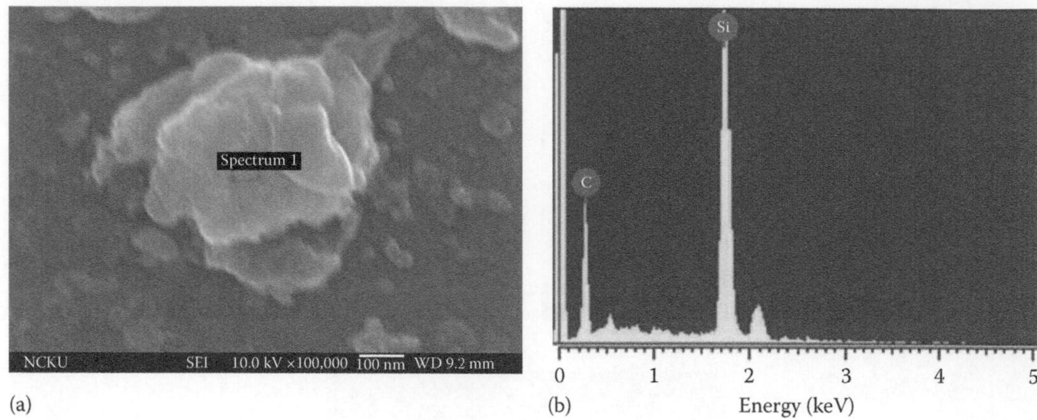

FIGURE 2.2 (a) A magnified FESEM micrograph of np-SiC with small particles surrounding a large one on the surface of Si/C/Si multilayers annealed at 900°C for 1.0 h and (b) an EDX spectrum for the measurement of Si/C composition ratio in the np-SiC area. The atomic composition ratio of Si/C is about 53/47. (From Chung, C.K. and Wu, B.H., *Nanotechnology*, 17, 3129, 2006. With permission.)

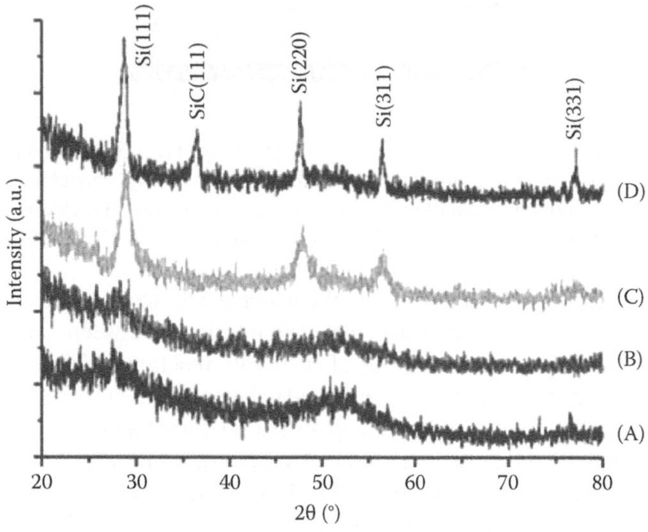

FIGURE 2.3 GIXRD patterns of the three-layer Si/C/Si films (A) as-deposited at RT and annealed at (B) 500°C, (C) 700°C, and (D) 900°C for 1.0 h, respectively. (From Chung, C.K. and Wu, B.H., *Nanotechnology*, 17, 3129, 2006. With permission.)

The FESEM and GIXRD results imply that the defects of stacking faults or disorders occur at the stage of SiC particle formation with the mixed stacking sequence of hexagonal close-packed (HCP) and face-centered cubic (FCC) structures along the normal direction of surface [67]. This stacking defect will induce extra interfacial free energy to the particle phase transformation. The increasing stacking defect energy may limit the size of large particle growth together with some small particles of tens of nanometers surrounding a large one without shrinking to dissipation. The particle size is concerned with the annealing time. Smaller SiC particles of 10–40 nm can be achieved by means of rapid thermal annealing at 900°C for 1 min, as shown in Figure 2.4. But the size of SiC particles at a longer furnace annealing time of 1.5 h is still about 300–600 nm, which doesn't vary much in comparison with 1 h. It supports the fact that the stacking defect energy limiting the size of large particles is reasonable.

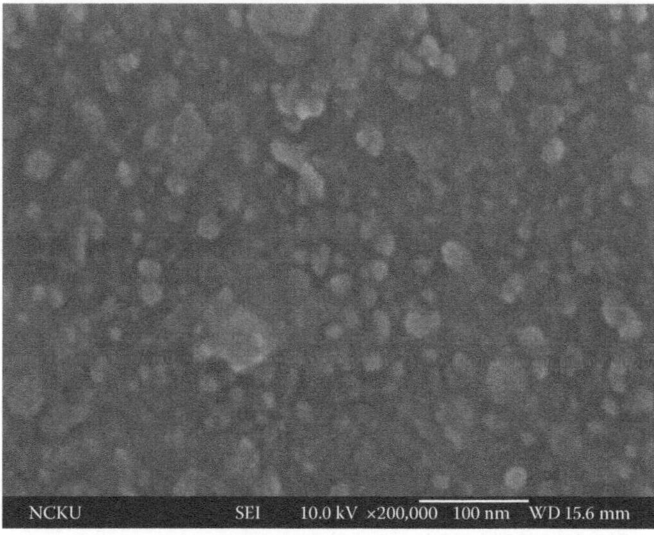

FIGURE 2.4 The FESEM micrograph of smaller np-SiCs formed by rapid thermal annealing (RTA) at 900°C for 1.0 min.

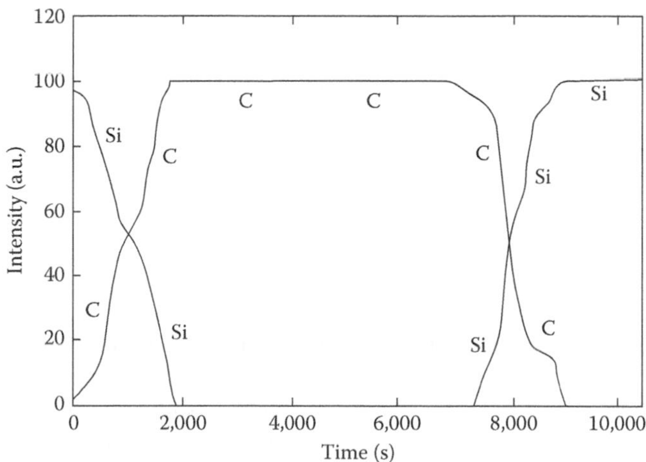

FIGURE 2.5 AES depth profile of three-layer Si/C/Si (75/150/75 nm) films annealed at 900°C for 1.0 h, respectively. (From Chung, C.K. and Wu, B.H., *Nanotechnology*, 17, 3129, 2006. With permission.)

Figure 2.5 shows the AES depth profile of three-layer Si/C/Si (75/150/75 nm) annealed at 900°C for 1.0 h to characterize the interdiffusion between C and Si. The *X*-axis is labeled by the sputter time of Ar^+ ions for AES depth profiling. Because the sputter rate of C is much lower than Si, about three to four times smaller, the time range of the thicker C layer is much larger than that of Si. The interdiffusion between the C and Si films becomes much more distinct at the two interfaces of top-layer Si/C and C/under-layer Si, which implies the formation of a new crystalline SiC phase at 900°C. The new SiC phase formation after interdiffusion of C and Si is in agreement with the HR-FESEM and the GIXRD results.

Curves A through D in Figure 2.6a and b show the Raman spectra with overlapping fitted curves at wavenumbers in the region of 400–1800 cm^{-1} and the magnified curves at 800–1800 cm^{-1} of three-layer Si/C/Si films, which are as-deposited at RT, and annealed at 500°C, 700°C, and 900°C for 1.0 h,

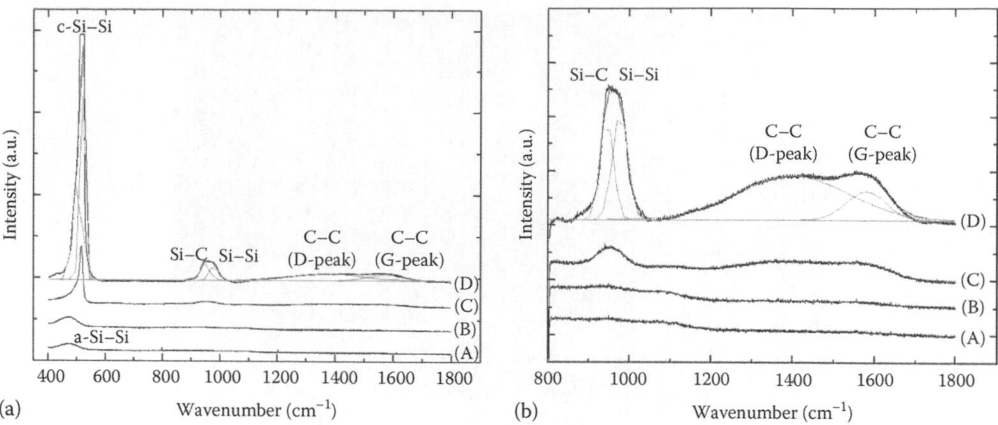

FIGURE 2.6 (a) Raman spectra with overlapping fitted curves of three-layer a-Si/C/a-Si films at wavenumbers in the regions from 400 to 1800 cm^{-1} and (b) the magnified curves from 800 to 1800 cm^{-1} for the samples deposited at RT and annealed at 500°C, 700°C, and 900°C for 1.0 h, respectively. (From Chung, C.K. and Wu, B.H., *Thin Solid Films*, 515, 1985, 2006. With permission.)

respectively. The bonding behavior of three-layer Si/C/Si for the obvious Si–Si bonding as well as Si–C bonding could be examined by Raman spectra over the wavenumbers of 400–1000 cm^{-1}, and for the C and Si bonding (Si–Si, Si–C, and C–C) over the wavenumbers of 800–1800 cm^{-1}. The peak intensity of crystalline Si–Si bonds (c-Si–Si) is found at a wavenumber of around 520 cm^{-1}, while that of amorphous Si–Si bonds (a-Si–Si) appears at around 480 cm^{-1} [68]. Also, the second-order Raman peak of c-Si–Si bonding and the longitudinal optical (LO) mode of crystalline Si–C bonding appear in the region of 950–1000 cm^{-1} [69–71]. From the variation of Si–Si bonding, the a-Si film is crystallized as the annealing temperature is increased to 700°C (curve C), and crystalline SiC is formed at 900°C annealing for 1.0 h (curve D). When the annealing temperature is increased to 900°C, the reaction is enhanced and the intensities of the c-Si–Si peak, the D-peak, and the G-peak in C–C bonding rise much higher. The second order of elemental silicon peak at 950–1000 cm^{-1} is in a flat band. As the SiC is formed in the film, the intensity of the SiC LO mode superimposed on the flat c-Si–Si band leads to the asymmetrical band at 950–1000 cm^{-1} (curve D). Because the Raman position of SiC LO and TO modes is of crystallographic orientation dependence [18], the absence of the TO peak at 796 cm^{-1} in this study may be attributed to the film defects, e.g., grain boundaries (GBs) and stacking faults for eliminating its intensity. The Raman result also indicates the intermixing of Si and C to form the crystalline SiC compound. The a-Si layers sandwiching the C layer in the C/Si multilayer under high temperature annealing at 900°C could enhance the interdiffusion and reaction between C and Si and form a nanocomposite structure with SiC, C, and Si phases due to the GBs of polycrystalline Si formed from both the a-Si layers at the top and the bottom of the C layer. The Raman result is also in agreement with the results of FESEM, GIXRD, and AES depth profile.

With regard to the layer thickness effect on the np-SiC formation, Figure 2.7a through c shows the FESEM micrographs of two-layer C/Si (200/50 nm), three-layer Si/C/Si (50/200/50 nm), and Si/C/Si (75/150/75 nm) multilayers on the Si(100) substrates annealed at 900°C for 1.0 h, respectively. No particle appears on two-layer specimens treated at 900°C, while many np-SiCs were found on both three-layer specimens at 900°C. This indicates that the three-layer Si/C/Si structure is beneficial for the formation of SiC nanoparticles compared to the two-layer structure. The particle sizes on both three-layer Si/C/Si (50/200/50 and 75/150/75 nm) were close and estimated to be about 300–600 nm in diameter. However, the density of particles on the former specimen has a uniform distribution of about 0.75×10^8 cm^{-2}, while that on the latter Si/C/Si (75/150/75 nm) is about 1.37×10^8 cm^{-2}, which is about 1.8 times higher compared to that of the thin a-Si structure (50/200/50 nm).

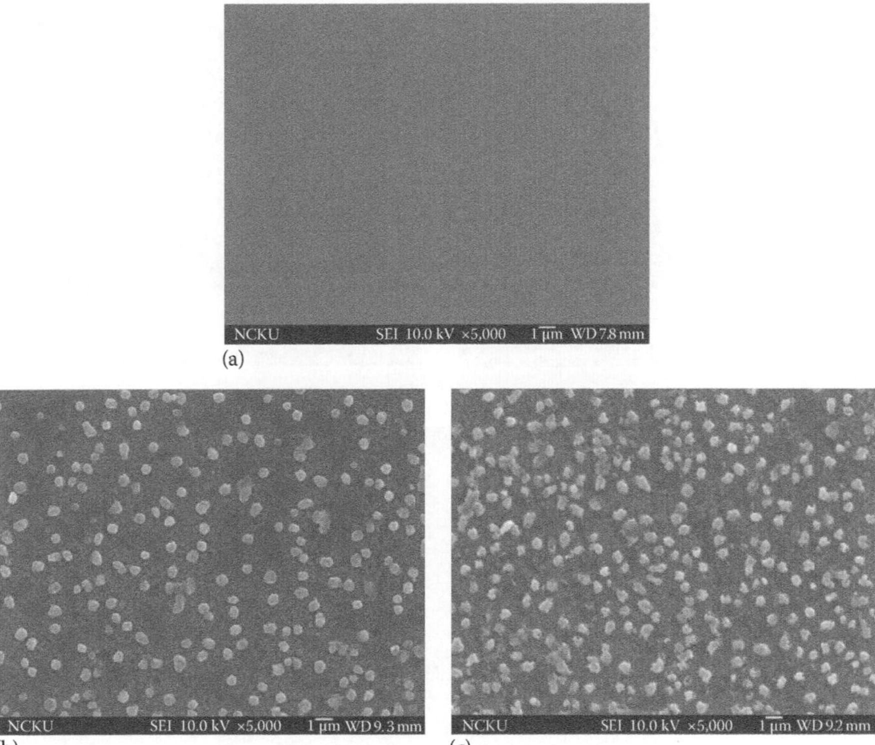

FIGURE 2.7 FESEM micrographs of (a) C/a-Si (200/50 nm), (b) a-Si/C/a-Si (50/200/50 nm) and (c) a-Si/C/a-Si (75/150/75 nm) films annealed at 900°C for 1.0 h, respectively. (From Chung, C.K. et al., *Thin Solid Films*, 517(20), 5867, 2009. With permission.)

The thicker a-Si and thinner C in a three-layer sandwiched structure are favorable for the formation of np-SiC through interdiffusion and reaction of C and Si to increase the particle density. It implies that a-Si/C/a-Si (75/150/75 nm) has a lower activation energy for SiC formation than the thin a-Si structure at the same total layer thickness. SiC formation is a thermally activated process. The np-SiC can start to appear at some lower temperature. Figure 2.8a through c shows the FESEM micrographs of both Si/C/Si (50/200/50 nm) and Si/C/Si (75/150/75 nm) on the Si(100) substrates annealed at 700°C and 500°C for 1.0 h, respectively. There are few particles on both three-layer specimens annealed at 700°C, and no particle appears at 500°C due to a low thermal energy. The particle size at 700°C is smaller than that at 900°C. The np-SiC density in a-Si/C/a-Si (75/150/75 nm) is also higher than that of the thin a-Si structure due to the lower formation temperature or activation energy. Compared to the two-layer C/a-Si structure, the three-layer a-Si/C/a-Si structure with more interfaces between the Si and C layers induces a larger driving force for interdiffusion and reaction of Si and C at high temperature due to the increased interfacial energy and GBs after Si polycrystallization [72].

Compared to Figure 2.5 for Si/C/Si (75/150/75 nm) structures, Figure 2.9a and b shows the AES depth profiles of C/a-Si (200/50 nm) and a-Si/C/a-Si (50/200/50 nm) structures annealed at 900°C for 1.0 h, respectively. No interdiffusion between Si and C is observed in the two-layer C/Si structure together with single-layer C on the Si(100) structure with a similar AES profile [72]. However, the pronounced interdiffusion occurs in both three-layer Si/C/Si structures. The surface concentration of Si in thinner a-Si is lower than that of the thicker one. The interdiffusion between the C and Si

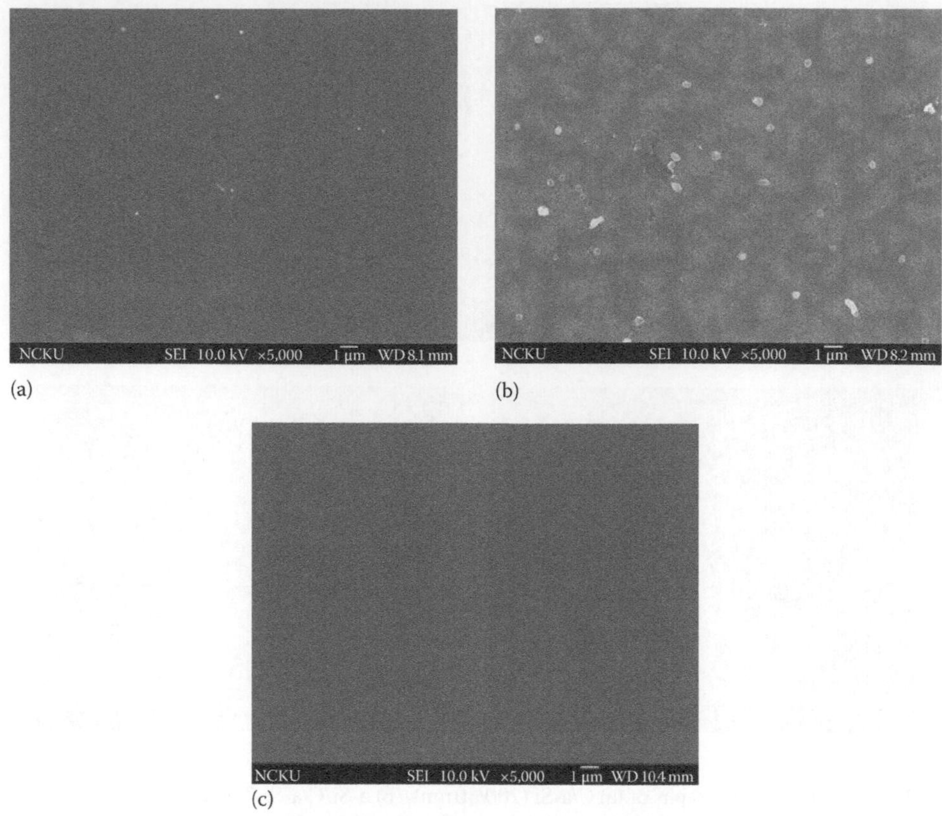

FIGURE 2.8 FESEM micrographs of (a) an a-Si/C/a-Si (50/200/50 nm) film annealed at 700°C and (b) a-Si/C/a-Si (75/150/75 nm) films at 700°C and (c) at 500°C for 1 h, respectively. (From Chung, C.K. et al., *Thin Solid Films*, 517(20), 5867, 2009. With permission.)

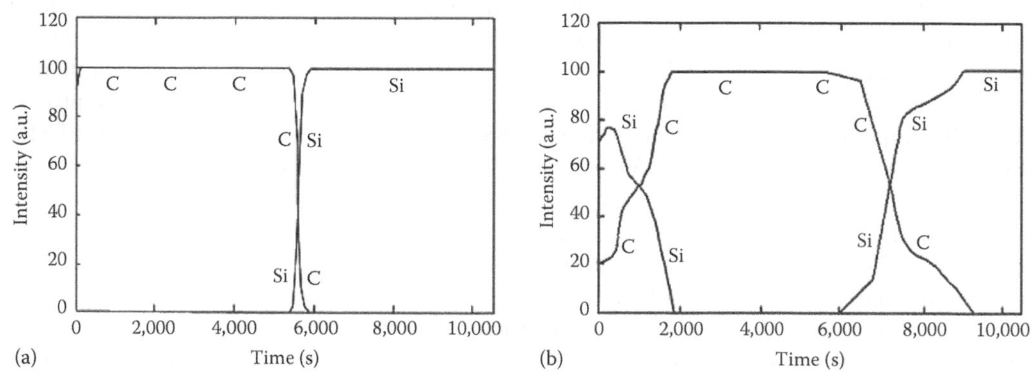

FIGURE 2.9 AES depth profiles of (a) C/Si (200/50 nm) and (b) Si/C/Si (50/200/50 nm) films annealed at 900°C for 1.0 h, respectively. (From Chung, C.K. et al., *Thin Solid Films*, 517(20), 5867, 2009. With permission.)

films becomes distinct at the two interfaces of the top-layer Si/C and C/under-layer Si, which implies the formation of a new crystalline SiC phase at 900°C. This is in agreement with the GIXRD results (Figure 2.3). From the results of FESEM, GIXRD, Raman, and AES depth profile, the annealing temperature plays an important role in the reaction mechanism of a-Si and C, to form np-SiC with a strong bonding and high surface energy.

2.3.2 Formation of Nanoweb SiC Films through In Situ High-Temperature Reaction of C Deposited on Si Substrate

Besides the post-annealing effect on the formation of nanostructured SiC films, the effect of the substrate temperature on the in situ formation of crystalline SiC (c-SiC) nanostructured films on Si(100) at 400°C–700°C using UHV IBS is stated in this section. In addition to GIXRD, FESEM, and AES depth profile, Raman spectroscopy is used for characterizing the sp^2–sp^3 bonding variation of C on the c-Si(100) substrate, corresponding to the reaction between C and c-Si. Curves A through E in Figure 2.10 show the GIXRD patterns of the C films deposited on c-Si(100) substrates at the temperatures of RT, 400°C, 500°C, 600°C, and 700°C, respectively. The broad peak centered at 2q about 52° indicates that the deposited C films at RT are amorphous. The amorphous microstructure is stable up to 500°C and transformed into a new phase of crystalline SiC at 600°C with diffraction peaks at 2θ of 35.7° from the SiC(111) plane. Higher intensities of the SiC(111) peak and one tiny diffraction peak from SiC(311) are observed at 700°C. This implies that the reaction temperature of C and c-Si for c-SiC starts at 500°C–600°C and more amount of c-SiC is formed at 700°C. The main driving force for the formation of c-SiC is the thermal energy that activates the large interdiffusion between C and c-Si. Although the amorphous C films exist from RT to 500°C, their disorder states are related to the temperature.

Curves A through E in Figure 2.11 show the Raman spectra of the C films deposited on c-Si substrates at the temperatures of RT, 400°C, 500°C, 600°C, and 700°C, respectively. In the region of Raman shift wavenumbers of 1200–1800 cm^{-1}, there are broad asymmetrical scattering bands for the G-peak and the D-peak of C–C bonding, which is a characteristic of the DLC film. The D-peak intensity increases from RT to 500°C, then decreases at 600°C, and then disappears at 700°C. It indicates that the disorder state of C increases with temperature before reaction, and SiC starts to form at around 600°C and reacts completely at 700°C. The new SiC phase can be observed from the Raman shift wavenumbers of 900–1100 cm^{-1}. The crystalline Si–Si (c-Si–Si) bonding from the c-Si wafer occurs in a broad band of 900–1050 cm^{-1} (curves A through C in Figure 2.11). The crystalline Si–C bonding in the LO mode appears in the region of 950–1000 cm^{-1} [69–71] and is superimposed to c-Si–Si spectra to make an asymmetrical spectrum (curves D and E in Figure 2.11). The more the c-SiC amount, the more asymmetrical is the band. Simultaneously, the intensities of the D-peak

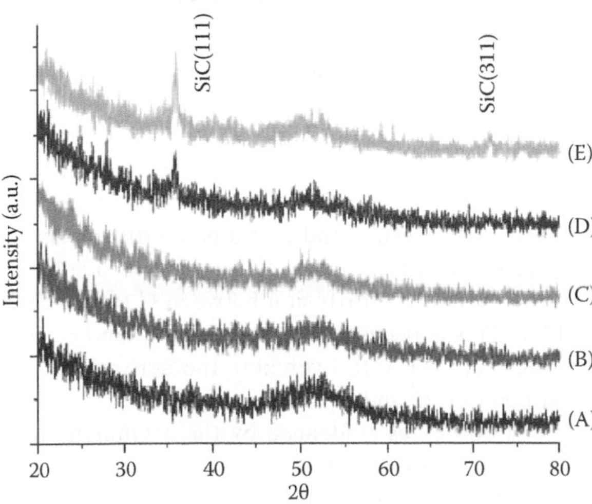

FIGURE 2.10 GIXRD patterns of the C films deposited on c-Si(100) substrates at the temperatures of (A) RT, (B) 400°C, (C) 500°C, (D) 600°C, and (E) 700°C, respectively. (Reprinted from Chung, C.K. and Wu, B.H., *J. Nanosci. Nanotechnol.*, 2009. With permission.)

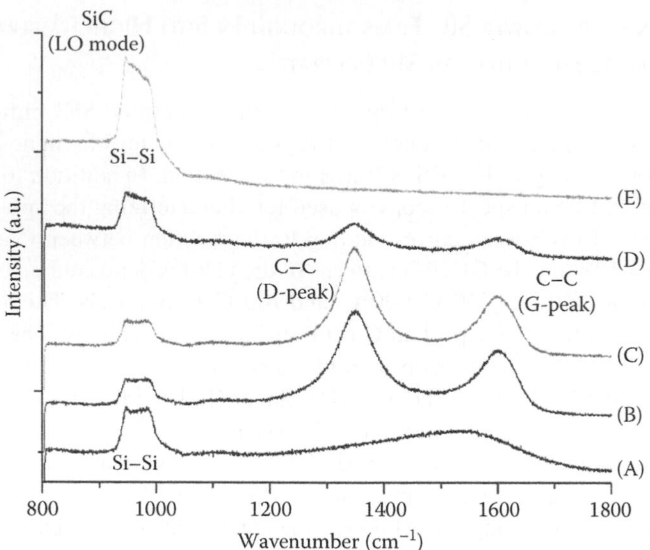

FIGURE 2.11 Raman spectra of C films deposited on c-Si substrates at the temperatures of (A) RT, (B) 400°C, (C) 500°C, (D) 600°C, and (E) 700°C, respectively. (Reprinted from Chung, C.K. and Wu, B.H., *J. Nanosci. Nanotechnol.*, 2009. With permission.)

and the G-peak in the C–C bonding decrease greatly at 600°C and disappear at 700°C due to the formation of c-SiC. This is consistent with the GIXRD result.

Figure 2.12a through c shows the FESEM micrographs of C films deposited on the c-Si substrates at temperatures of 500°C, 600°C, and 700°C, respectively. A smooth morphology of the amorphous C film is observed at 500°C, while high density of a nanoweb-like morphology appears at 600°C due to the formation of SiC. The nanoweb-like morphology becomes larger and thicker at 700°C due to more amount of SiC after the complete reaction. The nanoweb-like morphology is concerned with the amount of SiC. It is attributed to the high surface energy of SiC with a strong Si–C bonding to form an SiC nanoweb embedded in the matrix of the film. Figure 2.13 shows the EDX spectrum of an SiC nanoweb formed at 700°C. The atomic composition ratio of Si to C (Si/C) is analyzed from the integrated peak intensity of EDX to be around 51/49 close to the stoichiometry unity. Therefore, the formation of a new c-SiC phase between the C and Si reaction can be evidenced from the evolution of the microstructure, atomic bonding, and the morphology, from the analysis of GIXRD, Raman, and FESEM.

Figure 2.14a through c shows the AES depth profiles of the C films deposited on c-Si(100) substrates at temperatures of 500°C, 600°C, and 700°C, respectively. The interdiffusion behavior could be observed from the elements' transition at the interface between different layers. More sharp or symmetrical profiles at the interface indicate less interdiffusion. On the contrary, a large interface range indicates that interdiffusion happened, and a solid solution or a new phase was formed [72]. There was no detectable interdiffusion between C and Si elements at 500°C, which is the same as 400°C and RT. This is because the reaction of C and c-Si(100) substrate at 500°C or below has insufficient thermal energy to overcome the activation barrier for enhancing the interdiffusion and reaction between C and Si. However, an identifiable interdiffusion has been observed between C and Si at 600°C, as evidenced by the asymmetrical profile at the interface between the C layer and the c-Si substrate. And a layer with a nearly uniform Si/C composition ratio was obtained at 700°C. This indicates that C and Si can intermix to form polycrystalline SiC at 600°C and complete the reaction at 700°C. This is consistent with Raman and GIXRD results. In short, the in situ formation of c-SiC was through the interdiffusion and reaction between the sputtered carbon and the crystalline Si substrate at high temperature. The amorphous-like C

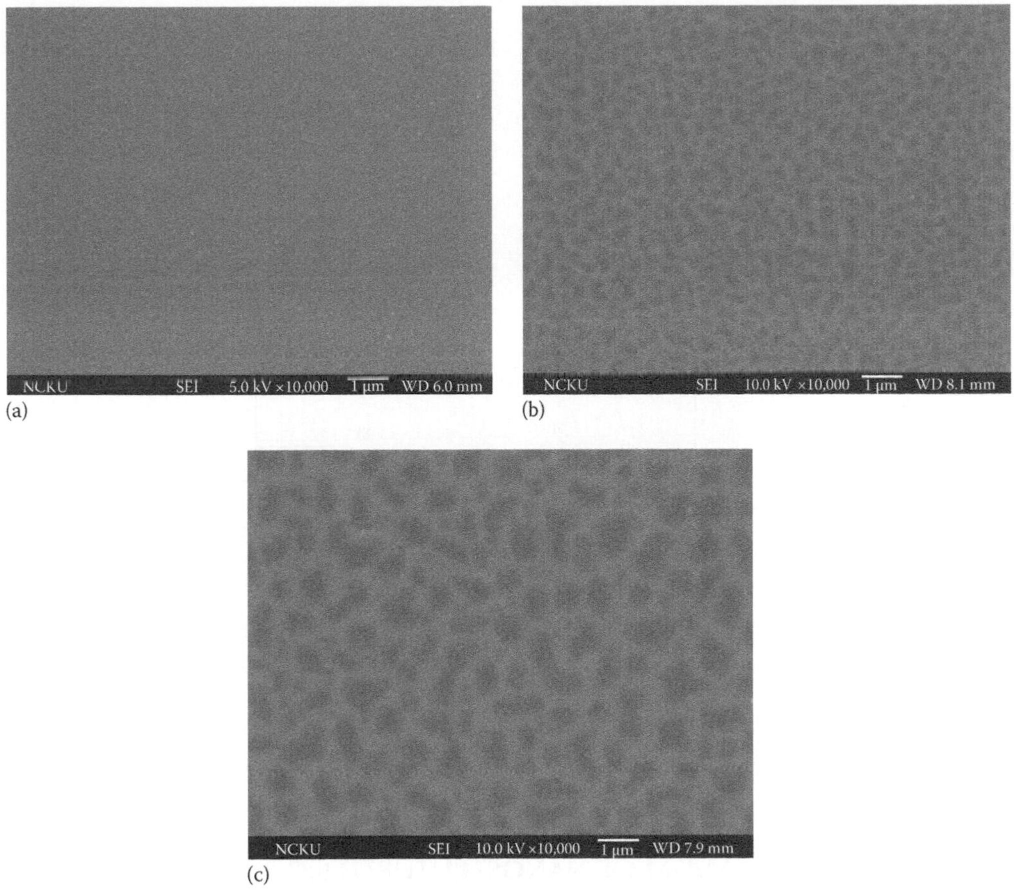

(a)

(b)

(c)

FIGURE 2.12 FESEM micrographs of C films deposited on c-Si substrates at the temperatures of (a) 500°C, (b) 600°C, and (c) 700°C, respectively. (Reprinted from Chung, C.K. and Wu, B.H., *J. Nanosci. Nanotechnol.*, 2009. With permission.)

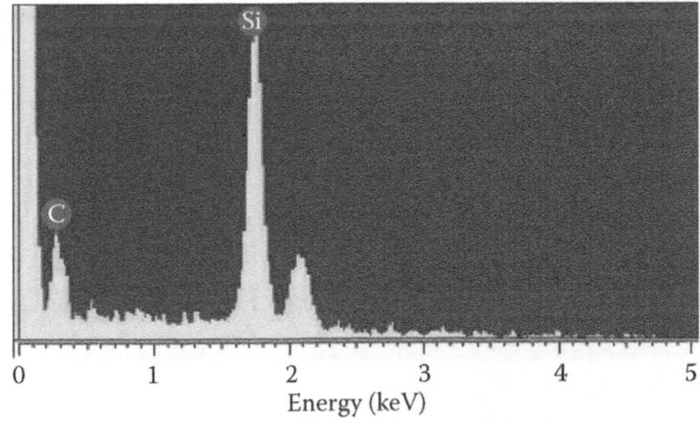

FIGURE 2.13 EDX spectrum for the measurement of the SiC nanoweb area. (Reprinted from Chung, C.K. and Wu, B.H., *J. Nanosci. Nanotechnol.*, 2009. With permission.)

microstructure is stable up to 500°C and transformed into a new phase of c-SiC at 600°C–700°C together with a nanoweb-like morphology. Therefore, the polycrystalline c-SiC nanostructured film can be obtained at 700°C using the in situ UHV IBS process that is much lower than in the conventional CVD epitaxial c-SiC. In addition, the nanoweb morphology is different from that of the post-annealing np-SiC.

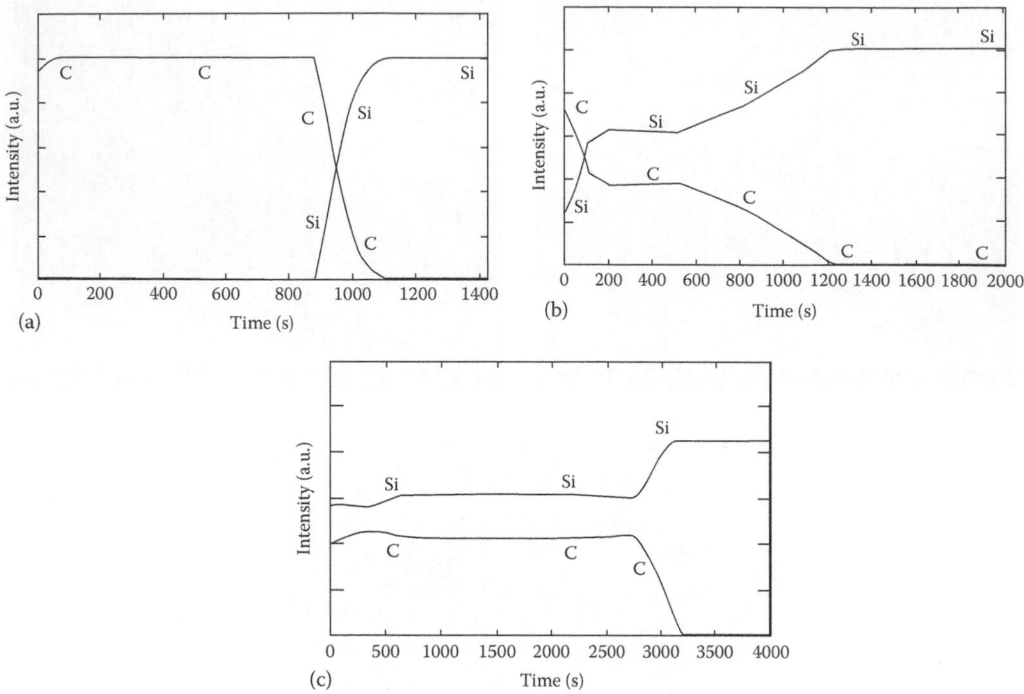

FIGURE 2.14 AES depth profiles of C films deposited on c-Si(100) substrates at the temperatures of (a) 500°C, (b) 600°C, and (c) 700°C, respectively. (Reprinted from Chung, C.K. and Wu, B.H., *J. Nanosci. Nanotechnol.*, 2009. With permission.)

2.4 REACTION MECHANISM OF NANOSTRUTURED SiC THIN FILMS BY USING THERMAL ANNEALING

2.4.1 Reaction Mechanism of np-SiC Films from Si/C/Si Multilayers by Using Post-Thermal Annealing

From the results of Section 2.3, the size, the density, and the distribution of np-SiC were significantly influenced by the annealing temperature, the a-Si thickness, and the layer number. The np-SiC formation is a thermally activated reaction. No nanoparticles are formed for Si/C/Si multilayers annealed at 500°C, while few particles start to appear at an annealing temperature of 700°C, and many small np-SiCs of several tens of nanometers surrounding larger submicron ones appear at 900°C with an about 16 times higher particle density than that observed at 700°C. With the higher annealing temperature, the nanoparticle size gets bigger, and the density gets higher. Such np-SiCs at 900°C have a density order about 10^8 cm^{-2} in both three-layer Si/C/Si (50/200/50 and 75/150/75 nm) structures, while no particles are formed in the single-layer C and two-layer C/Si structures at the same annealing temperature. The a-Si-layer position underlying or sandwiching the C layer influences the reaction temperature between the a-Si and C layers at constant annealing time. Also, the thick a-Si structure (75/150/75 nm) produces a particle density approximately 1.8 times higher than that of the thin structure (50/200/50 nm). The transformation of a-Si to polycrystalline Si is observed at 700°C, and significant interdiffusion occurs at 900°C as well as c-SiC formation in the multilayer. The interdiffusion between C and Si in the three-layer Si/C/Si is much more enhanced than in the two-layer C/Si. The two a-Si layers sandwiching the C layer in the three-layer a-Si/C/a-Si structure lead to the formation of the nanostructured SiC thin film at 900°C, which is lower than the conventional single-crystalline SiC formed at 1000°C–1200°C in a CVD process. The thick a-Si in an a-Si/C/a-Si structure exhibits a lower activation energy of SiC formation compared to the thin a-Si in an a-Si/C/a-Si structure at the same total three-layer thickness. A mechanism for the above formation of np-SiC films from thermodynamic and kinetic viewpoints is proposed.

In terms of thermodynamics, the Si and C in different films will react to form the SiC film at the same high temperature if the reaction time is long enough or infinite. But reaction kinetics related to the activation energy will dominate the transformation of films at a finite time and a specific temperature. In terms of the qualitative analysis of reaction kinetics, the reaction rate is related to the activation energy (Q) and the temperature (T) in the Arrehnius exponential relationship, i.e., proportional to $\exp(-Q/kT)$, where k is the Boltzmann constant. Easier diffusion paths, such as GBs, will speed up the interdiffusion between Si and C at a specific temperature to overcome the activation energy of polycrystalline SiC formation that is lower than in the conventional CVD ep-SiC process. In other words, the activation energy of the reaction will be reduced at a location where easy diffusion paths exist. The SiC formed is detected from the Raman analysis, and its presence reveals that the activation energy of the reaction between C and Si in the three-layer a-Si/C/a-Si structure on the Si substrate is smaller than single C or two-layer C/a-Si on the Si substrate.

Also, the Si/C/Si multilayer structure with interfaces between Si and C layers induces a large driving force due to the increased interfacial energy for reactions at high temperature. Thermal energy is needed for increasing the interdiffusion between Si and C layers for the formation of a SiC phase, together with the reduced interfacial energy. High temperature results in higher mobility or more rapid diffusion of the atoms together with the larger driving force of superheating for the formation of SiC. Due to the high surface energy of SiC films with a strong Si–C bonding [55], SiC nanoparticles, instead of films, were preferred to form on the surface in order to reduce the surface energy of films during post-annealing. More thermal energy or higher atomic mobility at a high temperature of 900°C enhances the particle growth to become larger in size. The larger superheating at 900°C increases the nucleation rate and results in a higher particle density. The particle growth at higher temperatures is meant to reduce the total surface energy of small particles.

From the theory of nucleation and growth of phase transformation in materials science [67], the radius of particles above a critical radius will grow, while the rest will shrink to dissipate below the critical radius. The critical nucleus radius (r^*) is related to the surface free energy (γ) (or interface energy or phase boundary energy, in a positive value) and the change of volume free energy (ΔG_v) (in a negative value), and expressed in Equation 2.1. A critical free energy (ΔG^*) occurs at the critical radius, corresponding to an activation free energy for the requirement of a stable nucleus, and is expressed in Equation 2.2. In heterogeneous nucleation, Equation 2.2 can be modified into Equation 2.3 by adding the shape-dependent $S(\theta)$ term as a function only of θ, i.e., the contact angle of the nucleus on the surface, whose value is between zero and unity due to a lower activation free energy than homogeneous nucleation.

$$r^* = -\frac{2\gamma}{\Delta G_v} \tag{2.1}$$

$$\Delta G^* = \frac{16\pi\gamma^3}{3(\Delta G_v)^2} \tag{2.2}$$

$$\Delta G^* = \frac{16\pi\gamma^3}{3(\Delta G_v)^2} S(\theta) \tag{2.3}$$

Equations 2.1 through 2.3 in qualitative analysis are suitable for different phase transformations with spherical (homogeneous nucleation) or nonspherical (heterogeneous nucleation) nuclei, including the liquid–solid (solidification), solid–vapor, and solid–solid examples. The change of volume free energy, ΔG_v, i.e., the driving force for phase transformation, increases with the increase in superheating (/temperature) or formation enthalpy in the case of the solid–solid transformation. This leads to a reduction in both the critical radius, r^*, and the activation free energy, ΔG^*. The magnitudes of ΔG_v and γ between various transformation types in real cases are related to the phase composition, the particle shape, the surface (/interface, phase boundary) energy between two phases, and the bulk defect energy, e.g., new phase–induced microscopic strains, dislocation, GB, and stacking faults. The change of volume free energy, ΔG_v, increases with the increase in superheating or temperature in this case of solid–solid transformation and leads to the reduction of the critical radius, r^*, and the activation free energy, ΔG^*. So, the activation free energy decreases with increasing temperature and results in an increased number and density of nuclei. The higher mobility of atoms at higher temperature leads to a higher growth rate for bigger particles together with more nuclei, which result in an about 16 times higher particle density at 900°C (1.37×10^8 cm^{-2}) than that at 700°C (8.33×10^6 cm^{-2}). The magnitudes of ΔG_v and γ among various transformation types in real cases will be related to the phase composition, the particle shape, the surface (/interface) energy or phase boundary energy between two phases, and the defect energy, e.g., new phase–induced microscopic strains, dislocation, and stacking faults. In our solid–solid transformation of multilayers, the surface free energy (γ) should be modified by the surface (/interface) energy between the particles and the substrates, and the stacking defect energy, in addition to the strain energy, and these will affect the magnitudes of ΔG^* and r^*. With regard to the thickness effect in our solid–solid transformation of multilayers, the SiC formation from three-layer Si/C/Si must go through the polycrystallization of a-Si, and the interdiffusion and reaction between C and Si (Figures 2.3, 2.5 through 2.10). Therefore, the number of GBs is concerned with the activation barrier of reaction. It is reasonable that a higher number of GBs with a larger driving force and a lower activation energy leads to a larger nucleation density. That is, the polycrystallization of the thicker a-Si structure in a-Si/C/a-Si (75/150/75 nm) at 900°C produces more GBs for a larger driving force (ΔG_v) to reduce the activation energy and increase the particle density, as compared to the thin a-Si (50/200/50 nm) structure. Moreover, in comparison with the AES depth profiles of Si/C/Si with thicker (Figure 2.5) and thinner a-Si layers (Figure 2.9b) after 1 h annealing, the former Si concentration is higher than the latter. It may be due to the limited elemental Si at the surface and a higher amount of np-SiC produced at the thicker a-Si top layer with a more important initial Si source.

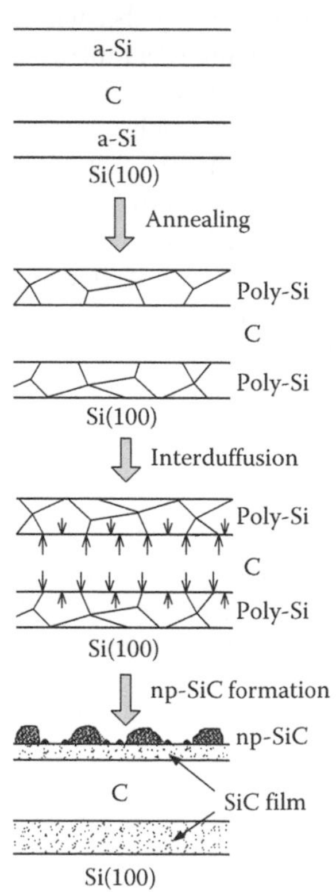

FIGURE 2.15 The schematic diagram of a proposed np-SiC formation mechanism via interdiffusion and reaction of C and Si in the a-Si/C/a-Si multilayer by means of thermal annealing. (From Chung, C.K. and Wu, B.H., *Nanotechnology*, 17, 3129, 2006. With permission.)

Figure 2.15 shows the schematic diagram of a proposed np-SiC formation mechanism via interdiffusion and reaction of C and Si in the a-Si/C/a-Si multilayer, by means of post-thermal annealing. According to the results of SEM, GIXRD, and AES depth profile, the annealing temperature plays an important role on the reaction of a-Si and C for the formation of a new SiC phase with a strong bonding and high surface energy. As annealing is performed at an exact temperature of 700°C, the crystallization of the two a-Si layers first takes place to form polycrystalline ones with many GBs, which are fast paths for interdiffusion between Si and C. Then, crystalline SiC with a high surface energy is formed. The SiC particles are preferred to appear on the surface in order to reduce the total free energy of the film. The grain growth of polycrystalline Si and interdiffusion between Si and C becomes more rapid at a higher temperature of 900°C than 700°C, owing to the higher thermal energy and atomic mobility. The particle density increases with the increasing temperature due to larger superheating at 900°C with more number of nuclei. The particle growth will dissipate the small particles below the critical radius. Some np-SiC particles of several tens of nanometers are retained to surround large ones due to the increase in the stacking fault energy during the growth of SiC with a mixed stacking sequence. The reduction of the activation energy is related to the small C elements with a radius of 0.077 nm, which diffuse much more easily through the fast pathway of GBs in both upward and downward directions toward the two polycrystalline Si layers sandwiching the C layer in the as-deposited three-layer a-Si/C/a-Si structure, than they do in only one diffusion direction in the as-deposited two-layer C/a-Si structure. As the C elements diffuse into the polycrystalline Si layers, the C layer leaves vacancies for the Si elements with a radius of 0.117 nm to diffuse in. Then the polycrystalline Si layers leave additional vacancies for C to diffuse in, and the bulk diffusion is accelerated. Cycling the interdiffusion between C and Si in the three-layer structure repeatedly, the C and Si can be intermixed to a large extent and their positions rearranged at a high temperature of 900°C for 1 h to form crystalline SiC, as detected in GIXRD (Figure 2.3). Because the interdiffusion region (Figure 2.9a) is small in the two-layer C/Si structure, the possible SiC formation under the C layer was not observed here. In fact, the interdiffusion and reaction mechanism of C and Si through the grain boundaries and grains of the polycrystalline Si layers transformed from a-Si could be very complex and not fully understood now. However, the more the diffusion paths and GBs, the faster the interdiffusion and reaction, and the lower the activation energy and the reaction temperature.

2.4.2 Reaction Mechanism of Nanoweb SiC Films through In Situ High-Temperature Reaction of C Deposited on Si

The morphology of the in situ formation of a nanoweb-SiC film through a high-temperature reaction of C deposited on Si is different from that of an np-SiC film by post-annealing the C film deposited on Si at RT. Both kinds of nanostructured SiC formations involve interdiffusion and reaction between C and Si atoms. But the former is concerned with the phase transformation of SiC from a vapor (C)–solid (C)–solid (Si) system, while the latter is related to a solid (Si)–solid (C)–solid (Si) system. That is, the in situ SiC formation at high temperature involves Si and C diffusion through the formed SiC layer, gas–solid reactions, and surface diffusion, while the in situ SiC formation is only related to solid–solid diffusion and reactions. However, the Si and C diffusion through the formed SiC layer in the former will dominate the in situ SiC formation at high temperature, because the much smaller C atoms have a higher diffusivity than Si atoms. The basic equations (Equations 2.1 through 2.3) from the theory of nucleation and growth of phase transformation can also be applied to the main discussion of the in situ high-temperature reaction of C deposited on the Si(100) substrate. With regard to the effect of substrate temperature on the reaction of the vapor (C)–solid (C)–solid (Si) system for the formation of c-SiC, the activation energy is attributed to the interdiffusion and reaction between C and Si atoms. Increasing substrate temperature leads to a reduction in the activation free energy, ΔG^*, for a higher reaction rate. Also, the atomic mobility and diffusion increase with increasing temperature. All above effects result in more amount of c-SiC formation and a thicker nanoweb at 700°C than at 600°C (Figures 2.10 and 2.11). In comparison with the morphologies by means of in situ annealing and post-annealing methods, the former are two-dimensional webs while the latter are sphere-like nanoparticles. This implies that non-spherical nuclei are preferred to occur and extend into a 2D web morphology in the in situ reaction of the vapor (C)–solid (C)–solid (Si) system at high temperature, while sphere-like nuclei can be formed during the post-annealing of solid (Si)–solid (C)–solid (Si) multilayers [65,66]. In solid (Si)–solid (C)–solid (Si) multilayers, the polycrystallization of Si offers fast diffusion paths of GBs for the interdiffusion and reaction between C and Si. In the in situ reaction of the vapor (C)–solid (C)–solid (Si) system, the GBs of the initially formed polycrystalline SiC at high temperatures of 600°C–700°C offer faster diffusion paths, which enhance the following interdiffusion between the C and Si elements for SiC formation. More amount of polycrystalline c-SiC and a lower activation free energy at a higher temperature of 700°C lead to the rapid and complete reaction between C and

Si, while some C remains on the surface of c-SiC at 600°C due to lower atom mobility. Compared to the conventional CVD c-SiC formation, the formation temperature of a new crystalline SiC phase by PVD is 600°C–700°C, much lower than that of CVD c-SiC, which is up to 1000°C–1200°C. But the deposits obtained in the two cases have completely different microstructures and crystallization. The PVD c-SiC is polycrystalline and has a nanowebbed morphology, while CVD c-SiC is smooth and single crystalline as the SiC buffer layer for 3C-SiC heteroepitaxy.

The detailed in situ reaction mechanism of nanoweb SiC films should be more complicated, and further investigation will be favorable for understanding and controlling the new nanostructured SiC films for emerging applications in future. For example, the interdiffusions and reactions between C and Si for both post-annealing and in situ annealing of the C film on Si are quite different for different morphologies of formation. There was no interdiffusion between the single-layer amorphous C and the c-Si(100) substrate using post-annealing at temperatures up to 900°C for 1.5 h [72], while the in situ deposited amorphous C microstructure stable at 500°C is transformed into a new phase of crystalline SiC (c-SiC), together with the remaining C at 600°C as well as the complete reaction of C and Si at 700°C (Figures 2.9 through 2.13). During post-annealing of Si/C/Si multilayers, the polycrystallization of a-Si layers in a-Si/C/a-Si can reduce the activation energy of c-SiC formation with the assistance of GB diffusion, and the morphology is formed by sphere-like SiC nanoparticles on the surface of the film, as stated in Section 2.3.1. Therefore, GB-enhanced interdiffusion to reduce the activation energy of reaction can also be applied to the in situ reaction between the C film and the Si substrate at high temperatures, as stated in Sections 2.3.2 and 2.4.2. The in situ SiC crystallization occurs at a high-enough substrate temperature to overcome the activation energy through the interdiffusion and intermixing mechanism. The SiC nanoweb morphology formed by in situ annealing may be attributed to the metastable reaction, which is different from post-annealing, inducing the formation of sphere-like SiC nanoparticle morphology. The stable state of c-SiC should be the single-crystalline SiC with the lowest free energy and smoother morphology ,and be achieved at a much high temperature. Therefore, our metastable polycrystalline nanoweb c-SiC can be achieved at lower temperature.

2.5 SUMMARY AND OUTLOOK

We have briefly reviewed the fabrication and properties of SiC and its nanostructures and nanocomposite thin films. The fabrication and characterization of SiC nanostructures and nanocomposite thin films using physical and chemical syntheses methods presented in many publications are also collected and introduced. The enhanced optical, optoelectronic, and mechanical properties can be obtained from the SiC nanostructures and nanocomposite thin films because of their nano-induced properties. Moreover, we have highlighted a new method of forming SiC nanoparticles and SiC nanoweb films using post-thermal annealing of Si/C/Si multilayers and in situ high-temperature deposition of C on Si(100), respectively, by means of the UHV IBS technology. The comparison between SiC nanoparticles and SiC nanoweb films in terms of fabrication, characterization, and the reaction mechanism model is addressed.

In future outlook [3], manipulating both carbon and silicon at the nanoscale will offer the possibility of the formation of novel materials for promising applications. One-dimensional NW systems can be synthesized for 3D structures with varied properties. These NW systems can also be used in composite materials to reinforce mechanical properties because of a high Young's modulus, as well as for cold cathodes in field-emission devices or for nanotips in near-field microscopy. The interesting 3D structures can be possibly formed by combining the properties of NWs and nanoporous SiCs for stimulating advanced investigation. In addition, the hydrogen storage may be enhanced by the hollow structures with tunable stoichiometries and nanotubes with unusual hybridization and chemical reactivity. The 3D complex architectures analogous to the zeolites can also be synthesized by means of the precursors of SiC nanocages. Overall, SiC chemistry and structures at the nanoscale should activate future experimental and theoretical exploration.

ACKNOWLEDGMENTS

Some figures in this chapter are extracted from our papers in [65,66,72], which are or will be published by Elsevier, the Institute of Physics, and American Scientific Publishers. We sincerely thank these publishers.

REFERENCES

1. Wright, N. G., Horsfall, A. B., and Vassilevski, K. 2008. Prospects for SiC electronics and sensors. *Mater. Today* 11: 16–21.
2. Mejregany, M. and Zorman, C. A. 1999. SiC MEMS: Opportunities and challenges for applications in harsh environments. *Thin Solid Films* 355: 518–524.
3. Mélinon, P., Masenelli, B., Tournus, F., and Perez, A. 2007. Playing with carbon and silicon at the nanoscale. *Nat. Mater*. 6: 479–490.
4. Harris, G. L. 1995. *Properties of Silicon Carbide*. INSPEC, London, U.K.
5. Kovacs, G. T. A. 1998. *Micromachined Transducers*. McGraw Hill, New York.
6. Woodell, T. R. 1936. *Electrochem. Soc*. 68: 111.
7. Wright, N. G. and Horsfall, A. B. 2007. SiC sensors: A review. *J. Phys. D: Appl. Phys*. 40: 6345–6354.
8. Cimalla, V., Pezoldt, J., and Ambacher, O. 2007. Group III nitride and SiC based MEMS and NEMS: Materials properties, technology and applications. *J. Phys. D: Appl. Phys*. 40: 6386–6434.
9. Zorman, C. A. and Parro, R. J. 2008. Micro- and nanomechanical structures for silicon carbide MEMS and NEMS. *Phys. Status Solidi (b)* 245: 1404–1424.
10. Baliga, B. J. 2006. *Silicon Carbide Power Devices*. World Scientific Press, Singapore.
11. Gupta, A. and Jacob, C. 2005. Selective epitaxy and lateral overgrowth of 3C-SiC on Si—A review. *Prog. Cryst. Growth Charact. Mater*. 51: 43–69.
12. Tabata, A., Kuno, Y., Suzuoki, Y., and Mizutani, T. 1997. Properties of hydrogenated amorphous silicon carbide films prepared by a separately excited plasma CVD method. *J. Phys. D* 30: 194–201.
13. Serre, C. 1996. Ion-beam synthesis of amorphous SiC films: Structural analysis and recrystallization. *J. Appl. Phys*. 79: 6907–6913.
14. Compagnini, G., Foti, G., and Makhtari, A. 1998. Vibrational analysis of compositional disorder in amorphous silicon carbon alloys. *Eur. Phys. Lett*. 41: 225–230.
15. Snead, L. L., Zinkle, S. J., Hay, J. C., and Osbome, M. C. 1997. Amorphization of SiC under ion and neutron irradiation. *Nucl. Instrum. Methods Phys. Res. B* 141: 123–132.
16. Shim, H. W. and Huang, H. 2007. Nanowebs and nanocables of silicon carbide. *Nanotechnology* 18: 335607.
17. Yang, W., Araki, H., Tang, C. et al. 2005. Single-crystal SiC nanowires with a thin carbon coating for stronger and tougher ceramic composites. *Adv. Mater*. 17: 1519–1523.
18. Bechelany, M., Brioude, A., Cornu, D., Ferro, G., and Miele, P. 2007. A Raman spectroscopy study of individual SiC nanowires. *Adv. Funct. Mater*. 17: 939–943.
19. Dai, H., Wang, E. W., Lu, Y. Z., Fan, S., and Lieber, C. M. 1995. Synthesis and characterization of carbide nanorods. *Nature* 375: 769–72.
20. Zhou, X. T., Zhang, R. Q., Peng, H. Y. et al. 2000. Highly efficient and stable photoluminescence from silicon nanowires coated with SiC. *Chem. Phys. Lett*. 332: 215–18.
21. Meng, G. W., Cui, Z., Zhang, L. D., and Philipp, F. 2000. Growth and characterization of nanostructured b-SiC via carbothermal reduction of SiO_2 xerogels containing carbon nanoparticles. *J. Cryst. Growth* 209: 801–806.
22. Li, Z. J., Li, H. J., Chen, X. L. et al. 2003. Large-scale synthesis of crystalline β-SiC nanowires. *Appl. Phys. A* 76: 637–640.
23. Zhang, H. F., Wang, C. M., and Wang, L. S. 2002. Helical crystalline SiC/SiO_2 core–shell nanowires. *Nano Lett*. 2: 941–944.
24. Zhang, Y., Suenaga, K., Colliex, C., and Iijima, S. 1998. Coaxial nanocable—Silicon carbide and silicon oxide sheathed with boron nitride and carbon. *Science* 281: 973–975.
25. Singjai, P., Womgjamras, A., Lu, L. D., and Tunkasiri, T. 2002. Production and characterization of beaded nanofibers from current heating of charcoal. *Chem. Phys. Lett*. 366: 51–55.
26. Saulig-Wenger, K., Cornu, D., Chassagneux, F., Epicier, T., and Miele, P. 2003. Direct synthesis of amorphous silicon dioxide nanowires and helical self-assembled nanostructures derived therefrom. *J. Mater. Chem*. 13: 3058–3061.

27. Saulig-Wenger, K., Cornu, D., Chassagneux, F. et al. 2002. Direct synthesis of β-SiC and h-BN coated β-SiC nanowires. *Solid State Commun*. 124: 157–161.
28. Saulig-Wenger, K., Bechelany, M., Cornu, D. et al. 2005. Synthesis and characterization of cubic silicon carbide (β-SiC) and trigonal silicon nitride (α-Si_3N_4) nanowires. *Ceram. Eng. Sci. Proc*. 26: 341–348.
29. Bechelany, M., Cornu, D., Chassagneux, F., Bernard, S., and Miele, P. 2006. Chemical surface transformation of SiC-based nanocables. *J. Opt. Adv. Mater*. 8: 638–642.
30. Zhou, X. T., Wang, N., Lai, H. L. et al. 1999. β-SiC nanorods synthesized by hot filament chemical vapor deposition. *Appl. Phys. Lett*. 74: 3942–3944.
31. Meng, G. W., Zhang, L. D., Mo, C. M. et al. 1998. Preparation of β-SiC nanorods with and without amorphous SiO_2 wrapping layers. *J. Mater. Res*. 13: 2533–2538.
32. He, X., Yi, J., Sun, Y., Xiao, P., and Zhao, X. 2007. Surface hardness enhancement in sp^3-bonded carbon doped SiC nanocomposite films. *Appl. Phys. Lett*. 91: 241914.
33. Spieß, L., Nennewitz, O., Weishart, H. et al. 1997. Aluminium implantation of *p*-SiC for ohmic contacts. *Diamond Relat. Mater*. 6: 1414–1419.
34. Kalish, R. 2001. The search for donors in diamond. *Diamond Relat. Mater*. 10: 1749–1755.
35. Pecz, B., Weishart, H., Heera, V., and Toth, L. 2003. Diamond formation in cubic silicon carbide. *Appl. Phys. Lett*. 82: 46–48.
36. Movchan, B. A. 2002. Functionally graded EB PVD coatings. *Surf. Coat. Technol*. 149: 252–262.
37. Kulikovsky, V., Vorlíček, V., Boháč, P. et al. 2008. Hardness and elastic modulus of amorphous and nanocrystalline SiC and Si films. *Surf. Coat. Technol*. 202: 1738–1745.
38. Zhao, Q. Z., Ciobanu, F., Malzer, S., and Wang, L. J. 2007. Enhancement of optical absorption and photocurrent of 6H-SiC by laser surface nanostructuring. *Appl. Phys. Lett*. 91: 121107.
39. Fedorenko, L., Medvid, A., Yusupov, M., Yukhimchuck, V., Krylyuk, S., and Evtukh, A. 2008. Nanostructures on SiC surface created by laser microablation. *Appl. Surf. Sci*. 254: 2031–2036.
40. Liu, X. M. and Yao, K. F. 2005. Large-scale synthesis and photoluminescence properties of SiC/SiOx nanocables. *Nanotechnology* 16: 2932–2935.
41. Nishino, S., Powell, J. A., and Will, H. A. 1983. Production of large-area single-crystal wafers of cubic SiC for semiconductor devices. *Appl. Phys. Lett*. 42: 460–462.
42. Nishino, S., Saraie, J. In: Rahman, M. M., Yang, C. Y., and Harris, G. L. (Eds.), 1989. *Amorphous and Crystalline Silicon Carbide II*. Springer, Berlin, Germany, p. 8.
43. Chiu, C. C., Desu, S. B., Chen, G., Tsai, C.Y., and Reynolds Jr., W. T. 1995. Deposition of epitaxial beta-SiC films on porous Si(100) from MTS in a hot wall LPCVD reactor. *J. Mater. Res*. 10: 1099–1107.
44. Hagiwara, C., Itoh, K. M., Muto, J. et al. 1997. *Proceedings of the 7th International Conference on Silicon Carbide, III-Nitrides and Related Materials*, Stockholm, Sweden, Vol. 5 p. 331.
45. Kamimura, K., Koike, K., Ono, H., Homma, T., Onuma, Y., and Yonekubo, S. 1992. In: Rahman, M. M., Yang, C.Y., and Harris, G. L. (Eds.), *Amorphous and Crystalline Silicon Carbide IV*. Springer, Berlin, Germany, p. 259.
46. Kobayashi, J., Yonekubo, S., Kamimura, K., and Onuma, Y. 1995. In: Nakashima, S., Matsunami, H., Yoshida, S., and Harima, H. (Eds.), *Silicon Carbide and Related Materials*. IOP Publishing Ltd., Bristol, U.K., p. 229.
47. Onuma, Y., Miyashita, S., Nishibe, Y., Kamimura, K., and Tezuka, K. 1989. In: Rahman, M. M., Yang, C. Y., and Harris, G. L. (Eds.), *Amorphous and Crystalline Silicon Carbide II*. Springer, Berlin, Germany, p. 212.
48. Nagasawa, H. and Yamaguchi, Y. 1993. Atomic level epitaxy of 3C-SiC by low pressure vapour deposition with alternating gas supply. *Thin Solid Films* 225: 230.
49. Fan, J. Y., Wu, X. L., Kong, F., Qiu, T., and Huang, G. S. 2005. Luminescent silicon carbide nanocrystallites in 3C-SiC/polystyrene films. *Appl. Phys. Lett*. 86: 171903.
50. Fan, J. Y., Wu, X. L., Li, H. X., Liu, H. W., Siu G. G., and Chu, P. K. 2006. Luminescence from colloidal 3C-SiC nanocrystals in different solvents. *Appl. Phys. Lett*. 88: 041909.
51. Ryu, Y., Tak, Y., and Yong, K. 2005. Direct growth of core–shell SiC–SiO_2 nanowires and field emission characteristics. *Nanotechnology* 16: S370–S374.
52. Sharma, R., Rao, D.V. S., and Vankar, V. D. 2008. Growth of nanocrystalline β-silicon carbide and nanocrystalline silicon oxide nanoparticles by sol gel technique. *Mater. Lett*. 62: 3174–77.
53. Zhu, H., Yang, H., Du, K. et al. 2007. Preparation of SiC and SiC/ZnO nanocomposites and its properties. *Mater. Lett*. 61: 4242–45.
54. Xu, M., Ng, V. M., Huang, S. Y., Long, J. D., and Xu, S. 2005. Growth of SiC nanoparticle films by means of RF magnetron sputtering. *IEEE Trans. Plasma Sci*. 33: 242–43.

55. Xu, Y., Narumi, K., Miyashita, K., and Naramoto, H. 2003. AFM observation of nanosized SiC dots prepared by ion beam deposition. *Surf. Interface Anal.* 35: 99–103.
56. Liu, X., Zhang, J., Yan, Z., Ma, S., and Wang, Y. 2001. Photoluminescence from SiC nanocrystals embedded in SiO_2. *Mater. Phys. Mech.* 4: 85–88.
57. Kametani, K., Sudoh, K., and Iwasaki, H. 2004. Growth of SiC nanodots on Si(111) by exposure to ferrocene and annealing studied by scanning tunneling microscopy. *Thin Solid Films* 467: 50–53.
58. Flores, M., Fuenzalida, V., and Häberle, P. 2005. Thermal effects in the size distribution of SiC nanodots on Si(111). *Phys. Status Solidi* 202: 1959–1966.
59. Lin, H. F., Gerbec, J. A., Sushchikh, M., and McFarland E. W. 2008. Synthesis of amorphous silicon carbide nanoparticles in a low temperature low pressure plasma reactor. *Nanotechnology* 19: 325601.
60. Takagi D., Hibino H., Suzuki S., Kobayashi Y., and Homma Y. 2007. Carbon nanotube growth from semiconductor nanoparticles. *Nano Lett.* 7: 2272–2275.
61. Kassiba, A., Makowska-Janusik, M., and Boucle, J. 2002. Photoluminescence features on the Raman spectra of quasistoichiometric SiC nanoparticles: Experimental and numerical simulations. *Physical Review B* 66: 155317.
62. Kassiba, A. et al. 2002. Stoichiometry and interface effects on the electronic and optical properties of SiC nanoparticles. *Diamond Relat. Mater.* 11:1243–1247.
63. Vicens, J., Herlin-Boime, N., Dufour, C. et al. 2004. Flame temperature effect on the structure of SiC nanoparticles grown by laser pyrolysis. *J. Nanoparticle Res.* 6: 63–70.
64. Clement, D., Mutschke, H., and Klein, R. 2003. New laboratory spectra of isolated β-SiC nanoparticles: Comparison with spectra taken by the infrared space observatory. *Astrophys. J.* 594: 642–650.
65. Chung, C. K. and Wu, B. H. 2006. Thermal induced formation of SiC nanoparticles from Si/C/Si multilayers deposited by ultra-high-vacuum ion beam sputtering. *Nanotechnology* 17: 3129–3133.
66. Chung, C. K., Wu, B. H., Chen, T. S., Peng, C. C., and Lai, C. W. 2009. Thickness effect on the formation of SiC nanoparticles in sandwiched Si/C/Si and C/Si multilayers. *Thin Solid Films* 517(20): 5867–5871.
67. Callister Jr., W. D. 2005. *Fundamentals of Materials Science and Engineering*, 2nd edn. John Wiley & Sons, Hoboken, NJ, pp. 420–427.
68. Kimura, Y. and Katoda, T. 1997. Effects of strain on crystallization of amorphous silicon characterized by laser Raman spectroscopy. *Appl. Surf. Sci.* 117/118: 790–793.
69. Kailer, A., Nickell, K. G., and Gogotsi, Y. G. 1999. Raman microspectroscopy of nanocrystalline and amorphous phases in hardness indentations. *J. Raman Spectrosc.* 30: 939–946.
70. Liu, J. and Vohra, Y. K. 1994. Raman mode of 6H polytype of silicon carbide to ultrahigh pressures: A comparison with silicon and diamond. *Phys. Rev. Lett.* 27: 4105–4108.
71. Nienhaus, H., Kampen, T. U., and Monch, W. 1995. Phonons in 3C-, 4H-, and 6H-SiC. *Surf. Rev. Lett.* 324: 328–332.
72. Chung, C. K. and Wu, B. H. 2006. Effect of amorphous Si layer on the reaction of carbon and silicon in the C/Si multilayer during vacuum annealing. *Thin Solid Films* 515: 1985–1991.
73. Chung, C. K. and Wu, B. H. Effect of substrate temperature in the in-situ formation of crystalline SiC nanostructured film using ultra-high-vacuum ion beam sputtering. *J. Nanosci. Nanotechnol.* doi: 10.1166/jnn.2010.1714 (in press).

3 Low-Dimensional Nanocomposite Fabrication and its Applications

Dong Kee Yi and Georgia C. Papaefthymiou

CONTENTS

3.1 INTRODUCTION

There is a need to develop general methodologies for an architecturally imposed order in low-dimensional nanomaterials composed of nanosystems of various shapes and sizes, such as spherical particles, rods, wires, combs, horns, and other nonspecified geometrical architectures. Dimensional and morphological tailoring is of importance, since useful and novel physicochemical properties can be decided upon by the geometry of the nanomaterials. To this end, a range of protocols have recently been introduced.

Numerous and varied zero- and one-dimensional nanostructures can be found in the literature. In terms of their physical and material properties, they can be classified into metallic, polymeric, and ceramic. Each type of material carries unique properties. Metal nanoparticles exhibit complex and interesting optical properties derived from the so-called surface plasmon resonance (SPR) caused by collective oscillations of the conduction electrons when the particles are exposed to electromagnetic radiation. Semiconductor nanoparticles, of the order of 10 nm diameter, show their characteristic quantum confinement effects as bright fluorescence when exposed to UV–vis electromagnetic energies. These novel properties have been adopted to develop timely useful devices in a wide range of applications, such as biosensors, bioimaging agents, and renewable solar cells.

For elongated nanoparticles of large aspect ratios compared to those of the spherical ones, the physicochemical properties can be fine-tuned by controlling the aspect ratios. For example, elongated Au nanoparticles could become available to exhibit SPR in the infrared. Likewise, metal-oxide nanowires are finding their use as gas or chemical sensors and piezoelectric nano-generators, applications not encountered in the spherical particulate domain. These low-dimensional nanomaterials can be integrated to create novel composite materials for applications to the next-generation devices, which will need to address the current energy crisis, environmental sustainability, and better performance requirements. While, at the atomic level, these are already nanocomposite, herewith the term "low-dimensional nanocomposite" is used to mean that the materials and device systems considered are nanocomposite at the molecular and/or higher levels.

First, the nanosized building blocks of the nanocomposites are considered. Thus, the beginning of this chapter is allocated to the description of the currently reported zero- and one-dimensional nanomaterials synthesis and applications. The recent overwhelming interest in nano-biotechnology has led to the inclusion of a section on bio-inspired nanocomposites. The development of size-controlled nanomaterials into useful devices necessitates the use of arraying technologies, such as the arraying of nanowires and nanospheres into thin-film nanomaterial composites and linear template-dot hybrids. Thus, the arraying technology itself can be incorporated as an important synthesis route to achieving nanocomposite fabrication at the molecular and higher levels.

Since the discovery, in the 1980s, of quantum confinement effects in semiconducting nanoparticles/nanocrystals, it is now understood how and why various low-dimensional nanomaterials show different physical/chemical properties compared to their bulk counterparts. New nanocomposites that use individual nanomaterials as basic components promise to exhibit additional novel properties and lead to new, useful applications. In this regard, the combination or integration of linear nanorods/nanowires and spherical nanoparticles to produce higher-level nanocomposites of greater complexity is an interesting theme, which we explore in this chapter.

Overall, the three topics will be introduced as follows:

1. Building blocks of nanocomposites
2. Chemical and physical routes to nanocomposites
3. Bio-inspired nanocomposites

3.2 BUILDING BLOCKS OF NANOCOMPOSITES

The building blocks of low-dimensional nanocomposites consist of linear and/or particulate nanomaterials that have contributed to various nanotechnological applications in combination or individually. Each type of nanomaterial, linear or particulate, has been designed for a specific implementation or for the study of novel nanoarchitectures. The application selection of nanocomposites comes only after the overall nanocomposite material is achieved, where linear and spherical nanomaterials comprise dominating portions of the low-dimensional nanocomposite. Preparation methods of two-dimensional nanomaterials and related property studies are reviewed; however, their detailed description is beyond the scope of this chapter. As building blocks of low-dimensional nanocomposites, individually standing nanomaterials play two distinct roles: (a) as platform materials, to evolve into advanced multifunctional nanocomposites with exceeding performance and (b) as component materials of the newly designed nanocomposite, within which their isolated component domain of nanoscale precision works synergistically with the platform [1].

3.2.1 Linear Nanomaterials

Linear or one-dimensional nanomaterials have been intensively studied worldwide over the past decade. Both inorganic and organic linear nanomaterials have been developed. The novel quantum confinement effect in nanowires opens up a new frontier in electronic device fabrication. Well-defined dimension controllability and reproducibility were the core challenges faced by the material chemists who focused on nanowire synthesis.

Inorganic nanowires include II-VI or III-V semiconductor, silicon and oxide compound nanowires. These inorganic nanowires can be obtained using various chemical and physical synthesis routes. Numerous specific preparation method protocols have been developed, which can be broadly classified as follows: (1) template-based methods, (2) vapor phase–based methods, (3) wet chemistry–based methods, and (4) miscellaneous methods. These methods are discussed below together with the physicochemical properties of the resulting nanowires.

3.2.1.1 Template-Based Method

Concave or edged sites prepared via mechanical treatment can be filled with nanowire precursor material followed by subsequent chemical treatment to convert the precursors to nanowires [2–6]. Solid semiconducting wafers were employed as concave templates for Al or Ga nanowire growth [2–4]. For example, novel metal wires were fabricated on a stepped structure of graphite by an electron deposition process [5,6], while metal-oxide nanowires were patterned onto a polymer groove template that was prepared using laser interference [7]. The resulting nanowires could be highly crystallized via thermal annealing at 900°C. The sol precursor of nanowires was spun onto the grooved pattern of a photo-responsive polymer and, subsequently, dried and thermally aged to produce linear patterns of oxide nanowires (see Figure 3.1). Regarding the dimensional regularity and versatility of material selection, pre-patterned porous membranes also serve as a strong tool.

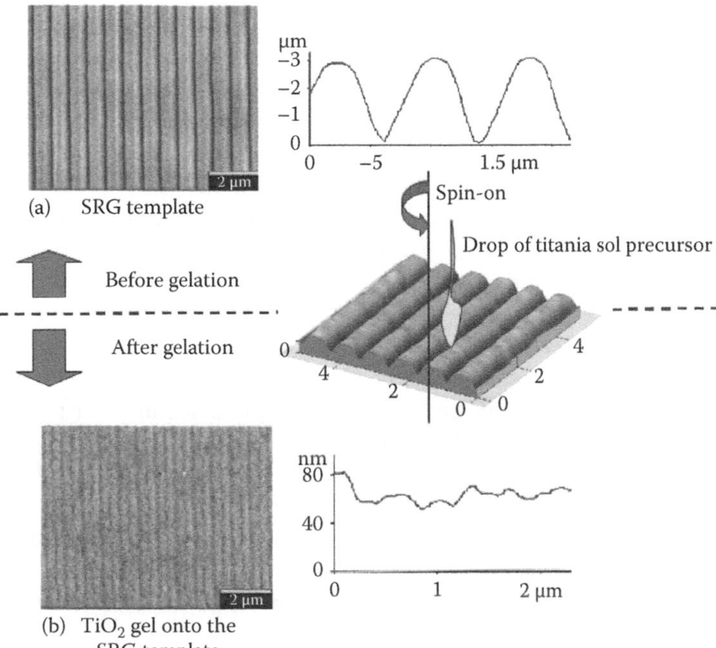

FIGURE 3.1 (a) SEM micrograph and surface undulations of the surface relief grating template. (b) SEM micrograph and surface undulations of cracked gel lines along the ridges. (Reproduced from Yi, D.K. et al., *Nano Lett.*, 2, 1101, 2002. With permission.)

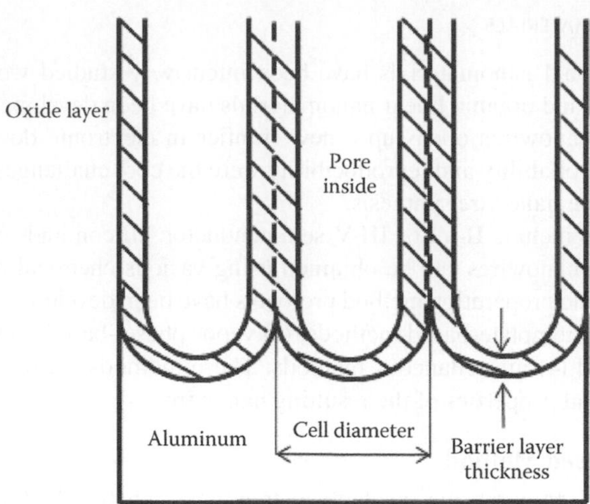

FIGURE 3.2 AAOs work as templates for nanotube and wire growth.

Anodized alumina oxides (AAOs) are reproducible, inorganic, porous, tubular tools for nanowire synthesis and arraying (see Figure 3.2) [8,9]. If a template with uniform pores is available, this method can provide for dimensions, especially height, and controlled nanorods and nanowires with a higher homogeneity, as compared to other template-based approaches. The AAO-based method can serve to fabricate metal nanorods and nanowires that have controlled crystallinity, width, and length, resulting in wires that are nanoscopic in width but macroscopic in length. In nanomaterials technology, morphological tuning is essential to meet industrial purposes. In this context, barrier layer thickness controllability can lead to dimension-controlled nanowires, using the AAO method. Nielsch et al. found that voltage variance works as a control parameter for the width of AAO templates [10]. The subsequent anodization with voltage control induces broadening or narrowing the AAO barrier width. AAOs are expanded to array nanoparticles to form linear architectures [11–14]. A similar porous template system is available as a polycarbonate membrane. Although the internal pore wall surfaces caused lengthwise dimensional irregularity in the resulting nanowires, an easy process ability and flexibility make it a strong polymeric template tool [15–20].

Mesoporous nanomaterials normally have uniform pores (2–50 nm diameter), which form continuous channels. Therefore, several approaches were tested for the fabrication of nanowires within the porous channels. Since the diameters of mesopores are so small, capillary force–driven nanowire precursor infiltration is quite challenging, and the resulting metal nanowires show polydispersity in wire length. However, due to their nanoscale dimensions and easy discarding of the template, mainly SiO_2-based, quite-small-diameter-dimensioned nanowires were produced [21–23]. Chemical vapor deposition (CVD) instead of liquid infiltration was also approached in order to avoid the capillary problem. However, the resulting nanowire length did not show distinguishing differences between the CVD and the solution infiltration methods (see Figure 3.3). Carbon nanotubes (CNTs) were also used as templates to regulate metal nanowire growth within their interior spaces. Solution-based metal salt was deposited onto the interior CNT wall. Subsequent drying and reduction led to metal nanowire–embedded CNT nanoarchitectures (see Figure 3.3) [24,25]. The exterior surface of the CNTs has also been used as a template to produce a nanowire [26]. This follows an extended work of the well-known DNA-templated nanowire fabrication [27,28].

3.2.1.2 Vapor Phase–Based Method

One-dimensional nanowires are unstable, nonequilibrium structures. Their growth was first explained in the case of silicon whisker synthesis by Ellis and Wagner [29], based on a vapor–liquid–solid

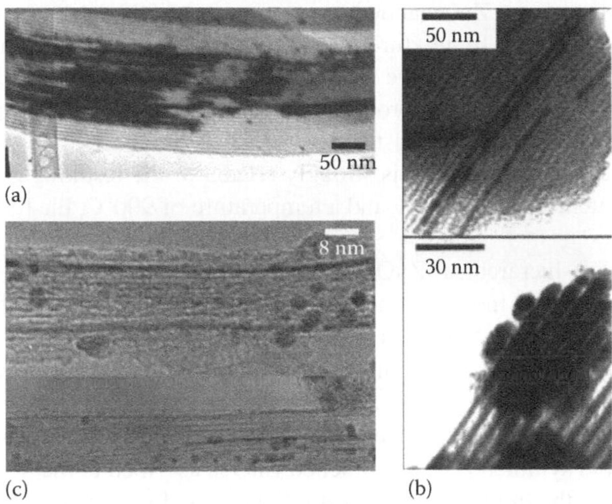

FIGURE 3.3 (a) Pt nanowire formation within the mesoporous templates using solution infiltration. (b) Pd within a mesopore wall via CVD. (c) Pt nanowires in CNT via solution method. (Reproduced from Han, Y.J. et al., *Chem. Mater.*, 12, 2068, 2000; Lee, K.B. et al., *Adv. Mater.*, 13, 517, 2001. With permission.)

(VLS) mechanism. They proposed a two-step growth process of Si nanowires, where, in the first step, nucleation and growth of liquid droplets is initiated, while, in the second step, nanowire growth is promoted via the VLS mechanism, each nanowire terminating at a liquid droplet on its tip (see Figure 3.4). Although the VLS mechanism resembles nanofiber growth using catalysts at a high temperature, CVD per se does not give a clear understanding of the metal-oxide wire growth, where phase reactions can result in nanowires of various compositions, for example, intrinsic semiconductor nanowires composed of Si. ZnO nanowires have also been extensively studied worldwide due to their unique optical and electrical properties [30]. Kong et al. [31] were the first to propose the synthesis of ZnO nanowires. Today we can see not only ZnO wire–type architectures but also belt, tube, and other categories with distinct geometrical features [32,33].

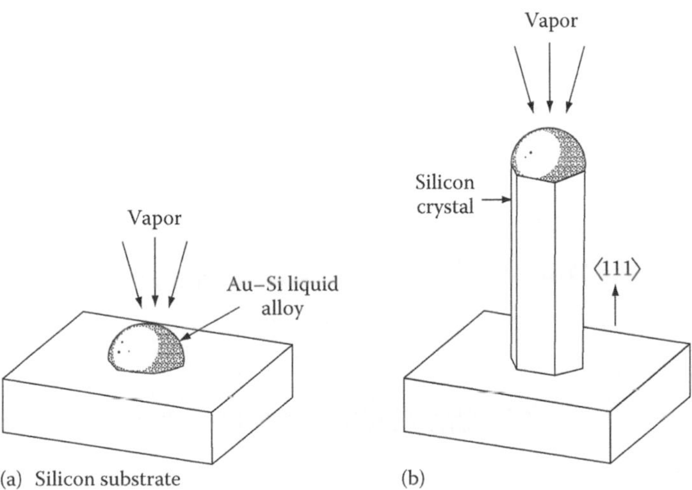

FIGURE 3.4 (a) Vapor condensed to liquid and goes through nucleation and growth. (b) Nanowires are grown with liquid drop at the tip following the VLS mechanism. (Reproduced from Wagner, R.S. et al., *Appl. Phys. Lett.*, 4, 89, 1964. With permission.)

Park et al. [34] have reported ZnO nanowire and nanosheet fabrication on an Au-coated silicon substrate prepared by heating a 1:1 mixture of ZnO and graphite under a constant flow of argon for 5–120min (see Figure 3.5). They have reported that there must be a critical limit in the partial pressure of oxygen gas required to produce ZnO nanowires. At least 1ppm of oxygen in the oxygen–argon combined gas was needed to produce ZnO nanowires. They have found that the temperature-and-oxygen-content window is strongly related to the resulting ZnO nanowire appearance. At an oxygen content of above 0.01% and a temperature of 900°C, the formation of combs and sheets was observed.

In another study [35], hierarchical ZnO nanostructures were synthesized on InO_3-core nanowires. Figure 3.6 shows three kinds of branch symmetries observed: sixfold, fourfold, and twofold symmetries. The main InO_3 core diameter ranges from 50 to 500nm, and the ZnO branches on the InO_3 cores reach up to a few micrometers in length, with diameters falling between 20 and 200nm. For small In_2O_3 nanowire core diameters, the branched ZnO nanorods grow in a single row (see Figure 3.6c), while for larger core diameters, multiple rows of branched ZnO are observed (see Figure 3.6d). The sixfold symmetry of the branched ZnO is ascribed to the hexagonal symmetry of the core. Similarly, when the In_2O_3-core wire is small, branched ZnO nanorods grow in a single row and the core forms multiple rows of ZnO rods with an angle in all four directions (Figure 3.6e and f). Twofold symmetry can be confirmed in Figure 3.6g. It was speculated that InO_x vaporizes first and then condenses on the collector due to the temperature gradient and oxygen partial pressure, and forms into hexagonal, tetragonal, or rectangular InO_3 nanowires. Subsequently, ZnO_x vaporizes and grows on the facets of the existing In_2O_3 nanowires so that 6-, 4-, and twofold symmetries

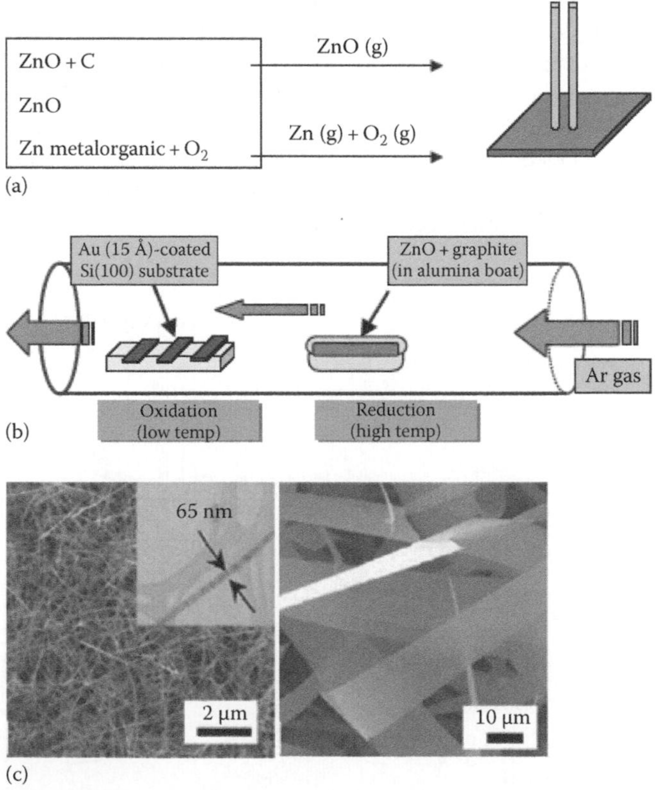

FIGURE 3.5 Typical synthetic route to ZnO nanowire and nanosheet using vapor-phase methods. (a) Oxidation mechanism of Zn, (b) schematic reactant gas flow, and (c) corresponding wire and sheet architectures. (Reproduced from Park, J.H. et al., *J. Cryst. Growth*, 280, 161, 2005. With permission.)

FIGURE 3.6 ZnO-folded architectures using vapor-phase transport and condensation method. (a) Low magnification image showing the abundance, (b) various structural symmetries, (c) 6s-fold symmetry, (d) 6m-fold symmetry, (e) 4s-fold symmetry, (f) 4m-fold symmetry, and (g) 2s-fold symmetry (s and m indicate the single and multiple rows). (Reproduced from Lao, J.Y. et al., *Nano Lett.*, 2, 1287, 2002. With permission.)

are completed. In the vapor phase–based method, using different vapor sources can lead to unique nanostructures. Yang et al. [36] synthesized comblike ZnO nanostructures, using Zn as the vapor source. Under different evaporation conditions, various types of nanostructures are produced, such as multi-pods and tapered nanowires. Reactant gasses of multicomponent mixtures were adopted to study tetrapod-like ZnO (T–ZnO) [37–40]. Recently, Dai et al. synthesized T–ZnO in high yield using 99.9% pure zinc powders that were oxidized to form T–ZnO without the presence of catalysts [40]. The structure was composed of four legs having a length of 2–3 μm, and the edge size of the central nucleus was 70–150 nm. The growth of T–ZnO nanorods could be controlled by the vapor–solid phase transition by changing the conditions under which the zinc vapor condenses onto the substrate during zinc vapor oxidation to form the solid phase. Most interestingly, these nanorod structures showed UV and vis emission at ~380 and ~495 nm, respectively.

Silicon and oxidized silicon (silica) nanowires can also be obtained by vapor-phase methods. Silicon is the paramount modern electronic material with high demands in memory devices. The formation of Si nanowires and their mass production can accelerate the fabrication of ultra highly integrated memory circuits and highly energy-efficient solar cells. Zhang [41] and Morales [42] independently developed mass production processes for Si nanowires, using oxide-assisted growth and laser-induced VLS growth methods, respectively.

High-energy laser ablation of a Si_xFe_x target results in a hot, highly dense vapor of elemental Fe and Si [42]. Subsequently, the hot vapor condenses into small clusters forming liquid droplets. Nanowire growth is initiated as a consequence of Si supersaturation (Figure 3c in Ref. [42]). Temperature quenching causes the solidification of the Si–Fe nanocluster and the termination of wire growth (Figure 3d in Ref. [42]). A transmission electron microscope (TEM) image shows a dark, contrasted core composed of Si, surrounded by a sheath of SiO_x. The spherical nanoclusters at the tip of nanowires can be seen in Figure 4 of Ref. [42], giving evidence that the process does, indeed, follow the vapor-based VLS growth mechanism.

Nitride nanowires can also be prepared using the vapor phase–based method. The well-known blue light–emitting GaN is synthesized by using $Ni(NO_3)$ and Ni [43–50] as catalysts. Most interestingly, ~4 nm width GaN nanowires have been mass-produced. This width lies within the exciton Bohr radius of GaN (~11 nm). Therefore, highly effective quantum confinement effects are expected in the resulting nanowires.

3.2.1.3 Wet Chemistry–Based Method

Solution-phase precursor molecules can be transformed into linear nanostructures via chemical reduction. Consequently, metal nanostructures are more readily obtainable via wet chemical approaches compared to AAO and vapor-phase methods. Linear, metallic nanomaterials have applications as gas and molecule sensors [5,6,51], due to their superior conducting properties for electrons and photons and dimension-dependent SPR properties. The latter can be modulated by tuning the aspect ratio, as demonstrated for Ag and Au [52–55] metal nanorods. The aspect ratio–dependent SPR nature can be potentially exploited in ultrasensitive RNA sensors, solar cell devices, and telecommunication systems. Metal precursors in a polyol reactant or solvent proceed to be reduced, and wire formation is readily achieved. Xia et al. have developed Se [56], Te [57], and Ag [58,59] nanowires using polyol as the reducing agent. Previously, polyol-based wet chemistry methods had been broadly used to produce nanoparticles. The introduction of exotic seeds into the polyol solvent can induce the formation of linear metal nanostructures. When metal precursors, like silver nitrate, are reduced in the presence of these exotic seeds, such as platinum, nucleation and Ostwald-ripening processes follow to form micrometer-length metal nanowires.

The aspect ratio–controlled metal nanorods can also be synthesized in an aqueous environment, using sodium borohydride and ascorbic acid as reducing agents together with molecular templates that regulate metal precursor arraying. Murphy and Wang [60,61], for example, have proposed that rodlike micelle molecular templates, such as cetyltrimethylammonium bromide (CTAB), can direct the growth of a gold salt in water. However, this method produces plates and prisms along with nanorods. Therefore, with respect to homogeneity, the wet chemistry approach does not yield monodispersed properties. Figure 3.7 shows the observed features of Au nanorods having different aspect

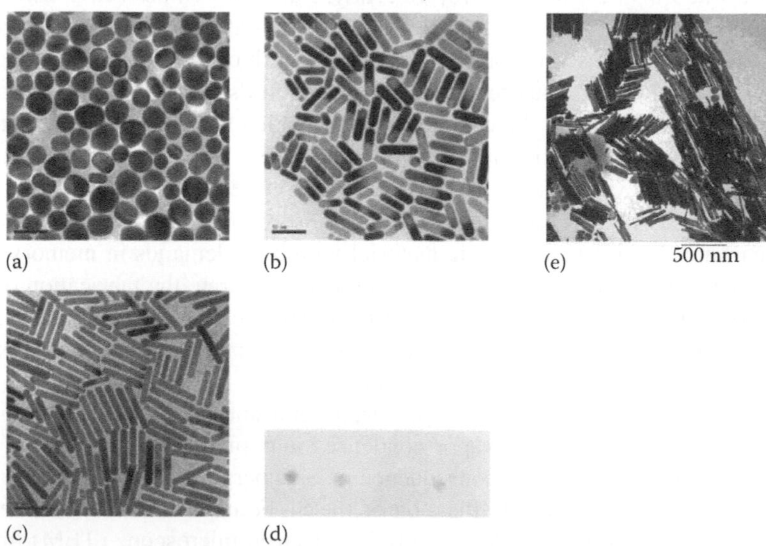

FIGURE 3.7 Au nanorod with different aspect ratios: (a) 1.2, (b) 3.9, and (c) 5.6. (d) Corresponding optical images. (e) TEM micrograph for aspect ratio 18. (Reproduced from Murphy, C. et al., *Adv. Mater.*, 14, 80, 2002; Chou, C.H. et al., *J. Phys. Chem. B.*, 109, 11135, 2005. With permission.)

ratios, and TEM images of nanorods with an aspect ratio of 18. Other groups, e.g., Link et al. [62] have tried similar approaches.

A representative wide-bandgap semiconductor, ZnO, can be prepared using wet chemical methods, specifically hydrothermal methods [63–67].

Conventional vapor-phase methods at high temperatures produce highly crystalline products, while crystal growth at low temperatures is believed to exhibit non-fully developed crystallinity. However, the hydrothermal method at a 90°C reaction temperature produces ZnO nanowires of high crystallinity grown onto a Si wafer [65]. The hydrothermal method using Zn acetate as the precursor material exhibits the highly crystalline nature of ZnO, and the pre-seeding of ZnO nanoparticles onto a wafer substrate with the resulting nanowires standing vertically onto the wafer. Overall, the hydrothermal method does necessitate a high-cost facility, and mass production of nanorods with high crystallinity is feasible. Therefore, there is a need for further exploration of various other semiconducting and metallic materials for the production of diverse nanorod systems.

3.2.1.4 Miscellaneous Methods

Polymer nanostructures in linear formation with nanometer precision can be obtained by electrospinning. A recent research on electrospinning, after an early 1980s patent [66], has been led by Reneker et al. [68–71]. Basic operating principles can be found in Figure 3.8. The polymer solution is contained in the base, which is composed of a capillary pipette maintained at a high voltage. The liquid spreads out as a jet from the tip of the pipette. The liquid jet dries and solidifies to form fibers. The resulting polymer nanofibers are collected onto a metallic collector. Using this method, various polymer nanofibers have been produced, including polyimides, polyamic acid, polyaramid, polyacrylonitrile, polyethylene terephthalate, nylon, polyaniline, and biopolymers such as DNA. Xia et al. have further applied electrospinning to fabricate hollow nanofibers and porous fiber bundles [72]. Figure 3.9 shows nanotubes made of hollow anatase and porous titania nanofibers. The hollow core was formed when mineral oil, corresponding to the core, and metal oxide, corresponding to the outer shell, were fed through a silica capillary together in the coaxial spinneret (see Figure 3.10).

3.2.2 Particulate Nanomaterials

Nanoparticulate materials have all three dimensions less than 100 nm. Well-known quantum-dot semiconducting nanocrystals and viruses can be classified as particulate nanomaterials. Many of them can be artificially synthesized and designed for specific end uses. Exact dimension controllability, monodispersity, and shape regulation are extremely important properties of modern nanoparticulate materials, since size and shape determine the properties, applications, and performance characteristics of

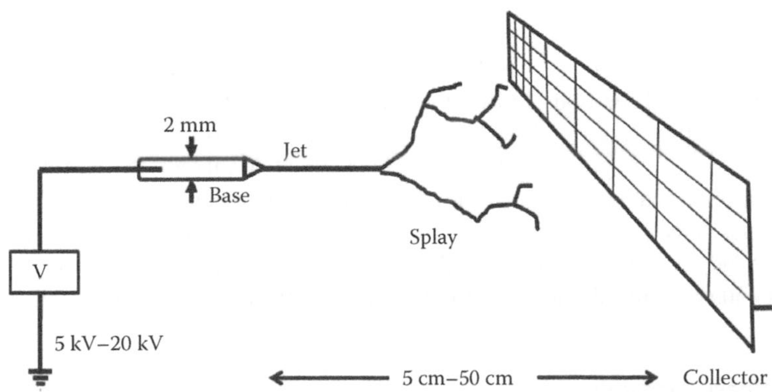

FIGURE 3.8 Diagram of an electrospinning apparatus. (Reproduced from Reneker, D.H. et al., *Nanotechnology*, 7, 216, 1996. With permission.)

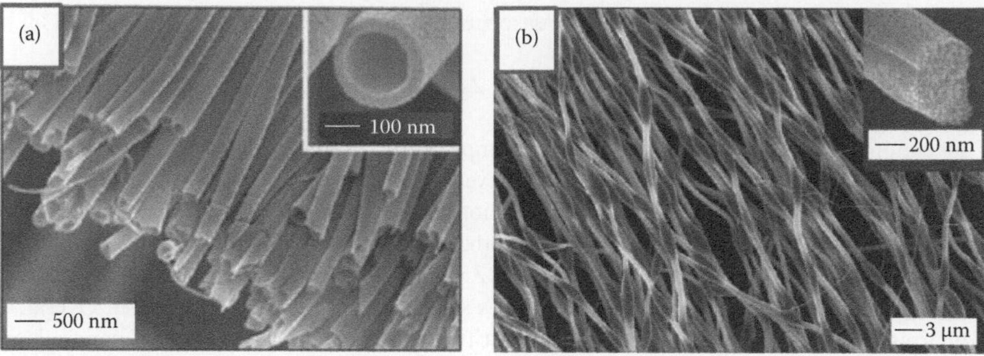

FIGURE 3.9 SEM images of a uniaxially aligned array of nanotubes made of (a) hollow anatase and (b) porous titania nanofibers. (Reproduced from Cann, J.T.M. et al., *J. Mater. Chem.*, 15, 735, 2005. With permission.)

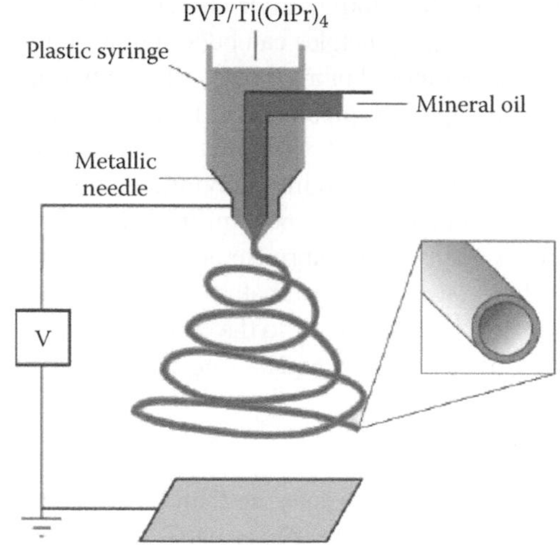

FIGURE 3.10 Schematic for fabrication of core-sheath nanofibers with a coaxial spinneret. (Reproduced from Cann, J.T.M. et al., *J. Mater. Chem.*, 15, 735, 2005. With permission.)

the nanomaterials. In this chapter, a brief review of the synthesis routes for particulate nanomaterials is given.

Nanoparticle synthesis routes can be broadly classified into (a) gas-phase and (b) liquid-phase approaches. Wet chemical methods in the liquid-phase approach are found in solgel, microemulsion, and sonochemical processes. Usually, gas-phase methods require high-cost facilities, and the resulting nanoparticles show polydispersity. The advantages of current gas-based methods include high mass productivity and stable processibility, although recent developments in the production of oxide nanoparticles have led to their mass production using liquid-phase processes [73,74]. Most liquid-based methods do not yield mass production. However, the broad window of feasible material classes and a narrow distribution of particle sizes are strong points in favor of the liquid-phase approach.

3.2.2.1 Gas-Phase Methods

The experimental or processing conditions in gas-phase synthesis of nanoparticles lead to thermodynamically unstable supersaturated vapors. Supersaturation can induce spontaneous particle nucleation and condensed phase formation. Subsequently, nucleation slows down and particle growth is initiated. Therefore, early density control of the supersaturated vapor is critical to achieving small and monodispersed nanoparticles.

The first step to nanoparticle preparation using a gas-phase approach is the formation of supersaturated vapors derived from solid [75–87], liquid, or gaseous sources. Metal sources can simply be heated into a gas and, subsequently, cooled to form a condensed phase [75,76,78]. Laser ablation of metals can also produce vapor sources appropriate for nanowire growing. Likewise, a laser can be used to vaporize a plume of semiconducting materials, such as silicon and titanium [83]. Arc discharge [85,86] and ion sputtering [87] methods can also be used to vaporize metal sources. The above-mentioned methods use different vaporizing protocols to prepare a condensed gas phase, leading to particle growth initiation. Among these methods, simple thermal heating is quite appropriate for achieving nanocomposite particles [76], such as in the PbS Ag system. Figure 3.11 shows that the gas-phase method can induce the formation of a dual system or a hybrid nanoparticle. Two different aerosols are fed via evaporated condensed phases and sintered prior to being mixed in an aggregation tube (see Figure 3.11a). The resulting nanocomposite has a diameter of about 20 nm.

Liquid and vapor phases can also be adopted to achieve the condensation and nucleation of supersaturated vapor [88–106]. The most versatile method is CVD, which is similar to that used for nanowire preparation. This method affords tremendous flexibility in producing a wide range of materials, e.g., oxide layer–passivated silicon nanoparticles, yttria, tungsten nanospheres, zirconia, and other metal oxide–coated or oxide-coated metal nanoparticles [88–91]. Furthermore, CVD expands its use into the fabrication of multicomponent nanoparticles, using several precursors in a consecutive order [92–94]. Kim et al. [96] have developed a monodispersed nanoparticle synthetic method by atomizing liquid sources to form fine droplets in the vapor phase. The vapor goes through the reactor furnace, and the resulting nanoparticles are screened by a mobility analyzer. This size-selection process is quite useful in obtaining the required monodispersed nanoproducts. As seen in the case of solid precursors, similar evaporation sources are available for liquid sources. For example, laser [97–100] and thermal plasmas [101] have been used for the preparation of a condensed vapor phase from liquid sources.

Ledoux et al. have prepared sub 10 nm silicon nanoparticles via pulsed CO_2 laser pyrolysis of silane in a gas flow reactor, equipped with a molecular-beam chopper that adds size-selection

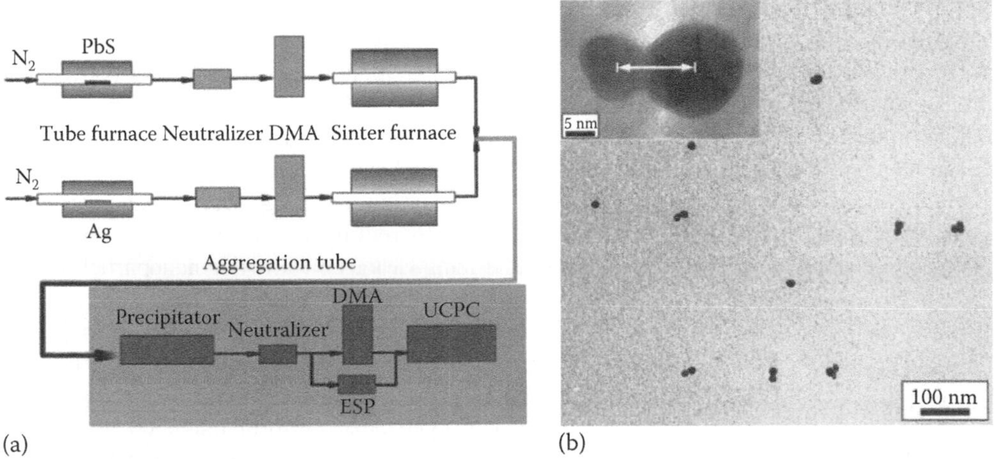

FIGURE 3.11 (a) Schematic drawing of the experimental setup for the gas-phase synthesis of composite nanoparticles. (b) TEM micrograph of Ag–PbS nanoaggregates composed of PbS and Ag nanoparticles having mobility diameters of 17.5 and 21.5 nm, respectively. (Reproduced from Maisels, A. et al., *Appl. Phys. Lett.*, 77, 4431, 2000. With permission.)

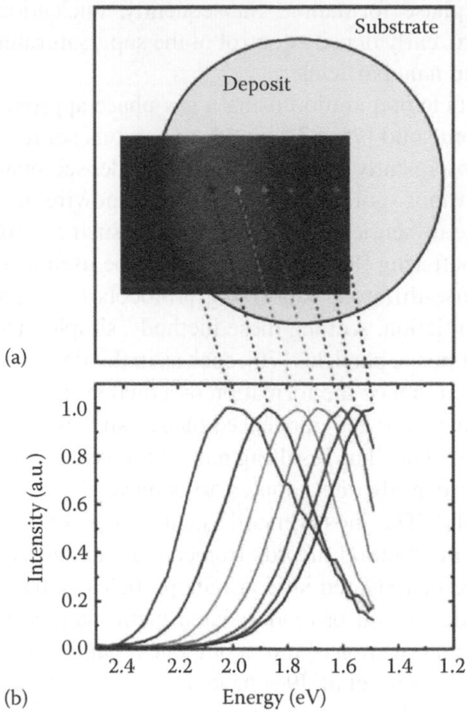

FIGURE 3.12 PL study of a specimen of size-screened Si nanoparticles. (a) Photograph of a part of the deposit illuminated by a simple UV lamp. (b) PL spectra show different energy values taken at different places on the deposit, as indicated by the arrows. (Reproduced from Ledoux, G. et al., *Appl. Phys. Lett.*, 80, 4834, 2002. With permission.)

capability to deposit screened nanocrystals onto a quartz substrate [99]. Figure 3.12 shows the photoluminescence (PL) spectra for the given nanocrystals.

3.2.2.2 Liquid-Phase Methods

Solgel processing in organic solvents, such as alcohols, has been developed over the past 50 years and still presents an important and widely accepted protocol to synthesize nano- and micron-scale particulate materials. In 1968, Stöber et al. [107] proposed that monodispersed SiO_2 nanospheres could be prepared via solgel processes in alcohol, using a silane precursor and a basic catalyst. As noted earlier, monodispersed nanoparticle preparations are extremely important in end-use applications of spherical nanoparticles, such as in photonic crystals, optical devices, and nano-biotechnology [108–111]. Liquid phase–based methods normally perform better in forming monodispersed nanoparticles compared to gas phase–based methods, which additionally require size-screening processes.

The solgel method has great versatility in synthesizing various metal-oxide nanoparticles, including ZrO_2, TiO_2, NbO_2, and WO_4. The well-known, general solgel chemistry undergoes two major steps: (1) hydrolysis of metal precursors, and (2) condensation of hydrous metal alcoxides. The ease and simplicity of the Stöber method influenced numerous solgel-based nanoparticle studies that followed.

Microemulsion techniques have been applied to synthesize uniform magnetic nanoparticles [112,113], CdS semiconducting nanoparticles or quantum dots [114] and silica-coated magnetic nanoparticles [115–117]. In the microemulsion process, emulsion stability can be controlled using an appropriate surfactant material that decreases the interfacial energy between the emulsion droplet and the major continuous phase of the reactant solution. Overall three types of surfactants are used to stabilize microemulsion systems: (1) anionic, (2) cationic, and (3) nonionic. Anionic and

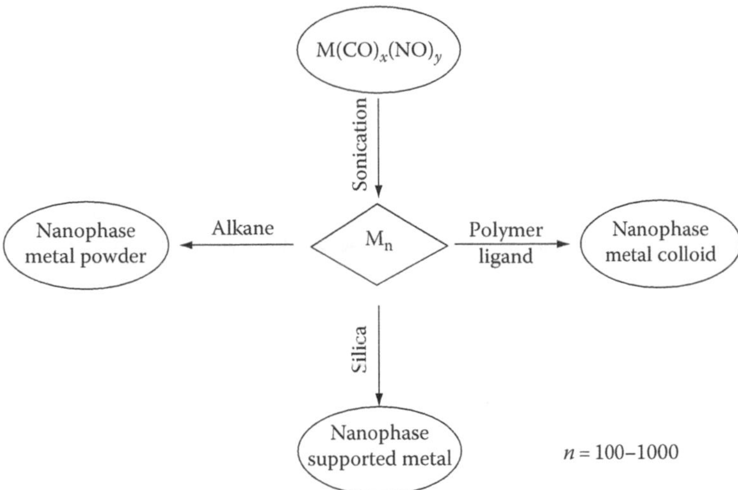

FIGURE 3.13 Sonochemical synthesis of various forms of nanostructured materials. (Reproduced from Suslick, K.S. et al., *Chem. Mater.*, 8, 2172, 1996. With permission.)

cationic surfactants have positive and negative counterions, respectively. Nonionic surfactants have a polyoxyethylene-based molecular backbone and hydroxyl terminals. Microemulsion droplets provide isolated spaces for the solgel chemical reaction to occur and form nanoparticles within the mini droplets. Therefore, the solgel chemistry combined with microemulsion methods has been proposed for the fabrication of metal-core and semiconductor-core oxide-shell nanocomposites. Further details on this subject are given in Section 3.3.4.

Semiconducting nanocrystals or quantum dots have been synthesized using thermal decomposition of the precursor materials followed by Ostwald ripening in order to grow up to 10 nm diameter nanocrystals [118]. What characterizes this method is the use of ligands to passivate the quantum-dot surface and impart a larger bandgap to the nanocrystal in order to increase quantum confinement efficiency [119–121]. For applications of quantum dots in biomedical imaging, ligand exchange and modification afford an essential and powerful approach. Current requirements for bioapplications of quantum dots are nontoxicity, biocompatibility, and highly durable fluorescent properties [122,123].

Sonochemical methods have also been used to synthesize ligated nanostructured particles [124,125]. Ultrasonic energy can generate a local hot zone with steep temperature variation and pressure. This abrupt change in pressure and temperature causes the decomposition of organometallic precursors and nanoparticle growth to be initiated. Figure 3.13 demonstrates that sonochemistry can be applied in the fabrication of particulate nanostructured materials.

3.3 CHEMICAL AND PHYSICAL ROUTES TO NANOCOMPOSITES

In the previous section, various types of 0D and 1D nanomaterials were introduced and the basic synthetic routes were reviewed. In this section, the preparation of nanocomposites via chemical and physical protocols is discussed.

Although the exact definition and classification of nanocomposites is quite controversial, we can broadly classify them into several groups based on shape: (1) 1D–0D nanocomposites, where 0D and 1D nanomaterials are conjugated, embedded into each other to synergistically conform to unique nanoarchitectures; (2) 1D–1D nanocomposites, where 1D nanomaterials comparable or differentiable in size are chemically bonded together; (3) 0D–0D nanocomposites, where hetero-junctioned nanocomponents are bound together by atomic lattice matching, polymeric wires, and molecular ligands; (4) 0D core-shell nanocomposites, where onion-type one-core/multishell or one-core/one-shell and

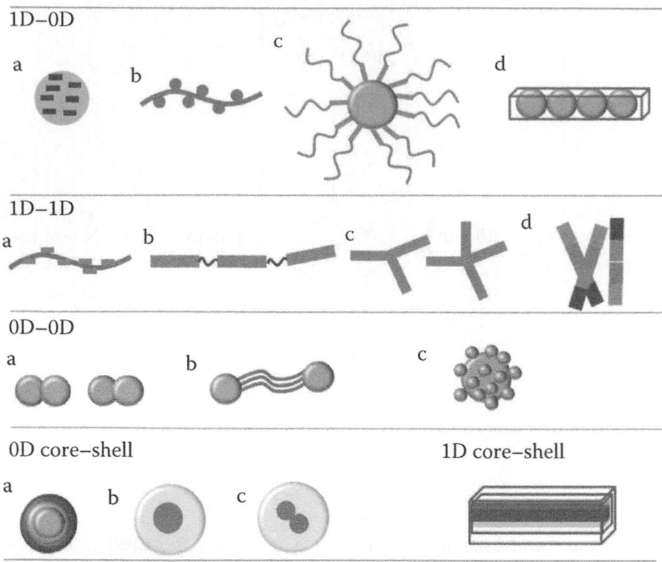

FIGURE 3.14 Various types of nanocomposites using low-dimensional nanomaterial building blocks.

multi-core/one-shell structures are available; and (5) 1D core-shell nanocomposites, forming hierarchical structures by downsizing the different shell components but keeping the same shape as the core. Figure 3.14 shows the overall schematic views of the various nanocomposites classified by shape.

3.3.1 1D–0D Nanocomposites

1D nanorods can be bonded onto the surface of polymeric microcapsules by the electro-attractive charge effect [126] or seed-mediated growth methods [127]. In addition to the use of different binding forces, these approaches utilize different schemes sourcing nanorods. The former makes use of readily available rods, while the latter utilizes an in situ growth of nanorods. Seed-mediated growth is performed in two different types of capsules: (1) polyelectrolyte multilayered (PEM) capsule and (2) silica layer, 50 nm, capsule, containing a composite layer of PEM. Au rods embedded into a polymeric or inorganic matrix are advantageous for fabricating optically stable devices and biosensors. In this context, the observed residual Au-rod stability increase in the Au rod–PEM composite is quite interesting and promising [127]. Gold rod–polymer capsule composites can open the way to novel drug delivery systems using the flexural volume-changing properties of the composite in different optical environments [126].

Long, extended 1D-wire nanomaterials can function as adsorption sites for 0D nanomaterials. CNTs [128,129], metal rods [130], and polymers [131] have been used as adsorption sites. As generally reported, ready-made metal and semiconducting nanoparticles show clean and typical features in 1D–0D nanocomposites in contrast to the in situ prepared nanoparticles using precursors [130], as shown in Figure 3.15. Most protocols bring individually standing components to form the nanocomposite. However, patterned arrays of block copolymers can form periodic assemblies of nanoparticle–polymer composites. Therefore, nanocomposite application schemes in devices can adopt these assembly and adsorption steps in a successive order [131–138], as shown in Figure 3.16. The nanoparticle adsorption area and pattern selection can be designed using nanoparticle surface–passivation with different chemicals. For example, a mixture of PS (polystyrene) and PVP (polystyrene-*b*-2 vinyl pyridine) can be coated over Au nanoparticles, which can then be placed at the interfaces between different blocks of polymers (see Figure 3.16a), but not within a specific block (see Figure 3.16b).

FIGURE 3.15 TEM micrographs. (a) DNA-modified CNT–Au nanoparticle composites using ready-made nanoparticles show more uniform structures compared to (b) the precursor method for Au rod/iron oxide nanoparticles. (Reproduced from Moghaddam, M.J. et al., *Nano Lett.*, 4, 89, 2004; Gole, A. et al., *Langmuir*, 24, 6232, 2008. With permission.)

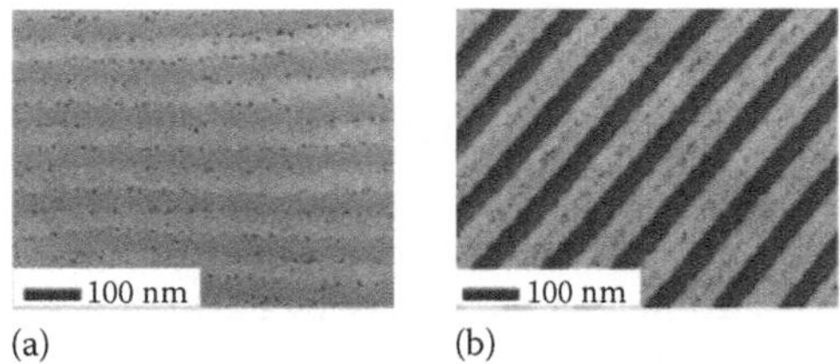

FIGURE 3.16 Selective nanoparticle array in the patterned block copolymers. Au nanoparticle array (a) at the interface and (b) within a block. (Reproduced from Chiu, J.J. et al., *J. Am. Chem. Soc.*, 127, 5036, 2005. With permission.)

Surface modification of nanoparticles can open up new passivation methods for nanoparticles synthesized in organic solvent environments, e.g., oleic acid–capped nanoparticles. Useful nanoparticle functions are not generally applicable to aqueous environments due to the as-synthesized nanoparticle hydrophobic nature. Several approaches were explored for the purpose of imparting hydrophilic passivation onto the nanoparticle surface [139,140]. Normally, the passivation molecules have long chains, and, therefore, ligand-exchanged nanoparticles can be considered as 1D–0D nanocomposites. Figure 3.17 shows schematically the conversion from hydrophobic oleic acid–passivated Fe_2O_3 nanoparticles to hydrophilic ones. In this experiment, oleic acid was replaced with α-cyclodextrin to disperse the nanoparticles in an aqueous medium [140].

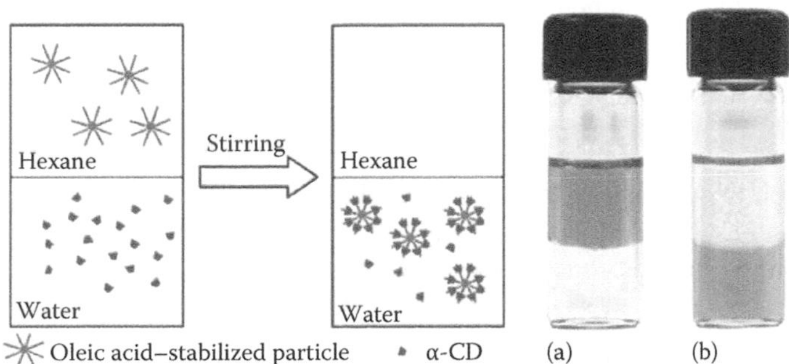

FIGURE 3.17 Ligand exchange for hydrophilic surface modification of synthesized iron oxide nanoparticle dispersion in hexane (a) above the aqueous solvent and (b) after surface modified. (Reproduced from Wang, Y. et al., *Nano Lett.*, 3, 1555, 2003. With permission.)

Tubular, hollow-containing linear porous materials can adsorb nanoparticles within their linear pores, that is, in 1D tube structures [141]. Since the entrapped nanoparticles present isolated nanoparticle formation, the composite can act as a novel nanocatalyst system (see Figure 3.18).

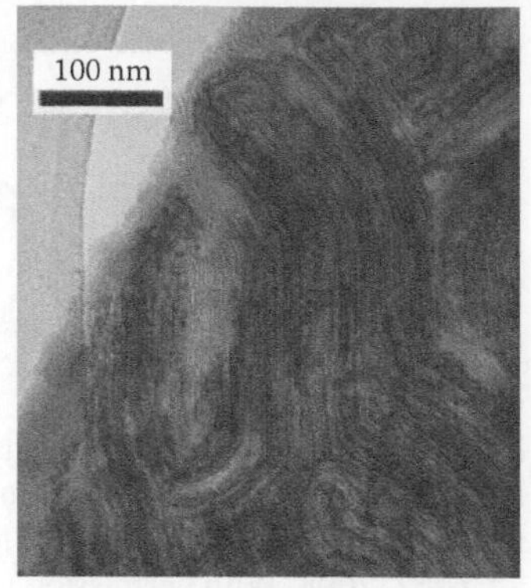

FIGURE 3.18 TEM micrograph. Pt nanoparticles in porous channels. (Also see Figure 3.3 for a reference.) (Reproduced from Fukuoka, A. et al., *Chem. Mater.*, 18, 337, 2006. With permission.)

3.3.2 1D–1D Nanocomposites

Just as 0D nanoparticle conjugation onto 1D linear rods or wires can be accomplished, 1D rods and wires can also be hybridized with other linear nanomaterials. For example, DNA-wrapped CNTs have been used to incorporate metal nanoparticles [128]. In this study, DNA was adsorbed onto the CNTs by physical adsorption (see Figure 3.19). And an early work of metal rod–CNT binding system can be found [142].

While the random adsorption of nanoparticles onto linear nanomaterials is the governing protocol in 0D–1D composites, the hybridization of linear materials along preferred directions is more challenging. A symmetric and tailored conjugation of Au-rod/Au-rod nanostructures in a parallel direction was studied by Caswell et al. [143] In their work, less than 10% of rod end-to-end (parallel) conjugation was achieved (see Figure 3.20). More selective and elaborate ways to derive preferred, one-directional conjugation of linear nanomaterials is expected to be found in the near future.

Flake or random-directional branching of rods is a consequence of nonequilibrium crystal growth processes. These classes of 1D–1D composites are available via wet chemical methods [144,145] and the vaporized synthesis method [146]. Figure 3.21 shows the nano-features of CdS and ZnO semiconducting nanopods and CoPt magnetic nanopods.

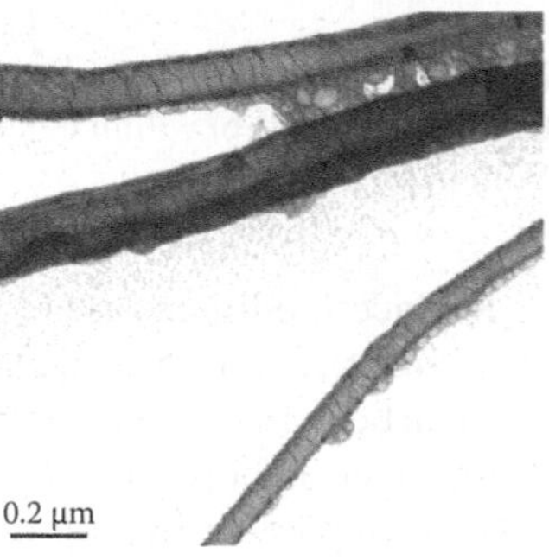

FIGURE 3.19 First CNTs were adsorbed onto a silica-mediated surface and then rod-shaped Au precursors grew on the surface. (Reproduced from Moghaddam, M.J. et al., *Nano Lett.*, 4, 89, 2004. With permission.)

As noted earlier, the AAO-based nanorod/wire protocol can serve to synthesize tailored rods and wires. Further studies of using AAO approaches have led to 1D–1D nanocomposite formation.

Ferromagnetic and insulating domains can form alternating nanoblock chains and overall long rods containing different components. Whitesides et al. prepared nanoblock chains with Ni forming the magnetic domain and Au the nonmagnetic section [147,148].

Figure 3.22 shows the basic scheme for this magnetic/nonmagnetic composite and the resulting feature. We note, however, that the demonstrated materials are not nanomaterials, as they are comprised of micrometer-sized components; but this novel scheme can presumably be downsized to the nanoscale. Nicewarner-Peña et al. [149] proposed a multicomponent rod using a conventional AAO method. In this scheme, a successive reduction of Au and Ag is used within the AAO membrane, and the resulting Au–Ag alternating blocks constitute a nanobar-coded material. The feasibility of bioassay applications using these metal-nanobar codes has also been demonstrated.

3.3.3 0D–0D Nanocomposites

The need for bifunctional and multifunctional nanocomposites gave the impetus to create heterojunctioned composite materials that are useful in optical imaging and biomedicine [150–152]. As individual components, oxides and metals have been combined and developed into hetero

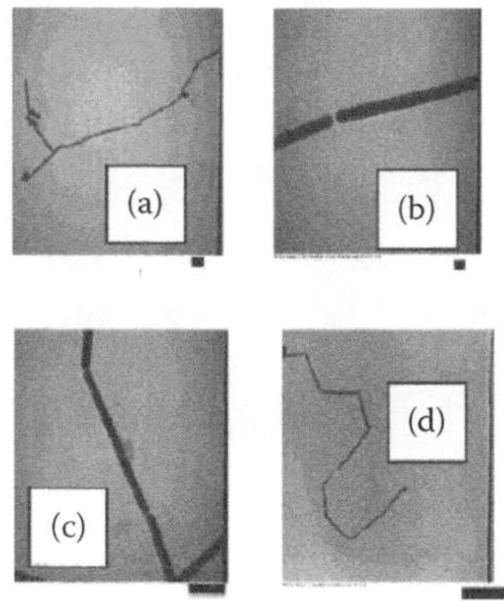

FIGURE 3.20 Rod–rod conjugation in parallel direction, scale bar: (a) 100 nm, (b) 20 nm, (c) 100 nm, and (d) 500 nm. (Reproduced from Caswell, K.K. et al., *J. Am. Chem. Soc.*, 125, 13914, 2003. With permission.)

composites. Hyeon et al. [73,74] have studied metal/metal-oxide composites using thermal decomposition methods, where thermal cracking induces metal-surfactant decomposition at a high temperature, ~300°C. Figure 3.23 shows the micrographs of metal/metal-oxide hetero composites [153].

Yet another type of heterostructure can be imaged: stretchable polymer brush-linked nanoparticles. Polyelectrolytes can be used to bond colloidal nanoparticles together. For example, polyvinyl pyridine, which shows pH-dependent stretching properties, has been used to bind Au nanoparticles at both ends, one at the glass slide [154]. At low pH, polyvinyl pyridine is more swollen than at high pH, with a concomitant change in the corresponding UV-absorbance data showing different traces (Figure 3.24). Furthermore, chemical ligands and polyelectrolytes can function as binders onto the central nanoparticles in order to adhere smaller-size nanocatalysts [155,156].

The central spherical nanoparticle could be comprised simply of PS [155] or more elaborate structures, such as magnetic functionalized silica [156]. The adsorbed smaller nanoparticles could be about 2 nm in diameter, and thus, of extremely high surface-to-volume ratio, making them good candidates for catalytic applications.

FIGURE 3.21 Electron micrographs. (a) CdS nano-bipod and (b) nano-tripod, scale bar, 10 nm. (c) CoPt nanopolypod. (d) ZnO nanocomb. (Reproduced from Jun, Y.W. et al., *J. Am. Chem. Soc.*, 123, 5150, 2001; Tzitzios, V. et al., *J. Am. Chem. Soc.*, 127, 13756, 2005; Yan, H.Q. et al., *J. Am. Chem. Soc.*, 125, 4728, 2003. With permission.)

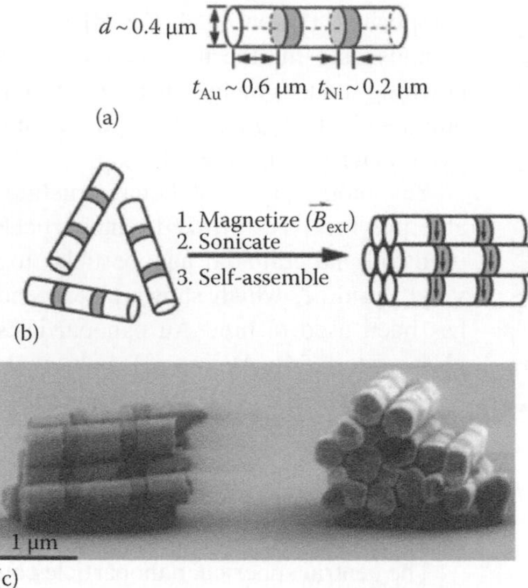

FIGURE 3.22 (a–c) Gray-colored block is for Ni magnetic domain and the white one is for Au. (b) Arrows in the Ni domain signify magnetic polarization. (Reproduced from Love, J.C. et al., *J. Am. Chem. Soc.*, 125, 12696, 2003; Urbach, A.R. et al., *J. Am. Chem. Soc.*, 125, 12704, 2003. With permission.)

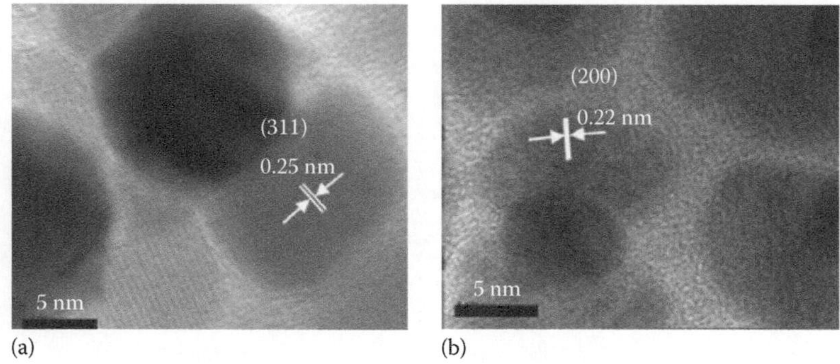

FIGURE 3.23 TEM micrographs of (a) Au–Fe_3O_4 and (b) Au–MnO heterostructures. (Reproduced from Choi, S.H. et al., *J. Am. Chem. Soc.*, 130, 15573, 2008. With permission.)

Yi et al. [156] demonstrated magnetic catalysts using the above described schemes. Chemical ligand functionalization was first performed onto a silica-coated magnetic nanocomposite. The ligand serves to attract nanocatalysts and fix them onto the silica-shell surface (see Figure 3.25). Amine and thiol terminal ligands exhibit different affinities for the Pd nanocatalysts (see Figure 3.26). The ligand's binding strength affects the morphology of the adhered nanocatalyst particles. A strong binding keeps the smaller catalytic particles bound to the surface isolated, while a weak binding leads to catalyst particle aggregation with deteriorating catalytic performance after repeated catalytic cycles.

Self-assembled nanoparticle arrays can open the way to novel particulate nanocomposites [53,157–160]. Sub 10 nm sized quantum dots [157] and less than 100 nm [53] diameter nanoparticles as well as mesoscale colloids [158–165] have been self-assembled to form monolayers. By self-assembly, nanoparticles are fabricated into device architectures showing film patterns. Self-assembled colloidal clusters can be effectively separated by density gradient centrifugation [160], highlighting the optimal packing mechanisms of colloids. Kim and Yi et al. [161–165] used a template approach

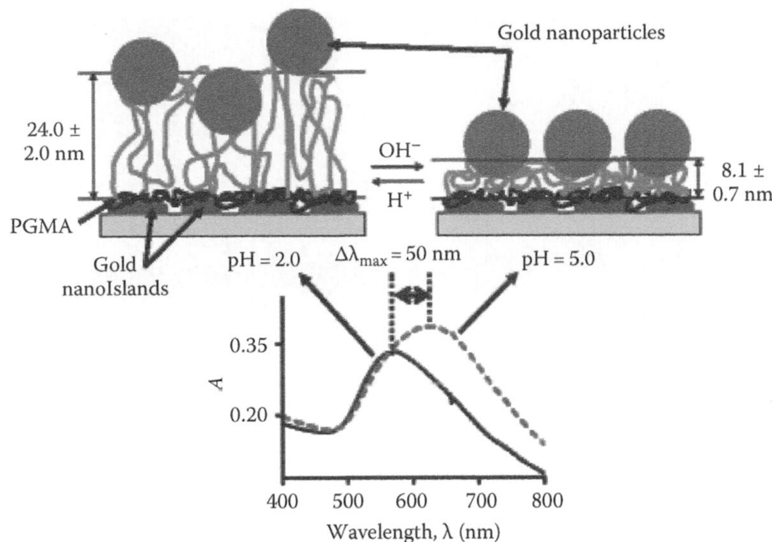

FIGURE 3.24 Scheme shows that the polymer binder between the Au island and the Au particles can be swollen or shrunk depending on the pH. And this effect affects the optical property of the nanocomposite film. (Reproduced from Tokareva, I. et al., *J. Am. Chem. Soc.*, 126, 15950, 2004. With permission.)

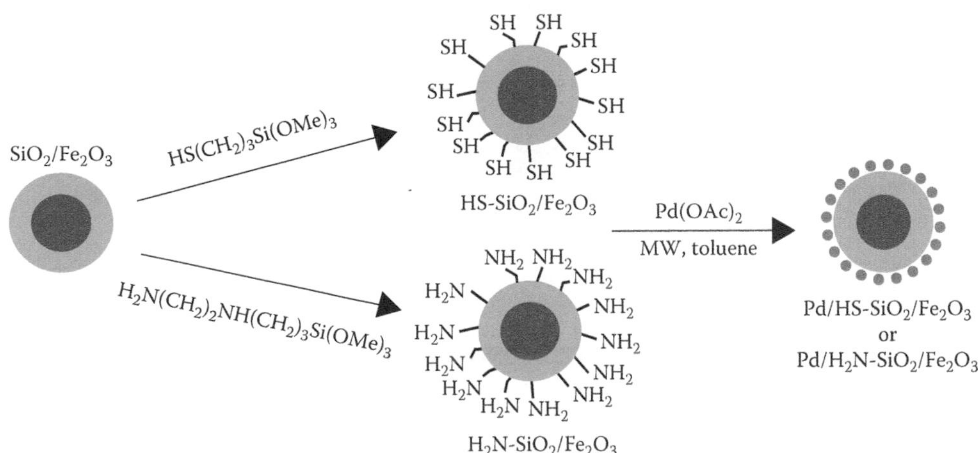

FIGURE 3.25 Silica-coated magnetic nanocomposite was further hybridized with metal nanocatalysts using a chemical ligand. (Reproduced from Yi, D.K. et al., *Chem. Mater.*, 18, 2459, 2006. With permission.)

to array self-assembled colloidal composites. Various architectures, such as necklace, square, and dumbbell, were achieved (see Figure 3.27). Even though various strategies have been proposed to fabricate defect-free crystal structures of colloids, nonuniformity of the wafer-scale area presents a current technical barrier.

3.3.4 0D Core-Shell Nanocomposites

Spherical nanoparticles can be dressed with additional shells, which can protect the core from environmental oxidation, high thermal stress, and optical and mechanical damage. Additionally, the shell can work synergistically with the core to improve the functionality of the nanocomposite: (1) the shell provides an organic functional platform, where useful biological and catalytic molecules

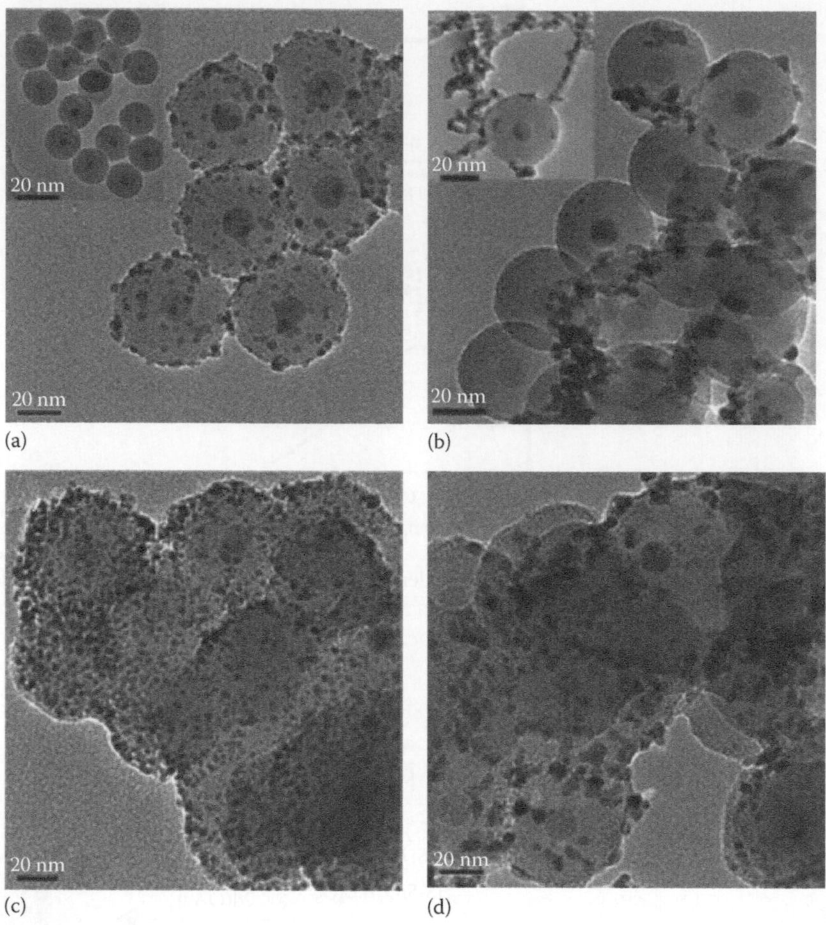

FIGURE 3.26 TEM micrographs of (a) HS ligand–treated Pd nanocrystal, silica-coated magnetic nanocomposites and (b) after the 14th recycle of catalytic hydrogenation reaction; (c) NH_2 ligand–treated Pd nanocrystal, silica-coated magnetic nanocomposites and (d) after the 14th recycle of catalytic hydrogenation reaction. (Reproduced from Yi, D.K. et al., *Chem. Mater.*, 18, 2459, 2006. With permission.)

can be conjugated; (2) the shell regulates the size and dimensionality of the nanocomposites; (3) well-controlled shell thickness tailors nanocomposites for specific optical, biomedical, and chemical applications; and (4) the shell imparts a secondary, but synergistic, function from the core to the nanocomposite.

For example, in the design of quantum dots, core-shell strategies have been used to passivate the semiconducting core bandgap crystals by coating them with higher-bandgap shell materials [118,166–174]. As a facile ligand detaching from the as-synthesized quantum dots is essential for biomedical applications, ligand surface engineering has been studied [167,168]. Optical applications, such as LED displays, require well-stabilized ligand/core-crystal bonding to ensure a stable light intensity from the quantum dots. Dendrimers and metal oxides have been studied to increase the binding stability between the nanocrystal surface and various ligands. Here, dendrimers can be adsorbed directly onto the nanocrystal surface, or, alternatively, a titania ligand can be associated with the present polymeric ligands over the nanocrystal surfaces [173,174].

In order to express the above-addressed synergistic shell-core functionality, the resulting 0D nanocomposites must be monodispersed and miscible within aqueous-based media, that is, they must be water soluble. To this end, shells made of silica have several advantages over other passivation

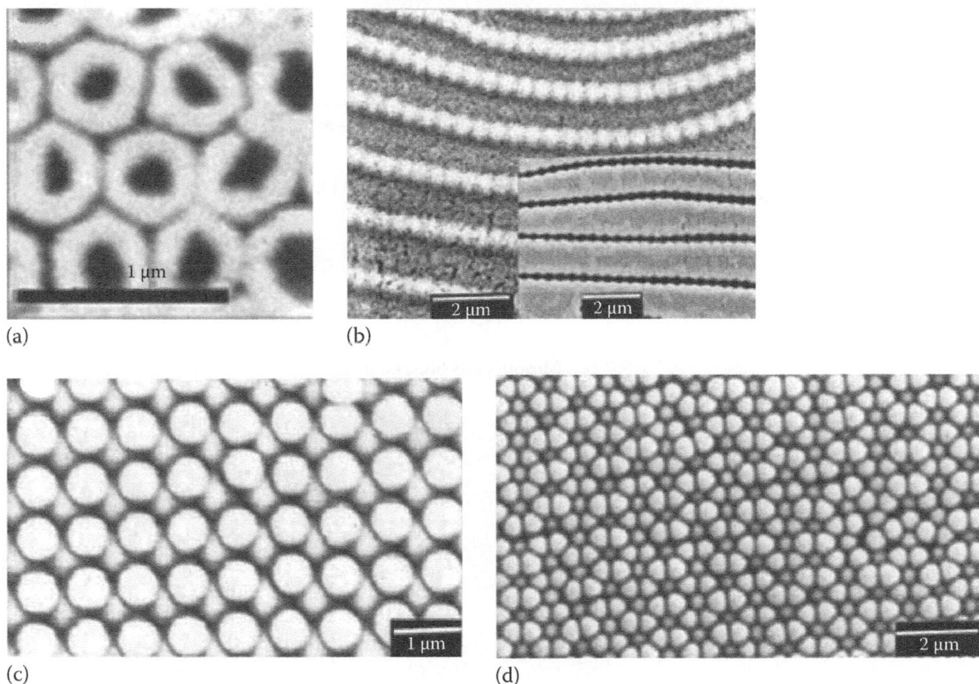

FIGURE 3.27 Self-assembled colloidal architectures: (a) rings, (b) necklace, (c) square symmetry, and (d) hexagonal symmetry. (Reproduced from Koo, H.Y. et al., *Adv. Mater.*, 16, 274, 2004; Yi, D.K. et al., *Chem. Commun.*, 8, 982, 2003; Yi, D.K. et al., *Appl. Phys. Lett.*, 80, 225, 2002; Yi, D.K. et al., *Langmuir*, 18, 5321, 2002; Yi, D.K. et al., *Nano Lett.*, 3, 207, 2003. With permission.)

materials. Silica advantages include (1) a facile synthesis, (2) miscibility within various dispersing media, and (3) an excellent platform for hybridization with biological and chemical molecules. Although numerous studies have addressed silica coating over materials [115–117,175–177], monodispersed silica-shell formation over nanoparticles had not been developed, till recently.

Yi et al. [178] prepared highly monodispersed maghemite-based magnetic nanoparticle/silica nanocomposites, having sub 1 nm scale silica and mesoporous silica-shell thickness modification control. This approach provides for controlled morphologies and tuning of magnetic behavior using different shell thicknesses and rattle-ball-type or core-free-type materials (see Figure 3.28). Most importantly, these systems show quite low cytotoxicity, making them excellent candidates for biomedical applications. In addition, these nanocomposites present themselves as excellent experimental systems for fundamental studies on interparticle magnetic interactions, an area of theoretical magnetism hindered by the lack of materials with tunable, homogeneous, and well-controlled magnetic dilution mechanisms over a large range of interparticle distances. Papaefthymiou et al. [179] carried out detailed interparticle magnetic interaction property studies for these nanocomposites. The results indicate that dipole–dipole interactions increase the superparamagnetic barrier of the nanoparticles and that their strength decreases rapidly with increasing interparticle distance.

Furthermore, fluorescent nanoparticles, i.e., quantum dots, can be embedded [180] within the silica shell. As mentioned above, core-free shells can also be prepared by etching away the core [181–184]. Therefore, the remaining shell can be further functionalized optically [180] and catalytically [181].

Multi-core/shell systems can also be found as proposed by Yi et al. [185] (see Figure 3.29). Since these nanocomposites have two cores, they have bifunctional flexibility and are more applicable to physicochemical and biomedical uses. Magnetic and fluorescent nanocomposites have been studied

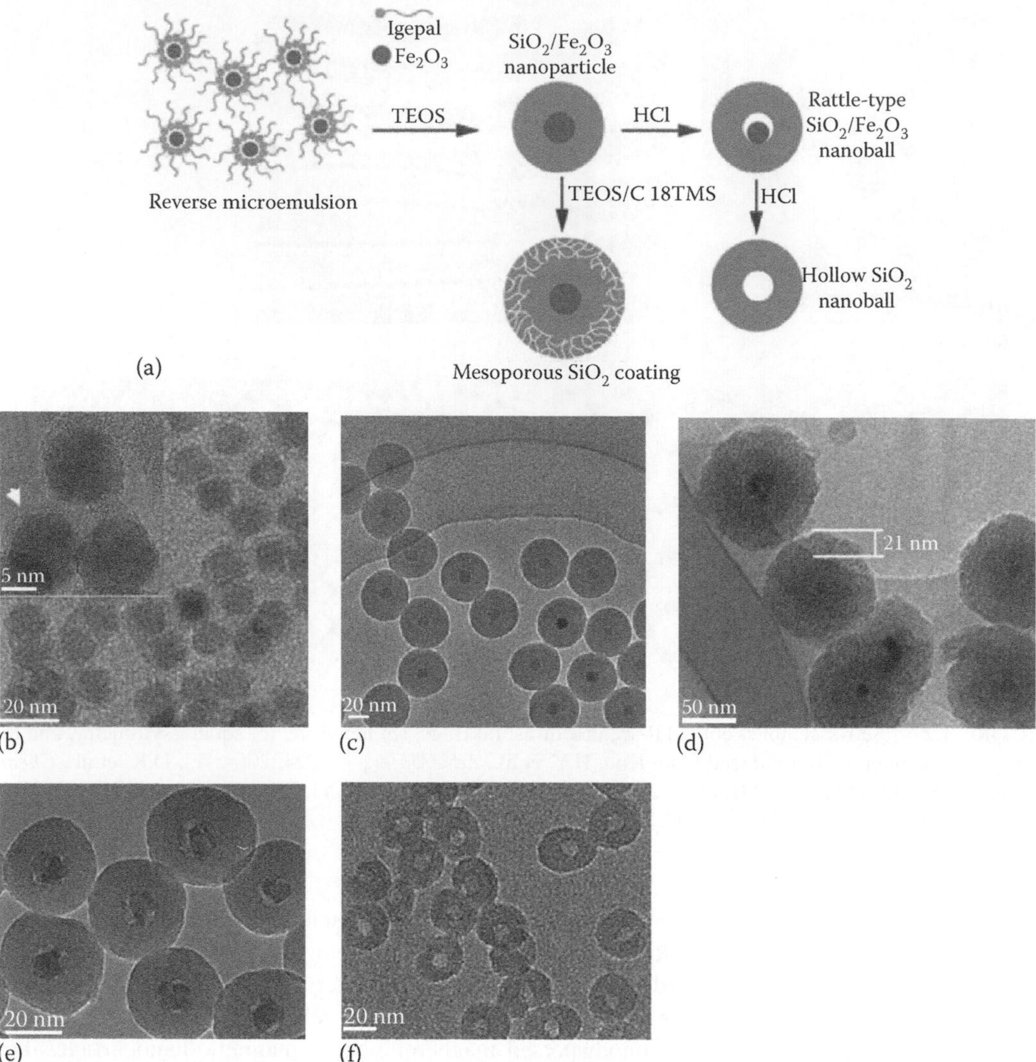

FIGURE 3.28 (a) Schematic for the synthesis of silica-coated magnetic nanocomposites. (b) Thin-shelled, (c) sub 1 nm (see inset) thick-shelled, (d) mesoporous-layered, (e) rattle-ball-type, and (f) core-free silica balls. (Reproduced from Yi, D.K. et al., *Chem. Mater.*, 18, 614, 2006. With permission.)

with high interest as potential in vivo bioimaging materials. The silica shell protects the bi-elemental cores and gives good wettability. However, it was observed that this magnetic/quantum-dot nanocomposite exhibited deteriorated optical properties as compared to the as-prepared oleic acid–covered fluorescent quantum dots. This was ascribed to (1) the removal of the oleic-acid passivation shell during the silica-coating process and consequent weak quantum confinement effects, (2) the introduction of quantum-dot crystal defect by exposure to a basic chemical, ammonium hydroxide, and (3) a destructive interference by the magnetic core in contact with the quantum dots.

Shells based on inorganic materials, including silica, are effectively water compatible, but exhibit weak interactions with gaseous acids, such as HCl. Therefore, for such applications, polymeric shells have also been studied as coatings for the preparation of functional nanoparticles [186,187]. PS and silane-conjugated PS have been studied as shell materials. It is expected that even more advanced core-shell nanosystems will be developed in the near future, in order to address requirements for better photostability, biocompatibility, and multifunctionality.

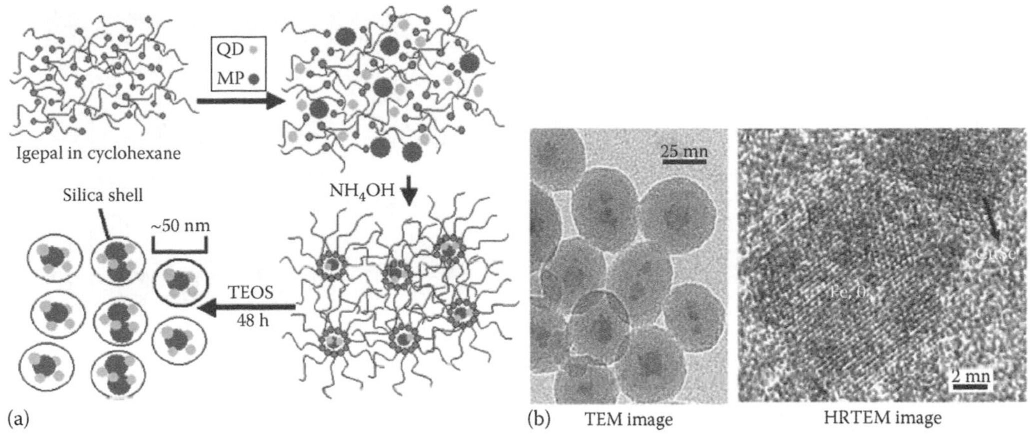

FIGURE 3.29 (a) Schematic view for magnetic quantum-dot composites and (b) TEM micrographs of the resulting materials. (Reproduced from Yi, D.K. et al., *J. Am.Chem. Soc.*, 127, 4990, 2005. With permission.)

3.3.5 1D Core-Shell Nanocomposites

Multiwalled or multishelled nanotubes can be considered as extended 1D analogue forms of multishelled nanospheres [188–190]. Under this category, materials containing homogeneous components, such as multiwalled carbon nanotubes (MWCNTs), and multispecies containing multishelled nanotubes can be found. Compared to the single-walled Au nanotubes, composite Au–Ag multiwalled nanotubes show optical absorbance-tuning properties, depending on the number of walls present (see Figure 3.30). MWCNTs should be noted as outstanding sub 10 nm scale nanosystems affording over 100 nm core-shell 1D nanocomposite structures.

3.4 BIO-INSPIRED NANOCOMPOSITES

In previous sections, we have reviewed how 0D and 1D nanomaterials have been used as building blocks, singly or in combination, in the fabrication of multifunctional nanocomposites. While materials scientists have only recently mastered such heterogeneous molecular nanoarchitectures in bottom-up synthesis schemes, nature has been at work integrating inorganic nanostructures within life's organic world over evolutionary times. For example, in the process of biomineralization, living organisms create the necessary chemical environmental conditions to control the nucleation and growth of unique mineral phases, forming exceptional nanocomposite materials, such as bones [191–193]; sea shells [194]; the bioinorganic mineral of the iron storage protein, ferritin [195–199]; and the nano-magnetite chains found in magnetotactic bacteria [200–202]. Figure 2 in Ref. [202] depicts this exquisite linear arrangement of magnetosomes, containing single-magnetic-domain magnetite encapsulated within membrane vesicles, as a startling example of a nanocomposite formation by design for end-use purposes, namely, the orientation of the bacterium in the Earth's magnetic field. In bones, the archetypal biological composite, nanoscale plate-like crystals of apatite are dispersed within a matrix of collagen fibers, while in ferritin, protein nanotemplates with an interior cavity of ~8 nm diameter are used as nanoreactors within which nanoparticles of ferrihydrite are formed. In solid-state chemistry and materials science, reproducing such processes in the laboratory under similarly mild and environmentally friendly conditions is an area of intense investigation. The key question for nano-engineers is whether such biomimetic processes can be used in the laboratory for creating novel biocompatible synthetic materials [203–205]. For example, while in vivo ferritin sequesters exclusively ferrihydrite, in vitro studies indicate that other iron-mineral phases can also be sequestered, depending on experimental conditions [206]. Figure 3.31 depicts the protein

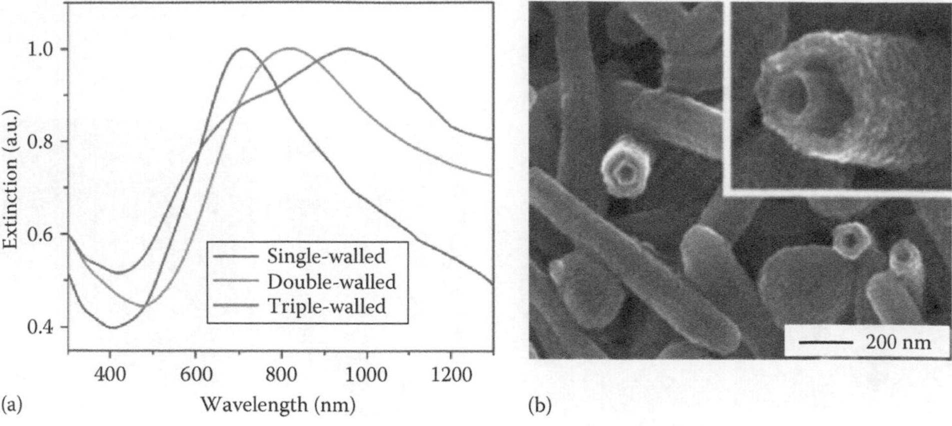

FIGURE 3.30 (a) Normalized UV–vis–NIR extinction spectra for multiwalled nanotubes composed of Au and Ag and (b) the corresponding TEM micrograph. (Reproduced from Sun, Y. et al., *J. Am. Chem. Soc.*, 126, 9399, 2004. With permission.)

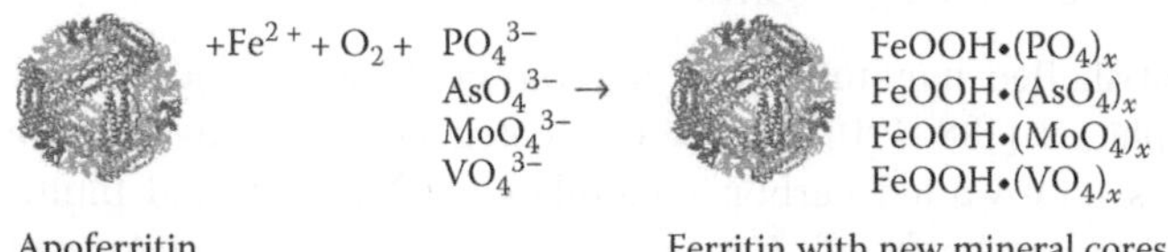

FIGURE 3.31 Reaction scheme for the formation of iron oxo-anions within the apoferritin cage. (Reproduced from Polanams, J. et al., *Inorg. Chem.*, 44, 3203, 2005. With permission.)

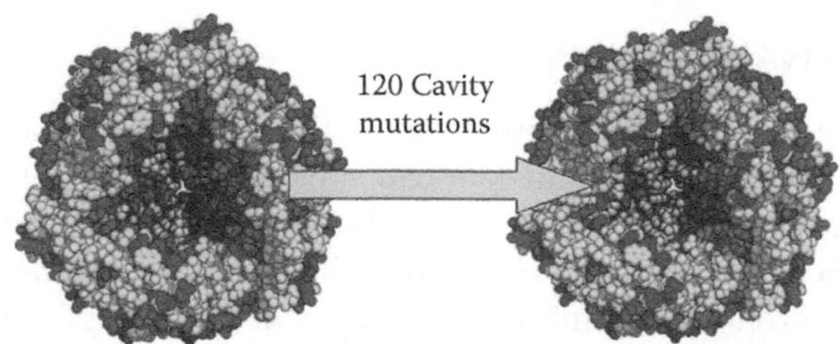

FIGURE 3.32 Designed interior surface of apoferritin shell via mutations introduced by genetic engineering. (Reproduced from Swift, J. et al., *J. Am. Chem. Soc.*, 128, 6611, 2006. With permission.)

shell and interior mineral core for the formation of alternate phases in the presence of oxo-anions. Ferritin presents a unique platform for nano-engineers, because the protein shell is extremely robust and remains assembled in the absence of a biomineral core in its interior. In addition, the protein is amenable to genetic engineering [207] for strategically altering the chemical characteristics of its interior surface (see Figure 3.32) for the purpose of promoting the nucleation and growth of elements other than iron, such as the formation of noble metal clusters [208]. Furthermore, the exterior surface of the protein cage can also be functionalized [209], as depicted in Figure 3.33, in order to interact with cell receptors of cancerous cells for bioimaging, targeted drug delivery, and hyperthermia cancer therapy purposes. Thus, a great deal of experimentation has been expended in the synthesis of nanoparticles, other than ferrihydrite, within the protein cage, with promising success in the manufacturing of magnetic resonance imaging (MRI)-enhancement agents [210,211], ecologically friendly nanocatalysts [212,213], and a variety of other phases including magnetic nanoparticles for

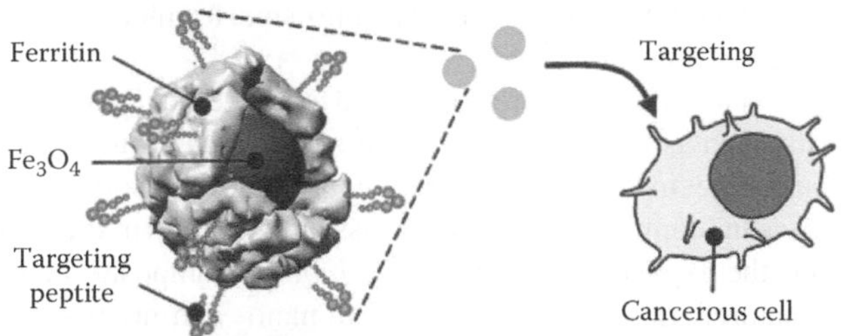

FIGURE 3.33 Magnetite synthesized within the interior cavity of a genetically engineered human H-chain ferritin (HFn). A cell-specific targeting peptide, RGD-4C, which binds $\alpha_v\beta_3$ integrins upregulated on tumor vasculature, was genetically incorporated on the exterior surface of HFn. (Reproduced from Uchida, M. et al., *J. Am. Chem. Soc.*, 128, 16626, 2006. With permission.)

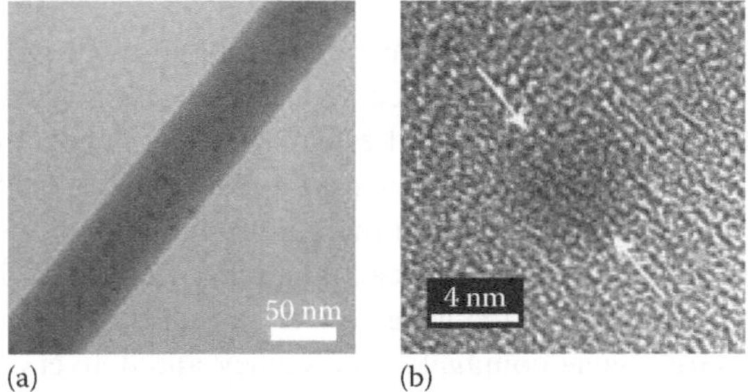

FIGURE 3.34 TEM images of PVA/ferritin nanofibers with a diameter of about 61 nm deposited on a carbon-coated grid. (a) A PVA/ferritin nanofiber containing uniformly dispersed ferritin and (b) a high magnification image of ferritin incorporated within the polymeric matrix. The arrows indicate the ferritin core diameter of approximately 6 nm. (Reproduced from Shin, M.K. et al., *Appl. Phys. Lett.*, 88, 193901-1, 2006. With permission.)

applications in nano-electronics [206,214–223]. In addition, ferritin has been applied in the functionalization of CNT-polymer composites [224] and also used for the reinforcement of polymeric nanofibers [225], as shown in Figure 3.34.

The above examples form a paradigm of the fact that biological nanomaterials, such as DNA, proteins, and viruses, can work as templates for the patterning of inorganic nanomaterials. The resulting nanocomposites contain the inorganic materials integrated within organic biomaterials, thus rendering them biocompatible, safe to be introduced within the human body and within the interior of a cell. Thus, these *bio*nanocomposites can be used as media for delivering chemicals to control biochemical events in cells, immuno-inflammation, cancer cell apoptosis, gene recognition, and other disease-causing or quenching phenomena. Thus, biological events can be investigated and differentiated from each other using *bio*nanocomposites, which can form nanoscale analogues of counter biochemical hosts in cells.

In cancer therapy, early detection of cancer cells is critical. *Bio*nanocomposites in combination with highly functional medical imaging instrumentation, such as, MRI, computer tomography (CT), positron emission tomography (PET), etc., have proven to be indispensable tools in fulfilling the need for early cancer detection. *Bio*nanocomposites promise to revolutionize medical diagnosis and therapy modalities in the near future, by introducing in situ, noninvasive methods for the diagnosis and treatment of diseases.

Conventional *bio*nanocomposites are designed to image cell events and must subsequently be discarded by metabolic processes in order to fulfill the safety requirement. In addition, modern *bio*-nanocomposites are required to perform the additional function of inducing cell lysis or apoptosis.

Ultimately, medical *bio*nanocomposites need evolve into smart nanodrugs having imaging functions and being metabolically processable.

The applications of *bio*nanocomposites are not limited to biomedical purposes. High-performance electronic nanomaterials, such as CNTs, can be hybridized with DNA to create novel *bio*nanocomposites that can be used for electronics and sensors, including biosensors. In many cases, for these physical applications, *bio*nanocomposites need to be assembled to form devices.

As mentioned earlier, the expanded definition of a *bio*nanocomposite includes bio-mimicking, that is, learning from "nature," in the effort to fabricate nano- and micro-systems that have superior performance compared to conventional artificial systems. These bio-inspired nanocomposites request more complex assembled processes and collegiate effects if they are to work systemically throughout the body.

In this section, specific details of these *bio*nanocomposites are discussed.

3.4.1 Biomedical Purposes

Since the 1970s, polymeric nanocompounds have been applied as drug-delivering agents [226]. Compared to the conventional viral vector–based gene delivery systems, *bio*nanocomposites can function near to target DNA plasmid by spatial effects. 1D–1D nanocomposites can be used as multicomponent gene delivery systems to introduce a supplementary gene to the damaged somatic cells [149,227]. However, the potential toxicity of inorganic components poses a major problem that must be overcome, before clinical applications can take place [228,229]. Therefore, less toxic Au nanorod–based and Au nanoparticle–based gene delivery systems have been developed [230,231]. Supplementary gene and target gene conjugation efficiency and delivery efficiency of nanocomposites are critical points to be addressed in the near future in relation to *bio*nanocomposite gene and drug targeting. Polymer-inorganic *bio*nanocomposite vector and gene binding, and controlled scission for delivery at cell body can be an emerging scheme for drug and gene delivery.

Bioligand-conjugated quantum dots, that is, semiconducting nanocrystal composites, have been studied as bioimaging materials [232,233]. The superior non-bleaching property and feasible biohybridization with interesting biological targets, such as genes and proteins, make these *bio*nanocomposites promising biolabel materials. Further studies on their biocompatibility and dispersity in various buffers are needed [234]. An additional functionalization of quantum dots by adhering them onto magnetic particles resulted in a new class of *bio*nanocomposites that led to in situ therapy and imaging (see Figure 3.35) [185,235–238]. Metal nanoparticles have also been shown to have the ability to be used as bioprobes for molecular imaging [239].

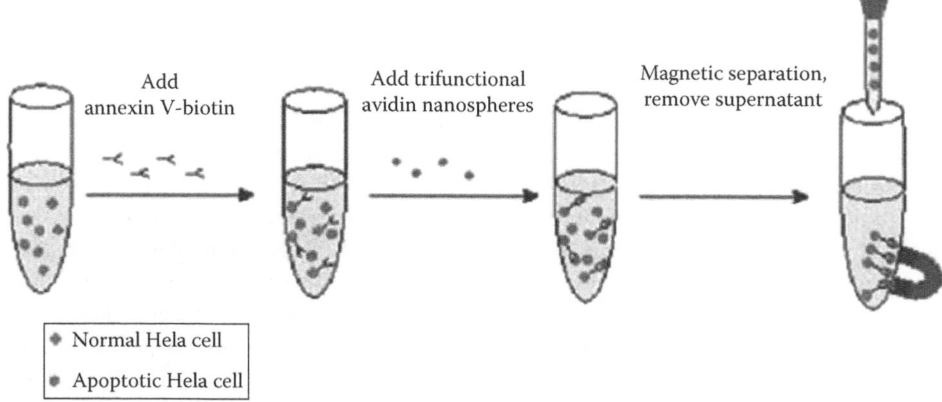

FIGURE 3.35 Cell separation and bioimaging using *bio*nanocomposites having magnetic and fluorescent properties. (Reproduced from Wang, G.P. et al., *Chem. Commun.*, 34, 4276, 2005. With permission.)

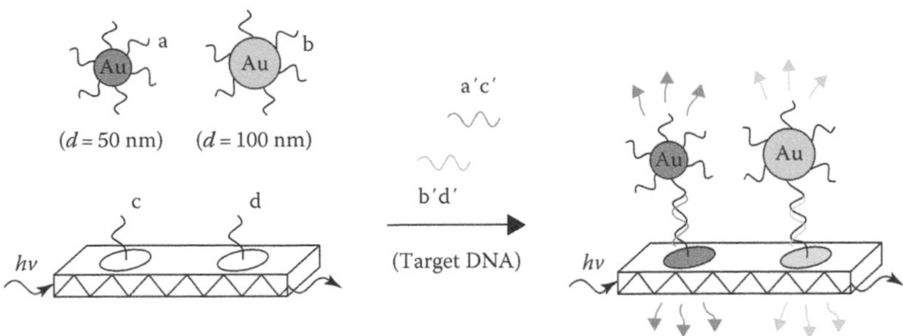

FIGURE 3.36 Oligonucleotide-functionalized bionanocomposite bioprobe. Hybridization is detected by simple color change. (Reproduced from Taton, T.A. et al., *J. Am. Chem. Soc.*, 123, 5164, 2001. With permission.)

Assembled *bio*nanocomposites can sense small amounts of biomolecules. The sensitivity and reproducibility of the sensing ability are dependent on the robustness of the assembly and the nature of *bio*nanocomposites. The detection of target DNA with paired DNA pieces is important for gene therapy, forensic work, and drug screening. Mirkin et al. have proposed to use the optical sensing property of Au nanoparticles for detecting when the appropriate DNA hybridization occurs (see Figure 3.36) [240]. This method serves as a simple and effective way for confirming the occurrence of hybridization events between DNAs. Then the CNT, placed in between two electrodes, will produce different current signals depending on the adsorbed biomaterials. Gruner et al. have used a field-effect transistor device in which the CNT was arrayed in between the source and the drain electrode (see Figure 3.37) [241]. This mechanism is as follows: when the interesting molecule, streptavidin, contacts the CNT *bio*nanocomposites, which are conjugates of CNTs and avidin, the current is varied according to the applied voltage, and the output current has different values in the presence or absence of streptavidin, respectively.

Recently, Yi and Cheong used Au nanorods to detect pathogens in a microfluidic chip apparatus. In this case, the optothermal property of Au nanorods was used to induce the lysis of bacterial cells and in situ amplify the DNA information by the interfaced polymerase chain reaction equipment (see Figure 3.38) [242]. The Au rod did not need to be assembled, but was confined in an 8 μL chamber.

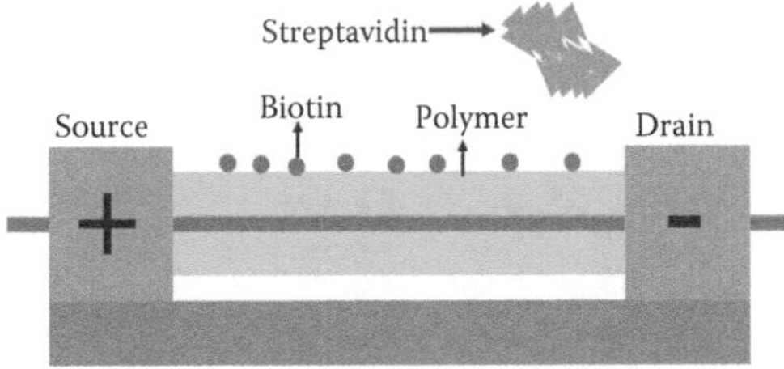

FIGURE 3.37 Schematic of the nanotube field-effect transistor using CNT as a linear biosensor that was conjugated with a molecular receptor, biotin, and recognizes streptavidin. (Reproduced from Star, A. et al., *Nano Lett.*, 3, 459, 2003. With permission.)

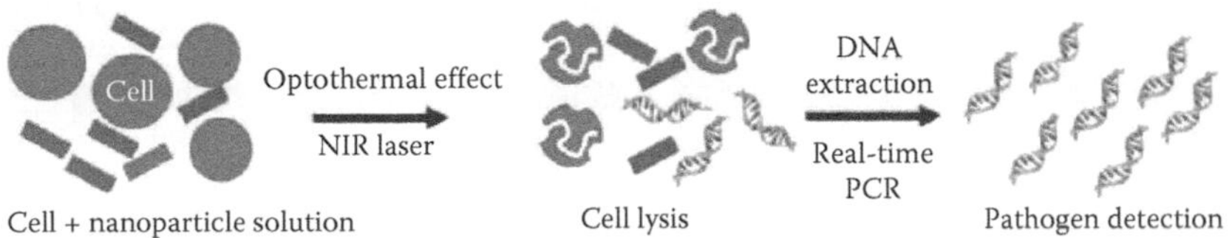

FIGURE 3.38 Overall scheme for pathogen detection using Au nanorods in a microfluidic chamber. Rectangular bars represent Au rods. (Reproduced from Cheong, K.H. et al., *Lab Chip*, 8, 810, 2008.)

3.4.2 Physical Applications and Bio-Mimicking

Biological systems possessing a maximized natural structure for living objects on earth are the best-designed *bio*nanocomposites produced by cellular processes with the highest energy efficiency. We, humans, macroscopically copied the bird in the invention of the aircraft and the fish in the construction of the submarine. However, the real core of biomimesis lies in the nanostructures of nature and their superb organization, which can make feasible the smart recognition of supramolecular structures and their consequent self-assembly within and outside the cell. Scientists have just started to understand this natural recognition process and assembly mechanism at the nanoscale. We are now able to copy DNA and make controlled sequences of DNA, but we cannot yet create any materials that can replace DNA. This proves that our understanding of nature is still evolving and far from complete.

DNA can actually function as a good template to pattern nanomaterials (see Figure 3.15). For example, mineralized DNA was used as a template to form a highly conducting Ag nanowire between two electrodes [28]. Furthermore, DNA templates can be made variable from junction to branching through the base sequence [243,244]. This more complex type of DNA assembly can be applied to make logic circuits [245]. DNA can be assembled over CNTs and can effectively separate CNTs having different electrical properties [142]. DNA-wrapped CNTs can also function as high-speed switching devices [246].

Protein-conjugated *bio*nanocomposites find physical applications as magnetic data storage. The iron mineral–containing protein, ferritin, can be used as a biotemplate to hold useful magnetic nanograins inside [219]. Viruses also have a good size analogy with nanomaterials; therefore, they have also been used as biomineralizing templates to engineer nanoporous materials [247] Mesoporous silica, thus prepared, shows self-assembled porous periodicity.

Assembled nanopillars can make artificial Gecko's foot structures. Gecko is a well-known reptile having sticky feet that adhere strongly onto the wall and the ceiling via van der Waals molecular forces. Flexible polymeric nanopillars have been suggested to be the artificial counterparts to physical adhesives. Geim et al. proposed to use e-beam lithography approaches, requiring high-cost steps, resulting in feet showing artificial hair effects clearly [248]. Kustandi et al. used a colloidal array as a mask to pattern vertical cavities where gaseous parylene was coated and developed into polymeric nanopillars that function as artificial feet (see Figure 3.39) [249].

FIGURE 3.39 (a) Parylene nanopillar for mimicking Gecko's foot. (b) Enlarged view. (Reproduced from Kustandi, T.S. et al., *Adv. Funct. Mater.*, 17, 2211, 2007.)

ACKNOWLEDGMENTS

D.K.Y. thanks Seonmi, Eunbin, Guanchen, and Xiaolin for their valuable contributions to this chapter, and the KRF grant (KRF-2007-331-D00585) for support. G.C.P. thanks the European Commission Marie Curie program for support at the NCSR Demokritos and the National Science Foundation, Grant No. DMR 0604049, for support at Villanova University.

REFERENCES

1. Shackelford, J.F. et al. 1996. *Introduction to Materials Science for Engineers*, 4th edn. Prentice Hall, Upper Saddle River, NJ.
2. Orritsma, J. et al. 1996. General technique for fabricating large arrays of nanowires. *Nanotechnology* 7:263–265.
3. Müller, T. et al. 2001. Formation of Ge nanowires in oxidized silicon V-grooves by ion beam synthesis. *Nucl. Instrum. Methods Phys. Res. B* 175–177:468–473.
4. Song, H.H. et al. 1999. Growth of Ag rows on Si (5 5 12). *J. Vac. Sci. Technol. A* 17:1696–1699.
5. Walter, E.C. et al. 2002. Palladium mesowire arrays for fast hydrogen sensors and hydrogen-actuated switches. *Anal. Chem.* 74:1546–1553.
6. Favier, F. et al. 2001. Hydrogen sensors and switches from electrodeposited palladium mesowire arrays. *Science* 293:2227–2231.
7. Yi, D.K. et al. 2002. Spin-on-based fabrication of titania nanowires using a sol-gel process. *Nano Lett.* 2:1101–1104.
8. Nishizawa, M. et al. 1995. Metal nanotubule membranes with electrochemically switchable ion-transport selectivity. *Science* 268:700–702.
9. Menon, V.P. et al. 1995. Fabrication and evaluation of nanoelectrode ensembles. *Anal. Chem.* 67:1920–1928.
10. Nielsch, K. et al. 2000. Uniform nickel deposition into ordered alumina pores by pulsed electrodeposition. *Adv. Mater.* 12:582–586.
11. Sandera, L. et al. 1973. Determination of the constitution of particles embedded in electrolytically dyed oxide films. *Aluminium* 49:533–539.
12. Kim, Y.S. et al. 2008. Patterned array of nanoparticles on substrates via contact printing method with CNTs/AAO stamp. *J. Nanosci. Nanotech.* 8:4803–4807.
13. Kim, Y.S. et al. 2008. The synthesis of MWNTs/SWNTs multiple phase nanowire arrays in porous anodic aluminum oxide templates. *Mater. Sci. Eng. A* 354:92–96.
14. Goad, D.G.W. et al. 1978. Colloidal metal in aluminum-oxide. *J. Appl. Phys.* 49:2929–2934.
15. Madsen, J.T. et al. 1985. Localization and electron-electron interaction effects in thin Pt wires. *Phys. Rev. B* 31:6395–6401.
16. Whitney, T.M. et al. 1993. Fabrication and magnetic properties of arrays of metallic nanowires. *Science* 261:1316–1319.
17. Piraux, L. et al. 1994. Giant magnetoresistance in magnetic multilayered nanowires. *Appl. Phys. Lett.* 65:2484.
18. Blondel, A. et al. 1994. Giant magnetoresistance of nanowires of multilayers. *Appl. Phys. Lett.* 65:3019.
19. Liu, K. et al. 1995. Perpendicular giant magnetoresistance of multilayered Co/Cu nanowires. *Phys. Rev. B* 51:7381–7384.
20. Chakarvarti, S.K. et al. 1991. Morphology of etched pores and microstructures fabricated from nuclear track filters. *Nucl. Instrum. Methods Phys. Res. B* 62:109–115.
21. Huang, M.H. et al. 2000. Ag nanowire formation within mesoporous silica. *Chem. Commun.* 12:1063–1064.
22. Han, Y.J. et al. 2000. Preparation of noble metal nanowires using hexagonal mesoporous silica SBA-15. *Chem. Mater.* 12:2068–2069.
23. Lee, K.B. et al. 2001. Size controlled synthesis of Pd nanowires using a mesoporous silica template via chemical vapor infiltration. *Adv. Mater.* 13:517–520.
24. Sloan, J. et al. 1999. Capillarity and silver nanowire formation observed in single walled carbon nanotubes. *Chem. Commun.* 8:699–700.
25. Govindaraj, A. et al. 2000. Metal nanowires and intercalated metal layers in single-walled carbon nanotube bundles. *Chem. Mater.* 12:202–205.

26. Zhang, Y. et al. 2000. Formation of metal nanowires on suspended single-walled carbon nanotubes. *Appl. Phys. Lett.* 77:3015.
27. Keren, K. et al. 2002. Sequence-specific molecular lithography on single DNA molecules. *Science* 297:72–75.
28. Braun, E. et al. 1998. DNA-templated assembly and electrode attachment of a conducting silver wire. *Nature* 391:775–778.
29. Wagner, R.S. et al. 1964. Vapor-liquid-solid mechanism of single crystal growth. *Appl. Phys. Lett.* 4:89–90.
30. Lieber, C.M. et al. 1998. One-dimensional nanostructures: Chemistry, physics & applications. *Solid State Commun.* 107:607–616.
31. Kong, Y.C. et al. 2001. UV-emitting ZnO nanowires synthesized by a PVD approach. *Appl. Phys. Lett.* 78:407–409.
32. Huang, M.H. et al. 2001. Communication catalytic growth of zinc oxide nanowires by vapor transport. *Adv. Mater.* 13:113–116.
33. Zheng, W.P. et al. 2001. Nanobelts of semiconducting oxides. *Science* 291:1947–1949.
34. Park, J.H. et al. 2005. Synthesis of ZnO nanowires and nanosheets by an O2-assisted carbothermal reduction process. *J. Cryst. Growth* 280:161–167.
35. Lao, J.Y. et al. 2002. Hierarchical ZnO nanostructures. *Nano Lett.* 2:1287–1291.
36. Yang, P. et al. 2002. Controlled growth of ZnO nanowires and their optical properties. *Adv. Funct. Mater.* 12:323–331.
37. Fuller, M.L. et al. 1944. Twinning in zinc oxide. *J. Appl. Phys.* 15:164–170.
38. Kitano, M. et al. 1990. Growth of large tetrapod-like ZnO crystals. *J. Cryst. Growth* 102:965–973.
39. Iwanaga, H. et al. 1993. Growth model of tetrapod zinc oxide particles. *J. Cryst. Growth* 134:275–280.
40. Dai, Y. et al. 2002. Synthesis and optical properties of tetrapod-like zinc oxide nanorods. *Chem. Phys. Lett.* 358:83–86.
41. Zhang, Y.F. et al. 1998. Silicon nanowires prepared by laser ablation at high temperature. *Appl. Phys. Lett.* 72:1835–1837.
42. Morales, A.M. et al. 1998. A laser ablation method for the synthesis of crystalline semiconductor nanowires. *Science* 279:208–211.
43. Chen, C.H. et al. 2001. Optical characterization of wurtzite gallium nitride nanowires. *Appl. Phys. Lett.* 79:3693–3695.
44. Cheng, G.S. et al. 1999. Large-scale synthesis of single crystalline gallium nitride nanowires. *Appl. Phys. Lett.* 75:2455–2457.
45. Kim, J.R. et al. 2002. Electrical transport properties of individual gallium nitride nanowires synthesized by chemical-vapor-deposition. *Appl. Phys. Lett.* 80:3548–3550.
46. He, M.Q. et al. 2001. Growth of GaN nanowires by direct reaction of Ga with NH3. *J. Cryst. Growth* 231:357–167.
47. Zhang, X.S. et al. 2001. Micro-Raman investigation of GaN nanowires prepared by direct reaction Ga with NH3. *Chem. Phys. Lett.* 345:372–376.
48. Shi, W.S. et al. 2001. Microstructures of gallium nitride nanowires synthesized by oxide-assisted method. *Chem. Phys. Lett.* 345:377–380.
49. Peng, H.Y. et al. 2002. Control of growth orientation of GaN nanowires. *Chem. Phys. Lett.* 359:241–245.
50. Chen, C.C. et al. 2000. Large-scale catalytic synthesis of crystalline gallium nitride nanowires. *Adv. Mater.* 12:738–741.
51. Zhou, H.X. et al. 2003. Study on sensing mechanism of CuO-SnO2 gas sensors. *Mater. Sci. Eng. B* 99:44–47.
52. Nath, N. et al. 2002. A colorimetric gold nanoparticle sensor to interrogate biomolecular interactions in real time on a surface. *Anal. Chem.* 74:504–509.
53. Wang, H. et al. 2005. Nanosphere arrays with controlled sub-10-nm gaps as surface-enhanced Raman spectroscopy substrates. *J. Am. Chem. Soc.* 127:14992–14993.
54. Liu, S.H. et al. 2005. Synthesis, functionalization, and bioconjugation of monodisperse, silica-coated gold nanoparticles: Robust bioprobes. *Adv Funct. Mater.* 15:961–967.
55. Daniel, M.C. et al. 2004. Gold nanoparticles: Assembly, supramolecular chemistry, quantum-size-related properties, and applications toward biology, catalysis, and nanotechnology. *Chem. Rev.* 104:293–346.
56. Xia, Y.N. et al. 2001. Large-scale synthesis of monodisperse nanorods of Se/Te alloys through a homogeneous nucleation and solution growth process. *Adv. Mater.* 13:1380–1384.

57. Xia, Y.N. et al. 2002. Formation of tellurium nanotubes through concentration depletion at the surfaces of seeds. *Adv. Mater.* 14:279–282.
58. Xia, Y.N. et al. 2002. Large-scale synthesis of uniform silver nanowires through a soft, self-seeding, polyol process. *Adv. Mater.* 14:833–837.
59. Xia, Y.N. et al. 2002. Crystalline silver nanowires by soft solution processing. *Nano Lett.* 2:165–168.
60. Murphy, C. et al. 2002. Controlling the aspect ratio of inorganic nanowire and nanorods. *Adv. Mater.* 14:80–82.
61. Chou, C.H. et al. 2005. Highly efficient, wavelength-tunable, gold nanoparticle based optothermal nanoconvertors. *J. Phys. Chem. B* 109:11135–11138.
62. Link, S. et al. 1999. Size and temperature dependence of the plasmon absorption of colloidal gold nanoparticles. *J. Phys. Chem. B* 103:4212–4217.
63. Zeng, H.C. et al. 2003. Hydrothermal synthesis of ZnO nanorods in the diameter regime of 50 nm. *J. Am. Chem. Soc.* 125:4430–4431.
64. Cho, S. et al. 2008. Morphology-controlled growth of ZnO nanostructures using microwave irradiation: From basic to complex structures. *J. Phys. Chem. C.* 112:12769–12776.
65. Greene, L.E. et al. 2002. Low temperature wafer-scale production of ZnO nanowire arrays. *Angew. Chem. Int. Ed.* 42:3031–3034.
66. Low, M. et al. 2005. Nanowire dye-sensitized solar cells. *Nat. Mater.* 4:455–459.
67. Tokumoto, M.S. et al. 2003. Catalysis and temperature dependence on the formation of ZnO nanoparticles and of zinc acetate derivatives prepared by the sol–gel route. *J. Phys. Chem. B.* 107:568–574.
68. Ormhals, A. 1934. U.S. Patent 1,975,504.
69. Reneker, D.H. et al. 1996. Nanometre diameter fibres of polymer, produced by electrospinning. *Nanotechnology* 7:216–223.
70. Reneker, D.H. et al. 2002. Nanofiber garland of polycaprolactone by electrospinning. *Polymer* 43:6785–6794.
71. Reneker, D.H. et al. 2000. Bending instability of electrically charged liquid jets of polymer solutions in electrospinning. *J. Appl. Phys.* 87:4531–4547.
72. Cann, J.T.M. et al. 2005. Electrospinning of nanofibers with core-sheath, hollow, or porous structures. *J. Mater. Chem.* 15:735–738.
73. Hyeon, T. et al. 2001. Synthesis of highly crystalline and monodisperse maghemite nanocrystallites without a size-selection process. *J. Am. Chem. Soc.* 123:12798–12801.
74. Jongnam, P. et al. 2004. Ultra-large-scale syntheses of monodisperse nanocrystals. *Nat. Mater.* 3:891–895.
75. Wegner, K. et al. 2002. Design of metal nanoparticle synthesis by vapor flow condensation. *Chem. Eng. Sci.* 57:1753–1762.
76. Maisels, A. et al. 2000. Synthesis of tailored composite nanoparticles in the gas phase. *Appl. Phys. Lett.* 77:4431–4433.
77. Nanda, K.K. et al. 2002. Band-gap tuning of PbS nanoparticles by in-flight sintering of size classified aerosols. *J. Appl. Phys.* 91:2315–2321.
78. Shinde, S.R. et al. 2000. Magnetic properties of nanosized powders of magnetic oxides synthesized by pulsed laser ablation. *J. Appl. Phys.* 88:1566–1575.
79. Nakaso, K. et al. 2002. Evaluation of the change in the morphology of gold nanoparticles during sintering. *J. Aerosol. Sci.* 33:1061–1074.
80. Ohno, T. et al. 2002. Morphology of composite nanoparticles of immiscible binary systems prepared by gas-evaporation technique and subsequent vapor condensation. *J. Nanoparticle Res.* 4:255–260.
81. Marine, W. et al. 2000. Strategy of nanocluster and nanostructure synthesis by conventional pulsed laser ablation. *Appl. Surf. Sci.* 154–155:345–352.
82. Nakata, Y. et al. 2002. Particle dynamics during nanoparticle synthesis by laser ablation in a background gas. *J. Appl. Phys.* 91:1640–1643.
83. Harano, A. et al. 2002. Crystal phases of TiO ultrafine particles prepared by laser ablation of solid rods. *J. Nanoparticle Res.* 4:215–219.
84. Makimura, T. et al. 2002. Laser ablation synthesis of hydrogenated silicon nanoparticles with green photoluminescence in the gas phase. *Jpn. J. Appl. Phys.* 41:L144–L146.
85. Weber, A.P. et al. 2001. Application of aerosol techniques to study the catalytic formation of methane on gasborne nickel nanoparticles. *J. Phys. Chem. A* 105:8958–8963.
86. Rexer, E.F. et al. 2000. Production of metal oxide thin films by pulsed arc molecular beamdeposition. *Rev. Sci. Instrum.* 71:2125–2130.

87. Urban, F.K. et al. 2002. Nanophase films deposited from a high-rate, nanoparticle beam. *J. Vac. Sci. Technol. B* 20:995–999.
88. Ostraat, M.L. et al. 2001. Ultraclean two-stage aerosol reactor for production of oxide-passivated silicon nanoparticles for novel memory devices. *J. Electrochem. Soc.* 148:G265–G270.
89. Magnusson, M.H. et al. 2000. Single-crystalline tungsten nanoparticles produced by thermal decomposition of tungsten hexacarbonyl. *J. Mater. Res.* 15:1564–1569.
90. Nasibulin, A.G. et al. 2002. Nanoparticle synthesis by copper(II) acetylacetonate vapor decomposition in the presence of oxygen. *Aerosol. Sci. Technol.* 36:899–911.
91. Schmechel, R. et al. 2001. Luminescence properties of nanocrystalline Y O:Eu3q in different host materials. *J. Appl. Phys.* 89:1679–1686.
92. Senter, R.A. et al. 2001. Synthesis of silicon nanocrystals with erbium-rich surface layers. *Nano Lett.* 1:383–386.
93. Winterer, M. et al. 2001. Nanocrystalline zirconia surface-doped with alumina: Chemical vapor synthesis, characterization, and properties. *J. Am. Ceram. Soc.* 84:2771–2776.
94. Ehrman, S.H. et al. 1999. Aquino-Class MI, Zachariah MR. Effect of temperature and vapor-phase encapsulation on particle growth and morphology. *J. Mater. Res.* 14:1664–1671.
95. Ahonen, P.P. et al. 2001. Mobility size development and the crystallization path during aerosol decomposition synthesis of TiO particles. *J. Aerosol. Sci.* 32:615–630.
96. Kim, J.H. et al. 2002. Size-monodisperse metal nanoparticles via hydrogen-free spray pyrolysis. *Adv. Mater.* 14:518–521.
97. Borsella, E. et al. 2001. MoS nanoparticles produced by laser induced synthesis from gaseous precursors. *J. Mater. Sci. Lett.* 20:187–191.
98. Kamlag, Y. et al. 2001. Laser CVD of cubic SiC nanocrystals. *Appl. Surf. Sci.* 184:118–122.
99. Ledoux, G. et al. 2002. Photoluminescence of size-separated silicon nanocrystals: Confirmation of quantum confinement. *Appl. Phys. Lett.* 80:4834–4836.
100. Ledoux, G. et al. 2002. Nanostructured films composed of silicon nanocrystals. *Mater. Sci. Eng. C* 19:215–218.
101. Heberlein, J. et al. 2001. Thermal plasma deposition of nanophase hard coatings. *Surf. Coat. Technol.* 142–144:265–271.
102. Janzen, C. et al. 2001. Formation of Fe O nano-particles in doped low-pressure H yO yAr flames. *Combust. Flame* 125:1150–1161.
103. Lee, D. et al. 2002. Coalescence enhanced synthesis of nanoparticles to control size, morphology, and crystalline phase at high concentrations. *J. Aerosol. Sci.* 33:1–16.
104. Wegner, K. et al. 2002. Flame-nozzle synthesis of nanoparticles with closely controlled size, morphology, and crystallinity. *Mater. Lett.* 55:318–321.
105. Kammler, H.K. et al. 2002. Flame temperature measurements during electrically assisted aerosol synthesis of nanoparticles. *Combust. Flame* 128:369–381.
106. Kammler, H.K. et al. 2002. Controlled synthesis of nanostructured particles by flame spray pyrolysis. *J. Aerosol. Sci.* 33:369–389.
107. Stöber, W. et al. 1968. Controlled growth of monodisperse silica spheres in the micron size range. *J. Colloid Interface Sci.* 26:62–69.
108. Vlasov, Y.A. et al. 2001. On-chip natural assembly of silicon photonic bandgap crystals. *Nature* 414:289–293.
109. Blanco, A. et al. 2000. Large-scale synthesis of a silicon photonic crystal with a complete three-dimensional bandgap near 1.5 micrometres. *Nature* 405:437–440.
110. Miguez, H. et al. 2002. Mechanical stability enhancement by pore size and connectivity control in colloidal crystals by layer-by-layer growth of oxide. *Chem. Commun.* 22:2736–2737.
111. Miguez, H. et al. 2003. A new synthetic approach to silicon colloidal photonic crystals with a novel topology and an omni-directional photonic bandgap: Micromolding in inverse silica opal (MISO). *Adv. Mater.* 15:597–600.
112. Gobe, M. et al. 1983. Preparation and characterization of monodisperse magnetite sols in W/O microemulsion. *J. Colloid Interface Sci.* 93:293–295.
113. Pillai, V. et al. 1995. Preparation of nanoparticles of silver halides, superconductors and magnetic materials using water-in-oil microemulsions as nano-reactors. *Adv. Colloid Interface Sci.* 55:241–269.
114. Pileni, M.P. et al. 1992. Synthesis of cadmium sulfide in situ in reverse micelles: Influence of the preparation modes on size, polydispersity, and photochemical reactions. *Chem. Mater.* 4:338–345.
115. Santra, S. et al. 2001. Synthesis and characterization of silica-coated iron oxide nanoparticles in microemulsion: The effect of nonionic surfactants. *Langmuir* 17:2900–2906.

116. Tago, T. et al. 2002. Novel synthesis of silica-coated ferrite nanoparticles prepared using water-in-oil microemulsion. *J. Am. Ceram. Soc.* 85:2188–2194.
117. Vestal, C.R. et al. 2003. Synthesis and magnetic characterization of Mn and Co spinel ferrite-silica nanoparticles with tunable magnetic core. *Nano Lett.* 3:1739–1743.
118. Peng, X.G. et al. 1998. Kinetics of II-VI and III-V colloidal semiconductor nanocrystal growth: "Focusing" of size distributions. *J. Am. Chem. Soc.* 120:5343–5344.
119. Wang, Y. et al. 1987. PbS in polymers. From molecules to bulk solids. *J. Chem. Phys.* 87:7315.
120. Trindade, T. et al. 1997. Synthesis of CdS and CdSe nanocrystallites using a novel single-molecule precursors approach. *Chem. Mater.* 9:523–530.
121. Ludolph, B. et al. 1998. Novel single molecule precursor routes for the direct synthesis of highly monodispersed quantum dots of cadmium or zinc sulfide or selenide. *Chem. Commun.* 1849–1850.
122. Bayer, I.S. et al. 2008. Biocompatible poly(vinylidene fluoride)/cyanoacrylate composite coatings with tunable hydrophobicity and bonding strength. *Appl. Phys. Lett.* 93:173902.
123. www. Technologyreview/read_article
124. Suslick, K.S. et al. 1986. Sonochemical hot spot. *J. Am. Chem. Soc.* 108:5641–5642.
125. Suslick, K.S. et al. 1996. Nanostructured materials generated by high-intensity ultrasound: Sonochemical synthesis and catalytic studies. *Chem. Mater.* 8:2172–2179.
126. Gorelikov, I. et al. 2004. Hybrid microgels photoresponsive in the near-infrared spectral range. *J. Am. Chem. Soc.* 126:15938–15939.
127. Koo, H.Y. et al. 2008. Direct growth of optically stable gold nanorods onto polyelectrolyte multilayered capsules. *Small* 4:742–745.
128. Moghaddam, M.J. et al. 2004. Highly efficient binding of DNA on the sidewalls and tips of carbon nanotubes using photochemistry. *Nano Lett.* 4:89–93.
129. Juaìrez, B.H. et al. 2008. Carbon supported CdSe nanocrystals. *J. Am. Chem. Soc.* 130:15282–15284.
130. Gole, A. et al. 2008. Iron oxide coated gold nanorods: Synthesis, characterization, and magnetic manipulation. *Langmuir* 24:6232–6237.
131. Cheyne, R.B. et al. 2005. Novel two-dimensional "ring and chain" morphologies in Langmuir–Blodgett monolayers of PS-b-PEO block copolymers: Effect of spreading solution concentration on self-assembly at the air–water interface. *Langmuir* 21:5453–5460.
132. Lauter-Pasyuk, V. et al. 1997. Effect of nanoparticle size on the internal structure of copolymer-nanoparticles composite thin films studied by neutron reflection. *Phys. B* 241:1092–1094.
133. Bronstein, L.M. et al. 1999. Induced micellization by interaction of poly(2-vinylpyridine)-block-poly(ethylene oxide) with metal compounds. Micelle characteristics and metal nanoparticle formation. *Langmuir* 15:6256–6262.
134. Tsutsumi, K. et al. 1999. Selective incorporation of palladium nanoparticles into microphase-separated domains of poly(2-vinylpyridine)-block-polyisoprene. *Langmuir* 15:5200–5203.
135. Lopes, W.A. et al. 2001. Hierarchical self-assembly of metal nanostructures on diblock copolymer scaffolds. *Nature* 414:735–738.
136. Horiuchi, S. et al. 2000. Nanoscale assembly of metal clusters in block copolymer films with vapor of a metal-acetylacetonato complex using a dry process. *Adv. Mater.* 12:1507–1511.
137. Zhang, Q.L. et al. 2005. Controlled placement of CdSe nanoparticles in diblock copolymer templates by electrophoretic deposition. *Nano Lett.* 5:357–361.
138. Chiu, J.J. et al. 2005. Control of nanoparticle location in block copolymers. *J. Am. Chem. Soc.* 127:5036–5037.
139. Euliss, L.E. et al. 2003. Cooperative assembly of magnetic nanoparticles and block copolypeptides in aqueous media. *Nano Lett.* 3:1489–1493.
140. Wang, Y. et al. 2003. "Pulling" nanoparticles into water: Phase transfer of oleic acid stabilized monodisperse nanoparticles into aqueous solutions of α-cyclodextrin. *Nano Lett.* 3:1555–1559.
141. Fukuoka, A. et al. 2006. Nanonecklaces of platinum and gold with high aspect ratios synthesized in mesoporous organosilica templates by wet hydrogen reduction. *Chem. Mater.* 18:337–343.
142. Zheng, M. et al. 2003. DNA-assisted dispersion and separation of carbon nanotubes. *Nat. Mater.* 2:338–342.
143. Caswell, K.K. et al. 2003. Preferential end-to-end assembly of gold nanorods by biotin–streptavidin connectors. *J. Am. Chem. Soc.* 125:13914–13915.
144. Jun, Y.W. et al. 2001. Controlled synthesis of multi-armed CdS nanorod architectures using monosurfactant system. *J. Am. Chem. Soc.* 123:5150–5151.
145. Tzitzios, V. et al. 2005. Synthesis and characterization of 3D CoPt nanostructures. *J. Am. Chem. Soc.* 127:13756–13757.

146. Yan, H.Q. et al. 2003. Dendritic nanowire ultraviolet laser array. *J. Am. Chem. Soc.* 125:4728–4729.
147. Love, J.C. et al. 2003. Three-dimensional self-assembly of metallic rods with submicron diameters using magnetic interactions. *J. Am. Chem. Soc.* 125:12696–12697.
148. Urbach, A.R. et al. 2003. Sub-100 nm confinement of magnetic nanoparticles using localized magnetic field gradients. *J. Am. Chem. Soc.* 125:12704–12705.
149. Nicewarner-Peña, S.R. et al. 2001. Submicrometer metallic barcodes. *Science* 294:137–141.
150. Yu, H. et al. 2002. Dumbbell-like bifunctional Au–Fe_3O_4 nanoparticles. *Nano Lett.* 5:379–382.
151. Li, Y.Q. et al. 2005. Enhanced magnetooptical response in dumbbell-like Ag–$CoFe_2O_4$ nanoparticle pairs. *Nano Lett.* 5:1689–1692.
152. Xu, C. et al. 2007. Au-Fe_3O_4 dumbbell nanoparticles as dual-functional probes. *Angew. Chem. Int. Ed.* 48:173–176.
153. Choi, S.H. et al. 2008. Simple and generalized synthesis of oxide-metal heterostructured nanoparticles and their applications in multimodal biomedical probes. *J. Am. Chem. Soc.* 130:15573–15580.
154. Tokareva, I. et al. 2004. Nanosensors based on responsive polymer brushes and gold nanoparticle enhanced transmission surface plasmon resonance spectroscopy. *J. Am. Chem. Soc.* 126:15950–15951.
155. Yu, M. et al. 2005. High catalytic activity of platinum nanoparticles immobilized on spherical polyelectrolyte brushes. *Langmuir* 21:12229–12234.
156. Yi, D.K. et al. 2006. Synthesis and applications of magnetic nanocomposite catalysts. *Chem. Mater.* 18:2459–2461.
157. Redl, F.X. et al. 2003. Three-dimensional binary superlattices of magnetic nanocrystals and semiconductor quantum dots. *Nature* 423:968–971.
158. Jiang, P. et al. 2005. Wafer-scale periodic nanohole arrays templated from two-dimensional nonclose-packed colloidal crystals. *J. Am. Chem. Soc.* 127:3710–3711.
159. Reculusa, S. et al. 2004. Synthesis of daisy-shaped and multipod-like silica/polystyrene nanocomposites. *Nano Lett.* 4:1677–1682.
160. Manoharan, V.N. et al. 2003. Dense packing and symmetry in small clusters of microspheres. *Science* 301:483–487.
161. Koo, H.Y. et al. 2004. A snowman-like array of colloidal dimers for antireflecting surfaces. *Adv. Mater.* 16:274–277.
162. Yi, D.K. et al. 2003. Polymer nanosphere lithography: Fabrication of an ordered trigonal polymeric nanostructure. *Chem. Commun.* 8:982–983.
163. Yi, D.K. et al. 2002. Surface-modulation-controlled three-dimensional colloidal crystals. *Appl. Phys. Lett.* 80:225.
164. Yi, D.K. et al. 2002. Fabrication of a mesoscale wire: Sintering of a polymer colloid arrayed inside a one-dimensional groove pattern. *Langmuir* 18:5321–5323.
165. Yi, D.K. et al. 2003. Novel approach to the fabrication of macroporous polymers and their use as a template for crystalline titania nanorings. *Nano Lett.* 3:207–211.
166. Wang, Y.A. et al. 2002. Stabilization of inorganic nanocrystals by organic dendrons. *J. Am. Chem. Soc.* 124:2293–2298.
167. Aldana, J. et al. 2001. Photochemical instability of CdSe nanocrystals coated by hydrophilic thiols. *J. Am. Chem. Soc.* 123:8844–8850.
168. Murray, C.B. et al. 1993. Synthesis and characterization of nearly monodisperse CdE (E = sulfur, selenium, tellurium) semiconductor nanocrystallites. *J. Am. Chem. Soc.* 115:8706–8715.
169. Peng, X. et al. 2000. Shape control of CdSe nanocrystals. *Nature* 404:59–61.
170. Qu, L. et al. 2001. Alternative routes toward high quality CdSe nanocrystals. *Nano Lett.* 1:333–337.
171. Qu, L. et al. 2002. Control of photoluminescence properties of CdSe nanocrystals in growth. *J. Am. Chem. Soc.* 124:2049–2055.
172. Yu, W.W. et al. 2002. Formation of high-quality CdS and other II-VI semiconductor nanocrystals in noncoordinating solvents: Tunable reactivity of monomers. *Angew. Chem., Inter. Ed.* 41:2368–2371.
173. Guo, X. et al. 2003. Luminescent CdSe/CdS Core/Shell nanocrystals in dendron boxes: Superior chemical, photochemical and thermal stability. *J. Am. Chem. Soc.* 125:3901–3909.
174. Petruska, M.A. et al. 2004. An amphiphilic approach to nanocrystal quantum dot–titania nanocomposites. *J. Am. Chem. Soc.* 126:714–715.
175. Phillipse, A.P. et al. 1994. Magnetic silica dispersions: Preparation and stability of surface-modified silica particles with a magnetic core. *Langmuir* 10:92–99.
176. Kobayashi, Y. et al. 2003. Preparation and properties of silica-coated cobalt nanoparticles. *J. Phys. Chem. B* 107:7420–7425.

177. Lu, Y. et al. 2002. Modifying the surface properties of superparamagnetic iron oxide nanoparticles through a sol–gel approach. *Nano Lett.* 2:183–186.
178. Yi, D.K. et al. 2006. Nanoparticle architectures templated by SiO_2/Fe_2O_3 nanocomposites. *Chem. Mater.* 18:614–619.
179. Papaefthymiou, G.C. et al. 2009. Interparticle interactions in magnetic core/shell nanoarchitectures. *Phys. Rev. B* 80:024406.
180. Gomez, D.E. et al. 2005. Tunable whispering gallery mode emission from quantum-dot-doped microspheres. *Small* 1:238–241.
181. Kim, S.W. et al. 2002. Fabrication of hollow palladium spheres and their successful application to the recyclable heterogeneous catalyst for suzuki coupling reactions. *J. Am. Chem. Soc.* 124:7642–7643.
182. Caruso, F. et al. 1998. Nanoengineering of inorganic and hybrid hollow spheres by colloidal templating. *Science* 282:1111–1114.
183. Caruso, F. et al. 1998. Electrostatic self-assembly of silica nanoparticle–polyelectrolyte multilayers on polystyrene latex particles. *J. Am. Chem.* 120.8523–8524.
184. Caruso, R.A. et al. 2001. Multilayered titania, silica, and laponite nanoparticle coatings on polystyrene colloidal templates and resulting inorganic hollow spheres. *Chem. Mater.* 13:400–409.
185. Yi, D.K. et al. 2005. Silica-coated nanocomposites of magnetic nanoparticles and quantum dots. *J. Am. Chem. Soc.* 127:4990–4991.
186. Deng, Y. et al. 2003. A novel approach for preparation of thermoresponsive polymer magnetic microspheres with core-shell structure. *Adv. Mater.* 15:1729–1732.
187. Ding, X. et al. 2004. Silica nanoparticles encapsulated by polystyrene via surface grafting and in situ emulsion polymerization. *Mater. Lett.* 58:3126–3130.
188. Sun, Y. et al. 2004. Synthesis and optical properties of nanorattles and multiple-walled nanoshells/nanotubes made of metal alloys. *J. Am.Chem. Soc.* 126:9399–9406.
189. Calderon Moreno, J.M. et al. 2001. Hydrothermal processing of high-quality multiwall nanotubes from amorphous carbon. *J. Am. Chem. Soc.* 123:741–742.
190. Libera, J. et al. 2001. Hydrothermal synthesis of graphite tubes using Ni catalyst. *Carbon* 39:1307–1318.
191. Tai, K. et al. 2006. Nanogranular origins of the strength of bone. *Nano Lett.* 6:2520–2525.
192. Gupta, S.H. et al. 2005. Nanoscale deformation mechanisms in bone. *Nano Lett.* 5:2108–2111.
193. Fratzl, P. et al. 2004. Structure and mechanical quality of the collagen-mineral nano-composite in bone. *J. Mater. Chem.* 14:2115–2123.
194. Li, X. et al. 2006. In situ observation of nanograin rotation and deformation in nacre. *Nano Lett.* 6:2301–2304.
195. Chasteen, N.D. and Harrison, P.M. 1999. Mineralization in ferritin: An efficient means of iron storage. *J. Struct. Biol.* 126:182.
196. Waldo, G.S. and Theil, E.C. 1996. Ferritin and iron biomineralization, in *Comprehensive Supramolecular Chemistry* (Suslick, K.S., Ed.), Vol. 5, pp. 65–89, Pergamon Press, Oxford, U.K.
197. Harrison, P.M. and Arosio, P. 1996. Ferritins: Molecular properties, iron storage function and cellular regulation. *Biochim. Biophys. Acta* 1275:161.
198. Chasteen, N.D. 1998. Ferritin. Uptake, storage, and release of iron, in *Metal Ions in Biological Systems* (Sigel, H. and Sigel, A., Eds.), Vol. 35, pp. 479–514, Marcel Dekker, New York.
199. Brooks, R.A., Vymazal, J., Goldfarb, R.B., Bulte, J.W., and Aisen, P. 1998. Relaxometry and magnetometry of ferritin. *Magn. Reson. Med.* 40: 227.
200. Blakemore, R.P. 1982. Magnetotactic bacteria. *Annu. Rev. Microbiol.* 36: 217.
201. Bazylinski, D.A., Frankel, R.B., and Jannasch, H.W. 1988. Anaerobic magnetite production by a marine, magnetotactic bacterium. *Nature* 334:518.
202. Dunin-Borkowski, R.E. et al. 1998. Magnetic microstructure of magnetotactic bacteria by electron holography. *Science* 282:1868–1870.
203. Mann, S. 2001. *Biomineralization: Principles and Concepts in Bioinorganic Materials Chemistry*, Oxford University Press, New York.
204. Mann, S. ed. 1996. *Biomimetic Materials Chemistry*, VCH Publishers, Inc., New York/Weinheim, Germany/Cambridge, U.K.
205. Mann, S. 1993. Molecular tectonics in biomineralization and biomimetic materials synthesis. *Nature* 365: 499.
206. Polanams, J., Ray, A.D., and Watt, R.K. 2005. Nanophase iron phosphate, iron arsenate, iron vanadate, and iron molybdate minerals synthesized within the protein cage of ferritin. *Inorg. Chem.* 44:3203–3209.

207. Swift, J. et al. 2006. Design of functional ferritin-like proteins with hydrophobic cavities. *J. Am. Chem. Soc.* 128:6611–6619.
208. Butts, C.A. et al. 2008. Directing noble metal ion chemistry within a designed ferritin protein. *Biochemistry* 47:12729–12739.
209. Uchida, M. et al. 2006. Targeting of cancer cells with ferrimagnetic ferritin cage nanoparticles. *J. Am. Chem. Soc.* 128:16626–16633.
210. Bulte, J.W.M., Douglas, T., Mann, S., Frankel, R.B., Moskovitz, B.M., Brooks, R.A., Baumgarner, C.D., Vymazel, J., Strub, M.-P., and Frank, J.A. 1994. Magnetoferritin: Characterization of a novel superparamagnetic magnetic resonance contrast agent. *J. Magn. Reson. Imaging* 4:497–505.
211. Bulte, J.W., Douglas, T., Mann, S., Frankel, R.B., Moskowitz, B.M., Brooks, R.A., Baumgarner, C.D., Vymazal, J., and Frank, J.A. 1994. Magnetoferritin. Biomineralization as a novel molecular approach in the design of iron-oxide-based magnetic resonance contrast agents. *Invest. Radiol.* 29(2):S214–S216.
212. Liu, G. et al. 2006. Characterization and surface reactivity of ferrihydrite nanoparticles assembled in ferritin. *Langmuir* 22:9313–9321.
213. Zhang, N. et al. 2000. Naturally occurring ferritin as a novel catalyst for selective hydroxylation of phenol. *React. Kinet. Catal. Lett.* 71:393–404.
214. Wong, K.K.W. and Mann, S. 1996. Biomimetic synthesis of cadmium sulfide-ferritin nanocomposites. *Adv. Mater.* 8:928.
215. Allen, M., Willits, D., Young, M., and Douglas, T. 2003. Constrained synthesis of cobalt oxide nanomaterials in the 12-subunit protein cage from *Listeria innocua*. *Inorg. Chem.* 42:6300.
216. Douglas, T., Dickson, D.P.E., Betteridge, S., Charnock, J., Garner, C.D., and Mann, S. 1995. Synthesis and structure of an iron(III) sulfide-Ferritin bioinorganic nanocomposite. *Science* 269:54.
217. Hainfeld, J.F. 1992. Uranium-loaded apoferritin with antibodies attached: Molecular design for uranium neutron-capture therapy. *Proc. Natl. Acad. Sci. USA* 89:11064.
218. Meldrum, F.C., Wade, V.J., Nimmo, D.L., Heywood, B.R., and Mann, S. 1991. Synthesis of inorganic nanophase materials in supramolecular protein cages. *Nature* 349:684.
219. Warne, B., Kasyutich, O., Mayes, E.L., Wiggins, J.A.L., and Wong, K.K.W. 2000. Self assembled nanoparticulate Co:Pt for data storage applications. *IEEE Trans. Magn.* 36:3009–3011.
220. Hoinville, J., Bewick, A., Gleeson, D., Jones, R., Kasyuitich, O., Mayes, E., Nartowiski, A., Warne, B., Wiggins, J., and Wong, K. 2003. Synthesis of inorganic nanophase materials in supramolecular protein cages. *J. Appl. Phys.* 93:7187.
221. Okuda, M., Iwahori, K., Yamashita, I., and Yoshimura, H. 2003. Fabrication of nickel and chromium nanoparticles using the protein cage of apoferritin. *Biotechnol. Bioeng.* 84:187.
222. Kramer, R.M., Li, C., Carter, D.C., Stone, M.O., and Naik, R.R. 2004. Engineered protein cages for nanomaterials synthesis. *J. Am. Chem. Soc.* 126:13282.
223. Yamashita, I., Hayashi, J., and Hara, M. 2004. Bio-template synthesis of uniform CdSe nanoparticles using cage-shaped protein, apoferritin. *Chem. Lett.* 33:1158.
224. Bhattacharyya, S. et al. 2005. Protein-functionalized carbon nanotube-polymer composites. *Appl. Phys. Lett.* 86:113104-1–113104-3.
225. Shin, M.K. et al. 2006. Reinforcement of polymeric nanofibers by ferritin nanoparticles. *Appl. Phys. Lett.* 88:193901-1–193901-3.
226. Kreuter, J. et al. 1976. In vitro studies of poly(methyl methacrylate) adjuvants. *J. Pharm. Sci.* 65:1624–1627.
227. Salemi, A.K. et al. 2003. Multifunctional nanorods for gene delivery. *Nat. Mater.* 2:668–671.
228. Kam, N.W.S. et al. 2005. Carbon nanotubes as multifunctional biological transporters and near-infrared agents for selective cancer cell destruction. *Proc. Natl. Acad. Sci. USA* 102:11600–11605.
229. Kam, N.W.S. et al. 2005. Carbon nanotubes as intracellular protein transporters: Generality and biological functionality. *J. Am. Chem. Soc.* 127:6021–6026.
230. Thomas, M. et al. 2003. Conjugation to gold nanoparticles enhances polyethylenimine's transfer of plasmid DNA into mammalian cells. *Proc. Natl. Acad. Sci. USA* 100:9138–9143.
231. Sudeep, P.K. et al. 2005. Selective detection of cysteine and glutathione using gold nanorods. *J. Am. Chem. Soc.* 127:6516–6517.
232. Wu, X.Y. et al. 2003. Immunofluorescent labeling of cancer marker Her2 and other cellular targets with semiconductor quantum dots. *Nat. Biotechnol.* 21:41–46.
233. Han, M.Y. et al. 2001. Quantum-dot-tagged microbeads for multiplexed optical coding of biomolecules. *Nat. Biotechnol.* 19:631–635.
234. Fan, H. et al. 2005. Surfactant-assisted synthesis of water-soluble and biocompatible semiconductor quantum dot micelles. *Nano Lett.* 5:645–648.

235. Xie, H.Y. et al. 2005. Cell-targeting multifunctional nanospheres with both fluorescence and magnetism. *Small* 1:506–509.
236. Gu, H. et al. 2004. Facile one-pot synthesis of bifunctional heterodimers of nanoparticles: A conjugate of quantum dot and magnetic nanoparticles. *J. Am. Chem. Soc.* 126:5664–5665.
237. Kim, H. et al. 2005. Synthesis and characterization of Co/CdSe Core/Shell nanocomposites: Bifunctional magnetic-optical nanocrystals. *J. Am. Chem. Soc.* 127:544–546.
238. Wang, G.P. et al. 2005. Biofunctionalization of fluorescent-magnetic-bifunctional nanospheres and their applications. *Chem. Commun.* 34:4276–4278.
239. Schultz, S. et al. 2000. Single-target molecule detection with nonbleaching multicolor optical immunolabels. *Proc. Natl. Acad. Sci. USA* 97:996–1001.
240. Taton, T.A. et al. 2001. Two-color labeling of oligonucleotide arrays via size-selective scattering of nanoparticle probes. *J. Am. Chem. Soc.* 123:5164–5165.
241. Star, A. et al. 2003. Electronic detection of specific protein binding using nanotube FET devices. *Nano Lett.* 3:459–463.
242. Cheong, K.H. et al. 2008. Gold nanoparticles for one step DNA extraction and real-time PCR of pathogens in a single chamber. *Lab Chip* 8:810–813.
243. Kallenbach, N.R. et al. 1983. An immobile nucleic acid junction constructed from oligonucleotides. *Nature* 305:829–831.
244. Seeman, N.C. et al. 2001. DNA nicks and nodes and nanotechnology. *Nano Lett.* 1:22–26.
245. Seeman, N.C. et al. 2000. Logical computation using algorithmic self-assembly of DNA triple-crossover molecules. *Nature* 407:493–496.
246. Dai, H. et al. 2006. DNA functionalization of carbon nanotubes for ultrathin atomic layer deposition of high K dielectrics for nanotube transistors with 60 mV/decade switching. *J. Am. Chem. Soc.* 128:3518–3519.
247. Mann, S. et al. 2001. Tobacco mosaic virus liquid crystals as templates for the interior design of silica mesophases and nanoparticles. *Adv. Mater.* 13:1266–1269.
248. Geim, A.K. et al. 2003. Microfabricated adhesive mimicking gecko foot-hair. *Nat. Mater.* 2:461–463.
249. Kustandi, T.S. et al. 2007. Self-assembled nanoparticles based fabrication of gecko foot-hair-inspired polymer nanofibers. *Adv. Funct. Mater.* 17:2211–2218.

4 Optical and Optoelectronic Properties of Silicon Nanocrystals Embedded in SiO_2 Matrix

T.P. Chen and L. Ding

CONTENTS

4.1 INTRODUCTION

4.1.1 BACKGROUND

Silicon has become the most important semiconductor material in microelectronic industry, since the invention of the first bipolar transistor in 1949 [1]. Nowadays, silicon devices constitute over 95% of the market of semiconductor devices [2]. On the other hand, as electronic device dimensions increasingly become smaller and smaller, the traditional electrical interconnects that were used for chip-to-chip communication become increasingly impractical due to the heat dissipation of the metal wires, which threatens the reliability of both the device and the system. When the device dimensions decrease to the nanoscale, the device will suffer a lot from the traditional electrical interconnects due to its propagation delay, high power consumption, and low bandwidth [3]. Fortunately, optical interconnects would probably provide us with a promising alternative strategy for overcoming these challenges. Since chip-to-chip communication via optical interconnects requires an on-chip emitter and detector, an important challenge on the materials and the integration of photonic devices into the mainstream Si process has recently triggered a new research subject, i.e., Si photonics. Most of the photonics components fabricated with Si technology have been demonstrated, such as optical modulator [4,5], switches [6,7], detectors [8,9], and low-loss waveguides [10,11]. However, bulk crystalline Si cannot be used for light-emitting diodes (LEDs) due to the reasons described below.

Silicon is an indirect bandgap material; thus, the probability for a radiative transition is very low. This is reflected in the very long time for radiative recombination. Due to the long radiative lifetime, excited free carriers have large probabilities to find non-radiative recombination centers and to recombine non-radiatively. Minimum energy transitions are vertical in k-space for a direct bandgap semiconductor materials whereas indirect bandgap materials undergo a non-vertical transition, as shown in Figure 4.1. In a direct bandgap material, any photon with energy equal to or larger than the energy bandgap (i.e., energy difference between the valence band maximum and the conduction band minimum) can participate in a direct transition and excite an electron from the top of the filled valence band to one of the states at the bottom of the conduction band. However, non-vertical absorptions in an indirect bandgap material occur via an intermediate virtual state, whereby photon absorption is accompanied by either the creation or annihilation of a phonon to achieve the momentum conservation, as shown in Figure 4.1b. The indirect transition requires a minimum photon energy of E_g+E_p to create an electron–hole pair and a phonon with the energy of E_p and the momentum of $\hbar(k_c-k_v)$. The alternative process is that a photon with the energy of E_g-E_p and a phonon with the energy of E_p and the momentum of $\hbar(k_c-k_v)$ are absorbed. In a perfect semiconductor, electron–hole pairs thermalize and accumulate at the conduction and valence band extremes. Fundamental radiative transitions in a semiconductor are those

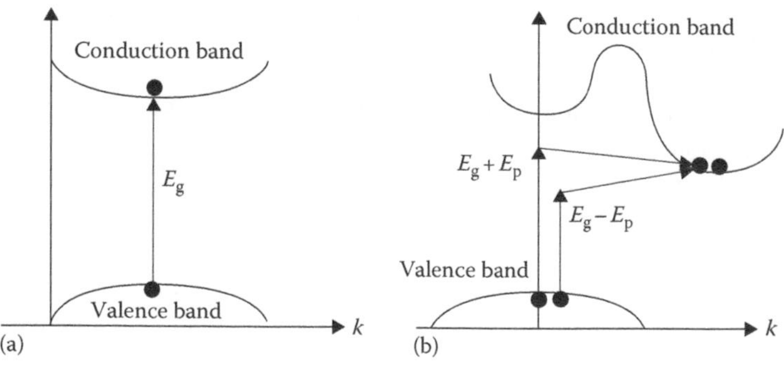

FIGURE 4.1 Schematic of the absorption of a photon in a direct bandgap (a) and indirect bandgap (b) semiconductor material.

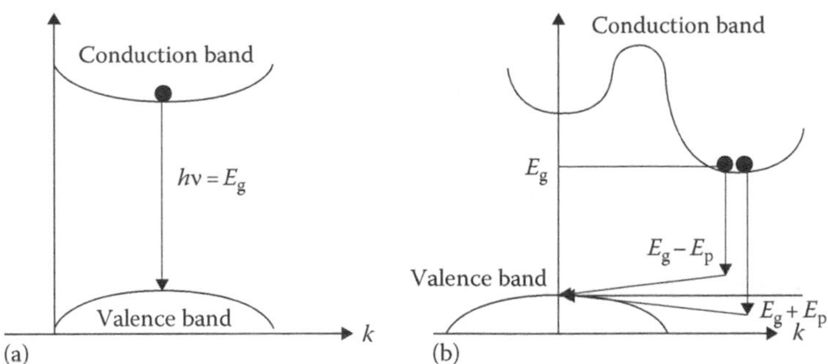

FIGURE 4.2 Schematic of direct band-to-band radiative transitions (a) and indirect band-to-band radiative transitions (b).

occurring at or near the band edges, namely, band-to-band or excitonic transitions. In a direct bandgap material, momentum-conserving transitions connect states with the same *k*-value. Accordingly, the emitted photon has a low-energy threshold at $h\nu = E_g$, as shown in Figure 4.2a. However, in an indirect bandgap semiconductor, since the momentum conservation requires electron–hole pairs to recombine radiatively only by phonon-assisted transitions, the probability of radiative transitions is smaller than that of competing non-radiative transitions. As shown in Figure 4.2b, indirect band-to-band radiative transitions could be realized by the assistance of phonon emission or absorption. With the involvement of a phonon, both the requirements of energy and momentum conservations are met for the bond-to-band transitions in the indirect bandgap semiconductor.

4.1.2 Silicon Nanocrystals

During the 1990s, many strategies have been employed to overcome these material limitations for fabricating Si-based LEDs. The most successful ones are based on the exploitation of low dimensional Si, where Si is nanoscaled and hence the optical and electronic properties of free carriers are modified by quantum confinement (QC) effects. Of all the types of low dimensional silicon, porous Si (PSi) has received the most intensive research attention in the early years of the exploitation of nanoscaled Si. The discovery of the visible photoluminescence (PL) from PSi in 1990 [12] has triggered a large research effort in the field of PSi and other Si nanostructures. It has been shown, as well, that luminescence wavelength can be tuned in a wide range and relatively high quantum efficiencies could be obtained, improving the prospect of using Si for LEDs. However, it is not reliable to utilize PSi in optoelectronic devices because of its instability in light emission [13], structural fragility [14], and incompatibility with conventional IC technology [15]. Recently, silicon nanocrystals (nc-Si) are considered to be the preferable strategy for overcoming these challenges [14,15]. In particular, nc-Si embedded in a dielectric matrix, such as SiO_2 and Si_3N_4 thin films, is a promising candidate for realizing Si LEDs with the advantages of chemical stability and full compatibility with silicon complementary metal-oxide-semiconductor (CMOS) processes [14,15].

In bulk crystalline silicon, since the radiative time of indirect transitions is extremely long and the transport of excitons is efficient, the main decay channel for free exciton is their capture in bound exciton states or non-radiative recombination. This results in a very low quantum yield of free exciton emission. The situation changes in a silicon nanocrystal. First, both absorbing and luminescing states shift to higher energies due to the spatial confinement, which would also affect the densities of electronic states due to the rising minimum kinetic energy. Second, the geometrical change in the nanocrystal causes a delocalization of carriers in *k*-space, thus allowing zero phonon optical transitions and significantly enhancing the oscillator strength of the zero phonon optical transitions in extremely small nc-Si. Third, due to the better overlap of the electron and hole envelope wavefunctions, a strong enhancement of the electron–hole (e-h) exchange interaction inducing a splitting of the exciton energy levels would be expected. In spite of the intensive research progress toward understanding the nature of efficient light emission from nc-Si, the mechanism is still controversial. This is because of the extreme complexity of the material system, as silicon nanocrystal ensembles are complicated heterogeneous systems that include various parameters, such as crystalline network properties, matrix properties, surface passivation conditions, nanocrystal size and shape distributions, etc., affecting the optical properties of nc-Si.

4.1.3 Synthesis of Silicon Nanocrystals

Nc-Si can be prepared by various methods. The earliest fabrication of nanoscaled silicon started in 1956 when Uhlir [16] made PSi for the first time by etching a *p*-type silicon wafer in a solution containing HF. PSi has been the most widely discussed and investigated nano-structured silicon system. It has attracted intensive research attention due to the simplicity of the preparation procedure and the high emission efficiency under optical excitation. Structural investigations have confirmed that it consists of Si nanocrystals with different sizes and shapes retaining the diamond lattice structure of bulk crystalline silicon [17]. PSi is usually formed by the electrochemical anodization of Si in an HF electrolyte. The choice for the substrate should naturally be *p*-type Si wafers because the anodization reaction at the interface of Si/electrolyte requires the presence of holes [18]. A most important factor that needs to be controlled in the fabrication is the porosity (i.e., the fraction of Si removed from the substrate) of PSi because it significantly influences the size of nc-Si. Furthermore, the value of porosity is directly related to the effective refractive index of the material. Indeed, as long as the typical structure size is much lower than the emission wavelength, the PSi layers appear as an effective medium, whose refractive index has an intermediate value between the refractive index of the Si (the host structures) and that of the air (the pores).

Both low-pressure chemical vapor deposition (LPCVD) [19–21] and plasma-enhanced chemical vapor deposition (PECVD) [22–24] have been frequently used in the fabrication of Si nanocrystals. CVD is a chemical process used to produce thin solid films on the surface of a substrate by the thermal decomposition and/or reaction of gaseous compounds. In a typical CVD process, the wafer (substrate) is exposed to one or more volatile precursors, which react and/or decompose on the substrate surface to produce the desired deposit. Figure 4.3 shows the schematic illustration of an LPCVD reactor. LPCVDs are CVD processes at subatmospheric pressures. Reduced pressures tend to reduce unwanted gas-phase reactions and improve film uniformity across the wafer. LPCVD is typically performed at ~800°C in a reactor. Nc-Si can be synthesized by controlling the early stages of LPCVD with pure SiH_4. In the PECVD technique, several reacting gases are activated by electric energy to react on the substrate, thus forming a thin solid film. Silicon nanocrystal films can be prepared using very high frequency (VHF) PECVD from a N_2O:SiH_4 mixture. After the mixture turns into plasma under high pressure and energy, the Si ions will be doped in situ with SiO_x ($x < 2$) film formation.

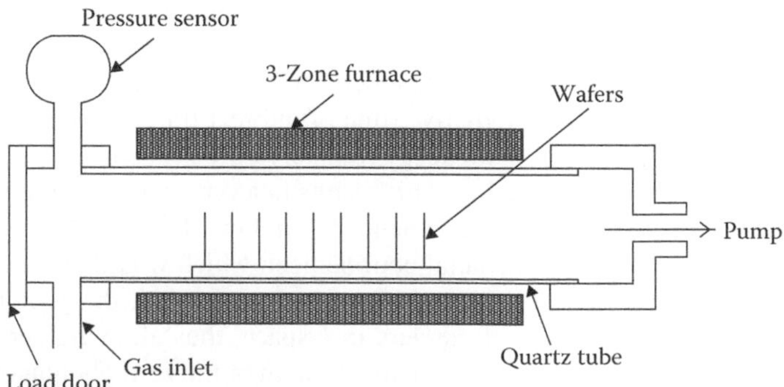

FIGURE 4.3 Schematic illustration of a typical LPCVD reactor.

Pulse laser deposition (PLD) has been recently used to successfully fabricate luminescent nc-Si. Strictly speaking, PLD is one type of sputtering technique. A pulsed laser beam is focused onto the surface of a solid target. The strong absorption of the electromagnetic radiation by the solid surface leads to the rapid evaporation of the target materials. The evaporated materials consist of highly excited and ionized species. They presented themselves as a glowing plasma plume immediately in front of the target surface, if the ablation is carried out in vacuum. Figure 4.4 shows the schematic illustration of a PLD system that was used to synthesize nc-Si [25]. Indeed, PLD is so straightforward that only a few parameters, such as laser energy density and pulse repetition rate, need to be adjusted during the process. The sizes of Si nanocrystals synthesized by PLD can vary from 2 to 10 nm depending on the background gas species and pressures [26].

Ion implantation is a versatile method to produce embedded semiconductor nanoclusters in dielectric films. Figure 4.5 presents the schematic illustration of a typical ion implanter. This technique combines the implantation of Si ions (implant energy is typically from a few to hundreds of keV; and typical implant dose is 10^{15}–10^{16} cm^{-2}) with a subsequent annealing at a high temperature (e.g., 1000°C) in N_2 gas for the duration of a few minutes to a couple of hours. Other annealing ambients, such as Ar and N_2+H_2, have also been used for the investigation on the PL properties of nc-Si embedded in dielectric films [27]. Figure 4.6 shows a typical cross-sectional HRTEM image of a Si nanocrystal embedded in an SiO_2 matrix fabricated by Si ion implantation at the implantation energy of 100 keV [28]. Wang et al. have carried out a study on the microstructural defects inside the nc-Si or around the interface using HRTEM [28,29]. The size of Si nanocrystal synthesized with this technique is usually ~2–~7 nm, and it does not depend much on the annealing time and annealing temperature. In addition, annealing does not change very much the profile of the implanted Si in the oxide. This is due to the fact that Si atoms have an extremely low diffusion coefficient in SiO_2. The advantages of this technique include its simple process and full compatibility with mainstream CMOS process. Ion implantation also offers many advantages in the fabrication of Si nanocrystals embedded in dielectric films. An ion implanter provides a high-voltage accelerating particles to form a high-velocity beam of Si ions, which can penetrate the surface of as-grown dielectric films. In a typical process, Si ions are implanted into an as-grown SiO_2 or Si_3N_4 film. The concentration and depth distribution of nc-Si in the film are decided by the implantation energy and the implantation dose. By this technique, the distribution of excess Si can be controlled through adjusting the implantation energy and the dose of implanted Si ions.

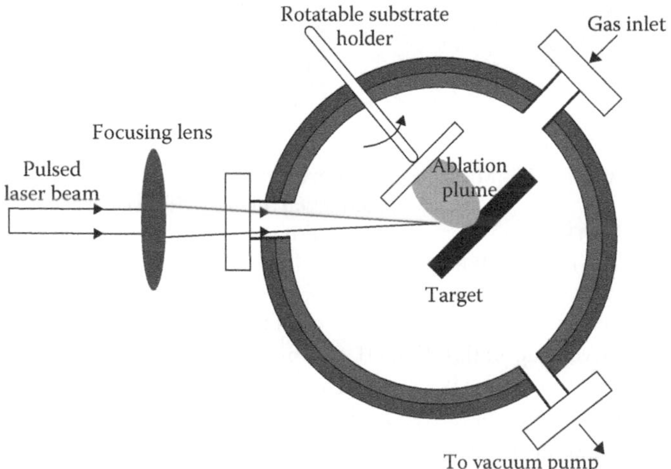

FIGURE 4.4 Schematic of a PLD system. (Modified from Chen, X.Y. et al., *J. Appl. Phys.*, 93, 6311, 2003. With permission.)

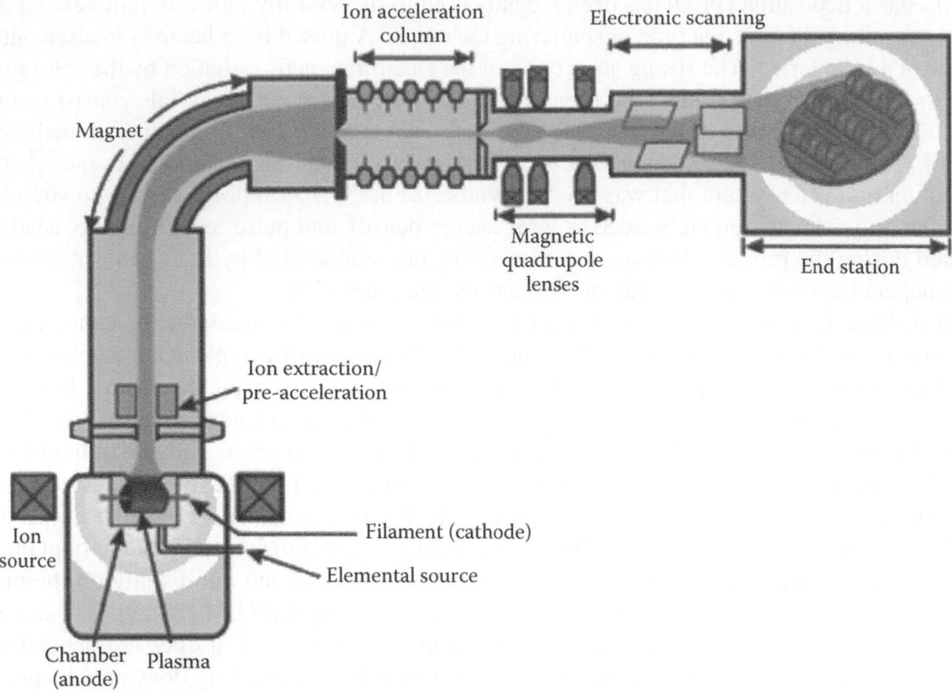

FIGURE 4.5 Schematic illustration of a typical ion implanter.

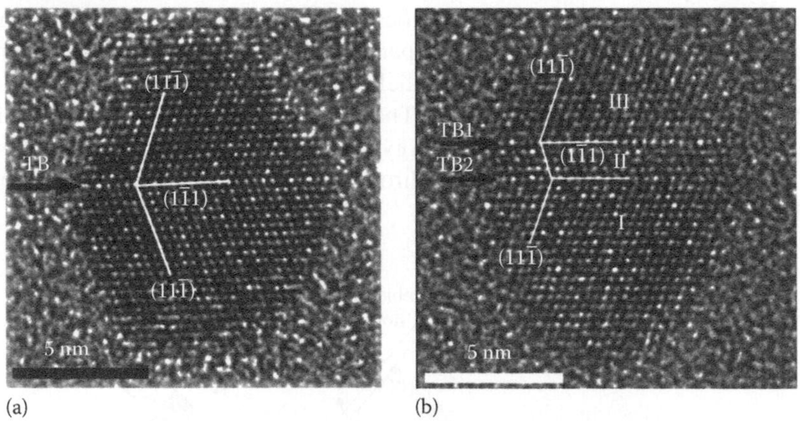

FIGURE 4.6 Cross-sectional HRTEM images of a Si nanocrystal embedded in SiO_2 synthesized by Si ion implantation and subsequent high temperature annealing: (a) Single-twin structure and (b) double-twin structure [28]. The implantation was carried out at the energy of 100 keV with the Si ion dose of 3×10^{17} cm^{-2}. (From Wang, Y.Q. *Nano Lett.*, 4, 2041, 2004. With permission.)

There are still many other techniques that have been used to synthesize Si nanocrystals, such as sputtering [30–32], reactive evaporation [33,34], mechanical ball milling followed by spin-coating [35] and aerosol synthesis [36–38]. All the fabrication techniques have their own advantages and disadvantages, though the purpose of all synthesizing methods is to achieve size-controlled Si nanocrystals. Except for the size of nc-Si, from the application point of view, researchers need to pay special attention to the size distribution and depth distribution of nc-Si in dielectric films. Methods mentioned above normally result in a relatively broad size distribution of synthesized nc-Si. For the ion implantation technique, size control is usually realized by shrinking the entire size distribution by varying the ion implant energy and dose [39,40] or by the subsequent oxidation of nanocrystals [20,41]. Note that a broad size distribution complicates the characterization of QC effects. In the case of optical characterization, this problem can be partially resolved by resonant excitation [42], which results in the narrowing of the size distribution of nc-Si that are optically excited.

4.2 DIELECTRIC FUNCTIONS AND OPTICAL CONSTANTS OF SILICON NANOCRYSTALS EMBEDDED IN A SiO_2 MATRIX

4.2.1 Introduction

In the material system of dielectric films embedded with nc-Si, the dielectric functions or optical constants of isolated nc-Si should be different from those of bulk crystalline silicon due to the size effect, and should be also different from those of a continuous Si nanocrystal film. Therefore, it would be interesting to examine the optical properties of isolated nc-Si embedded in SiO_2 matrix. It is obviously important to the fundamental physics as it is concerned with a system of quantum dots isolated by a dielectric matrix, and it is also necessary to the optoelectronic and photonic applications of the nc-Si.

A lot of investigations on optical properties of nc-Si have been carried out by luminescence experiments [12,14,22,30,43–52] and theoretical calculations [53–59]. However, there have been few optical studies so far to experimentally determine the dielectric functions or optical constants of nc-Si embedded in a dielectric matrix in a wide range of photon energy due to the complexity of the material system. Some experimental studies of the optical properties of a continued Si nanocrystal thin film [60] and SiO_2/nanocrystalline Si multilayers [61] have been reported recently. Nevertheless, a comprehensive experimental study of the optical properties in a wider photon energy range and a proper modeling to the optical properties are still lacking. Furthermore, few studies have been focused on optical properties of nc-Si prepared under different annealing conditions. Due to the reduced dimensionality, Si nanocrystals are expected to present size-dependent dielectric functions. However, only a few theoretical studies have been carried out on this significant issue [62–67]. In this section, a comprehensive study on the dielectric properties of nc-Si embedded in SiO_2 matrix by spectroscopic ellipsometry (SE) is presented [68].

4.2.2 Dielectric Functions of Dispersedly Distributed nc-Si Embedded in SiO_2

4.2.2.1 Experiment

To elaborate the methodology to determine the dielectric functions of nc-Si using SE, we take an example of nc-Si embedded in a SiO_2 matrix that is synthesized by Si^+-implantation with a dose of 1×10^{17} atoms/cm^2 at the energy of 100 keV into a 550-nm-thick SiO_2 film thermally grown on a *p*-type Si substrate. A thermal annealing at 1000°C for 30 min in nitrogen gas was carried out for the formation of Si nanocrystals. Figure 4.7 shows the high-resolution transmission electron microscopy (HRTEM) image of the nc-Si embedded in the SiO_2 matrix [68]. The average size of nc-Si determined from the broadening of the Bragg peak in x-ray diffraction (XRD) spectrum is ~4.2 nm, which is consistent with the HRTEM measurement. The SE measurements were carried out in the wavelength range of 250–1100 nm with a step of 5 nm, and the incidence angle was set to 75°. The nc-Si distribution in the

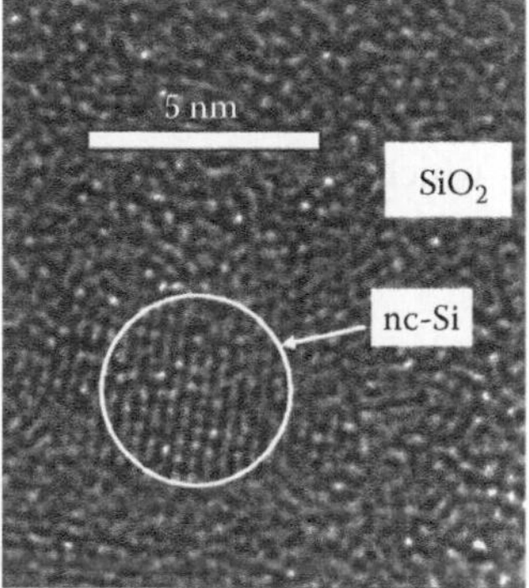

FIGURE 4.7 Cross-sectional HRTEM image of nc-Si embedded in SiO_2 matrix [68]. The sample was fabricated by Si ion implantation with the dose of 1×10^{17} cm^{-2} at 100 keV followed by annealing at 1000°C at N_2 for 30 min. (From Ding, L. et al., *Phys. Rev. B*, 72, 125419, 2005. With permission.)

SiO_2 thin film can be determined from secondary ion mass spectroscopy (SIMS) measurement. The SIMS intensity $I(x)$ due to the excess silicon in the Si^+-implanted region at a given depth x can be obtained by deducting the Si SIMS signal of the pure SiO_2 region from the measured total Si SIMS signal (which is from both the excess Si and the SiO_2) at the depth. The amount of excess Si at depth x should be proportional to the intensity $I(x)$. The volume fraction of nc-Si at the depth should be proportional to the amount of the excess Si at the depth. Thus the volume fraction $f(x)$ of the nc-Si embedded in SiO_2 at depth x can be expressed as

$$f(x) = \frac{QI(x)}{N_{Si}\int_0^{d_{max}} I(x)dx}, \tag{4.1}$$

where

Q is the dose of implanted Si ions in the unit of atoms/cm^2
d_{max} is the maximum depth in SiO_2 beyond which no excess Si can be detected
N_{Si} is the Si density in the unit of atoms/cm^3

In this example, Q is equal to 1×10^{17} atoms/cm^2 and N_{Si} is 5×10^{22} atoms/cm^3

4.2.2.2 SE Analysis

The volume fraction of nc-Si in the SiO_2 calculated from the SIMS measurement as a function of the depth is shown in Figure 4.8. As can be seen in this figure, the nc-Si distributes from the surface of the SiO_2 film to a depth of 250 nm and there is almost no nc-Si in the SiO_2 film beyond a depth of 250 nm. Therefore the thin film system can be divided into two layers, namely, the first layer ($0 \leq$ depth ≤ 250 nm) with nc-Si distributing in SiO_2 and the second layer (depth > 250 nm), which is just a basically pure SiO_2 layer without nc-Si. In the first layer, the optical properties vary with the depth as the volume fraction of the nc-Si varies with the depth. In order to model the optical properties of the first layer, it is divided into m sub-layers with equal thickness d_0 ($m=25$ and $d_0=10$ nm in this study). Each sub-layer has an nc-Si volume fraction (f_i, $i=1, 2,\ldots,m$), which can be calculated from the SIMS measurement, and the nc-Si volume fraction is considered to be constant within each sub-layer. Each sub-layer has its own effective dielectric function ε_i ($i=1, 2,\ldots,m$) due to its own nc-Si volume fraction. As such, the optical system in the SE analysis can be described with the multilayer model shown in Figure 4.8. Each sub-layer can be optically schematized as an effective medium in which the SiO_2 is the host matrix while the nc-Si is an inclusion embedded in the SiO_2 matrix, represented by the Maxwell-Garnett effective medium approximation (EMA)(68)

$$\frac{\varepsilon_i - \varepsilon_{SiO_2}}{\varepsilon_i + 2\varepsilon_{SiO_2}} = \frac{\varepsilon_{nc\text{-}Si} - \varepsilon_{SiO_2}}{\varepsilon_{nc\text{-}Si} + 2\varepsilon_{SiO_2}} f_i, \tag{4.2}$$

where

ε_i ($= N_i^2$ where N_i is the complex refractive index of the ith sub-layer) is the effective complex dielectric function of the ith sub-layer
ε_{SiO_2} is the dielectric function of SiO_2 matrix
$\varepsilon_{nc\text{-}Si}$ ($=(n_{nc\text{-}Si} - ik_{nc\text{-}Si})^2$ where $n_{nc\text{-}Si}$ and $k_{nc\text{-}Si}$ are the refractive index and extinction coefficient of the nc-Si, respectively) is the complex dielectric function of the nc-Si
f_i is the volume fraction of nc-Si in the ith sub-layer

As the volume fraction (f_i) and ε_{SiO_2} are known, from Equation 4.2 the effective complex dielectric functions ε_i (and thus the effective complex refractive index N_i) for the ith sub-layer ($i=1, 2,\ldots,m$) can be expressed in terms of $\varepsilon_{nc\text{-}Si}$ (or the refractive index and extinction coefficient of nc-Si). Therefore, in the SE analysis, the ellipsometric angles (Ψ and Δ) can be expressed as functions of the optical

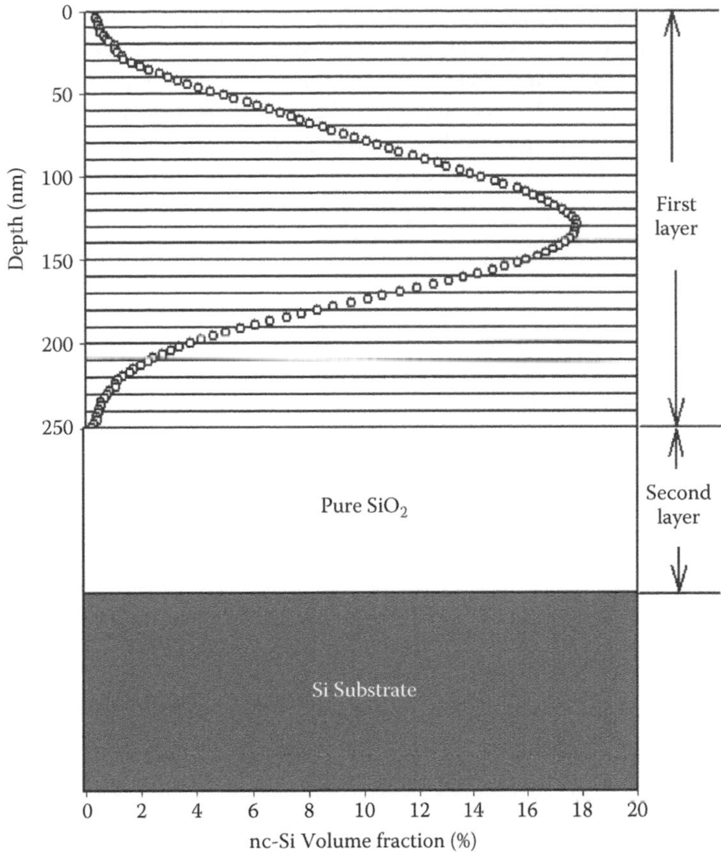

FIGURE 4.8 Multilayer model used in the SE analysis [68]. The nc-Si volume faction is calculated from SIMS measurement. (From Ding, L. et al., *Phys. Rev. B*, 72, 125419, 2005. With permission.)

constants of the nc-Si, although these functions cannot be displayed with analytical formulas due to their complexities. Based on these functions, a spectral fitting to the experimental data of Ψ and Δ can yield the dielectric function and optical constants of the nc-Si at various wavelengths.

In the spectral fitting, an appropriate optical dispersion model should be used to describe the spectral dependence of dielectric functions of the nc-Si. Here two optical dispersion models, the Lorentz oscillator model [69] and the four-term Forouhi–Bloomer (FB) model [70], were used to carry out the spectral fitting.

It is found that a combination of four Lorentz oscillators with different resonant energies is necessary to describe the spectral dependence of optical properties of the nc-Si embedded in SiO_2 over the wavelength range of 250–1100 nm. The Lorentz oscillator model can be expressed as [69]

$$\varepsilon(E) = \varepsilon_1(\infty) + \sum_{i=1}^{4} \frac{A_i}{E_i^2 - E^2 - i\Gamma_i E}, \tag{4.3}$$

where

$\varepsilon(\infty)$ refers to the dielectric constant at very large photon energies
A_i is the amplitude of the ith oscillator with the unit of $(eV)^2$
Γ_i is the damping factor of the ith oscillator with the unit of eV
E_i is the resonant energy with the unit of eV

Besides the Lorentz oscillator model, the four-term FB model is found to be another choice to get a reasonable spectral fitting. It should be pointed out that the FB model can yield not only the information of optical constants and dielectric functions but also the bandgap of the isolated nc-Si embedded in the SiO_2 matrix. The information of the nc-Si bandgap is very important because a bandgap expansion (if any) is a direct evidence of the QC effect. Based on the four-term FB model, the optical constants including the refractive index (n) and extinction coefficient (k) of the nc-Si are given by [70]

$$k(E) = \left(\sum_{i=1}^{4} \frac{A_i}{E^2 - B_i E + C_i} \right) \left(E - E_g\right)^2, \tag{4.4}$$

$$n(E) = n(\infty) + \sum_{i=1}^{4} \frac{B_{0_i} E + C_{0_i}}{E^2 - B_i E + C_i}, \tag{4.5}$$

where

$$B_{0_i} = \frac{A_i}{Q_i} \left(-\frac{B_i^2}{2} + E_g B_i - E_g^2 + C_i \right), \tag{4.6}$$

$$C_{0_i} = \frac{A_i}{Q_i} \left(\left(E_g^2 + C_i\right) \frac{B_i^2}{2} - 2 E_g C_i \right), \tag{4.7}$$

$$Q_i = \frac{1}{2} \left(4C_i - B_i^2\right)^{\frac{1}{2}}, \tag{4.8}$$

where
- A_i, B_i, and C_i (i = 1, 2, 3, and 4) are some parameters related to electron transition
- $n(\infty)$ is the refractive index when photon energy $E \rightarrow \infty$
- E_g is the energy bandgap of the nc-Si

The spectral fitting is carried out by freely varying the parameters of the models to minimize the following mean-square error (MSE)

$$\text{MSE} = \frac{1}{2N - M} \sum_{i=1}^{N} \left[\left(\frac{\psi_i^{cal} - \psi_i^{exp}}{\sigma \psi_i^{exp}} \right)^2 + \left(\frac{\Delta_i^{cal} - \Delta_i^{exp}}{\sigma \Delta_i^{exp}} \right)^2 \right], \tag{4.9}$$

where
- N is the number of data points in the spectra
- M is the number of variable parameters in the model
- σ is the standard deviations on the experimental data points
- ψ^{exp} and Δ^{exp} are the measured values of the ψ and Δ while ψ^{cal} and Δ^{cal} are the corresponding calculated values [70]

For an efficient spectral fitting, the initial values of parameters of the two models are taken to be equal to that of bulk crystalline silicon listed in Tables 4.1 and 4.2. An excellent spectral fitting

TABLE 4.1
Values of the Parameters A_i, E_i, Γ_i (i=1, 2, 3, 4), $\varepsilon_1(\infty)$ of the Lorentz Oscillator Model for Both Bulk Crystalline Silicon and the nc-Si Embedded in SiO_2

	A_i	E_i (eV)	Γ_i (eV^2)	$\varepsilon_1(\infty)$
Bulk crystalline silicon	10.1124	3.4423	0.1407	3.803
	45.9013	3.7005	0.4823	
	92.2078	4.3172	0.5289	
	11.5679	5.3233	0.3823	
Si nanocrystals embedded in SiO_2	30.8123	3.5560	0.4313	6.209
	13.7089	4.0151	0.3681	
	5.42090	4.0529	0.2242	
	10.3012	5.16135	0.8173	

Source: Ding, L. et al., *Phys. Rev. B*, 72, 125419, 2005. With permission.

TABLE 4.2
Values of the Parameters A_i, B_i, and C_i (i=1, 2, 3, 4), $n(\infty)$ and E_g of the FB Model for Both Bulk Crystalline Silicon and the nc-Si Embedded in SiO_2

	A_i	B_i (eV)	C_i (eV^2)	$n(\infty)$	E_g (eV)
Bulk crystalline silicon	0.0036	6.8811	11.8486	2.3688	1.12
	0.0140	7.4010	13.7473		
	0.0683	8.6340	18.7952		
	0.0496	10.234	26.5029		
Si nanocrystals embedded in SiO_2	0.0538	7.1119	12.7176	2.8237	1.7369
	0.0056	8.0157	16.0797		
	0.0603	8.0300	18.7101		
	0.0003	10.323	33.6447		

Source: Ding, L. et al., *Phys. Rev. B*, 72, 125419, 2005. With permission.

based on the above approach in a wide wavelength range of 250–1100 nm has been obtained, as shown in Figure 4.9, in which the best-fit spectra based on the FB model and the Lorentz oscillator model are both included. As can be seen in this figure, all the complicated spectral features of both Ψ and Δ are fitted excellently.

4.2.2.3 Dielectric Functions and Optical Constants of nc-Si

Using the values of the parameters of the nc-Si shown in Tables 4.1 and 4.2, the complex dielectric function and the optical constants of the nc-Si are calculated with Equation 4.3 for the Lorentz oscillator model and Equations 4.4 and 4.5 for the FB model, respectively. The dielectric functions and optical constants of the nc-Si are shown in Figures 4.10 and 4.11, respectively, in which the results calculated with the above two optical dispersion models are both included. The dielectric function and the optical constants of bulk crystalline Si are also included in the two figures for comparison. As can be seen in Figures 4.10 and 4.11, the overall spectral features of

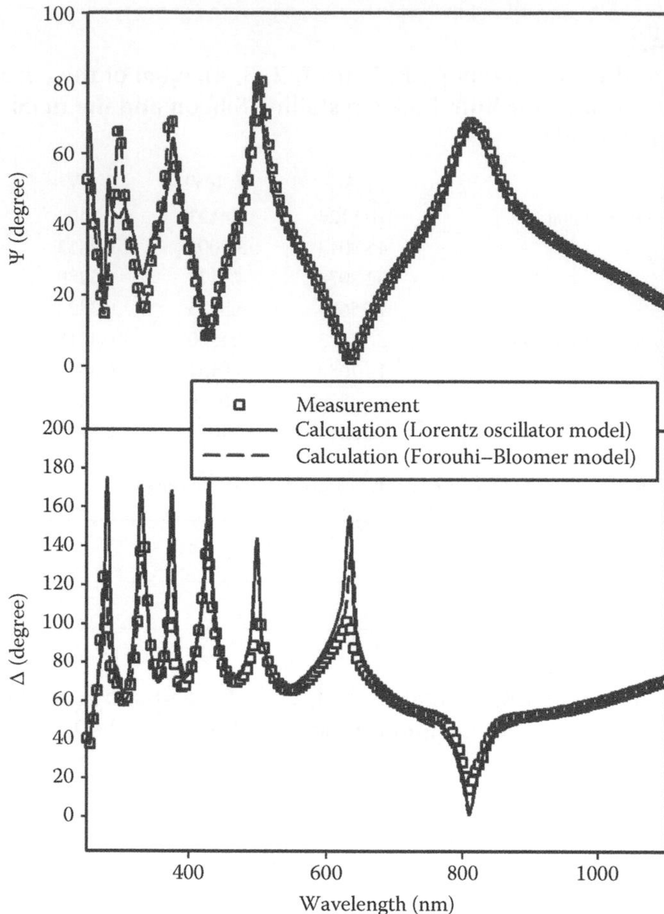

FIGURE 4.9 Best spectral fittings of ψ and Δ based on the Lorentz oscillator model and the four-term FB model with the approach described in the text. (From Ding, L. et al., *Phys. Rev. B*, 72, 125419, 2005. With permission.)

optical properties of nc-Si are similar to that of bulk crystalline Si. However, the nc-Si shows a significant reduction in the dielectric function and optical constants as compared with bulk crystalline Si. This is also indicated by the parameters of Lorentz oscillator modeling for the dielectric functions of nc-Si. In general, the dielectric functions of bulk crystalline Si show main peaks at the transition energies E_1 (~3.4 eV), E_2 (~3.7 eV), E_3 (~4.3 eV), and E_4 (~5.3 eV) as its critical points as listed in Table 4.1. However, in the case of nc-Si, the spectral peaks below 4 eV, i.e., the two lower-frequency transition peaks, undergo slight blueshifts in E_1 (~3.6 eV) and E_2 (4.0 eV), while the two higher-frequency transition peaks show slight redshift in E_3 (4.1 eV) and E_4 (5.2 eV). These peak positions are almost in coincidence with the corresponding values of $B_i/2$ ($i = 1, 2, 3$, and 4) of the four-term FB model. This good agreement in the peak energy positions between the two models underlines the validity of the modeling. The fact of the blueshift of the lower transition energies and the redshift of the higher transition energies could be considered as a trend that the different transition energies tend to merge as the size of Si becomes nanoscale. Such consideration is supported by the studies of Amans [60] and Lee [61] in which the dielectric function and optical constants of Si nanocrystals have only single peak in the spectral ranges of interest. The tendency of transition energies to merge is possibly due to the large broadening and significant weakening of strength of some oscillators. The trend of single-peak structure is

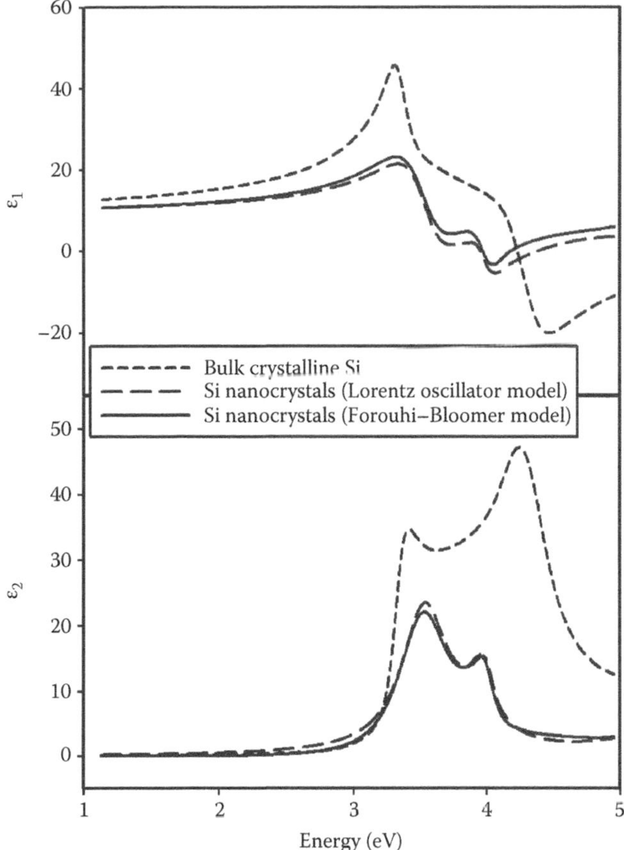

FIGURE 4.10 Real (ε_1) and imaginary (ε_2) parts of the complex dielectric functions of the nc-Si obtained from the spectral fitting based on the Lorentz oscillator model and the FB model [68]. The dielectric functions of bulk crystalline are also included for comparison. (From Ding, L. et al., *Phys. Rev. B*, 72, 125419, 2005. With permission.)

also reported by Rossow et al. [71] and Nguyen et al. [72], for porous silicon and nc-Si layer, respectively.

4.2.2.4 Bandgap of nc-Si

On the other hand, based on the plot of $(\alpha E)^\gamma$ versus E, where α is the absorption coefficient of the nc-Si and E is the photon energy, one can examine whether the nc-Si is a direct ($\gamma=2$) or indirect ($\gamma=1/2$) bandgap semiconductor [73]. As an example, Figure 4.12a shows the Tauc plot ($\gamma=1/2$) of nc-Si annealed at 1000°C for 20 min, and the plot for bulk crystalline Si is also included for comparison [74]. With $\gamma=1/2$, a linear relationship in the photon energy range near the absorption edge is observed for both the nc-Si and bulk crystalline Si, as shown in Figure 4.12a. In contrast, as shown in Figure 4.12b, the linearity is much poorer when $\gamma=2$ for both nc-Si and bulk crystalline silicon. This suggests that the nc-Si has an indirect bandgap structure. As shown in Figure 4.12a, the bandgap of the nc-Si obtained by extrapolating the linear portion of the plot is 1.75 eV, which is almost the same as the value (1.74 eV) given in Table 4.2 obtained from the SE analysis based on the FB model. This means that the nc-Si embedded in SiO_2 has a bandgap expansion of ~0.6 eV as compared to bulk crystalline Si. The bandgap obtained in this work is in good agreement with the first-principle calculation of the optical gap of Si nanocrystals based on QC [58]. A fit to the calculation shown in Figure 3 of the paper by Ogut et al. [58] yields

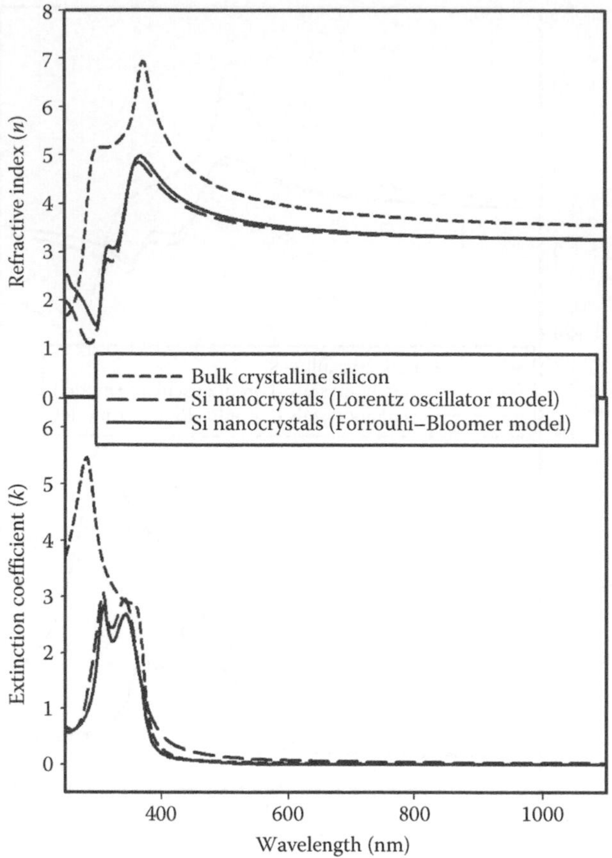

FIGURE 4.11 Refractive index (n) and extinction coefficient (k) of the nc-Si and bulk crystalline Si as functions of wavelength. (From Ding, L. et al., *Phys. Rev. B*, 72, 125419, 2005. With permission.)

$$E_g(D) = E_{g0} + C/D^n, \tag{4.10}$$

where

- D is the nanocrystal size in nm
- $E_g(D)$ is the bandgap in eV of the nanocrystal
- $E_{g0} = 1.12\,\text{eV}$ is the bandgap of bulk crystalline Si, C=3.9, and n=1.22

For the nc-Si size of ~4.2 nm of this study, Equation 4.10 gives a bandgap of ~1.79 eV, which is close to the bandgap values mentioned above.

4.2.2.5 Dielectric Suppression

It has been well established that a reduction of the static dielectric constant becomes significant as the size of the quantum confined physical systems, such as quantum dots and wires, approaches the nano-metric range [62,75–77]. However, the origin of the reduction in static dielectric constant with the size is still not fully understood. It is often attributed to the opening of the gap that should lower the polarizability, but it is also shown that the reduction is due to the breaking of polarizable bonds at the surface and is not due to the opening of the bandgap induced by the confinement [78].

Nevertheless, the reduction in the static dielectric constant is also observed in the above SE study. The static dielectric constant of the dispersedly distributed nc-Si embedded in SiO_2 matrix is found to be 9.7, which is obtained from the SE analysis based on the four-term FB model by setting the photon energy to zero. Taking the screening effect by the medium into account, Wang et al.

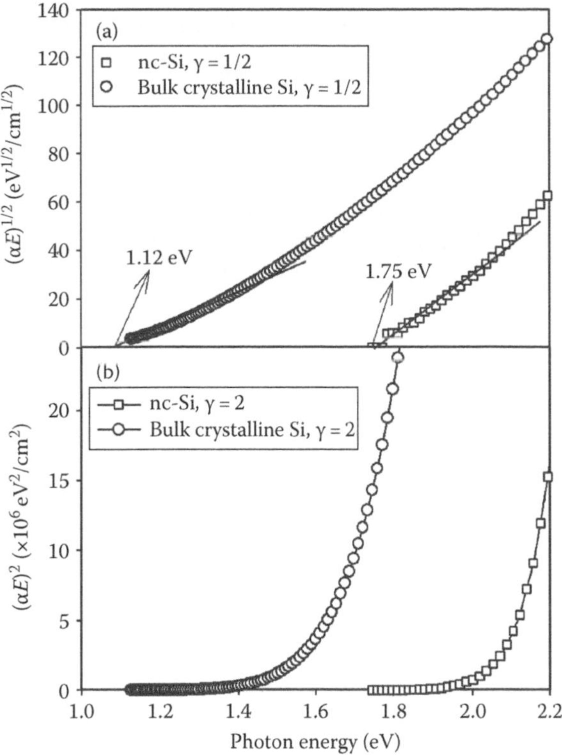

FIGURE 4.12 Plots of $(\alpha E)^{\gamma}$ versus photon energy (E) for both the nc-Si synthesized at 1000°C and bulk crystalline silicon. (a) $\gamma = 1/2$ and (b) $\gamma = 2$. (From Ding, L. et al., *J. Nanosci. Nanotechnol.*, 8, 3555, 2008. With permission.)

pointed out that the static dielectric constant of the nc-Si as a function of the nanocrystal size could be expressed as follows [62]

$$\varepsilon_r(D) = 1 + \frac{\varepsilon_r(\infty) - 1}{1 + \left(\frac{6.9}{R}\right)^{1.37}}, \tag{4.11}$$

where

$\varepsilon_r(\infty)$ (=11.4) is the static dielectric constant of bulk crystalline silicon
R is the radius of nc-Si with the unit of angstrom

The static dielectric constant of nc-Si with the diameter of 4.5 nm calculated with Equation 4.11 is 9.7, which is equal to the value obtained from the SE analysis based on the FB model.

4.2.3 Depth Profile of Optical Constant of SiO_2 Films Containing nc-Si

In optoelectronic device applications of nc-Si, it is necessary to know the depth profiles of optical constants of SiO_2 films containing Si nanocrystals synthesized with the ion implantation technique in order to perform reliable device modeling. As the dielectric functions and optical constants of nc-Si embedded in SiO_2 matrix have been determined with the SE analysis, it is now convenient to calculate the optical constants of each sub-layer based on the EMA theory. Therefore, the depth profiles of optical constants of the film can be determined [79].

Based on Equation 4.2, the refractive index n_i and extinction coefficient k_i for the ith (i=1, 2, …, 25) sub-layer can be obtained as

$$n_i = n_{SiO_2}\left[\frac{\sqrt{(A^2+B^2)(C^2+D^2)}+(AC+BD)}{2(C^2+D^2)}\right]^{1/2}, \tag{4.12}$$

and

$$k_i = n_{SiO_2}\left[\frac{\sqrt{(A^2+B^2)(C^2+D^2)}-(AC+BD)}{2(C^2+D^2)}\right]^{1/2}, \tag{4.13}$$

where

$$A = (1+2f_i)(n_{\text{nc-Si}}^2 - k_{\text{nc-Si}}^2) + 2(1-f_i)n_{SiO_2}^2, \tag{4.14}$$

$$B = 2(1+2f_i)n_{\text{nc-Si}}k_{\text{nc-Si}}, \tag{4.15}$$

$$C = (1-f_i)(n_{\text{nc-Si}}^2 - k_{\text{nc-Si}}^2) + (2+f_i)n_{SiO_2}^2, \tag{4.16}$$

$$D = 2(1-f_i)n_{\text{nc-Si}}k_{\text{nc-Si}}. \tag{4.17}$$

In the above equations, $n_{\text{nc-Si}}$ and $k_{\text{nc-Si}}$ are the mean refractive index and extinction coefficient of the nc-Si, and n_{SiO_2} is the refractive index of SiO_2. Note that all the parameters are dependent on wavelength. By using the nc-Si optical constants obtained in Section 4.2.2.2, the refractive index n_i and extinction coefficient k_i for the ith ($i=1, 2,\ldots, 25$) sub-layer can be calculated with Equations 4.12 and 4.13, respectively, and thus the depth profiles of optical constants (n and k) for SiO_2 thin film containing Si nanocrystals are obtained.

Figure 4.13 shows the three-dimensional view of the refractive index and extinction coefficient as functions of both depth and wavelength for the sample used in Section 4.2.2.2. As shown in this figure, the peaks of both the refractive index and extinction coefficient are located at the depth of ~120 nm, and in the region deeper than 250 nm, the optical constants are basically the

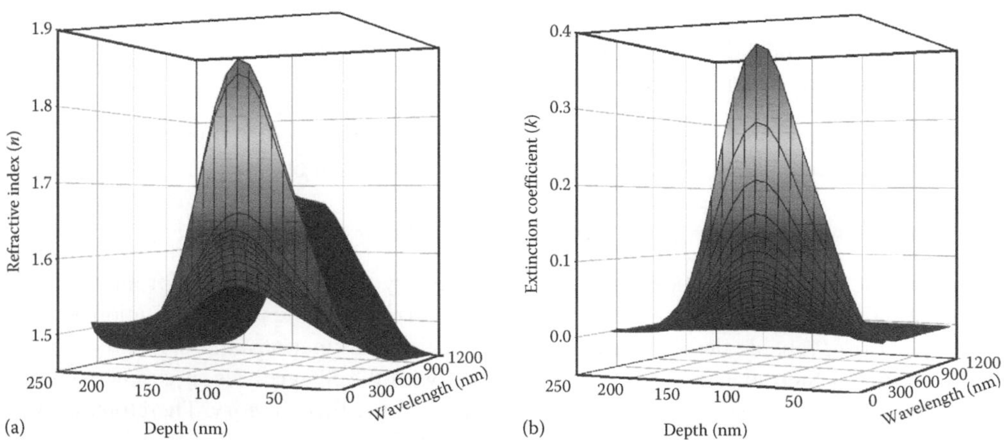

FIGURE 4.13 Depth profiles of optical constants of a SiO_2 film containing Si nanocrystals synthesized with Si ion implantation.

same as that of SiO_2. It can be expected that such a structure containing nc-Si in SiO_2 film with a proper nc-Si distribution has a graded refractive-index profile that can offer light-trapping ability and thus the possibility of waveguiding. For a clear demonstration of the light-trapping ability of this structure, we replot the figure in a two-dimensional style as shown in Figure 4.14 [79]. Figure 4.14 shows the spectral dependence of refractive index and extinction coefficient at the depths of 20, 50, 90, and 120 nm. For comparison, the optical constants of pure SiO_2 are also included in the figure. As can be seen in this figure, for any wavelength both the refractive index and extinction coefficient increase with the depth. However, they decrease when the depth > ~120 nm (for simplicity, the optical constants beyond the depth of 120 nm are not shown in Figure 4.14). The sharp change of the optical constants in the wavelength range of 250–350 nm is due to the spectral features of the nc-Si optical constants in this wavelength range. Figure 4.15 shows the depth dependence of $\Delta n/n_{SiO_2}$ (where Δn is the difference between the refractive index at a given depth and that of pure SiO_2) for the wavelengths of 250, 300, 350, 400, 600 and 900 nm [79]. As shown in this figure, for all the wavelengths, the maximum refractive index is located at the depth of about 120 nm due to the maximum nc-Si concentration at this depth, and the refractive indices are basically the same as that of SiO_2 in the region deeper than 250 nm. It is clear from Figure 4.15 that the maximum change of refractive index due to the existence

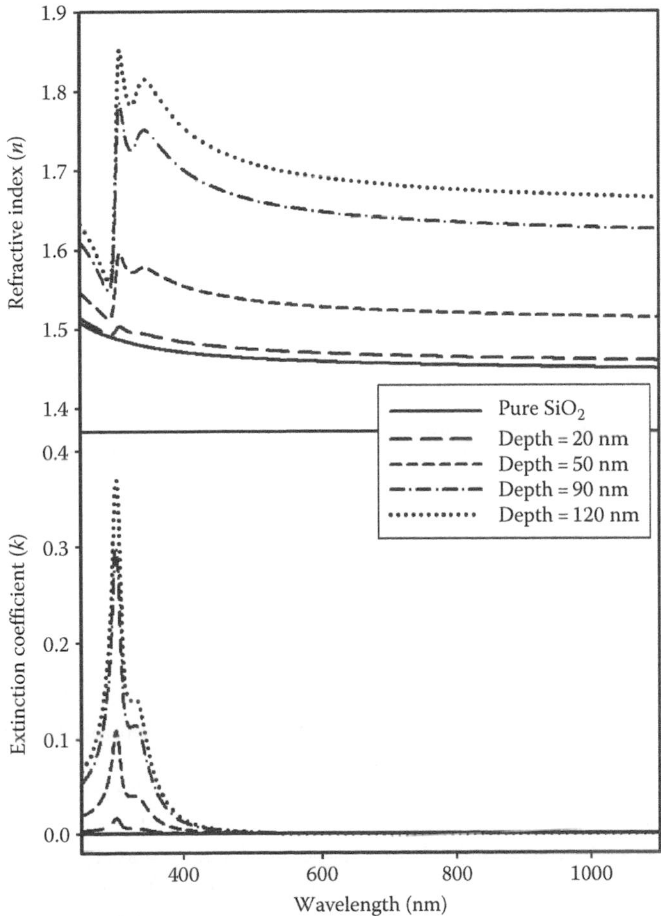

FIGURE 4.14 Optical constants of the film structure as functions of wavelength at the depths of 20, 50, 90, and 120 nm. The optical constants of pure SiO_2 are also included for comparison. (From Ding, L. et al., *Nanotechnology*, 16, 2657, 2005. With permission.)

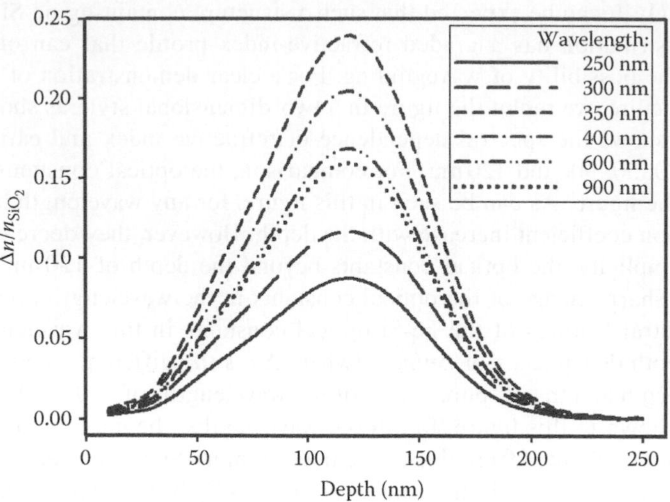

FIGURE 4.15 Depth profile of $\Delta n/n_{SiO_2}$ for various wavelengths. (From Ding, L. et al., *Nanotechnology*, 16, 2657, 2005. With permission.)

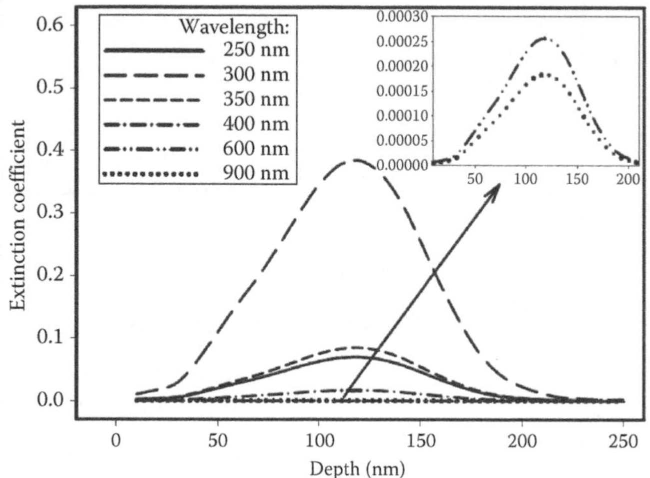

FIGURE 4.16 Depth profile of extinction coefficient for various wavelengths. (From Ding, L. et al., *Nanotechnology*, 16, 2657, 2005. With permission.)

of nc-Si occurs at the wavelength of ~350 nm. This is because of the peak structure of nc-Si refractive index at the wavelength of about 350 nm. For the wavelengths longer than 350 nm, the changes in refractive index due to the nc-Si decrease with wavelength because of the decrease in the nc-Si refractive index with wavelength. Figure 4.16 shows the depth profile of extinction coefficient for the same wavelengths as in Figure 4.15 [79]. As can be seen in this figure, for short wavelengths (< ~400 nm), absorption occurs only in the region of 0 < depth < 250 nm with the strongest absorption at the depth of ~120 nm. However, for long wavelengths (> ~400 nm), there is almost no absorption everywhere including in the region of 0 < depth < 250 nm. Therefore, it can be concluded from Figures 4.15 and 4.16 that for long wavelengths (> ~400 nm) there is a large refractive index but almost no absorption in the region of 0 < depth < 250 nm. This suggests that the thin film structure has the capability of waveguiding with low optical loss in the visible to infrared range.

4.2.4 Thermal Annealing Effect on Dielectric Functions of Si Nanocrystals

Very few studies have been focused on dielectric functions of dispersedly distributed nc-Si embedded in SiO_2 synthesized with Si ion implantation under different annealing conditions. In this section, the effect of post-implantation thermal annealing on both the bandgap and the dielectric functions of the nc-Si embedded in SiO2 is discussed [84]. Post-implantation thermal annealings are carried out in nitrogen ambient at various temperatures for different annealing times. For simplicity, here only the results of annealing at 1000°C for various annealing durations ranging from 0 to 100 min (0, 20, 30, 40, 60, 80 and 100 min) and the results of annealing for 20 min at different temperatures (500°C, 700°C, 850°C, 900°C and 1000°C) are presented.

The average size of nc-Si can be determined from the broadening of the Bragg peak in XRD measurement. The nc-Si size obtained is from ~3.3 nm for the as-implanted sample to ~4.5 nm for the samples annealed at 1000°C for different durations, and from ~3.3 nm for the as-implanted sample to ~4.6 nm for the samples annealed for 20 min at different annealing temperatures. The nc-Si sizes for various annealing conditions obtained from the XRD measurements are shown in Table 4.3. A very small increase in the nc-Si size with annealing time and temperature is observed. The slow nc-Si size evolution with thermal annealing is similar to the results of the paper by Lopez et al. [80] who reported that the nc-Si size increases slightly from 2.5 nm after 1 min annealing to 3 nm after annealing for 16 h at the temperature of 1100°C. This slow evolution in the size is due to the very low diffusion coefficient (10^{-16} to 10^{-18} cm/s) of Si in SiO_2 film [81–83]. The implanted Si distributes from the surface to the depth of ~250 nm, and the subsequent annealing does not change the profile significantly because of the extremely low diffusion coefficient of Si in SiO_2 films.

Based on the method discussed in Section 4.2.2, the dielectric functions of nc-Si have been obtained from the best spectral fitting. The nc-Si bandgap (E_g) obtained from the spectral fittings for various annealing durations is given in Table 4.3. The nc-Si shows a large expansion in the bandgap as compared to the bulk crystalline Si. The nc Si bandgap obtained in this work is in good agreement with the values calculated with Equation 4.10, which is obtained from the fitting result of the

TABLE 4.3
Size, Bandgap, and Static Dielectric Constant of nc-Si Embedded in SiO_2 Matrix Annealed for Different Durations

Annealing Conditions	Size (nm)	E_g^{cal} (eV)	E_g^{exp} (eV)	ε_s^{cal}	ε_s^{exp}
As-implanted	3.3±0.4	2.03±0.15	1.78	8.98±0.34	8.82
1000°C for 20 min	4.6±0.2	1.73±0.03	1.74	9.72±0.08	9.67
1000°C for 30 min	4.2±0.2	1.79±0.05	1.74	9.54±0.10	9.52
1000°C for 40 min	4.5±0.2	1.74±0.04	1.74	9.68±0.09	9.73
1000°C for 30 min	4.3±0.2	1.78±0.04	1.75	9.59±0.10	9.73
1000°C for 30 min	4.5±0.2	1.74±0.04	1.74	9.68±0.10	9.61
1000°C for 30 min	4.5±0.2	1.74±0.04	1.72	9.68±0.10	9.73
500°C for 20 min	3.3±0.4	2.03±0.15	1.78	8.98±0.34	8.82
700°C for 20 min	3.1±0.4	2.10±0.16	1.84	8.82±0.32	8.78
850°C for 20 min	3.9±0.2	1.86±0.07	1.76	9.38±0.12	9.29
900°C for 20 min	4.1±0.2	1.82±0.06	1.76	9.46±0.11	9.41

Note: E_g^{exp} and ε_s^{exp} are the bandgap and the static dielectric constant of nc-Si obtained from the spectral fittings based on the four-term FB model, respectively. E_g^{cal} is the nc-Si bandgap obtained from the calculation with Equation 4.10, while ε_s^{cal} is the nc-Si static dielectric constant obtained from the calculation with Equation 4.11.

first-principle calculation of the nc-Si optical gap based on QC. For comparison, the bandgap of nc-Si calculated with Equation 4.10 is presented in Table 4.3. The good agreement shown in Table 4.3 suggests that the bandgap expansion is a result of quantum size effect. It can be also concluded from this table that the annealing does not cause a significant change in the nc-Si bandgap because there is no large change in the nc-Si size.

As examples, the dielectric functions of nc-Si obtained after 0 (non-annealing), 30, and 80 min of annealing are shown in Figure 4.17 [84]. The dielectric functions of bulk crystalline Si are included in the figure for comparison. Obviously, the nc-Si exhibits a significant reduction in the dielectric functions with respect to bulk crystalline silicon. It is observed that the dielectric-function spectra of the nc-Si of different annealing durations except the non-annealing sample are very similar. This is not surprising, and it can be explained in terms of quantum size effect on the dielectric functions [62,85]. As shown in Table 4.3, except the non-annealing sample that shows the nc-Si size of ~3.3 nm, all other samples with different annealing durations have almost the same nc-Si size (~4.5 nm). In general, the dielectric functions of bulk crystalline Si show main peaks at the transition energies E_1 (~3.4 eV) and E_2 (~4.3 eV) as its critical points. As for the case of nc-Si

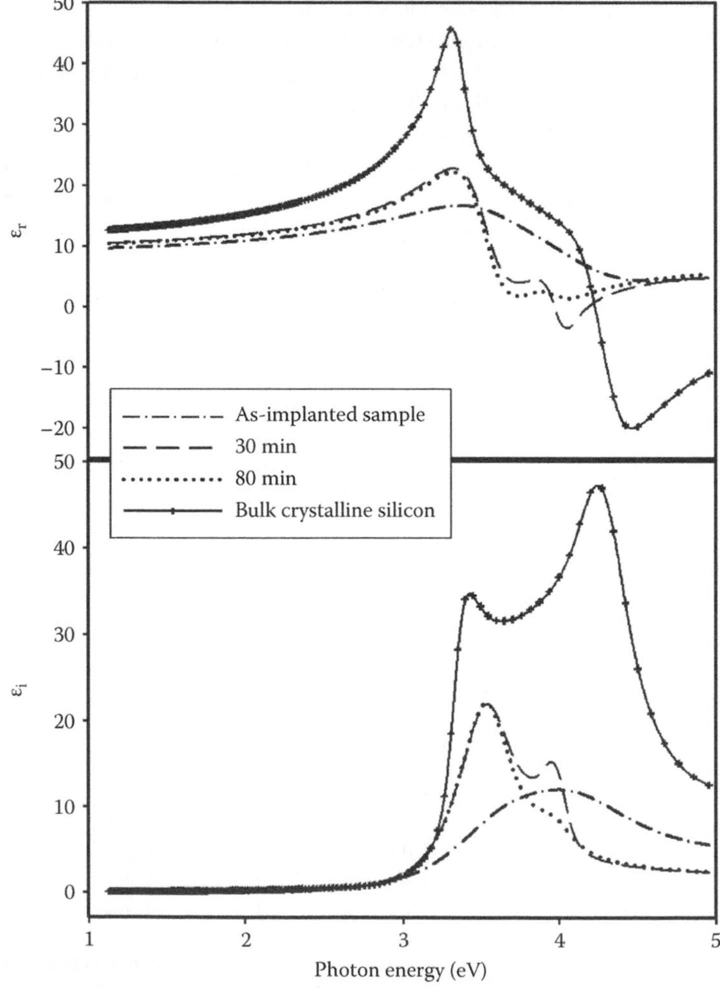

FIGURE 4.17 Real (ε_r) and imaginary (ε_i) parts of the complex dielectric functions of both the nc-Si annealed for 0, 30, and 80 min at 1000°C and the bulk crystalline silicon as functions of photon energy. (From Ding, L. et al., *Appl. Phys. Lett.*, 87, 121903, 2005. With permission.)

formed after high temperature annealing, the dielectric spectra show a slight blueshift in the main transition energy E_1 (~3.6 eV) position, and a slight redshift in E_2 (~4.0 eV) position. However, in the case of as-implanted (i.e., non-annealing) sample, the dielectric functions of nc-Si show only a single peak in the experimental photon-energy range, possibly because of larger broadenings of E_1 and E_2 due to a significant change of surface area to volume ratio when the size of nc-Si is being close to 3 nm.

In order to investigate the effect of annealing temperature on dielectric functions of nc-Si, the samples annealed for 20 min at various temperatures (500°C, 700°C, 800°C, 850°C, 900°C and 1000°C) were also studied with SE. For simplicity, only the dielectric functions of nc-Si for the as-implanted sample and the samples annealed at 700°C and 900°C are shown in Figure 4.18. As can be seen in Table 4.3, the average size of nc-Si for 500°C annealing sample is approximately the same as that of the as-implanted sample. When the annealing temperature goes up greater than 500°C, e.g., 700°C, 900°C, and 1000°C, the dielectric functions and optical constants will have a distinct change due to the slightly increasing size with annealing, as shown in Figure 4.18. This indicates that a high temperature (>500°C) annealing is needed for the evolution of nc-Si size. On the

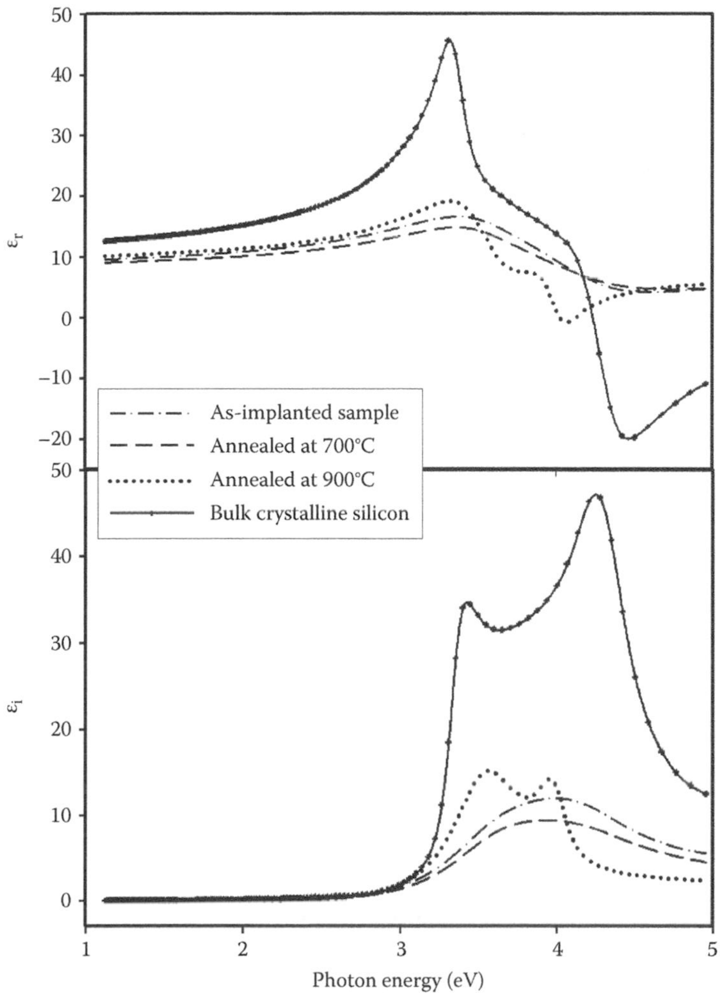

FIGURE 4.18 Real (ε_r) and imaginary (ε_i) parts of the complex dielectric functions of both the nc-Si of as-implanted sample and annealed at 700°C and 900°C, for 20 min and the bulk crystalline silicon as functions of photon energy.

other hand, the static dielectric constant (ε_s^{exp}) of nc-Si is obtained by setting the photon energy to zero in the FB model, and the influence of annealing on the static dielectric constant is also shown in Table 4.3.

4.2.5 Size Effect on Dielectric Functions and Bandgap of Silicon Nanocrystals

Si nanocrystals with different sizes ranging from 4.6 to 7.8 nm embedded in SiO_2 are synthesized with various implantation recipes (i.e., different implantation energies and doses). The influence of nanocrystal size on dielectric functions and bandgap of the nc-Si is discussed in this section.

SiO_2 films with thickness of 550 nm were grown on Si substrate by wet oxidation at 1000°C. Afterward, a high dosage of Si ions (10^{16}–10^{17} atoms/cm^2) was implanted into the SiO_2 films at various energies ranging from 100 to 3 keV (100, 50, 18, 5, and 3 keV) at room temperature, and the samples are denoted as sample 1, sample 2,…, and sample 5, respectively. Implantations were conducted at 7° off axis to reduce channeling effect. Nc-Si were formed after high temperature annealing at 1000°C for 20 min in N_2 ambient. The average size of nc-Si was determined from the broadening of Bragg peak in the XRD spectrum. Table 4.4 gives the sizes of nc-Si for all the

TABLE 4.4
Values of the Parameters A_i, B_i, and C_i (i = 1, 2, 3, 4), $n(\infty)$ and E_g of the Four-Term FB Model for the Five Samples

	nc-Si Size (nm)	A_i	B_i (eV)	C_i (eV^2)	$n(\infty)$	E_g (eV)
Sample 1	4.6	0.0534	7.1121	12.7223	2.8143	1.7371
		0.0057	8.0148	16.0834		
		0.0605	8.0312	18.7121		
		0.0003	11.3674	33.6447		
Sample 2	5.3	0.0458	7.1119	12.7176	2.8237	1.6864
		0.0089	8.0157	16.0797		
		0.0843	8.3300	18.7101		
		0.0113	10.3227	33.6447		
Sample 3	5.8	0.0043	6.8850	11.8641	2.0179	1.6312
		0.0154	7.5102	14.1610		
		0.0652	8.7143	19.1612		
		0.1854	10.5212	29.2021		
Sample 4	6.3	0.0145	7.0851	12.7275	2.0211	1.5312
		0.0146	7.4057	13.7881		
		0.0653	8.7143	19.1601		
		0.2154	10.5312	29.2021		
Sample 5	7.8	0.0041	6.8854	11.8654	1.9498	1.4503
		0.0149	7.4013	13.7545		
		0.0767	8.6341	18.8134		
		0.2117	10.6517	29.8456		
Bulk crystalline silicon		0.0036	6.8811	11.8486	2.3688	1.12
		0.014	7.4013	13.7473		
		0.0683	8.6348	18.7952		
		0.0496	10.2339	26.5029		

Source: From Ding, L. et al., *J. Appl. Phys.*, 101, 103525, 2007. With permission.
Note: The corresponding values of bulk crystalline Si are also included.

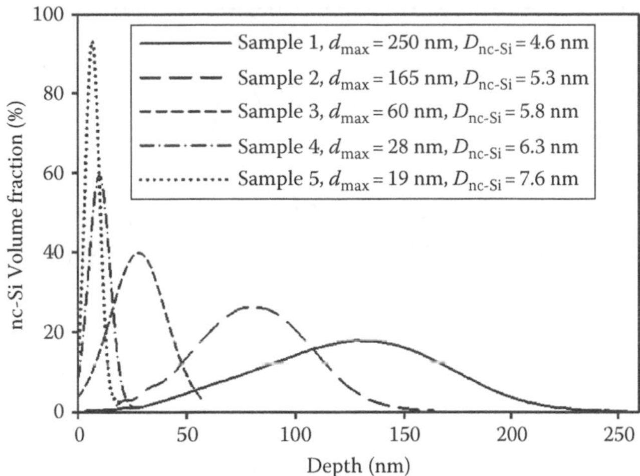

FIGURE 4.19 Volume fractions of nc-Si in SiO_2 as a function of depth for the five samples obtained from SRIM simulation. (From Ding, L. et al., *J. Appl. Phys.*, 101, 103525, 2007. With permission.)

samples (i.e., samples 1 to 5). Figure 4.19 shows the depth profiles of the volume fraction of excess Si in SiO_2 for the five samples [86].

Using the methodology described in Section 4.2.2.2, the dielectric functions of the nc-Si with various sizes embedded in SiO_2 are obtained for the five samples, and they are shown in Figure 4.20 [86]. The parameters including A_i, B_i, C_i (i=1, 2, 3, and 4), $n(\infty)$ and E_g of the four-term FB model for the nc-Si with various sizes are given in Table 4.4. Table 4.4 clearly shows that the nc-Si exhibits a large expansion in the bandgap as compared to that of the bulk crystalline Si and the bandgap of the nc-Si increases when the nc-Si size is reduced. For example, for the nc-Si with a size of 4.6 nm (sample 1), it has a bandgap of 1.74 eV that is significantly larger than the bandgap (1.1 eV) of the bulk crystalline silicon. The bandgap expansion is the most direct evidence of QC effect of nc-Si. The bandgap expansion of nc-Si and its dependence on the nanocrystal size have been demonstrated by some theoretical calculations of the bandgap of nc-Si [58,62]. A calculation of the bandgap expansion with Equation 4.10 is shown in Figure 4.21, and the calculation is also compared with the bandgap expansions obtained from the SE analysis discussed above [86]. A good agreement in the comparison can be seen in the figure.

4.2.6 Dielectric Functions of Densely Stacked nc-Si Layer Embedded in SiO_2

With the Si^+-implantation technique, isolated Si nanocrystals dispersedly distributed in a SiO_2 matrix [43,87,88] or densely stacked Si nanocrystal layers [89–92] embedded in SiO_2 can be formed depending on the implantation recipe. The former could be used for Si optoelectronic applications while the latter could be used in memory devices. The optical properties of the densely stacked nanocrystal layer should be different from that of both the isolated nc-Si and the continuous nc-Si film. It is necessary to investigate the optical properties of the densely stacked nanocrystal layer because of the importance of such nanoscale structure in both fundamental physics and applications. In practice, the Si ion implantation into a thin SiO_2 film with a high dose (typically in the range of ~10^{16} to ~10^{17} cm^{-2}) at a very low energy (≤2 keV) can form such densely stacked nanocrystal layers embedded in SiO_2 [89–92]. In this section, the dielectric functions of the densely stacked Si nanocrystal layer embedded in SiO_2 are determined by SE fittings, and the influence of thermal annealing on the dielectric functions is also examined [93].

SiO_2 thin films of 30 nm were thermally grown in dry oxygen at 950°C on a *p*-type Si (100) wafer. Si ions with a dose of 8×10^{16} cm^{-2} were then implanted into the SiO_2 thin films at 1 keV.

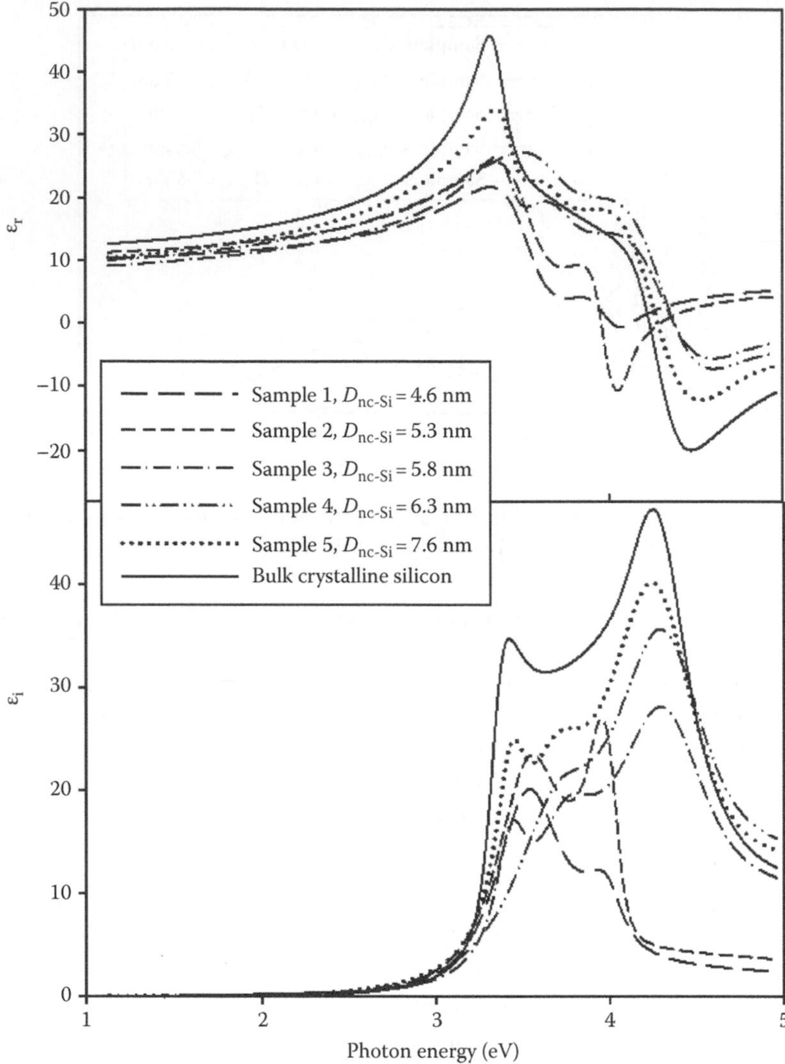

FIGURE 4.20 Real (ε_1) and imaginary (ε_2) parts of the complex dielectric function of the nc-Si with various sizes obtained from the spectral fittings. The dielectric function of bulk crystalline silicon is also included for comparison. (From Ding, L. et al., *J. Appl. Phys.*, 101, 103525, 2007. With permission.)

Post-implantation thermal annealing was carried out in N_2 at various annealing temperatures for different durations. The mean size of nc-Si was estimated from full width at half-maximum (FWHM) of the Bragg in the XRD measurement. The average size of nanocrystals was estimated to be ~3–4 nm. It was found that the nc-Si size changes little with annealing. This observation is consistent with those reported in literature [80,91]. The cross-sectional transmission electron microscopy (TEM) measurement reveals the existence of a densely stacked Si nanocrystal layer with a thickness of ~16 nm embedded in SiO_2, as shown in Figure 4.22 [93]. The sample is annealed at 1000°C for 20 min (i.e., sample *a*).

For the SE analysis, an appropriate optical model is required. As a densely stacked Si nanocrystal layer is formed in the SiO_2 thin film, the model developed for the isolated nc-Si embedded in a SiO_2 matrix that has been developed in the above sections is inadequate for this study. In the present study, a single layer of densely stacked nc-Si is embedded in the SiO_2 thin film as revealed by the TEM image shown in Figure 4.22, and the situation is different from that

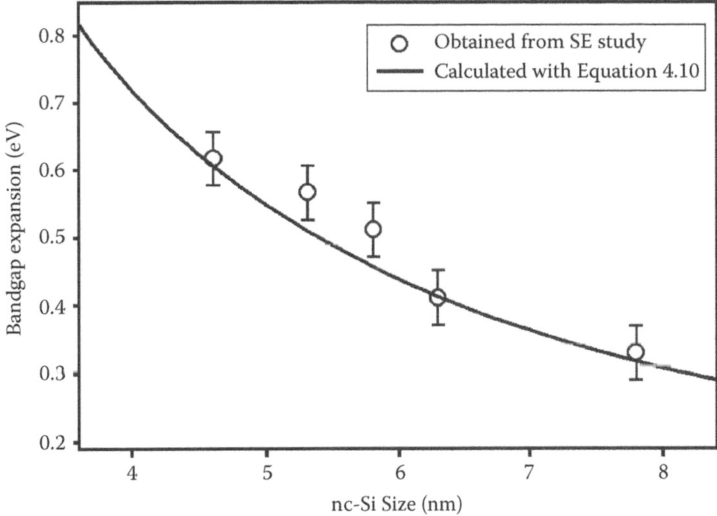

FIGURE 4.21 Bandgap expansion of nc-Si as a function of nc-Si size. (From Ding, L. et al., *J. Appl. Phys.*, 101, 103525, 2007. With permission.)

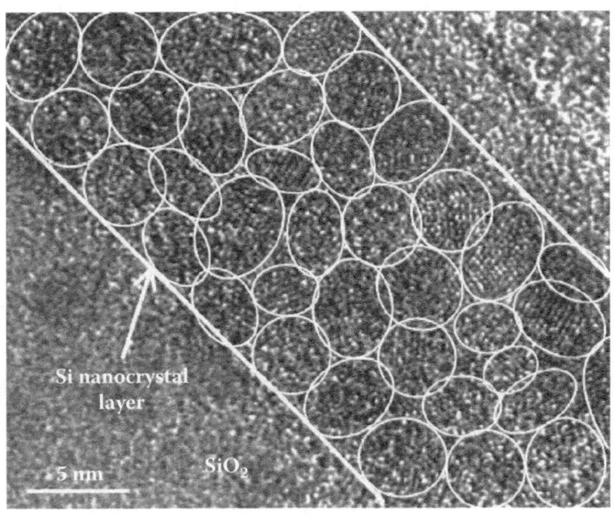

FIGURE 4.22 Cross-sectional TEM image of the densely stacked Si nanocrystal layer embedded in SiO_2 thin film for sample *a*. (From Ding, L. et al., *Appl. Phys. Lett.*, 89, 251910, 2006. With permission.)

of the isolated nc-Si dispersedly distributed in a SiO_2 matrix. Obviously, this densely stacked nc-Si layer should be treated as a phase in the ellipsometric analysis. Therefore, a five-phase model, namely, air/SiO_2 layer/densely stacked nc-Si layer/SiO_2 layer/Si substrate, which is shown in Figure 4.23a, was employed to carry out the SE fittings in the present study [93]. Note that in this model, the ellipsometric angles (Ψ and Δ) are functions of the thicknesses and dielectric functions of all the layers. Excellent fittings with meaningful outputs have been achieved with this model. Note that no optical dispersion model was required for the SE fittings with the model. As an example, Figure 4.23b shows the spectral fitting of Ψ and Δ for sample *a* (i.e., the sample annealed at 1000°C for 20 min) [93]. As can be seen in this figure, all the complicated spectral features of both Ψ and Δ can be fitted excellently, indicating that the model shown in Figure 4.23a is effective.

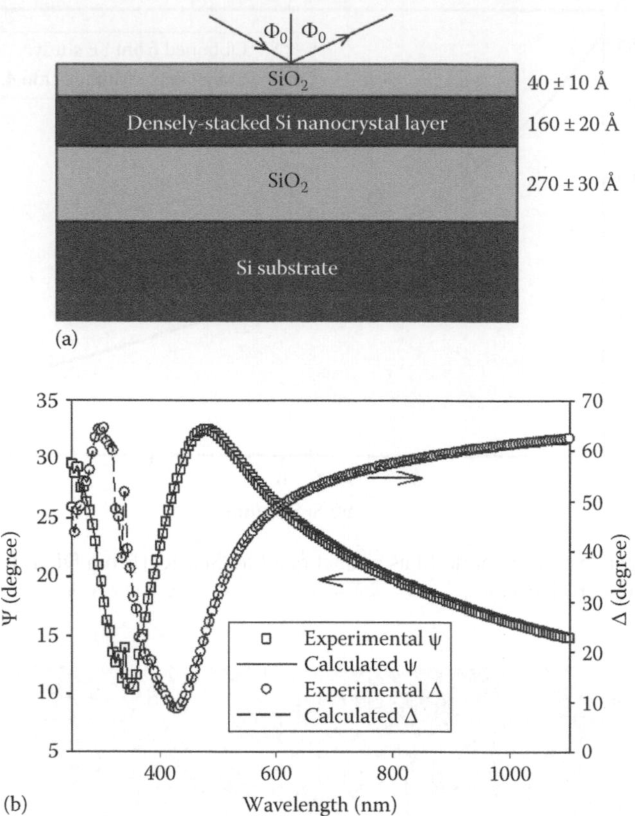

FIGURE 4.23 (a) Five-phase model used in the SE analysis. (b) Spectral fittings of Ψ and Δ for sample *a*. (From Ding, L. et al., *Appl. Phys. Lett.*, 89, 251910, 2006. With permission.)

Figure 4.24 shows the dielectric functions of the densely stacked Si nanocrystal layer for sample *a* [93]. For comparison, the dielectric functions of other Si materials are also included in Figure 4.24. As can be seen in the figure, the dielectric spectra of the densely stacked Si nanocrystal layer are similar to that of amorphous Si but different from those of both bulk crystalline Si and the isolated nc-Si whose dielectric functions have been investigated in Section 4.2.2. For the imaginary part of dielectric functions, the densely stacked Si nanocrystal layer and amorphous Si show a single broadened peak; in contrast, bulk crystalline Si and the isolated nc-Si exhibit a two-peak structure. The peak structures in the dielectric spectra are believed to originate from singularities in the joint density of states (DOS). Essentially, the DOS in the amorphous state is a broadened version of crystalline state, which leads to a single broad peak in the dielectric spectra of amorphous semiconductors. Therefore, the single-peak structure may suggest that the nanocrystal layer is in an amorphous state to certain extent. On the other hand, the nanocrystal layer also shows significant reductions in the amplitudes of dielectric functions as compared with bulk crystalline Si and amorphous Si, which is related to the size effect of the nanocrystals.

Furthermore, a significant influence of the annealing on the dielectric functions of the densely stacked Si nanocrystal layer is observed. Figure 4.25 shows the imaginary part of dielectric functions of the densely stacked silicon nanocrystal layer synthesized at a fixed annealing temperature of 1000°C for different annealing durations [93]. The dielectric function changes with the annealing duration, showing a trend toward bulk crystalline Si. Firstly, the dielectric function increases with the annealing duration. Secondly, the dielectric function exhibits a clear two-peak structure when the annealing duration is 30 min, and the structure is further enhanced when the annealing duration is 80 min. The structure is analogue to that of bulk Si that has two peaks at the transition energies

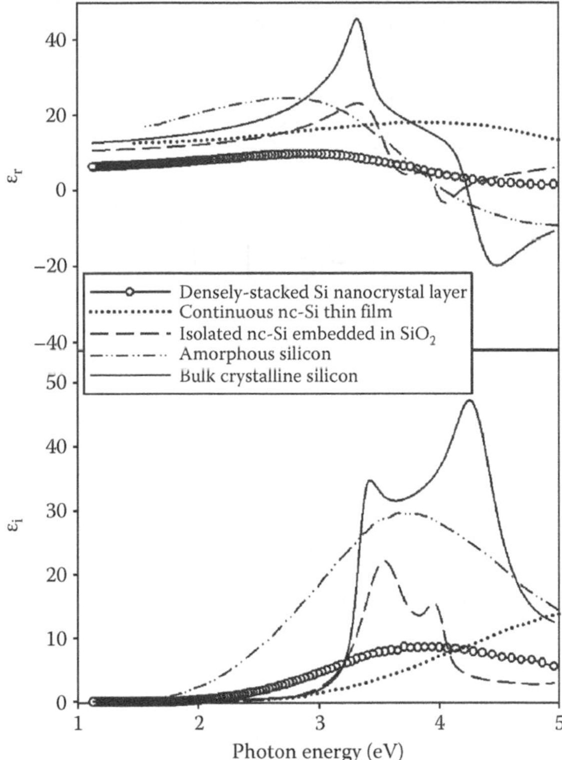

FIGURE 4.24 Real (ε_1) and imaginary (ε_2) parts of the complex dielectric functions of the densely stacked Si nanocrystal layer for sample *a*. For comparison, the dielectric functions of the continuous nc-Si thin film [60], the isolated nc-Si dispersedly distributed in a SiO_2 matrix, amorphous Si and bulk crystalline Si are also included in this figure. (From Ding, L. et al., *Appl. Phys. Lett.*, 89, 251910, 2006. With permission.)

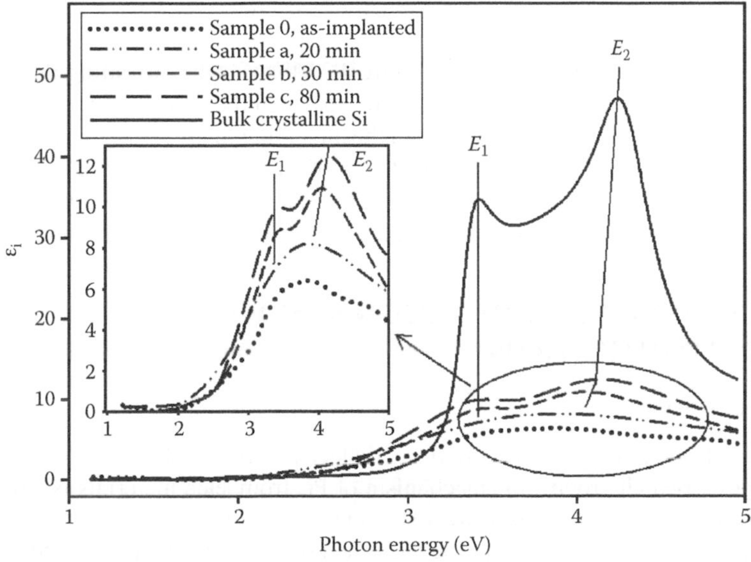

FIGURE 4.25 Influence of annealing duration on the imaginary (ε_i) part of the complex dielectric functions of the densely stacked Si nanocrystal layer. The annealing temperature is fixed at 1000°C. (From Ding, L. et al., *Appl. Phys. Lett.*, 89, 251910, 2006. With permission.)

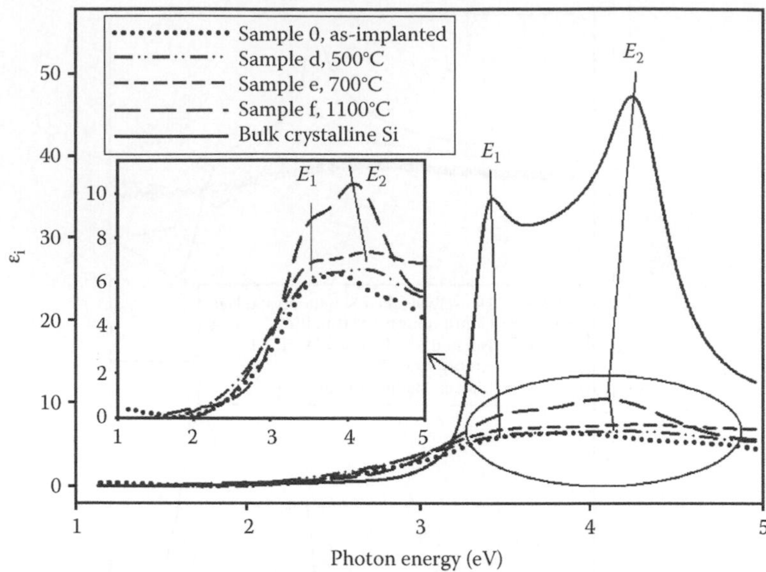

FIGURE 4.26 Influence of annealing temperature on the imaginary (ε_i) part of the densely stacked Si nanocrystal layer. The annealing duration is 20 min. (From Ding, L. et al., *Appl. Phys. Lett.*, 89, 251910, 2006. With permission.)

E_1 (~3.4 eV) and E_2 (~4.3 eV) as its critical points. E_1 of the nanocrystal layer is about the same as that of bulk crystalline Si, and it changes very little with the annealing duration. However, E_2 of the nanocrystal layer is slightly lower than that of bulk Si, but it shows a small blueshift toward that of bulk Si with the increase in annealing duration. However, it should be pointed out that extending the annealing duration should not result in a dielectric function that is approaching that of bulk crystalline Si because of the quantum size effect of the nc-Si and the difference in the structure between the two materials. A similar annealing effect on the dielectric functions is also observed from the experiment of various annealing temperatures. Figure 4.26 shows the evolution of the dielectric functions with annealing temperature for the fixed annealing duration of 20 min [93]. One can observe from Figure 4.26 together with Figure 4.25 that for the annealing duration of 20 min, only a single broad peak in the dielectric spectra exists when the annealing temperature is 1000°C or lower, but a two-peak structure emerges when the annealing temperature is 1100°C. Therefore, it is clear that thermal annealing promotes the evolution of the dielectric functions toward that of bulk crystalline Si. The mechanism for the annealing effect is not known yet, but one may attribute it to the crystallization and or the growth of the nanocrystals caused by the annealing.

4.3 PHOTOLUMINESCENCE OF SILICON NANOCRYSTALS EMBEDDED IN SiO_2 FILMS

4.3.1 Introduction

Although efficient room-temperature PL from dielectric films embedded with nc-Si has been reported by many research groups, the mechanism of PL from such materials is still under debate. For example, for the system synthesized with Si ion implantation, the visible-light (in particular the red-light) luminescence was attributed to the QC effect of the nc-Si [14,43,47,49,51,52,57,87,88,94–96]. However, it was also suggested that the short-wavelength luminescence was related to the presence of the oxygen vacancies in the Si^+-implanted dielectric matrix [97–103]. Thanks to the large amount of previous studies, it is certain that there are multiple mechanisms responsible for the visible luminescence. However, it is still controversial on the evolution of PL mechanisms with thermal annealing. This section presents an investigation of PL properties of nc-Si embedded in SiO_2 films [74].

4.3.2 Experiments

550-nm SiO_2 films were grown on *p*-type Si wafers with (100) orientation by thermal oxidation in dry oxygen. A dose of 1×10^{17} cm^{-2} of Si ions was implanted into the SiO_2 films at the energy of 100 keV at room temperature. Afterward, post-implantation thermal annealing was carried out in nitrogen ambient at various temperatures ranging from 500°C to 1100°C (500°C, 600°C, 700°C, 850°C, 900°C, 1000°C, 1100°C) for 20 min. No significant change in the nc-Si size under different annealing temperatures was observed from the XRD measurement. This could be due to the very low diffusion coefficient of Si in SiO_2 films [80], and/or the relatively short annealing duration. Both the XRD and TEM measurements show the formation of nc-Si with a mean size of ~4.5 nm embedded in the SiO_2 matrix. For the study of size effect, various nc-Si sizes were synthesized using different implantation recipes: (a) 3×10^{16} ions/cm^2 at 120 keV, (b) 1×10^{17} ions/cm^2 at 100 keV, (c) multiple implantations with 5×10^{15} ions/cm^2 at 20 keV, 1×10^{16} ions/cm^2 at 40 keV, and 2×10^{16} ions/cm^2 at 80 keV, (d) 4×10^{16} at 10 keV, and (e) multiple implantations with 3×10^{15} ions/cm^2 at 1 keV, 8×10^{15} ions/cm^2 at 5 keV, and 4×10^{16} ions/cm^2 at 16 keV. The PL spectra of the samples were measured using the 325-nm line of a He-Cd laser as the excitation source. The spectra were recorded at room temperature with a photomultiplier tube (PMT) detector.

4.3.3 Annealing Effect

For the study of annealing effect on the PL properties of Si^+-implanted SiO_2 films, the samples implanted at 100 keV with 1×10^{17} cm^{-2} are used. In the study of different annealing temperatures, the samples were annealed for 20 min, and the annealing temperature was varied from 500°C to 1100°C as given above. For the study on the effect of different annealing times on the PL spectra, the annealing temperature was fixed at 1100°C, and the annealing time was varied from 20 to 150 min. The bandgaps of nc-Si synthesized at various annealing temperatures were obtained from the SE analysis. It can be observed in Figure 4.27 that the nc-Si bandgap, which is related to the

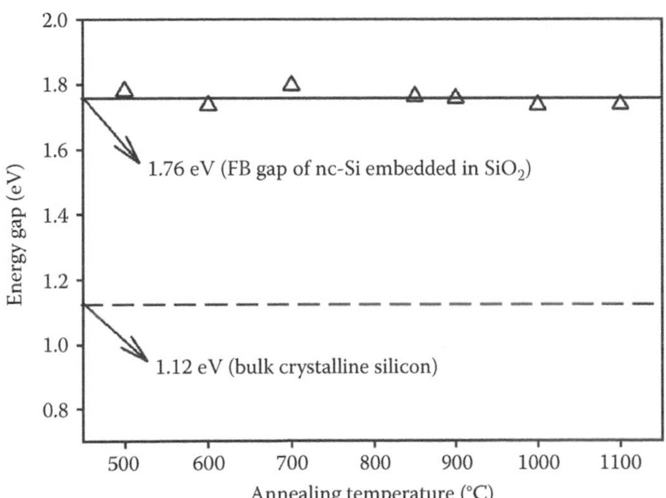

FIGURE 4.27 Bandgap of the nc-Si embedded in SiO_2 versus annealing temperature [74]. The bandgap was extracted from the FB model in the SE analysis. (From Ding, L. et al., *J. Nanosci. Nanotechnol.*, 8, 3555, 2008. With permission.)

optical transitions, is not significantly affected by the annealing [74]. As discussed early, the Tauc plots indicate that nc-Si has an indirect bandgap structure and the annealing does not change the band structure and bandgap of nc-Si.

Visible light emission is observed with the excitation wavelength of 325 nm at room temperature. All the samples show a very broad PL spectrum, suggesting that it may be composed of several different PL bands. Figure 4.28 shows the PL spectra for the samples annealed at various temperatures ranging from 500°C to 1100°C [74]. The PL spectrum changes significantly when the annealing temperature increases from 500°C to 1100°C. It is found that the PL spectra of the samples annealing at 500°C, 600°C, and 700°C have a similar shape while the PL spectra of the samples annealing at above 700°C show some different traits. Particularly, the PL peak intensity of the sample annealed at 1100°C is almost 30 times stronger than that of the samples annealed at temperatures below 1100°C. The PL spectra of all the samples can be well represented by six Gaussian-shaped peaks centered at the wavelengths of ~415, ~460, ~520, ~630, ~760, and ~845 nm, respectively, which have been frequently reported and widely accepted in the literature. The full widths at half maximum (FWHM) of the six bands are ~50, ~64, ~125, ~149, ~118, and ~50 nm, respectively. The FWHMs of the six PL bands do not show a significant change with the annealing temperature. The decomposition of the PL spectra into the six Gaussian-shaped peaks is shown in Figure 4.28 [74]. Figure 4.29 shows the dependence of annealing temperature on the integrated PL intensity for each band. As can be observed in this figure, the ~415-, ~460-, and ~520-nm PL bands have a similar behavior, i.e., the integrated intensity first increases slightly and then decreases dramatically with the increase of the annealing temperature. The integrated intensity for the ~415-, ~460-, and ~520-nm bands reaches a maximum at 700°C, 600°C, and 700°C, respectively. The above three bands almost disappear when the annealing temperature exceeds 900°C. The ~630-nm band always exists for annealing temperatures up to 1000°C, and no significant change in the intensity is observed with increasing annealing temperature. In contrast, the integrated PL intensity of both the ~760- and ~845-nm bands continuously increases with the annealing temperature. Their integrated intensities increase slightly with the annealing temperature up to 1000°C and then experience a dramatic increase by approximately 30 times when the annealing temperature reaches 1100°C.

As discussed early, the annealing in the temperature range of 500°C–1100°C does not change both the indirect band structure and the bandgap of the nc-Si. This contrasts with the very large changes in the PL discussed above. Therefore, one may conclude that not all the PL bands observed in this study originate from the QC effect of the nc-Si and some PL bands can be ascribed to the oxide matrix. It is well known that various defects in the oxide matrix can serve as visible luminescent centers. In the following discussions, the PL bands at ~415, ~460, ~520, and ~630 nm are attributed to the defects in the oxide matrix. As for the disappearance or steep decrease of these four PL bands when the annealing temperature is approaching 1100°C, a possible explanation is that the high annealing temperature leads to the elimination of most of the luminescent defects in the oxide.

To study the effect of annealing time on the PL spectra, the annealing temperature is fixed at 1100°C. The PL spectra for the samples annealed at 1100°C for 20, 40, and 120 min are presented in Figure 4.30. As clearly seen in the figure, the PL peak position does not change with the annealing time, but the intensity increases with the annealing time. Figure 4.31 shows the integrated PL intensity as a function of annealing time. As can be seen in this figure, the dependence of the PL intensity on the annealing time has two distinct regimes. The intensity first increases rapidly with the annealing time when the annealing time is shorter than 60 min, and then it increases slowly until its saturation after a certain annealing time. This behavior can be understood based on the theory of Ostwald ripening of nanocrystal population [81,104]. When the annealing time is sufficiently long, the ripening stage of crystal growth and elimination of non-radiative defects is reached leading to a much slower increasing rate or even constant PL intensity.

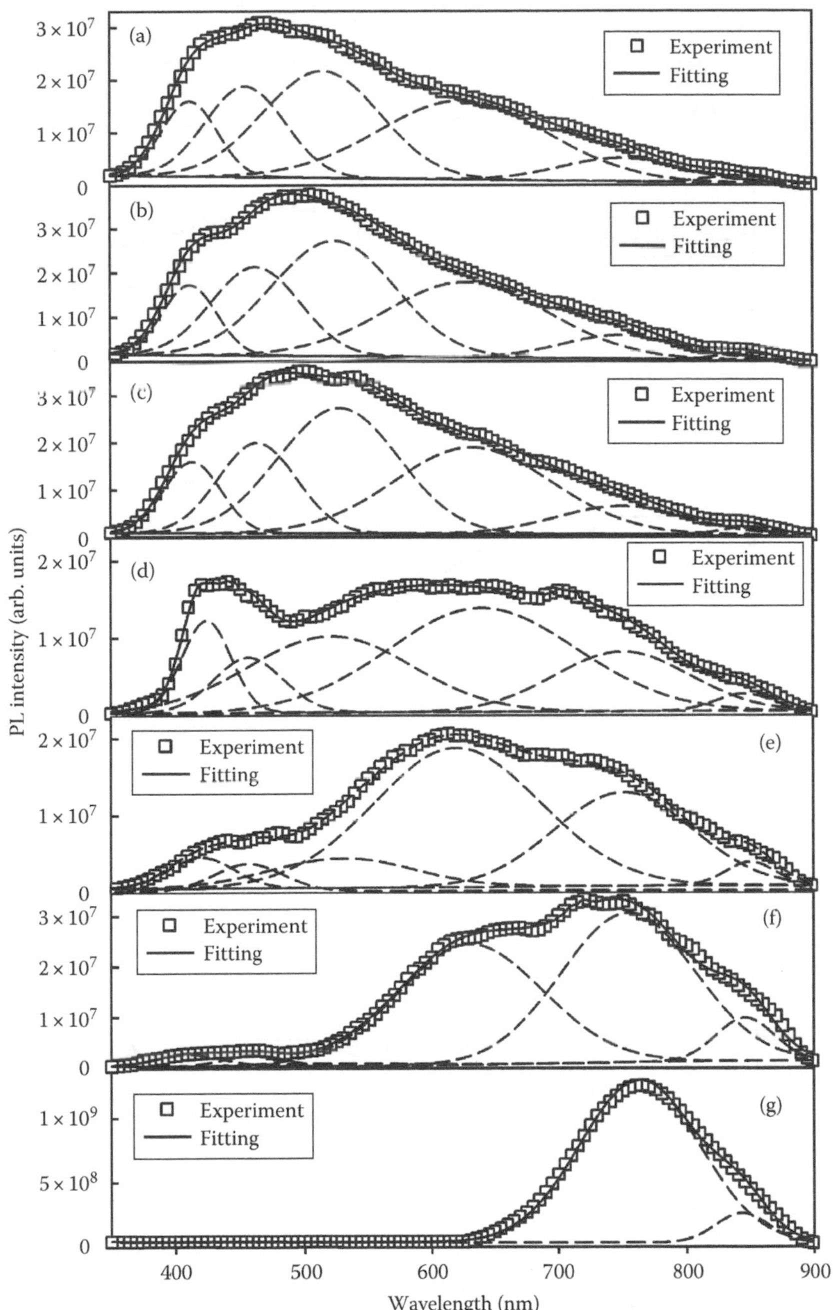

FIGURE 4.28 Decomposition of PL spectra for the samples of Si$^+$-implanted SiO$_2$ annealed at (a) 500°C, (b) 600°C, (c) 700°C, (d) 850°C, (e) 900°C, (f) 1000°C, and (g) 1100°C for 20 min. (From Ding, L. et al., *J. Nanosci. Nanotechnol.*, 8, 3555, 2008. With permission.)

4.3.4 PL Mechanisms

The 415-nm (~3 eV) PL band can be attributed to the weak oxygen bond (WOB) defects in silicon oxide [105–107]. The WOB defects are generated in the Si$^+$-implanted SiO$_2$ layers after a suitable thermal annealing. Oxygen vacancies are induced by the displacement of oxygen caused by the Si$^+$-implantation into a SiO$_2$ network, and the oxygen interstitials, which are

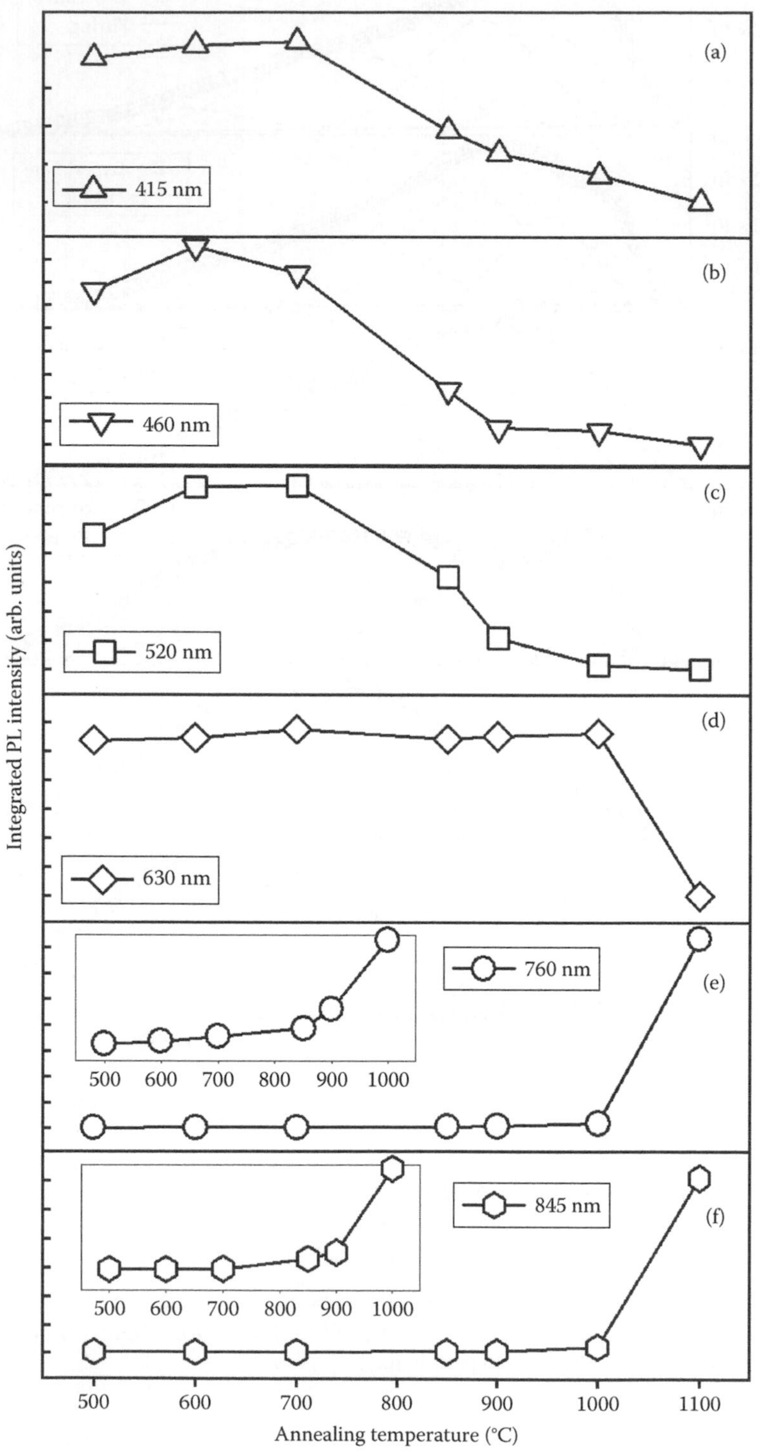

FIGURE 4.29 Integrated intensity of each PL band as a function of annealing temperature. (From Ding, L. et al., *J. Nanosci. Nanotechnol.*, 8, 3555, 2008. With permission.)

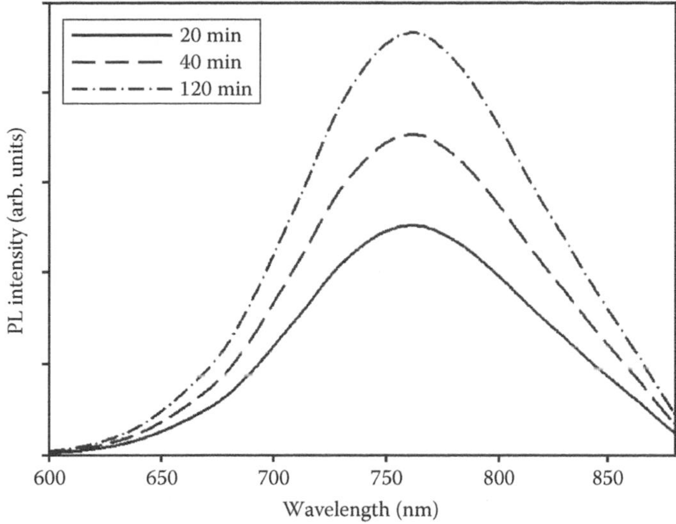

FIGURE 4.30 PL spectra for the samples annealed at 1100°C for various durations.

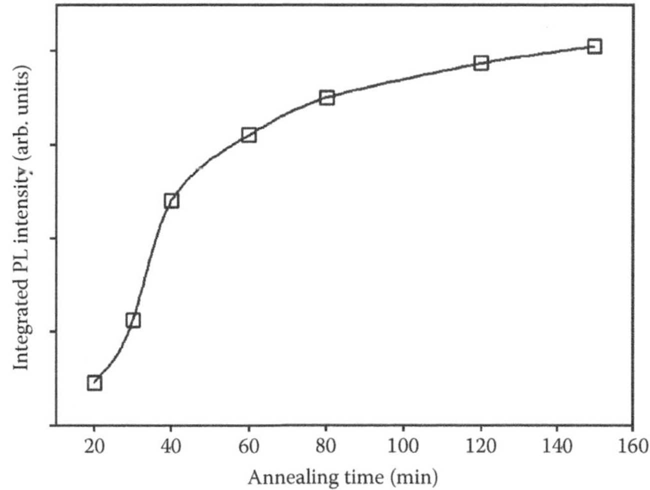

FIGURE 4.31 Integrated PL intensity as a function of annealing time. The annealing temperature is 1100°C.

considered as the precursors of the WOB defects, are created concurrently. The reaction can be represented by $O_3 \equiv Si - O - Si \equiv O_3 \rightarrow O_3 \equiv Si - Si \equiv O_3 + O_{interstitial}$. The oxygen interstitials will change into the WOB defects immediately after annealing, as described by the reaction $O_{interstitial} + O_{interstitial} \rightarrow$ O–O, which is a reversible reaction subject to an excessive thermal annealing energy. The reverse reaction is indicated by the decrease of the 415-nm PL band with the increase of annealing temperature when the temperature is higher than 700°C. The ~460-nm (~2.7 eV) PL band has been observed from pure silica glass [108] and Si-rich silicon oxides [53,56,101]. The ~460-nm PL band was shown to be associated with the neutral oxygen vacancy (NOV) defect represented by $O_3 \equiv Si - Si \equiv O_3$ [101,105,109]. The NOV defect could be observed in ion-implanted or radiation-damaged SiO_2 networks. As for the Si^+-implanted SiO_2

film, the NOV defects are actually generated in the same process as for the WOB defects, as described above. Therefore, it is not surprising that the ~415- (WOB) and ~460-nm (NOV) PL bands have a very similar dependence on the annealing temperature. The origin of the ~520-nm band is thought to be associated with the E'_δ defect [105,110]. In Si^+-implanted SiO_2, E'_δ defects can be generated by both extrinsic ion-implantation-induced dissociation process [105,106] and intrinsic UV photon absorption during the PL measurement [111]. The generation of E'_δ defects caused by ion implantation is a transformation of an NOV defect to a E'_δ center, as described by $O_3 \equiv Si - Si \equiv O_3 \rightarrow O_3 \equiv Si \bullet^+ Si \equiv O_3 + e^-$ [105,110]. As can be seen in Figure 4.29, the WOB-related 415-nm band, the NOV-related ~460-nm band, and the E'_δ-related 520-nm band are observed in the same temperature range, and they have a similar dependence on annealing temperature. This indicates that the extrinsic ion-implantation-induced dissociation process plays an important role in the generation of E'_δ defects. Another radiative defect, the non-bridging oxygen hole center (NBOHC) that has been observed in both pure silica glass and ion-irradiated SiO_2 [106,112–115], could be responsible for the ~630-nm (~2.0 eV) PL band observed in this study. As shown in Figure 4.29d, the intensity of the ~630-nm band changes little when the annealing temperature is increased until 1000°C, indicating that the NBOHC defects are stable during the thermal annealing when the annealing temperature is below 1100°C. However, the steep decrease in the intensity of the ~630-nm PL band for the annealing at 1100°C suggests that the NBOHC defects have been largely reduced at this annealing temperature.

In contrast to the above PL bands, the ~760 and ~845-nm PL bands both show a drastic increase in the peak intensities when the annealing temperature is 1100°C, suggesting that these two bands have luminescence mechanisms different from the defects in the oxide matrix discussed above. As the peak energy (~1.63 eV) of the 760-nm band is close to the FB gap (~1.76 eV) of the nc-Si, one may easily link this PL band to the QC effect. Indeed, the QC effect has been frequently used to explain the PL from nc-Si, although there are many debates. However, the difference of ~0.13 eV (i.e., the Stokes shift) between the energy of the ~760-nm PL peak and the bandgap has prompted us to rethink the QC mechanism for the ~760-nm band. As discussed above, the nc-Si embedded in the SiO_2 matrix has an indirect band structure. The probability of direct transitions is extremely low because of the indirect band structure of nc-Si. On the other hand, the role of the nc-Si/SiO_2 interface should not be overlooked in such a system in which the nanocrystals are embedded in the oxide matrix. It is interesting to note that the Stokes shift is about the same as the energy (~0.13 eV) of the Si-O vibration with a stretching frequency of ~1083 cm^{-1} in the system of nc-Si embedded in SiO_2 [116]. This could imply the important role of the phonons associated with the nc-Si/SiO_2 interface in the light emission of the ~760-nm band. The important role of the nc-Si/SiO_2 interface was highlighted in a previous study, which showed that the coupling of the confined excitons and the Si-O stretch vibrations dominates the PL process in nc-Si embedded in SiO_2 [117]. Considering both the energy and momentum conservations and the indirect band structure of the nc-Si, we therefore are inclined to believe that the Si-O vibration at the nc-Si/SiO_2 interface provides the mean required for both the energy dissipation due to the energy conservation requirement and the momentum conservation in the PL process. In other words, the ~760-nm band can be attributed to the band-to-band transition of the nc-Si assisted by the emission of a phonon, which is actually the Si-O vibration at the nc-Si/SiO_2 interface. For Si^+-implanted SiO_2, the 1100°C annealing could lead to both ultrafine nc-Si and the desirable nc-Si/SiO_2 interface such that the probability of the indirect exciton recombination assisted by the Si-O vibration is large leading to a strong light emission at 760 nm. As for the 845-nm band, it probably originates from the localized luminescent centers at the nc-Si/SiO_2 interface. In the study of Allan et al., the 845-nm (~1.5-eV) PL was attributed to the luminescent surface state of nc-Si [118]. This argument was later supported by Zhuravlev et al. who also observed the 1.5-eV PL band from Si^+-implanted SiO_2 after high temperature treatment [119].

4.3.5 Size Effect

In order to clarify the relationship between the nc-Si size and the PL properties as well as to further confirm the origin of the main PL peak for the sample annealed at 1100°C, other samples with different nc-Si sizes fabricated with Si ion implantation and subsequent annealing are investigated. The implantation details of the five samples are given in the following: (a) 3×10^{16} cm^{-2} at 120 keV, (b) 1×10^{17} cm^{-2} at 100 keV, (c) multiple implantations with 5×10^{15} cm^{-2} at 20 keV, 1×10^{16} cm^{-2} at 40 keV, and 2×10^{16} cm^{-2} at 80 keV, (d) 4×10^{16} cm^{-2} at 10 keV, and (e) multiple implantations with 3×10^{15} cm^{-2} at 1 keV, 8×10^{15} cm^{-2} at 5 keV, and 4×10^{16} cm^{-2} at 16 keV. After implantation, the samples were annealed at 1100°C for 20 min in the atmosphere of N_2 to induce nanocrystallization. XRD measurements show that the average nc-Si sizes for the four samples are ~4 nm, ~4.5 nm, ~4.9 nm, ~6.0 nm, and ~7.2 nm, respectively. This result can be explained by the concentration of excess Si in SiO_2. The average volume fractions of excess Si over the regions containing the nc-Si were calculated to be 3%, 8%, 11%, 15%, and 30%, respectively. Excess Si with a higher concentration in SiO_2 forms a larger nanocrystal size after a high temperature annealing. The bandgaps of the samples obtained from the SE study are 1.85, 1.76, 1.70, 1.56, and 1.47 eV, respectively. The room-temperature PL spectra were taken using the same excitation source as mentioned previously. The PL of the five samples is similar in their spectrum shapes, but the main PL peak positions are different, as shown in Figure 4.32 [74]. The main peaks of the five samples are located at ~720 nm (1.72 eV), ~760 nm (1.63 eV), ~785 nm (1.58 eV), ~870 nm (1.43), and ~925 nm (1.34 eV), respectively. This means that the PL peak shifts to a lower energy when the concentration of excess Si increases, which is consistent with a previous study [120]. For all the five samples, the difference between the energy gap and the energy of the corresponding main PL peak is always 0.13 ± 0.01 eV, as shown in Figure 4.33. This energy difference is actually equal to the Si-O vibration energy. Although both the bandgap and the PL peak energy change with the nc-Si size, the energy difference remains unchanged. This strongly supports the suggestion that the main PL peaks of the samples annealed at 1100°C are due to the band-to-band transition of the nc-Si assisted by the Si-O vibration at the nc-Si/SiO_2 interface.

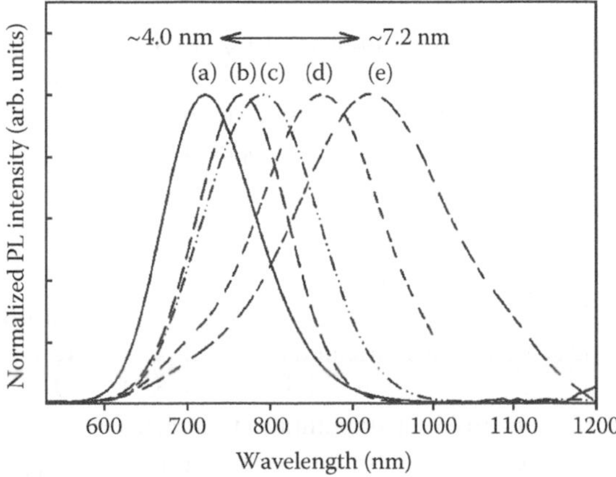

FIGURE 4.32 Normalized PL spectra for the following samples annealed at 1100°C for 20 min: (a) 3×10^{16} ions/cm^2 at 120 keV, (b) 1×10^{17} ions/cm^2 at 100 keV, (c) multiple implantations with 5×10^{15} ions/cm^2 at 20 keV, 1×10^{16} ions/cm^2 at 40 keV, and 2×10^{16} ions/cm^2 at 80 keV, (d) 4×10^{16} at 10 keV, and (e) multiple implantations with 3×10^{15} ions/cm^2 at 1 keV, 8×10^{15} ions/cm^2 at 5 keV, and 4×10^{16} ions/cm^2 at 16 keV. (From Ding, L. et al., *J. Nanosci. Nanotechnol.*, 8, 3555, 2008. With permission.)

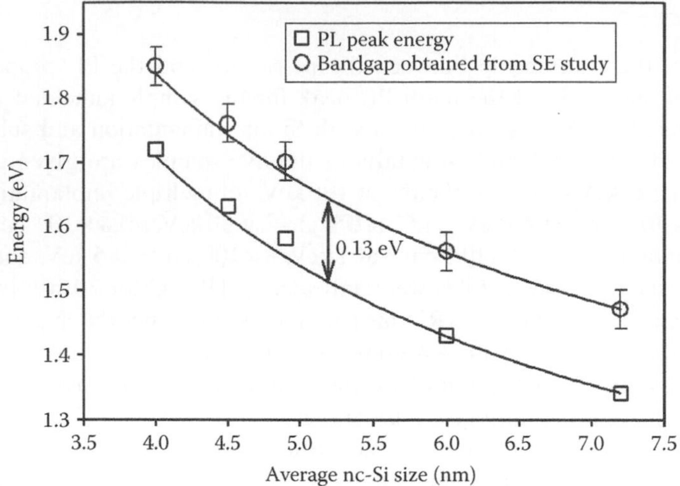

FIGURE 4.33 Bandgap and PL peak energy as functions of nc-Si diameter.

4.4 ELECTROLUMINESCENCE FROM SILICON NANOCRYSTALS EMBEDDED IN SiO_2 THIN FILMS

4.4.1 Introduction

Not long after the discovery of visible PL from PSi by Canham [12], Koshida et al. reported electroluminescence (EL) from PSi [121]. Visible EL has also been observed from Si^+-implanted SiO_2 films [98,122–131], and even electrically driven ultraviolet emission from Si^+-implanted SiO_2 films was predicted [128]. In previously reported studies on the light emission properties of Si^+-implanted SiO_2 films, usually medium- or high-energy Si ions were implanted into SiO_2 films with the thickness of hundreds of nanometers. With such a thick SiO_2 film a high voltage (~20 – ~60 V) is required for a measurable EL emission. For practical applications, the thickness of the SiO_2 thin films must be reduced to achieve higher carrier injection and lower turn-on voltage for the EL emission. In this section, we will discuss the EL from nc-Si light-emitting devices (LEDs) with a 30-nm SiO_2 thin film embedded with nc-Si fabricated with low-energy Si ion implantation. Different nanocrystal distributions in the SiO_2 films are achieved by varying the Si ion dose and the implantation energy. The influence of nanocrystal distribution on the EL properties is investigated. The EL mechanisms are also discussed based on the experimental results.

4.4.2 Device Fabrication and Typical EL Spectra

The active region of the LED structures is fabricated by the implantation of various doses of Si ions into thermally grown 30-nm SiO_2 thin films on *p*-type Si wafers at different implantation energies. For the study of influence of nc-Si concentration, three samples are fabricated with the Si ion dose of 1×10^{16}, 2×10^{15}, and 3×10^{14} cm^{-2}, respectively, at the implantation energy of 5 keV; and for the study of the influence of nc-Si distribution, the other three samples are fabricated with the Si ion dose of 1×10^{16} cm^{-2} at the implantation energies of 8, 5, and 2 keV, respectively. Table 4.5 summarizes the implantation recipes of all the samples under investigation (each sample is denoted with a sample number) [132].

TABLE 4.5
Summary of Implantation Recipes for Different EL Devices

Sample No.	Implantation Energy (keV)	Si Ion Dose (cm^{-2})
5a	5	1×10^{16}
5b	5	2×10^{15}
5c	5	3×10^{14}
8a	8	1×10^{16}
2a	2	1×10^{16}

After the ion implantation, thermal annealing was conducted at 1000°C in N_2 for 1 h to induce the nanocrystallization of excess Si in SiO_2. Afterward, the backside of the samples was coated with 1-μm Al layer as the back ohmic contact using an electron beam evaporator system. Indium tin oxide (ITO) with a thickness of 130 nm was sputtered onto the surface of the Si^+-implanted SiO_2 film through a hard shadow mask with a pad radius of 1.2 mm. Because of its conductivity and high transparency over the visible to infrared range, the ITO film acts as both the gate electrode and the EL emission window. Figure 4.34a shows the schematic cross section of the MOS-like LED structure with the Si^+-implanted SiO_2 as the gate oxide. The cross-sectional transmission electron microscopy (TEM) image confirms the formation of nc-Si embedded in the SiO_2 matrix [132]. As an example, Figure 4.34b shows the TEM image of the sample before implantation, and Figure 4.34c presents the TEM image of nc-Si with the size of ~4 nm after the ion implantation followed by the high temperature annealing in N_2. As can be seen in the figures, the thermally grown SiO_2 is still amorphous after high temperature annealing. The EL measurements were carried out with a PDS-1 PMT detector together with a monochromator. The measurements were performed at room temperature.

The implanted Si depth profiles in the SiO_2 films were calculated with SRIM simulations. Figure 4.35 shows the calculated depth profiles for the five samples [132]. As can be seen in Figure 4.35a, for the samples implanted at 5 keV, the excess Si distributes from the oxide surface to the depth of ~23 nm, and the concentration of excess Si reaches its maximum at the depth of ~8 nm. The peak concentrations of excess Si for samples 5a, 5b, and 5c are 1×10^{22} cm^{-3}, 2×10^{21} cm^{-3}, and 3×10^{20} cm^{-3}, respectively. The corresponding peak nc-Si volume fractions are 20%, 4%, and 0.6%, respectively. The depth profiles of excess Si for the samples fabricated at different implanted energies with

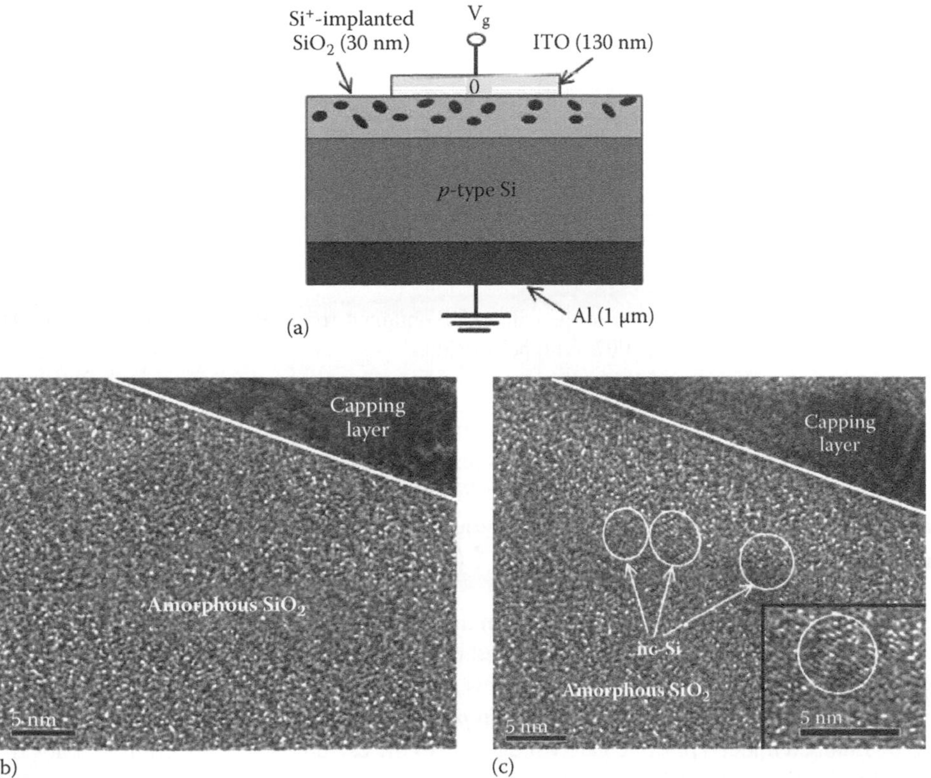

FIGURE 4.34 (a) Schematic illustration of the LED structure. (b) Cross-sectional TEM image of sample 2a before implantation. (c) Cross-sectional TEM image of nc-Si embedded in SiO_2 of sample 2a. (From Ding, L. et al., *Nanotechnology*, 18, 455306, 2007. With permission.)

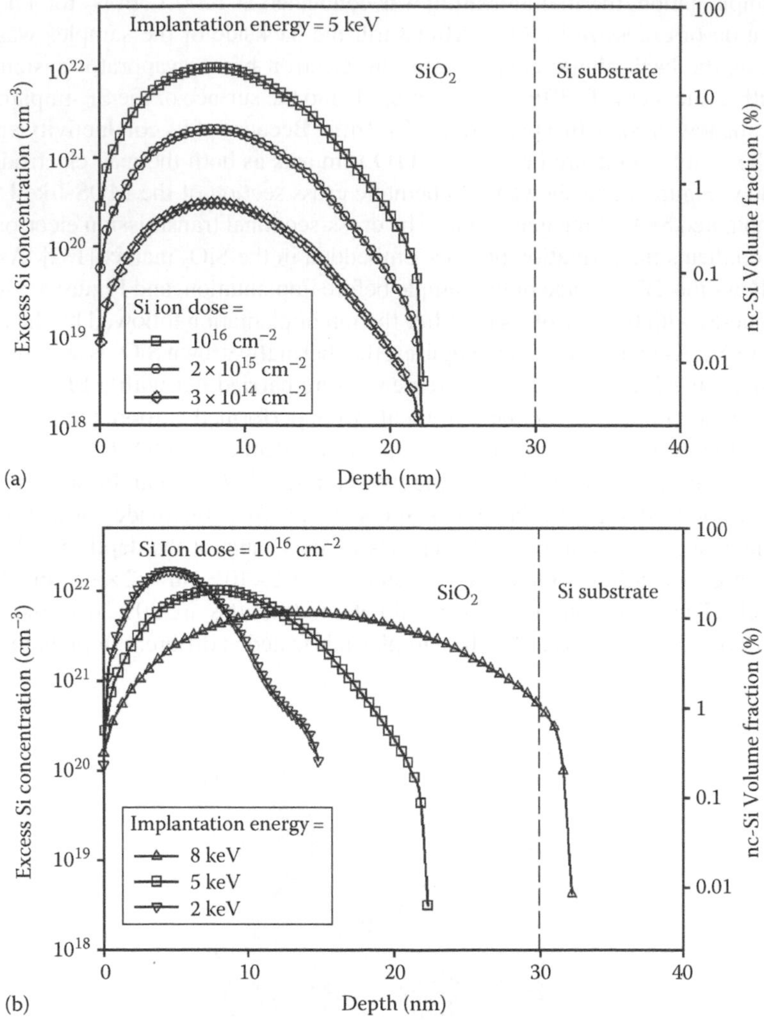

FIGURE 4.35 Distributions of nc-Si in the gate oxide obtained from SRIM simulations. (From Ding, L. et al., *Nanotechnology*, 18, 455306, 2007. With permission.)

the same implantation dose (i.e., samples 8a, 5a, and 2a) are presented in Figure 4.35b. As shown in this figure, for sample 8a, the excess Si distributes throughout the entire oxide layer with the peak concentration of 5.7×10^{21} cm^{-3} (or the peak nc-Si volume fraction of 11%) located at the depth of ~15 nm from the oxide surface. For sample 2a, the excess Si distributes to the depth of ~15 nm and reaches the peak concentration of 1.7×10^{22} cm^{-3} (the corresponding nc-Si volume fraction is 33%) at the depth of ~4.5 nm.

Visible and infrared EL can be observed from the MOS-like LED structures when a negative voltage is applied to the ITO gate. No EL is detected under a positive gate voltage due to the extremely low gate current. When a positive gate voltage is applied, the holes can hardly be provided by the ITO contact, and the electrons are minorities in *p*-type Si substrate. Therefore, a strong EL is difficult to be induced under a positive gate voltage due to the extremely low injection of holes from the positively biased ITO gate and electrons from *p*-type Si substrate. As examples, Figure 4.36 shows the EL spectra for sample 5a (implant dose = 1×10^{16} cm^{-2}; implant energy = 5 keV) under different gate voltages. As shown in this figure, one can see that the EL intensity increases with the applied gate voltage, but the spectral shape changes little under different gate voltages [132]. The injected

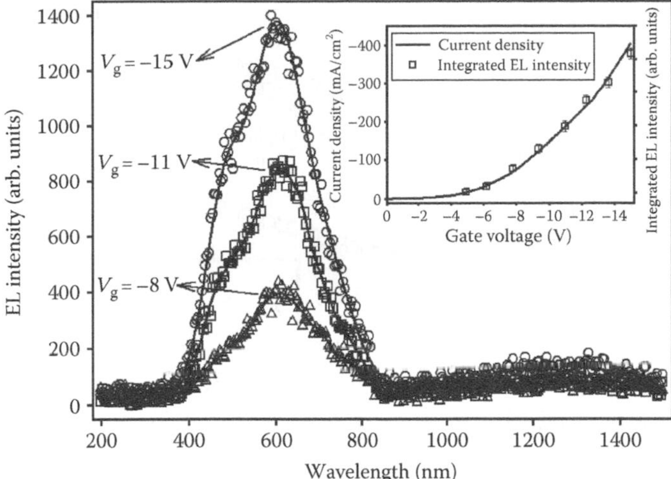

FIGURE 4.36 EL spectra for sample 5a under the gate voltages of −15, −11, and −8 V, respectively. The inset shows the *I–V* characteristic and the integrated EL intensity as a function of the gate voltage. (From Ding, L. et al., *Nanotechnology*, 18, 455306, 2007. With permission.)

current and the integrated EL intensity versus the gate voltage are shown in the inset of Figure 4.36. The current and the integrated EL intensity have a similar dependence on the gate voltage, showing a linear relationship between the integrated EL intensity and the current.

4.4.3 Effect of Implantation Recipe

First, the influence of implant dose on both the current transport and the EL is examined. The samples used for this study are fabricated with the same implantation energy (i.e., 5 keV), but they have different implant doses (i.e., 1×10^{16}, 2×10^{15}, and 3×10^{14} cm^{-2}). Therefore, they have the same distribution range of excess Si, namely, the excess Si is distributed from the SiO_2 surface to a depth of ~22 nm in the SiO_2 thin film. This means that there is no excess Si in the oxide in the region from the depth of ~22 nm to the SiO_2/substrate interface. In other words, there is a tunnel oxide (i.e., the pure SiO_2 region without the excess Si) with the thickness of ~8 nm for all the three samples. The Log-Log plots of current–voltage (*I–V*) characteristics for the three samples are shown in Figure 4.37 [133]. The tunnel current through the pure SiO_2 region and the current conduction in the excess-Si-distributed region play roles in the current transport. However, when the voltage is sufficiently high, the tunneling across the pure oxide is sufficient enough such that the current transport and the light emission are limited only by the conduction in the excess-Si-distributed region. In the EL study, the voltage is at least ~5 V to induce measurable EL emission. Therefore, the current conduction in the Si^+-implanted region rules both the EL and the current transport in the voltage range of our interest. The current conduction in the excess-Si-distributed region could be explained with the percolation concept [134]. Similarly to the situation of tunneling paths formed by neutral oxide traps in SiO_2 thin films [134], carriers can be transported through tunneling via the defects, nc-Si, and Si nanoclusters that have been confirmed to coexist in the Si suboxide films after the annealing at 1000°C [135–141]. With the formation of many conductive percolation paths in the excess-Si-distributed region, the conduction of the system is enhanced. This explains why the current conduction increases with the concentration of excess Si as shown in Figure 4.37. As a result of the increase in the current conduction, more electrons from the ITO gate and more holes from the *p*-type Si substrate are injected into the Si-implanted region, leading to an increase in the radiative recombination of the injected electrons and holes and thus an increase in the EL intensity. Indeed, as shown in Figure 4.37, the integrated EL intensity increases with the concentration of excess Si following the same trend of the current conduction.

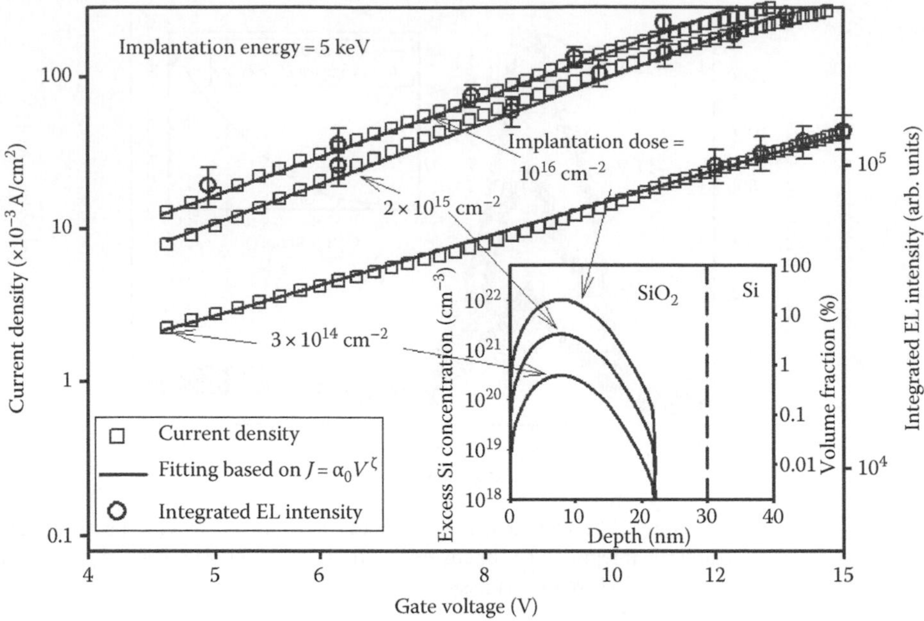

FIGURE 4.37 Dependence of the current and the integrated EL intensity on the voltage as a function of the implanted Si ion dose (samples 5a, 5b, and 5c). The implantation energy is fixed at 5 keV. The inset shows the concentrations and distributions of the excess Si in the oxide thin film obtained from the SRIM simulations. (From Ding, L. et al., *IEEE Trans. Electron Dev.*, submitted for publication.)

Now the influence of the distribution of excess Si in SiO_2 thin film is investigated. Various distributions of excess Si are achieved by varying the implantation energy [133]. The inset of Figure 4.38 shows the concentrations of excess Si and distribution profiles for samples 2a, 5a, and 8a that were fabricated at the energies of 2, 5, and 8 keV, respectively, with the fixed Si ion dose of 1×10^{16} cm^{-2}. For sample 2a, the excess Si is distributed from the SiO_2 surface to a depth of ~14 nm in the SiO_2 thin film, forming a tunnel oxide of ~16 nm; for sample 5a, the excess Si is distributed from the SiO_2 surface to a depth of ~22 nm, forming a tunnel oxide of ~8 nm; and for sample 8a, the excess Si is distributed from the SiO_2 surface until the SiO_2/substrate interface leading to no tunnel oxide formed. As a constant implantation dose is used, the peak concentration of excess Si decreases with the extension of the Si^+-implanted region. With higher implantation energy, the excess-Si-distributed region extends wider while the pure oxide becomes thinner; thus, the voltage drop in the excess-Si-distributed region increases for a given applied voltage. Therefore, for a given applied voltage the current will be higher if the conduction in the excess-Si-distributed region plays a dominant role in the current transport. On the other hand, when the excess Si is distributed throughout the entire SiO_2 thin film (i.e., the situation of sample 8a), conductive tunneling paths connecting the substrate to the gate will be formed, leading to a large increase in the current conduction. As a result of the two scenarios, the current conduction increases with the implantation energy, as shown in Figure 4.38. As the EL intensity reflects the transport of the injected carriers, the increase of the current conduction is translated to an increase in the EL intensity with the implantation energy, as shown in Figure 4.38 also.

4.4.4 EL Mechanisms

It has been found that the EL spectra can be decomposed into four Gaussian-shaped EL bands. As an example, the EL spectral decomposition for sample 5a under the gate voltage of −15 V is shown in Figure 4.39 [132]. As can be seen in this figure, the ~610-nm luminescence band dominates the

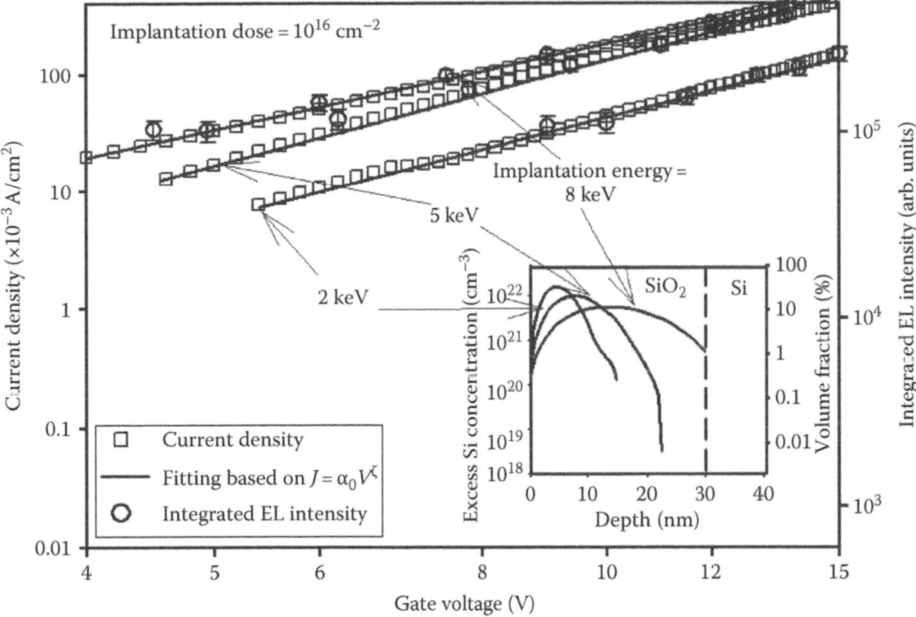

FIGURE 4.38 Dependence of the current and the integrated EL intensity on the voltage as a function of the implantation energy (samples 2a, 5a, and 8a). The implanted Si ion dose is fixed at 1×10^{16} cm^{-2}. The inset shows the concentrations and distributions of the excess Si in the oxide thin film obtained from the SRIM simulations. (From Ding, L. et al., *IEEE Trans. Electron Dev.*, submitted for publication.)

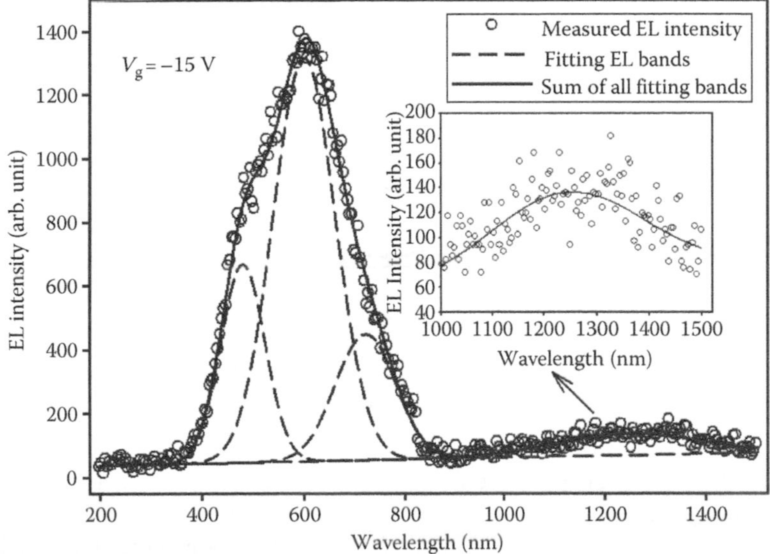

FIGURE 4.39 Decomposition of the EL spectrum for sample 5a under the gate voltage of −15 V into the following four EL bands: ~460 nm, ~610 nm, ~740 nm, and ~1260 nm bands. (From Ding, L. et al., *Nanotechnology*, 18, 455306, 2007. With permission.)

EL spectrum with two shoulder bands centered at ~460 nm and ~740 nm, respectively. In addition, a luminescence band located at ~1260 nm was observed in the near-infrared region. Figure 4.40 shows the peak wavelengths of the four EL bands obtained at different gate voltages [132]. It can be seen from the figure that the peak wavelengths change little with the gate voltage. On the other hand, the contribution (in percentage) of each EL band, which is the ratio of the integrated intensity of each

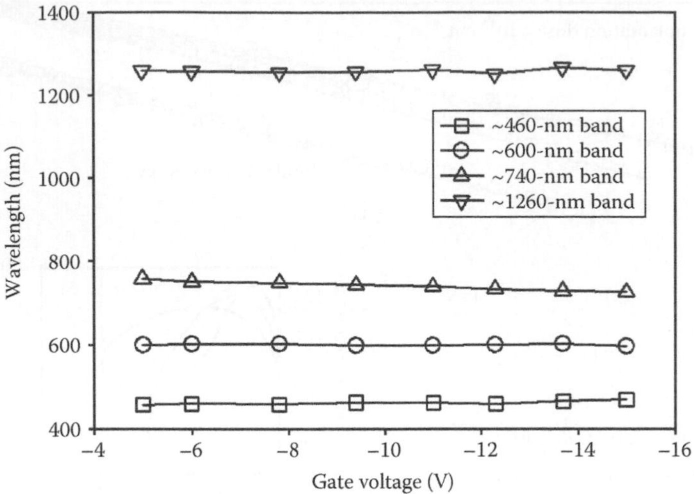

FIGURE 4.40 Evolution of the peak wavelength of each EL band with gate voltage for sample 5a. (From Ding, L. et al., *Nanotechnology*, 18, 455306, 2007. With permission.)

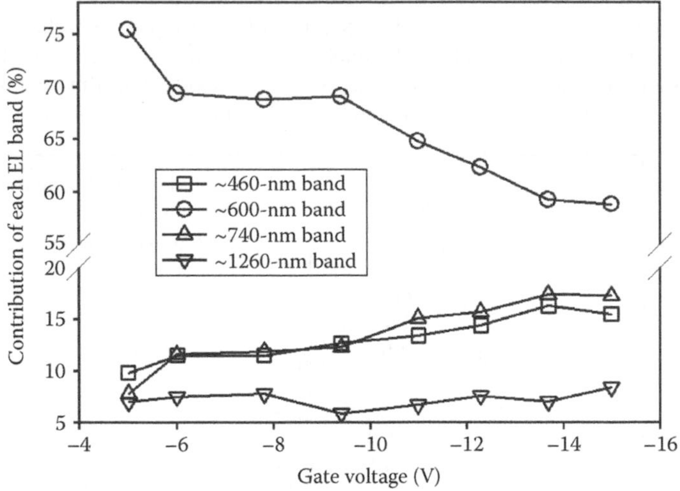

FIGURE 4.41 Gate-voltage dependence of the contribution of each EL band for sample 5a. (From Ding, L. et al., *Nanotechnology*, 18, 455306, 2007. With permission.)

EL band to the integrated intensity of the EL spectrum, changes with the gate voltage. Figure 4.41 shows the gate-voltage dependence of the contribution of each EL band for sample 5a [132]. As can be seen in the figure, when the gate voltage increases, the contribution of the dominant EL band (i.e., the ~610-nm band) decreases, while the contributions of the ~460- and ~740-nm bands increase but the contribution of the ~1260-nm band shows no change.

Figure 4.42 shows the EL spectra under the gate voltage of −15 V for the samples fabricated with different implantation doses (i.e., samples 5a, 5b, and 5c) [132]. As shown in Figure 4.42, the integrated EL intensity increases with the implantation dose. Figure 4.43 shows the EL spectra for the samples fabricated at different implantation energies with the same Si ion dose (i.e., samples 8a, 5a, and 2a). It can be observed that the EL intensity also increases with the implantation energy. The increase in the EL intensity with either the implanted Si ion dose or the implantation energy can be explained by the enhancement in the current conduction of the Si^+-implanted oxide, as discussed above.

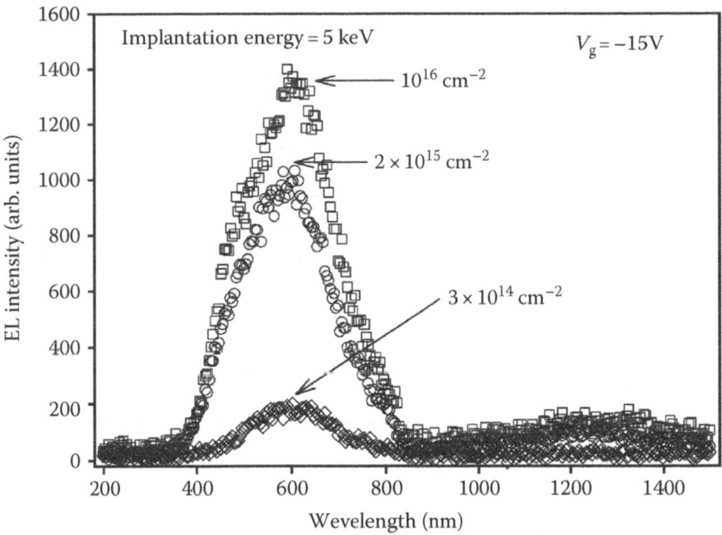

FIGURE 4.42 Influence of the implanted Si ion dose on the EL spectrum [132]. The EL spectra shown in the figure are from samples 5a, 5b, and 5c. (From Ding, L. et al., *Nanotechnology*, 18, 455306, 2007. With permission.)

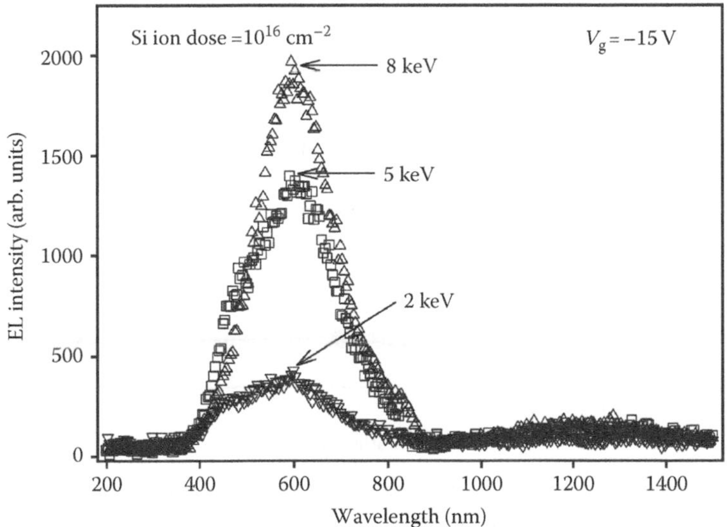

FIGURE 4.43 Influence of the implantation energy on the EL spectrum [132]. The EL spectra shown in the figure are from samples 8a, 5a, and 2a. (From Ding, L. et al., *Nanotechnology*, 18, 455306, 2007. With permission.)

Although the integrated EL intensity increases with both the implanted Si ion dose and implantation energy, the contribution of each EL band shows a different picture of its dependence on the implanted Si ion dose and implantation energy. Figure 4.44 shows the contribution of each EL band as a function of the implanted Si ion dose under the gate voltage of −15 V [132]. As shown in Figure 4.44, for the samples with the same implantation energy (5 keV), the contribution of each EL band is almost independent of the Si ion dose. However, the contribution of each EL band changes with the implantation energy. Figure 4.45 shows the dependence of the contribution of each EL band on the implantation energy for the samples fabricated with the same Si ion dose (10^{16} cm^{-2}) [132]. As can

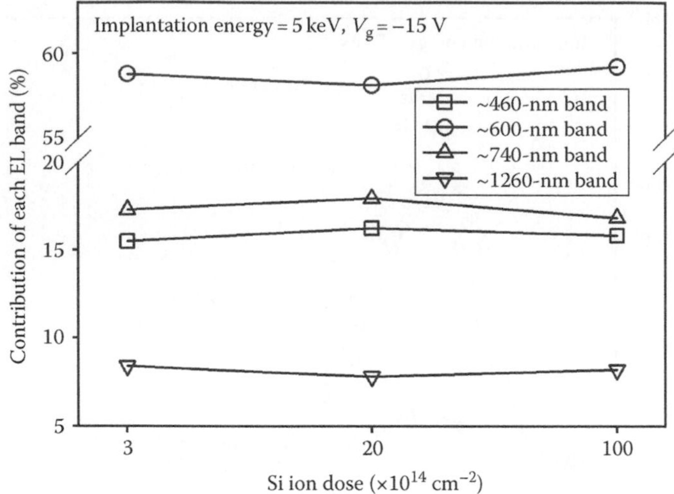

FIGURE 4.44 Contribution of each EL band as a function of the implanted Si ion dose. (From Ding, L. et al., *Nanotechnology*, 18, 455306, 2007. With permission.)

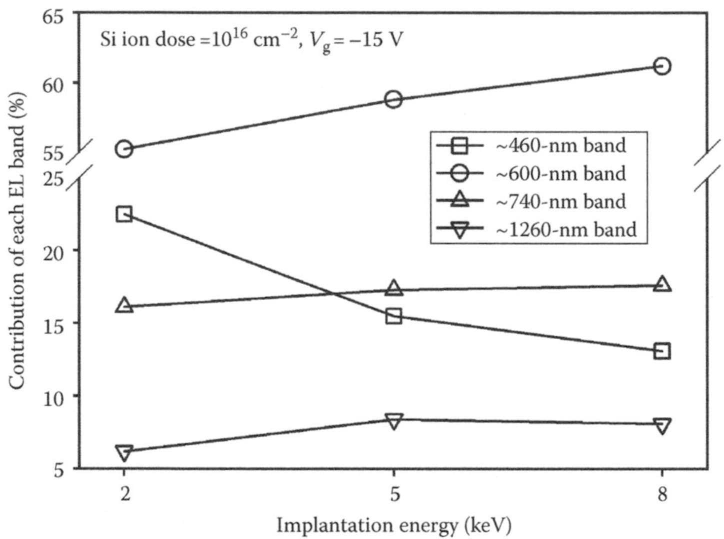

FIGURE 4.45 Contribution of each EL band as a function of the implantation energy. (From Ding, L. et al., Nanotechnology, 18, 455306, 2007. With permission.)

be seen in the figure, when the implantation energy increases from 2 to 8 keV, the contribution of the ~600-nm band significantly increases from ~ 55% to ~ 61%, the contributions of both the ~740-nm band and the ~1260-nm band show almost no changes, and the contribution of the ~460-nm EL band significantly decreases from ~23% to ~13%.

The EL bands observed in the present study are very similar to those reported by Song et al. [131]. In their study, they also observed the ~460-nm, ~600-nm, and ~740-nm EL bands with the ~600-nm band as the dominant one from the Si^+-implanted SiO_2 films. Many kinds of luminescent defects, such as weak oxygen bond (WOB) (O-O) defects [98,105], NOV ($O_3\equiv$Si-Si$\equiv O_3$) [45,98,101, 105,115,128,130,131], NBOHC ($O_3\equiv$Si-Si-O•) [40,45,98,105,115,131], *D* center [45], and *E′* center [14,41,45,50,98,105,131,142] have been proposed for the mechanisms of luminescence from Si^+-implanted SiO_2 films. Although the origins for the above-mentioned EL bands are still under debate, the following mechanisms are popularly accepted in the literature, and thus our discussions will be also based on these mechanisms. The ~460 nm (~2.7 eV) band is ascribed to the NOV defect that was proposed for the EL band observed from pure silica glass [108] and Si^+-implanted SiO_2 films [45,98,101,105,115,128,130,131]. It has been known that the NBOHC defects in SiO_2 can emit light at around 2 eV (~600 nm) [40,45,98,105,112,115,131]. Therefore, following Song et al. [131], we attribute the ~600-nm band to the NBOHC defects.

As regard to the ~740-nm (~1.7 eV) EL band, first we assume that it is probably due to the existence of nc-Si in the oxide. As mentioned above, the average diameter of nc-Si formed in the SiO_2 films in this study is ~4 nm. The ~740-nm PL band was observed, and it was believed to be due to the formation of nc-Si with a mean size of ~4 nm [115]. For a nc-Si with a diameter of ~4 nm, its bandgap is estimated to be ~1.85 eV. The difference between the bandgap and the emission energy indicates that the ~740-nm EL band should not originate from the direct band-to-band transition within the nc-Si. In fact, it has been suggested that the nc-Si with the size larger than ~1 nm still has an indirect band structure [118,143,144]. The energy difference (~0.15 eV) between the bandgap and the emission energy is approximately equal to the energy (~0.134 eV) of Si-O vibration with a stretching frequency of $1083\,cm^{-1}$, implying the important role of Si-O vibration at the nc-Si/SiO_2 interface in the 740-nm band luminescence. Considering the energy conservation and the indirect band structure of the nc-Si, we therefore are inclined to believe that the Si-O vibration at the nc-Si/SiO_2 interface provides the mean required for both the energy dissipation due to the energy conservation requirement and the momentum conservation in the EL process. In other words, the ~740-nm band can be attributed to the recombination of carries within the nc-Si assisted by the Si-O vibration at the nc-Si/SiO_2 interface.

As for the infrared EL band center at ~1260 nm, it mostly originates from the Si substrate. In order to confirm whether the virgin oxide and the substrate have any contribution to the observed EL spectra, we measured the EL spectra of the LEDs fabricated with pure thermal grown SiO_2 and SiO_2 films implanted with different elements such as Ge and Al. Although the EL spectra in the visible range are different, all the samples exhibit infrared EL centered at ~1260 nm. The electrical-driven infrared luminescence was also reported by Kulakci et al. recently [125]. Although the authors could not determine the peak wavelength of the infrared EL due to the detector limitation, the light emission above 900 nm was indeed observed. In our study, it is noticeable that the photon energy of the peak of the ~1260 nm band is ~0.985 eV that has a difference of ~0.135 eV as compared to the bandgap of bulk crystalline silicon. The energy difference is approximately the same as the Si-O vibration energy (0.134 eV). We therefore suggest that the ~1260 nm band is a result of electron–hole recombination in the accumulation region formed beneath the oxide layer near the interface between the Si substrate and the SiO_2. Such recombination is assisted by the Si-O vibration at the interface between the SiO_2 and the Si substrate. In other words, the electron–hole recombination near the interface emits a photon with the energy of ~0.985 eV and a phonon of ~0.135 eV that is actually the Si-O vibration.

The excitation of the NOV defect as a luminescence center can be enhanced by a high electric filed in the oxide (E_{ox}) [98,131,145]. This agrees with the result shown in Figure 4.41 that the contribution of the ~460-nm band increases with the applied gate voltage. On the other hand, as discussed early, the nc-Si distributed region extends with increasing the implantation energy, leading to an increase in the gate current. As a result, for a given gate voltage, the voltage drops in the ITO/oxide junction and the oxide/substrate junction increases, and thus the voltage drop in the oxide decreases. Thus, the electric field in the oxide decreases weakening the excitation of the NOV defect. Therefore, the contribution of the ~460-nm band decreases with the implantation energy, as shown in Figure 4.45.

In our experiments, the ~600-nm band always dominates the EL spectra for all the samples, indicating that the excitation energy of the ~600-nm band should be much lower than that of other luminescent defects. The energy distribution of injected carriers can easily satisfy the requirement for the excitation of the NBOHC defects. The excitation of the NBOHC defects becomes saturated at a gate voltage less than 5 V. As the contributions of the ~460-nm band and the ~740-nm band increase with the gate voltage, the contribution of the ~600-nm band decreases with the gate voltage due to the saturation in the excitation of the NBOHC defects (note that the contribution is a relative contribution and the sum of all the contributions is 100%), as shown in Figure 4.41. On the other hand, as shown in Figure 4.45, the contribution of the ~600-nm band increases with the implantation energy. This could suggest that the generation of the NBOHC defects increases with the implantation energy.

As the carrier transport via the nc-Si is enhanced by the electric field in the nc-Si distributed region, the light emission from the nc-Si should increase with the gate voltage. Indeed, the contribution of the ~740-nm band that is attributed to the nc-Si increases with the gate voltage, as shown in Figure 4.41. In contrast to the contributions of the ~460-nm band and the ~600-nm band that change significantly with the implantation energy, the contribution of the ~740-nm band shows only a small increase with the implantation energy, indicating that the light emission from the nc-Si is not very sensitive to the implantation energy. For all the EL bands, their contributions are insensitive to the implanted Si dose, as shown in Figure 4.44. This indicates that the increase in the implantation dose does not change the composition of the EL spectra.

4.5 SUMMARY

The dielectric function and optical constants of nc-Si embedded in SiO_2 matrix have been obtained with SE based on the Maxwell-Garnett effective medium approximation. The nc-Si shows a significant reduction in the dielectric function and optical constants as compared with bulk crystalline silicon. The bandgap expansion of nc-Si with reducing nc-Si size is also observed, and it is in a good agreement with the first-principles calculation of the optical gap of nc-Si based on QC. In addition, thermal annealing does not change the indirect band structure and the energy gap of the nc-Si.

PL from nc-Si embedded in SiO_2 matrix has been studied. The PL shows a continuous evolution with annealing. Six PL bands located at 415, 460, 520, 630, 760, and 845 nm are observed depending on the annealing temperature. The annealing at 1100°C yields the strongest PL band (i.e., the main PL peak) at 760 nm (~1.63 eV) with intensity much higher than that of all other PL bands. Both the bandgap energy and the PL peak energy change with the nc-Si size. However, the difference between the PL peak energy and the energy gap related to the optical transition is always 0.13 ± 0.01 eV, being equal to the Si-O vibration energy. This strongly suggests that the main PL peak is due to the indirect band-to-band transition of the nc-Si assisted by the Si-O vibration at the nc-Si/SiO_2 interface. On the other hand, the ~415-, ~460-, ~520-, and 630-nm bands are ascribed to the WOB, NOV, E'_{δ}, and NBOHC defects, respectively, while the 845-nm band is proposed to be related to the interfacial luminescent centers.

Both visible and near-infrared EL has been observed from MOS-like LED structures with nc-Si embedded in the gate oxide synthesized by low-energy ion implantation. It is observed that the EL intensity increases with the Si ion dose and/or the implantation energy, due to the enhanced current transport. The EL spectra can be decomposed into four EL bands peaked at ~460, ~600, ~740, and ~1260 nm, respectively. The first two EL bands are ascribed to the NOV and NBOHC defects, respectively. The ~740-nm band is believed to be associated with the formation of the nc-Si embedded in SiO_2. The ~1260-nm EL band is assigned to the electron–hole recombination in the accumulation region in the Si substrate underneath the oxide.

ACKNOWLEDGMENT

This work has been financially supported by the Ministry of Education of Singapore under project No. ARC 1/04 and the National Research Foundation of Singapore under project No. NRF-G-CRP 2007-01. The authors wish to thank Y. Liu, C. Y. Ng, M. Yang, J. I. Wong, and Z. H. Cen for their collaboration in the research.

ABBREVIATIONS

CMOS	Complementary metal-oxide-semiconductor
CVD	Chemical vapor deposition
DOS	Density of states
DRAM	Dynamic random access memory
EL	Electroluminescence
EMA	Effective medium approximation
FB	Forouhi–Bloomer
HRTEM	High resolution transmission electron microscopy
ITO	Indium tin oxide
I–V	Current–voltage
LED	Light emitting diode
LPCVD	Low pressure chemical vapor deposition
MOS	Metal-oxide-semiconductor
MSE	Mean-square Error
NBOHC	Non-bridging oxygen hole center
nc-Si	Si nanocrystals
NOV	Neutral oxygen vacancy
PECVD	Plasma-enhanced chemical vapor deposition
PL	Photoluminescence
PLD	Pulse laser deposition
PMT	Photomultiplier tube
PSi	Porous silicon
QC	Quantum confinement
SE	Spectroscopic ellipsometry
SIMS	Secondary ion mass spectroscopy
SRIM	Stopping and range of ions in matter
TEM	Transmission electron microscopy
WOB	Weak oxygen bond
XRD	X-ray diffraction

SYMBOLS

E	Photon energy
f	Volume fraction
k	Extinction coefficient
n	Refractive index
N_{si}	Atom density of Si
Q	Implantation dose
α	Absorption coefficient
ε	Dielectric function
Ψ and Δ	Ellipsometric angles

REFERENCES

1. Shockley, W. 1949. The theory of p-n junctions in semiconductors and p-n junction transistors. *Bell System Technical Journal* 28:435–439.
2. Grimmeiss, H. G. 1999. Silicon-germanium: A promise into the future. *Semiconductors* 33:939–941.
3. Pavesi, L. and Lockwood, D. J. 2004. *Silicon Photonics*. New York: Springer.

4. Cutolo, A., Lodice, M., Spirito, P., and Zeni, L. 1997. Silicon electro-optic modulator based on a three terminal device integrated in a low-loss single-mode SOI waveguide. *Journal of Lightwave Technology* 15:505–518.
5. Liu, A., Jones, R., Liao, L. et al. 2004. A high-speed silicon optical modulator based on a metal-oxide-semiconductor capacitor. *Nature* 427:615–618.
6. Liu, Y., Liu, E., Li, G. et al. 1994. Novel silicon waveguide switch based on total internal reflection. *Applied Physics Letters* 64:2079–2080.
7. Zhao, C. Z., Chen, A. H., Liu, E. K., and Li, G. Z. 1997. Silicon-on-insulator asymmetric optical switch based on total internal reflection. *IEEE Photonics Technology Letters* 9:1113–1115.
8. Ghioni, M., Zappa, F., Kesan, V. P., and Warnock, J. 1996. A VLSI-compatible high-speed silicon photodetector for optical data link applications. *IEEE Transactions on Electron Devices* 43:1054–1060.
9. Hawkins, A. R., Wu, W., Abraham, P., Streubel, K., and Bowers, J. E. 1997. High gain-bandwidth-product silicon heterointerface photodetector. *Applied Physics Letters* 70:303–305.
10. Pellegrino, P., Garrido, B., Garcia, C. et al. 2005. Low-loss rib waveguides containing Si nanocrystals embedded in SiO_2. *Journal of Applied Physics* 97:074312–8.
11. Valenta, J., Pelant, I., and Linnros, J. 2002. Waveguiding effects in the measurement of optical gain in a layer of Si nanocrystals. *Applied Physics Letters* 81:1396–1398.
12. Canham, L. T. 1990. Silicon quantum wire array fabrication by electrochemical and chemical dissolution of wafers. *Applied Physics Letters* 57:1046–1048.
13. Tischler, M. A., Collins, R. T., Stathis, J. H., and Tsang, J. C. 1992. Luminescence degradation in porous silicon. *Applied Physics Letters* 60:639–641.
14. Mutti, P., Ghislotti, G., Bertoni, S. et al. 1995. Room-temperature visible luminescence from silicon nanocrystals in silicon implanted SiO_2 layers. *Applied Physics Letters* 66:851–853.
15. Park, N.-M., Kim, T.-S., and Park, S.-J. 2001. Band gap engineering of amorphous silicon quantum dots for light-emitting diodes. *Applied Physics Letters* 78:2575–2577.
16. Uhlir, A. Jr. 1956. Electrolytic shaping of germanium and silicon. *Bell System Technical Journal* 35:333–347.
17. Cullis, A. G. and Canham, L. T. 1991. Visible light emission due to quantum size effects in highly porous crystalline silicon. *Nature* 353:335–338.
18. Pickering, C., Beale, M. I. J., Robbins, D. J., Pearson, P. J., and Greef, R. 1984. Optical studies of the structure of porous silicon films formed in p-type degenerate and non-degenerate silicon. *Journal of Physics C: Solid State Physics* 17:6535–6552.
19. Fukuda, M., Nakagawa, K., Miyazaki, S., and Hirose, M. 1997. Resonant tunneling through a self-assembled Si quantum dot. *Applied Physics Letters* 70:2291–2293.
20. Nakajima, A., Sugita, Y., Kawamura, K., Tomita, H., and Yokoyama, N. 1996. Microstructure and optical absorption properties of Si nanocrystals fabricated with low-pressure chemical-vapor deposition. *Journal of Applied Physics* 80:4006–4011.
21. Darma, Y., Murakami, H., and Miyazaki, S. 2003. Formation of nanometer silicon dots with germanium core by highly-selective low-pressure chemical vapor deposition. *Japanese Journal of Applied Physics (Part 1: Regular Papers, Short Notes and Review Papers)* 42:4129–4133.
22. Tong, J. F., Hsiao, H. L., and Hwang, H. L. 1999. Adjustable emissions from silicon-rich oxide films prepared by plasma-enhanced chemical-vapor deposition. *Applied Physics Letters* 74:2316–2318.
23. Yang, C. S., Lin, C. J., Kuei, P. Y. et al. 1997. Quantum size effects on photoluminescence from Si nanocrystals in PECVD silicon-rich-oxide. *Applied Surface Science* 113–114:116–120.
24. Wang, Y. Q., Kong, G. L., Chen, W. D. et al. 2002. Getting high-efficiency photoluminescence from Si nanocrystals in SiO_2 matrix. *Applied Physics Letters* 81:4174–4176.
25. Chen, X. Y., Lu, Y. F., Wu, Y. H. et al. 2003. Mechanisms of photoluminescence from silicon nanocrystals formed by pulsed-laser deposition in argon and oxygen ambient. *Journal of Applied Physics* 93:6311–6319.
26. Suzuki, N., Makino, T., Yamada, Y., Yoshida, T., and Seto, T. 2001. Monodispersed, nonagglomerated silicon nanocrystallites. *Applied Physics Letters* 78:2043–2045.
27. Wilkinson, A. R. and Elliman, R. G. 2004. The effect of annealing environment on the luminescence of silicon nanocrystals in silica. *Journal of Applied Physics* 96:4018–4020.
28. Wang, Y. Q., Smirani, R., and Ross, G. G. 2004. Nanotwinning in silicon nanocrystals produced by ion implantation. *Nano Letters* 4:2041–2045.
29. Wang, Y. Q., Smirani, R., Ross, G. G., and Schiettekatte, F. 2005. Ordered coalescence of Si nanocrystals in SiO_2. *Physical Review B* 71:161310–4.

30. Allegrini, M., Ciofi, C., Diligenti, A. et al. 1996. Photoluminescence from ion-beam cosputtered Si/SiO_2 thin films. *Solid State Communications* 100:403–406.
31. Baru, V. G., Chernushich, A. P., Luzanov, V. A. et al. 1996. Optical properties of Si nanocrystals prepared by magnetron sputtering. *Applied Physics Letters* 69:4148–4150.
32. He, Y., Bi, L., Feng, J. Y., and Wu, Q. L. 2005. Properties of Si-rich SiO_2 films by RF magnetron sputtering. *Journal of Crystal Growth* 280:352–356.
33. Kahler, U. and Hofmeister, H. 1999. Silicon nanocrystallites in buried SiO_x layers via direct wafer bonding. *Applied Physics Letters* 75:641–643.
34. Kahler, U. and Hofmeister, H. 2001. Visible light emission from Si nanocrystalline composites via reactive evaporation of SiO. *Optical Materials* 17:83–86.
35. Lau, H. W., Tan, O. K., Liu, Y. et al. 2005. Defect-induced photoluminescence from tetraethylorthosilicate thin films containing mechanically milled silicon nanocrystals. *Journal of Applied Physics* 97:104307–4.
36. Littau, K. A., Szajowski, P. J., Muller, A. J., Kortan, A. R., and Brus, L. E. 1993. A luminescent silicon nanocrystal colloid via a high-temperature aerosol reaction. *Journal of Physical Chemistry* 97:1224–1230.
37. Ostraat, M. L., De Blauwe, J. W., Green, M. L. et al. 2001. Ultraclean two-stage aerosol reactor for production of oxide-passivated silicon nanoparticles for novel memory devices. *Journal of the Electrochemical Society* 148:G265–G270.
38. Ostraat, M. L., De Blauwe, J. W., Green, M. L. et al. 2001. Synthesis and characterization of aerosol silicon nanocrystal nonvolatile floating-gate memory devices. *Applied Physics Letters* 79:433–435.
39. Fischer, T., Petrova-Koch, V., Shcheglov, K., Brandt, M. S., and Koch, F. 1996. Continuously tunable photoluminescence from Si+-implanted and thermally annealed SiO_2 films. *Thin Solid Films* 276:100–103.
40. Muller, T., Heinig, K.-H., and Moller, W. 2002. Size and location control of Si nanocrystals at ion beam synthesis in thin SiO_2 films. *Applied Physics Letters* 81:3049–3051.
41. Nakajima, A., Sugita, Y., Kawamura, K., Tomita, H., and Yokoyama, N. 1996. Si quantum dot formation with low-pressure chemical vapor deposition. *Japanese Journal of Applied Physics (Part 2: Letters)* 35:L189–L191.
42. Kanemitsu, Y. and Okamoto, S. 1998. Phonon structures and Stokes shift in resonantly excited luminescence of silicon nanocrystals. *Physical Review B* 58:9652–9655.
43. Guha, S., Pace, M. D., Dunn, D. N., and Singer, I. L. 1997. Visible light emission from Si nanocrystals grown by ion implantation and subsequent annealing. *Applied Physics Letters* 70:1207–1209.
44. Guha, S. 1998. Characterization of Si^+ ion-implanted SiO_2 films and silica glasses. *Journal of Applied Physics* 84:5210–5217.
45. Song, H. Z. and Bao, X. M. 1997. Visible photoluminescence from silicon-ion-implanted SiO_2 film and its multiple mechanisms. *Physical Review B* 55:6988–6993.
46. Dinh, L. N., Chase, L. L., Balooch, M., Siekhaus, W. J., and Wooten, F. 1996. Optical properties of passivated Si nanocrystals and SiO_x nanostructures. *Physical Review B* 54:5029–5037.
47. Takeoka, S., Fujii, M., and Hayashi, S. 2000. Size-dependent photoluminescence from surface-oxidized Si nanocrystals in a weak confinement regime. *Physical Review B* 62:16820–16825.
48. Cho, K. S., Park, N.-M., Kim, T.-Y. et al. 2005. High efficiency visible electroluminescence from silicon nanocrystals embedded in silicon nitride using a transparent doping layer. *Applied Physics Letters* 86:071909–3.
49. Vial, J. C., Bsiesy, A., Gaspard, F. et al. 1992. Mechanisms of visible-light emission from electro-oxidized porous silicon. *Physical Review B* 45:14171–14176.
50. Shimizu-Iwayama, T., Fujita, K., Nakao, S. et al. 1994. Visible photoluminescence in Si^+-implanted silica glass. *Journal of Applied Physics* 75:7779–7783.
51. Takagi, H., Ogawa, H., Yamazaki, Y., Ishizaki, A., and Nakagiri, T. 1990. Quantum size effects on photoluminescence in ultrafine Si particles. *Applied Physics Letters* 56:2379–2380.
52. Deak, P., Rosenbauer, M., Stutzmann, M., Weber, J., and Brandt, M. S. 1992. Siloxene: Chemical quantum confinement due to oxygen in a silicon matrix. *Physical Review Letters* 69:2531–2534.
53. Hill, N. A. and Whaley, K. B. 1995. Size dependence of excitons in silicon nanocrystals. *Physical Review Letters* 75:1130–1133.
54. Carrier, P., Lewis, L. J., and Dharma-wardana, M. W. C. 2001. Electron confinement and optical enhancement in Si/SiO_2 superlattices. *Physical Review B* 64:195330–7.
55. Cichos, F., Martin, J., and von Borczyskowski, C. 2004. Emission intermittency in silicon nanocrystals. *Physical Review B* 70:115314–9.

56. Leung, K. and Whaley, K. B. 1997. Electron-hole interactions in silicon nanocrystals. *Physical Review B* 56:7455–7468.
57. Proot, J. P., Delerue, C., and Allan, G. 1992. Electronic structure and optical properties of silicon crystallites: Application to porous silicon. *Applied Physics Letters* 61:1948–1950.
58. Ogut, S., Chelikowsky, J. R., and Louie, S. G. 1997. Quantum confinement and optical gaps in si nanocrystals. *Physical Review Letters* 79:1770–1173.
59. Bulutay, C. 2007. Interband, intraband, and excited-state direct photon absorption of silicon and germanium nanocrystals embedded in a wide band-gap lattice. *Physical Review B* 76:205321–14.
60. Amans, D., Callard, S., Gagnaire, A. et al. 2003. Ellipsometric study of silicon nanocrystal optical constants. *Journal of Applied Physics* 93:4173–4179.
61. Lee, K.-J., Kang, T.-D., Lee, H. et al. 2005. Optical properties of SiO_2/nanocrystalline Si multilayers studied using spectroscopic ellipsometry. *Thin Solid Films* 476:196–200.
62. Wang, L.-W. and Zunger, A. 1994. Dielectric constants of silicon quantum dots. *Physical Review Letters* 73:1039–1042.
63. Delerue, C., Allan, G., and Lannoo, M. 1993. Theoretical aspects of the luminescence of porous silicon. *Physical Review B* 48:11024–11036.
64. Vasiliev, I., Ogut, S., and Chelikowsky, J. R. 2001. Ab Initio absorption spectra and optical gaps in nanocrystalline silicon. *Physical Review Letters* 86:1813–1816.
65. Weissker, H. C., Furthmuller, J., and Bechstedt, F. 2002. Optical properties of Ge and Si nanocrystallites from ab initio calculations. I. Embedded nanocrystallites. *Physical Review B* 65:155327–9.
66. Weissker, H. C., Furthmuller, J., and Bechstedt, F. 2002. Optical properties of Ge and Si nanocrystallites from ab initio calculations. II. Hydrogenated nanocrystallites. *Physical Review B* 65:155328–7.
67. Weissker, H. C., Furthmller, J., and Bechstedt, F. 2003. Validity of effective-medium theory for optical properties of embedded nanocrystallites from ab initio supercell calculations. *Physical Review B* 67:165322–5.
68. Ding, L., Chen, T. P., Liu, Y., Ng, C. Y., and Fung, S. 2005. Optical properties of silicon nanocrystals embedded in a SiO_2 matrix. *Physical Review B* 72:125419–7.
69. Azzam, R. M. A. and Basharra, N. M. 1977. *Ellipsometry and Polarized Light*. Amsterdam, the Netherlands: North-Holland.
70. Forouhi, A. R. and Bloomer, I. 1988. Optical properties of crystalline semiconductors and dielectrics. *Physical Review B* 38:1865–1874.
71. Rossow, U., Frotscher, U., Pietryga, C., Richter, W., and Aspnes, D. E. 1996. Interpretation of the dielectric function of porous silicon layers. *Applied Surface Science* 102:413–416.
72. Nguyen, H. V., Lu, Y., Kim, S., Wakagi, M., and Collins, R. W. 1995. Optical properties of ultrathin crystalline and amorphous silicon films. *Physical Review Letters* 74:3880–3883.
73. Pankove, J. I. 1975. *Optical Process in Semiconductors*. New York: Dover.
74. Ding, L., Chen, T. P., Liu, Y., et al. 2008. Evolution of photoluminescence mechanisms of Si^+-implanted SiO_2 films with thermal annealing. *Journal of Nanoscience and Nanotechnology* 8:3555–3560.
75. Penn, D. R. 1962. Wave-number-dependent dielectric function of semiconductors. *Physical Review* 128:2093–2097.
76. Wang, L.-W. and Zunger, A. 1996. Pseudopotential calculations of nanoscale CdSe quantum dots. *Physical Review B* 53:9579–9582.
77. Tsu, R., Babic, D., and Ioriatti, J. L. 1997. Simple model for the dielectric constant of nanoscale silicon particle. *Journal of Applied Physics* 82:1327–1329.
78. Delerue, C., Lannoo, M., and Allan, G. 2003. Concept of dielectric constant for nanosized systems. *Physical Review B* 68:115411–4.
79. Ding, L., Chen, T. P., Liu, Y., Ng, C. Y., and Fung, S. 2005. An approach to optical-property profiling of a planar-waveguide structure of Si nanocrystals embedded in SiO_2. *Nanotechnology* 16:2657–2660.
80. Lopez, M., Garrido, B., Bonafos, C., Perez-Rodriguez, A., and Morante, J. R. 2001. Optical and structural characterization of Si nanocrystals ion beam synthesized in SiO_2: Correlation between the surface passivation and the photoluminescence emission. *Solid-State Electronics* 45:1495–1504.
81. Garrido, B., Lopez, M., Gonzalez, O. et al. 2000. Correlation between structural and optical properties of Si nanocrystals embedded in SiO_2: The mechanism of visible light emission. *Applied Physics Letters* 77:3143–3145.
82. Nesbit, L. A. 1985. Annealing characteristics of Si-rich SiO_2 films. *Applied Physics Letters* 46:38–40.
83. Agarwal, A. M. and Dunham, S. T. 1995. Consistent quantitative model for the spatial extent of point defect interactions in silicon. *Journal of Applied Physics* 78:5313–5319.

84. Ding, L., Chen, T. P., Liu, Y. et al. 2005. Thermal annealing effect on the band gap and dielectric functions of silicon nanocrystals embedded in SiO_2 matrix. *Applied Physics Letters* 87:121903–3.
85. Chen, T. P., Liu, Y., Tse, M. S. et al. 2003. Dielectric functions of Si nanocrystals embedded in a SiO_2 matrix. *Physical Review B* 68:153301–4.
86. Ding, L., Chen, T. P., Liu, Y. et al. 2007. Influence of nanocrystal size on optical properties of Si nanocrystals embedded in SiO_2 synthesized by Si ion implantation. *Journal of Applied Physics* 101:103525–6.
87. Guha, S., Qadri, S. B., Musket, R. G., Wall, M. A., and Shimizu-Iwayama, T. 2000. Characterization of Si nanocrystals grown by annealing SiO_2 films with uniform concentrations of implanted Si. *Journal of Applied Physics* 88:3954–3961.
88. Cheylan, S. and Elliman, R. G. 2001. Effect of particle size on the photoluminescence from hydrogen passivated Si nanocrystals in SiO_2. *Applied Physics Letters* 78:1912–1914.
89. Bonafos, C., Carrada, M., Cherkashin, N. et al. 2004. Manipulation of two-dimensional arrays of Si nanocrystals embedded in thin SiO_2 layers by low energy ion implantation. *Journal of Applied Physics* 95:5696–5702.
90. Ng, C. Y., Chen, T. P., Zhao, P. et al. 2006. Electrical characteristics of Si nanocrystal distributed in a narrow layer in the gate oxide near the gate synthesized with very-low-energy ion beams. *Journal of Applied Physics* 99:106105–3.
91. Carrada, M., Cherkashin, N., Bonafos, C. et al. 2003. Effect of ion energy and dose on the positioning of 2D-arrays of Si nanocrystals ion beam synthesised in thin SiO_2 layers. *Materials Science and Engineering B* 101:204–207.
92. Normand, P., Beltsios, K., Kapetanakis, E. et al. 2001. Formation of 2-D arrays of semiconductor nanocrystals or semiconductor-rich nanolayers by very low-energy Si or Ge ion implantation in silicon oxide films. *Nuclear Instruments and Methods in Physics Research B* 178:74–77.
93. Ding, L., Chen, T. P., Wong, J. I. et al. 2006. Dielectric functions of densely stacked Si nanocrystal layer embedded in SiO_2 thin films. *Applied Physics Letters* 89:251910–3.
94. Kim, T.-Y., Park, N.-M., Kim, K.-H. et al. 2004. Quantum confinement effect of silicon nanocrystals in situ grown in silicon nitride films. *Applied Physics Letters* 85:5355–5357.
95. Zacharias, M., Heitmann, J., Scholz, R. et al. 2002. Size-controlled highly luminescent silicon nanocrystals: A SiO/SiO_2 superlattice approach. *Applied Physics Letters* 80:661–663.
96. Sham, T. K., Jiang, D. T., Coulthard, I. et al. 1993. Origin of luminescence from porous silicon deduced by synchrotron-light-induced optical luminescence. *Nature* 363:331–334.
97. Ghislotti, G., Nielsen, B., Asoka-Kumar, P. et al. 1996. Effect of different preparation conditions on light emission from silicon implanted SiO_2 layers. *Journal of Applied Physics* 79:8660–8663.
98. Lin, C. J. and Lin, G. R. 2005. Defect-enhanced visible electroluminescence of multi-energy silicon-implanted silicon dioxide film. *IEEE Journal of Quantum Electronics* 41:441–447.
99. Qin, G. G., Liu, X. S., Ma, S. Y. et al. 1997. Photoluminescence mechanism for blue-light-emitting porous silicon. *Physical Review B* 55:12876–12879.
100. Kobitski, A. Y., Zhuravlev, K. S., Wagner, H. P., and Zahn, D. R. T. 2001. Self-trapped exciton recombination in silicon nanocrystals. *Physical Review B* 63:115423–5.
101. Liao, L.-S., Bao, X.-M., Zheng, X.-Q., Li, N.-S., and Min, N.-B. 1996. Blue luminescence from Si^+-implanted SiO_2 films thermally grown on crystalline silicon. *Applied Physics Letters* 68:850–852.
102. Qin, G. G., Li, A. P., Zhang, B. R., and Li, B.-C. 1995. Visible electroluminescence from semitransparent Au film/extra thin Si-rich silicon oxide film/p-Si structure. *Journal of Applied Physics* 78:2006–2009.
103. Tong, S., Liu, X.-N., Gao, T., and Bao, X.-M. 1997. Intense violet-blue photoluminescence in as-deposited amorphous Si:H:O films. *Applied Physics Letters* 71:698–700.
104. Bonafos, C., Garrido, B., Lopez, M. et al. 1999. Structural and optical characterization of Mn doped ZnS nanocrystals elaborated by ion implantation in SiO_2. *Nuclear Instruments and Methods in Physics Research B* 147:373–377.
105. Lin, C.-J., Lee, C.-K., Diau, E. W.-G., and Lin, G.-R. 2006. Time-resolved photoluminescence analysis of multidose Si-ion-implanted SiO_2. *Journal of the Electrochemical Society* 153:E25–E32.
106. Nishikawa, H., Watanabe, E., Ito, D. et al. 1995. Photoluminescence study of defects in ion-implanted thermal SiO_2 films. *Journal of Applied Physics* 78:842–846.
107. Cheang-Wong, J. C., Oliver, A., Roiz, J. et al. 2001. Optical properties of Ir_2^+-implanted silica glass. *Nuclear Instruments and Methods in Physics Research B* 175–177:490–494.
108. Tohmon, R., Shimogaichi, Y., Mizuno, H. et al. 1989. 2.7-eV luminescence in as-manufactured high-purity silica glass. *Physical Review Letters* 62:1388–1391.

109. Bae, H. S., Kim, T. G., Whang, C. N. et al. 2002. Electroluminescence mechanism in SiO_x layers containing radiative centers. *Journal of Applied Physics* 91:4078–4081.
110. Nishikawa, H., Stahlbush, R. E., and Stathis, J. H. 1999. Oxygen-deficient centers and excess Si in buried oxide using photoluminescence spectroscopy. *Physical Review B* 60:15910–15918.
111. Stathis, J. H. and Kastner, M. A. 1984. Photoinduced paramagnetic defects in amorphous silicon dioxide. *Physical Review B* 29:7079–7081.
112. Skuja, L. 1992. Time-resolved low temperature luminescence of non-bridging oxygen hole centers in silica glass. *Solid State Communications* 84:613–616.
113. Munekuni, S., Yamanaka, T., Shimogaichi, Y. et al. 1990. Various types of nonbridging oxygen hole center in high-purity silica glass. *Journal of Applied Physics* 68:1212–1217.
114. Bakos, T., Rashkeev, S. N., and Pantelides, S. T. 2002. The origin of photoluminescence lines in irradiated amorphous SiO_2. *IEEE Transactions on Nuclear Science* 49:2713–2717.
115. Valakh, M. Y., Yukhimchuk, V. A., Bratus, V. Y. et al. 1999. Optical and electron paramagnetic resonance study of light-emitting Si^+ ion implanted silicon dioxide layers. *Journal of Applied Physics* 85:168–173.
116. Liu, Y., Chen, T. P., Fu, Y. Q. et al. 2003. A study on Si nanocrystal formation in Si-implanted SiO 2 films by x-ray photoelectron spectroscopy. *Journal of Physics D: Applied Physics* 36:L97–L100.
117. Kanemitsu, Y., Shimizu, N., Komoda, T., Hemment, P. L. F., and Sealy, B. J. 1996. Photoluminescent spectrum and dynamics of Si^+-ion-implanted and thermally annealed SiO_2 glasses. *Physical Review B* 54:14329–14332.
118. Allan, G., Delerue, C., and Lannoo, M. 1996. Nature of luminescent surface states of semiconductor nanocrystallites. *Physical Review Letters* 76:2961–2964.
119. Zhuravlev, K. S., Gilinsky, A. M., and Kobitsky, A. Y. 1998. Mechanism of photoluminescence of Si nanocrystals fabricated in a SiO_2 matrix. *Applied Physics Letters* 73:2962–2964.
120. Komoda, T., Kelly, J., Cristiano, F. et al. 1995. Visible photoluminescence at room temperature from microcrystalline silicon precipitates in SiO_2 formed by ion implantation. *Nuclear Instruments and Methods in Physics Research B* 96:387–391.
121. Koshida, N. and Koyama, H. 1992. Visible electroluminescence from porous silicon. *Applied Physics Letters* 60:347–349.
122. Fojtik, A., Valenta, J., Stuchlikova, T. H. et al. 2006. Electroluminescence of silicon nanocrystals in p-i-n diode structures. *Thin Solid Films* 515:775–777.
123. Irrera, A., Iacona, F., Franzo, G. et al. 2005. Correlation between electroluminescence and structural properties of Si nanoclusters. *Optical Materials* 27:1031–1040.
124. Irrera, A., Pacifici, D., Miritello, M. et al. 2003. Electroluminescence properties of light emitting devices based on silicon nanocrystals. *Physica E: Low-dimensional Systems and Nanostructures* 16:395–399.
125. Kulakci, M., Serincan, U., and Turan, R. 2006. Electroluminescence generated by a metal oxide semiconductor light emitting diode (MOS-LED) with Si nanocrystals embedded in SiO_2 layers by ion implantation. *Semiconductor Science and Technology* 21:1527–1532.
126. Lalic, N. and Linnros, J. 1999. Light emitting diode structure based on Si nanocrystals formed by implantation into thermal oxide. *Journal of Luminescence* 80:263–267.
127. Luterova, K., Pelant, I., Valenta, J. et al. 2000. Red electroluminescence in Si^+-implanted sol-gel-derived SiO_2 films. *Applied Physics Letters* 77:2952–2954.
128. Matsuda, T., Nishihara, K., Kawabe, M. et al. 2004. Blue electroluminescence from MOS capacitors with Si-implanted SiO_2. *Solid-State Electronics* 48:1933–1941.
129. Muller, D., Knapek, P., Faure, J. et al. 1999. Blue electroluminescence from high dose Si^+ implantation in SiO_2. *Nuclear Instruments and Methods in Physics Research B* 148:997–1001.
130. Rebohle, L., von Borany, J., Yankov, R. A. et al. 1997. Strong blue and violet photoluminescence and electroluminescence from germanium-implanted and silicon-implanted silicon-dioxide layers. *Applied Physics Letters* 71:2809–2811.
131. Song, H.-Z., Bao, X.-M., Li, N.-S., and Zhang, J.-Y. 1997. Relation between electroluminescence and photoluminescence of Si^+-implanted SiO_2. *Journal of Applied Physics* 82:4028–4032.
132. Ding, L., Chen, T. P., Liu, Y. et al. 2007. The influence of the implantation dose and energy on the electroluminescence of Si^+-implanted amorphous SiO_2 thin films. *Nanotechnology* 18:455306–6.
133. Ding, L., Chen, T. P., Yang, M. et al. Unpublished. Relationship between current transport and electroluminescence in Si^+-implanted SiO_2 thin films. *IEEE Transaction on Electron Devices* (submitted).
134. Chen, T. P., Tse, M. S., and Zeng, X. 2001. Snapback behavior of the postbreakdown *I–V* characteristics in ultrathin SiO_2 films. *Applied Physics Letters* 78:492–494.
135. Min, K. S., Shcheglov, K. V., Yang, C. M. et al. 1996. Defect-related versus excitonic visible light emission from ion beam synthesized Si nanocrystals in SiO_2. *Applied Physics Letters* 69:2033–2035.

136. Yi, L. X., Heitmann, J., Scholz, R., and Zacharias, M. 2002. Si rings, Si clusters, and Si nanocrystals—Different states of ultrathin SiO_x layers. *Applied Physics Letters* 81:4248–4250.
137. Yi, L. X., Heitmann, J., Scholz, R., and Zacharias, M. 2003. Phase separation of thin SiO layers in amorphous SiO/SiO 2 superlattices during annealing. *Journal of Physics: Condensed Matter* 15:S2887–S2895.
138. Fazio, B., Vulpio, M., Gerardi, C. et al. 2002. Residual crystalline silicon phase in silicon-rich-oxide films subjected to high temperature annealing. *Journal of the Electrochemical Society* 149:G376–G378.
139. Chen, X. Y., Lu, Y. F., Wu, Y. H. et al. 2006. Correlation between optical properties and Si nanocrystal formation of Si-rich Si oxide films prepared by plasma-enhanced chemical vapor deposition. *Applied Surface Science* 253:2718–2726.
140. Cheylan, S., Langford, N., and Elliman, R. G. 2000. The effect of ion-irradiation and annealing on the luminescence of Si nanocrystals in SiO_2. *Nuclear Instruments and Methods in Physics Research B* 166–167:851–856.
141. Garrido, B., Lopez, M., Perez-Rodriguez, A. et al. 2004. Optical and electrical properties of Si-nanocrystals ion beam synthesized in SiO_2. *Nuclear Instruments and Methods in Physics Research B* 216:213–221.
142. Shimizu-Iwayama, T., Hole, D. E., and Ian, W. B. 1999. Mechanism of photoluminescence of Si nanocrystals in SiO_2 fabricated by ion implantation: The role of interactions of nanocrystals and oxygen. *Journal of Physics: Condensed Matter* 11:6595–6604.
143. Nayfeh, M. H., Rigakis, N., and Yamani, Z. 1997. Photoexcitation of Si-Si surface states in nanocrystallites. *Physical Review B* 56:2079–2084.
144. Hybertsen, M. S. 1994. Absorption and emission of light in nanoscale silicon structures. *Physical Review Letters* 72:1514–1517.
145. Sahoo, P. K., Gasiorek, S., Dhar, S., Lieb, K. P., and Schaaf, P. 2006. Cathodoluminescence and epitaxy after laser annealing of Cs^+-irradiated α-quartz. *Applied Surface Science* 252:4477–4480.

5 Electrical Properties of Silicon Nanocrystals Embedded in Amorphous SiO_2 Films

Wali Zhang, Sam Zhang, Zhen Liu, and T.P. Chen

CONTENTS

5.1 INTRODUCTION

Nonvolatile memory chips with low power consumption and low cost have attracted increasing attention due to the booming portable electronic devices market, particularly devices such as cellular phones and digital cameras. There are mainly four types of nonvolatile memory technologies: flash memory, ferro-electric random access memory, magnetic random access memory, and phase change memory. Among these, flash memory can achieve the highest chip density and possesses multi-bit per cell storage capability [1–3]. Therefore, it has become the mainstream nonvolatile

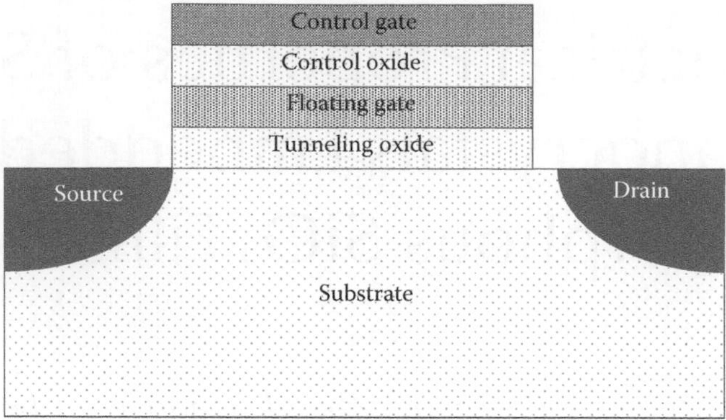

FIGURE 5.1 Schematic cross-section of a floating-gate memory device, in which the tunneling oxide must be thicker than 8 nm to maintain 10 years of retention time at 85°C.

memory technology nowadays. Furthermore, flash memory enjoys a fabrication process compatible with the current complementary metal-oxide-semiconductor (CMOS) process, and thus has become a suitable solution for embedded memory applications.

A floating-gate flash memory structure is shown in Figure 5.1. The gate stack consists of a thermal oxide as the tunneling layer, a floating polysilicon gate as the charge storage layer, and an inter-poly oxide layer as the control oxide layer. Flash memory devices with this kind of structure usually need a programming voltage higher than 10 V and a tunneling oxide thicker than 8 nm for a required retention time in industry (around 10 years) [4].

5.1.1 Quantum Dot Nonvolatile Flash Memory

The market demand of high density, low cost, and low power consumption in flash memory technology results in aggressive scaling of semiconductor memory cells and a dramatic increase in the density of the memory array. The scaling down of conventional nonvolatile flash memory can be achieved by continuously thinning the control and tunneling oxide [5]. To improve the write/erase speed of a floating-gate memory device, the thickness of the tunneling oxide must be reduced further [5,6], for example, the tunneling oxide must be less than 2.5 nm in order to achieve a write/erase time of 100 ns for a reasonable programming voltage (less than 10 V) [5]. Unfortunately, certain defects in the tunneling oxide can lead to a huge leakage current that greatly reduces the retention time. For a conventional floating-gate flash memory device, if there is a defect in the tunneling oxide, a tunneling path could be formed through the defect under a high electric field, and all of the charges stored on the floating gate will gradually leak back to either the channel or the source/drain through the tunneling path, leading to the failure of the device. Therefore, the floating-gate memory requires a thick tunneling oxide to reduce the defect-related charge loss. This limits the scaling down of the floating-gate flash memory device. Currently, commercial flash memory devices use a tunneling oxide thicker than 8 nm to guarantee a long retention time, which results in high programming voltage and slow operation speed [7].

The utilization of discrete-trap storage nodes in flash memory devices has been proposed as one of the key concepts for the continuous survival of existing flash memory technology. The basic idea of the "discrete traps" mechanism is to replace the continuous floating gate in nonvolatile memory devices by a layer of discrete charge trapping centers [1–4,8–11], such as the quantum dots, as shown in Figure 5.2. Quantum dot memory can significantly reduce the leakage problem during retention. In a quantum dot flash memory device, a single leakage path due to a defect in the tunneling oxide can only discharge the charges stored in the particular dot near the defect since the

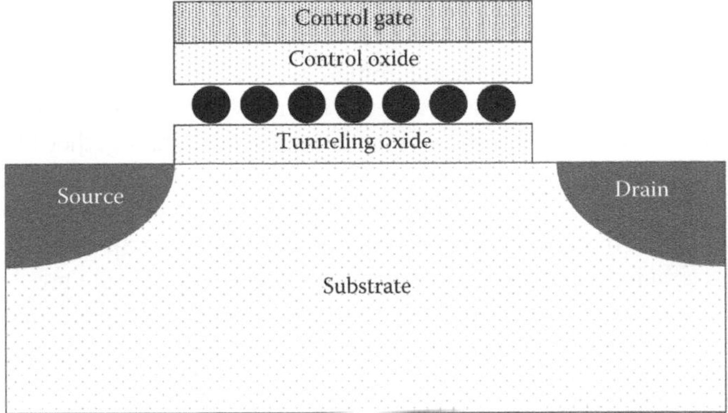

FIGURE 5.2 Schematic of a quantum dot nonvolatile memory device.

dots are separated from each other within the gate dielectric. Hence the tunneling oxide thickness in the quantum dot memory can be reduced. The reduction in thickness enables direct tunneling, hence, faster write/erase operations are achieved compared to the conventional flash memory devices (mainly the Fowler–Nordheim tunneling). The thinner tunneling oxide also allows for a lower voltage operation and less power consumption. Many types of quantum dot memories have been demonstrated in the literature. Among these, Si nanocrystal (nc-Si)–based memories have been considered as one of the most promising candidates because their fabrication process is fully compatible with the mainstream CMOS process.

The write and erase processes for an n-channel quantum dot memory device are illustrated in Figure 5.3. During the write process, a positive gate voltage is applied to inject the electrons from the inversion layer of the Si substrate into the nanocrystals. During the erase process, a negative gate bias is applied to cause the electrons in the nc-Si to tunnel back into the source region. If the quantum dots are well separated, the lateral tunneling between the quantum dots can be neglected; this is essential in enhancing the retention time and reliability when compared with the conventional floating-gate flash memory.

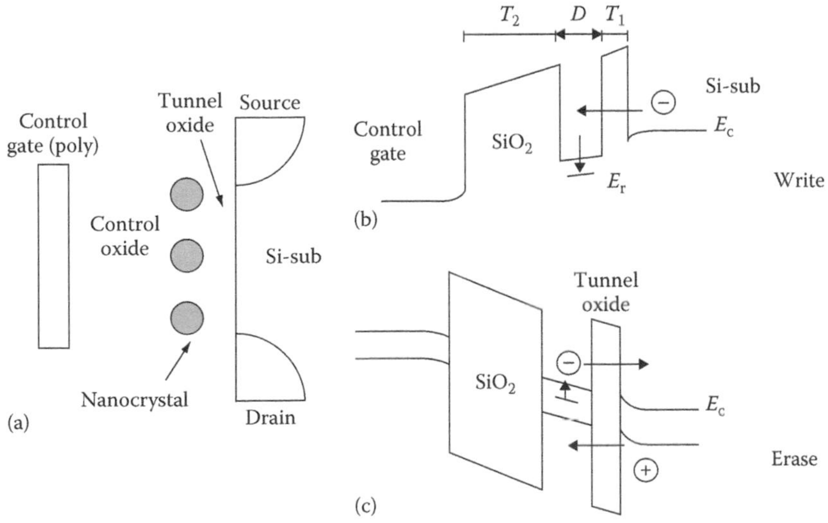

FIGURE 5.3 (a) Schematic cross-section of a nanocrystal memory device structure; (b) illustration of write process; (c) illustration of erase process.

5.1.2 Band Structure of Silicon

A silicon atom has 14 electrons with a configuration of $1s^2 2s^2 2p^6 3s^2 3p^2$. In other words, these 14 electrons are distributed such that there are two each in the 1s and 2s states, six in the 2p states, and two each in the 3s and 3p states. When the silicon atoms are brought together to form a crystal, the discrete levels are broadened into bands so that all the electrons originally in a given energy state have energies slightly different from each other. Because there are so many atoms per unit volume, these bands may be regarded as continuous. In the crystal, the 3s and 3p states intermingle to form two bands separated by a large energy gap without the energy states. There are four quantum states per atom in the lower band, called the valence band, and four quantum states per atom in the upper band, called the conduction band. The gap between these two bands is called the forbidden energy gap, or simply the forbidden gap.

The band structure of the silicon is quite complex. Both experimental and theoretical works have been conducted to establish the shapes of the conduction and valence bands. An important qualitative aspect of the band structure is as follows: the minimum of the conduction band is displaced from the maximum of the valence band so that a momentum change is required when the electrons move across the gap from the valence band to the conduction band [12].

The band structure of silicon is schematically shown in Figure 5.4 [13], where the energy is plotted as a function of the wave vector k, along the main crystallographic directions in the crystal. This figure shows the lowest energy of the bands above the forbidden gap and the highest energy of the bands below the gap as a function of wave vector k, along two low index directions. The energy gap is the difference between the valence band at $k=0$ and the lowest point in the conduction band.

When the lowest point occurs at $k=0$, the semiconductor is that of a direct gap type because a transition can occur at the zone center with both the initial and final states having the same momentum vector, $k=0$. While, for the silicon the minimum energy in the conduction band is shifted by a k-vector relative to the valence band. The k-vector difference represents a difference in momentum, resulting in an indirect band gap. The bulk silicon is an indirect band gap semiconductor, with a band gap of 1.12 eV at room temperature.

5.1.3 The Band Gap in nc-Si

Quantum dots are a unique class of material because they are so small, ranging from 2 to 10 nm in diameter. At the size of this range, materials behave differently, giving the quantum

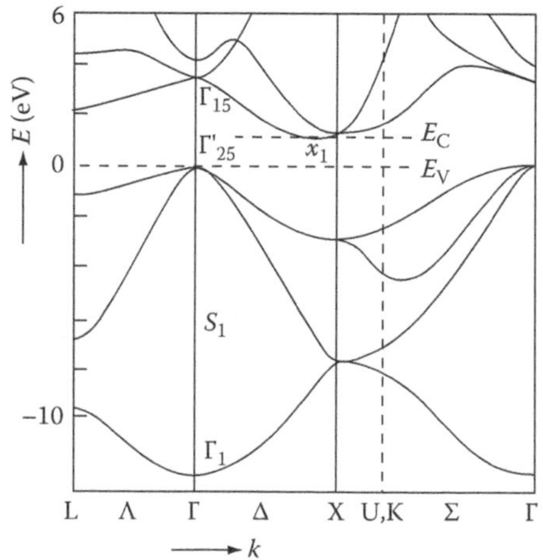

FIGURE 5.4 Energy band diagram of silicon. (Reprinted from Chelikowsky, J.R. and Cohen, M.L., *Phys. Rev. B*, 10, 5095, 1974. With permission.)

dot an unprecedented tenability. In bulk materials, the energy levels are very close together, so close that they are described as continuous, meaning there is almost no energy difference between them. For the quantum dots, the concepts of energy level, band gap, conduction band, and valence band still apply. However, the electron energy levels have become discrete instead of being continuous. Excitons have an average physical separation between the electron and hole, referred to as the exciton Bohr radius. In bulk, the dimensions of the semiconductor crystals are much larger than the exciton Bohr radius, allowing the exciton to extend to its natural limit. However, for a quantum dot small enough whose size approaches that of the material's exciton Bohr radius, the electron energy levels can no longer be treated as continuous. They must be treated as discrete, meaning that there is a small and finite separation between the energy levels.

Because the quantum dot's electron energy levels are discrete rather than continuous, the addition or subtraction of just a few atoms to the quantum dot has the effect of altering the boundaries of the band gap. The size dependence of the energy band gap of the Si quantum dot has been studied widely, both in theoretical and experimental researches. The theoretical works concerning the size dependence of the band gap of quantum dots were first reported in the early 1980s [14]. Most of the theoretical works reported were based on the effective mass theory and tight-binding semi-empirical approaches [14–20]. The relationship between the band gap of the nc-Si and its size can be calculated using (Equation 5.1 [14])

$$E_g = E_g^0 + \frac{C}{d^n} \tag{5.1}$$

where
- d is the diameter of the nanocrystal
- E_g is the band gap of nc-Si
- E_g^0 is the band gap of bulk silicon at room temperature
- C is an appropriately dimensioned (energy × (length)^n) constant
- n is the exponent related to the material

This size-dependent band gap of the semiconductor nanoclusters has been directly observed by light emission studies. A size-dependent photoluminescence (PL) has been observed by many researchers from the semiconductor quantum dots. Ledoux et al. [21] have synthesized Si nanoparticles with size ranging from 2 to 8 nm by pulsed CO_2 laser pyrolysis of silane in a gas flow reactor. The correlation between the average diameter and the PL peak energy is shown in Figure 5.5 [21]. It can be seen that the size dependence of the PL peak energy agrees well with (Equation 5.1).

However, the predication of Equation 5.1 shows somewhat of a disagreement with the experiments shown in Figure 5.6. One of the reasons for this is that the theoretical calculations were performed for a single nanocrystal, whereas experiments were performed on an ensemble of quantum dots of various sizes. Ranjan et al. have improved the theoretical calculation by taking the size distribution into account [14]. The band gap of the nc-Si is given by [14]

$$E_g = E_g^0 + \frac{C}{d^n}\left(\frac{d_m}{d_0}\right)^{n(2n+5)/3} \tag{5.2}$$

where
- d_0 is the mean size of the nanocrystals
- d_m is the size at which the maximum occurs in the normal distribution

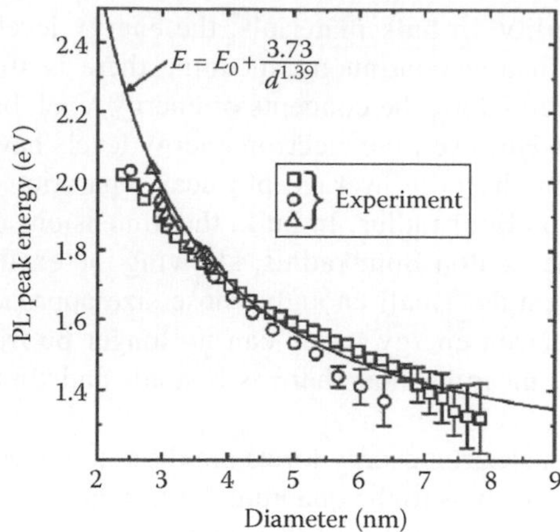

FIGURE 5.5 Size dependence of PL peak energy. (Reprinted from Ledoux, G. et al., *Appl. Phys. Lett.*, 80, 4834, 2002. With permission.)

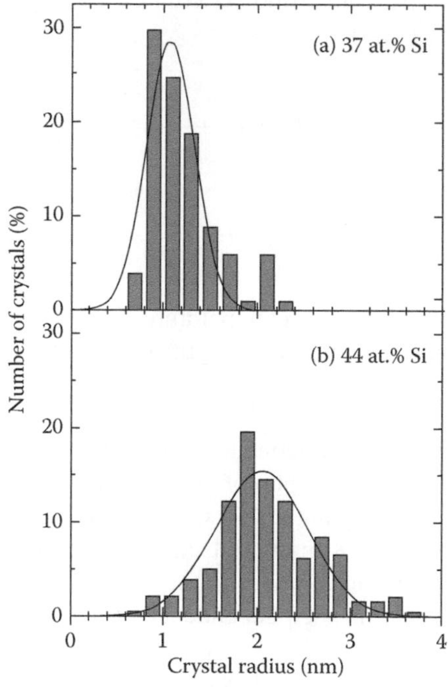

FIGURE 5.6 nc-Si size distribution for the SiO_x samples having a Si concentration of 37 at.% (a) and 44 at.% (b), after annealing at 1250°C. (From Franzo G. et al., *Mater. Sci. Eng. B*, 69–70, 454, 2000. With permission.)

5.2 MICROSTRUCTURE OF nc-Si/a-SiO_2 NANOCOMPOSITES

The nanocomposite nc-Si/a-SiO_2 has drawn much attraction for its potential applications in quantum dots memory [5,22–24]. The memory effect is realized by the charging/discharging in the nc-Si that can lead to flat band voltage shifts of MOS capacitors based on nc-Si/a-SiO_2 films. The nc-Si/a-SiO_2 can be synthesized by ion implantation of Si ions into an amorphous SiO_2 matrix followed by a thermally induced crystallization process [25–40], deposition of sub-stoichiometric oxide films using chemical vapor deposition [41–54], sputtering process [55–64], or by reactive evaporation followed by a thermally induced phase separation and crystallization of the excess Si [65–70]. This process is compatible with the modern CMOS process.

The electrical and optical properties of the nc-Si/a-SiO_2 nanocomposite films are strongly correlated with their microstructures. Thus, a clear understanding on the microstructure of the nc-Si/a-SiO_2 nanocomposites is mandatory to realize all kinds of novel applications.

5.2.1 Types of nc-Si/a-SiO_2 Nanocomposites

The simplest nc-Si/a-SiO_2 nanocomposite is a single layer of Si-rich oxide film. The nc-Si is distributed throughout the oxide layer. Conventional synthesis of the nc-Si by ion implantation, sputtering, CVD, or evaporation results in a relatively wide size distribution of the nc-Si [7,46,71–76] as shown in Figure 5.6. The wide size distribution of nc-Si strongly affects the optical and electrical properties due to the size-dependent band gap nature of the quantum dots [7,52,77]. Although the research on room-temperature PL from nc-Si/a-SiO_2 nanocomposites has been conducted for more than 10 years, the commercial Si-based light-emitting diode (LED) is still not available. One of the critical issues is the broad size distribution and random spatial arrangement of the nc-Si in the SiO_2 matrix.

Size control of the nc-Si is normally realized by shrinking the entire size distribution by varying the silicon content within the SiO_2 matrix, by changing the etching conditions in the case of porous silicon, or by subsequent oxidation of the nanocrystals [78–80]. A broad size distribution complicates the characterization of the quantum confinement effects and is undesirable for potential applications. In recent years, new ways of narrowing the size distribution of the nc-Si have been developed. Among them, the utilization of the Si/SiO_2 superlattice nanocomposite is a popular method [55,62,81–86]. In this methodology, the Si/SiO_2-multilayer structures can be synthesized by alternative deposition of the Si and SiO2 layers by CVD, sputtering, evaporation, etc. For such a structure, the nc-Si average size is determined by the thickness of the Si layers. On the other hand, since the growth of the nc-Si is restricted by the SiO_2 layers, the size uniformity can be controlled. However, there are some disadvantages of the nc-Si related to this method. Firstly, the nc-Si thus produced are generally brick-shaped rather than spherical. Secondly, the brick-shaped nc-Si are not well passivated by the SiO_2 matrix. Grain boundaries exist between the different brick-like nc-Si that are undesirable because numerous non-radiative recombination centers may form.

Very recently, a new approach for the tight control of the size distribution of nc-Si in the SiO_2 matrix was reported that employs the phase separation of the ultrathin SiO_x layers within SiO_x/a-SiO_2 superlattice nanocomposites, where $1<x<2$ [11,67,68,87–91]. In this approach, the Si-rich oxide layers replaced the Si layers in the Si/a-SiO_2 multilayer films during fabrication. The phase separation of the SiO_x layer results in the formation of amorphous silicon clusters in the SiO_2 matrix during a low-temperature annealing, and the amorphous Si clusters crystallize at the subsequent high-temperature annealing. In such a structure, the average size of the nanocrystals is controlled by the SiO_x layer thickness, and the size uniformity can be controlled by the encapsulation of the SiO_2 layers. Such a structure can overcome the disadvantages encountered in the Si/SiO_2-multilayer structure. The nc-Si thus fabricated are spherical and well passivated by the newly segregated SiO_2 and initial SiO_2 layers, as shown in Figure 5.7 [68].

Besides the size uniformity, tight control of the density of the nc-Si can also be achieved by the SiO_x/a-SiO_2 superlattice structure. The density of nc-Si can be controlled by the concentration of the Si in the SiO_x layers. Lu et al. have reported that they have successfully synthesized nc-Si with a very narrow size distribution and different densities by controlling the concentration of the Si in the SiO_x layer as shown in Figure 5.8 [11]. The stoichiometry of the SiO_x layer in Figure 5.8a is $SiO_{0.9}$, and the average nc-Si density is calculated to be $2.79\times10^{12}/cm^2$; while the stoichiometry of the SiO_x layer in Figure 5.8b is $SiO_{1.63}$, and the average nc-Si density is $0.93\times10^{12}/cm^2$. The average sizes of nc-Si in these two samples are both around 3.8 nm.

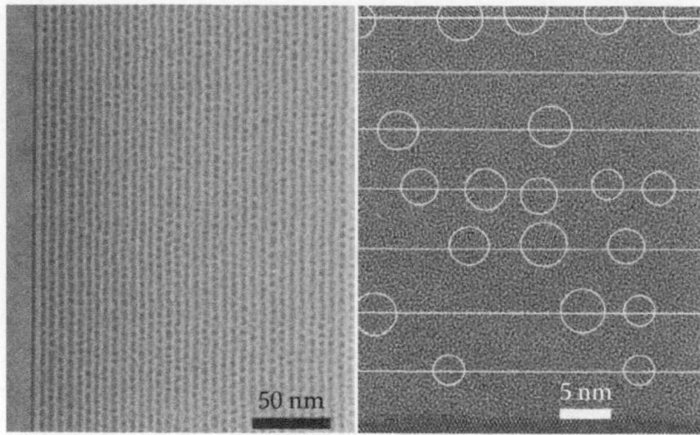

FIGURE 5.7 (a) TEM image of SiO/SiO_2 superlattice with 3 nm thick SiO layer after annealing at 1100°C under N_2 ambient, (b) HRTEM image of the same sample. White circles indicate well-oriented nanocrystals, and white lines are parallel to the substrate surface. (Reprinted from Heitmann, J. et al., *J. Non-Cryst. Solids*, 299, 1075, 2002. With permission.)

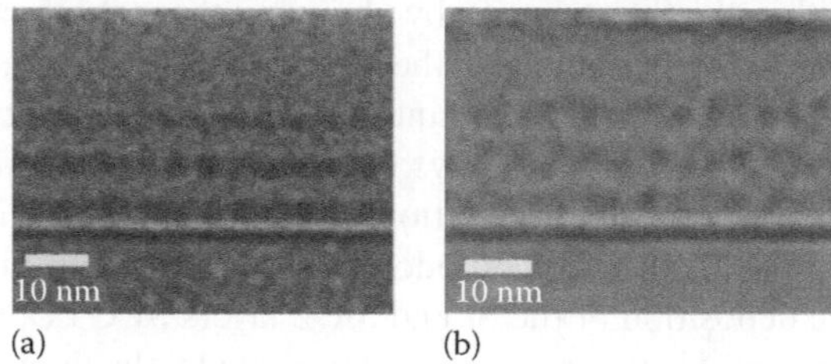

FIGURE 5.8 SiO_x/SiO_2 superlattices with $x=0.9$ (a) and $x=1.64$ (b). (Reprinted from Lu, T.Z. et al., *J. Appl. Phys.*, 100, 5, 2006. With permission.)

5.2.2 Local Atomic Structure of the Amorphous SiO_x Films

Si nanocrystals are generally formed by thermally annealing the as-deposited amorphous SiO_x films. Thus, the microstructure and the chemical structure of the as-deposited films will play a critical role in the growth of nc-Si. The local atomic structure of SiO_x was first brought to the attention of researchers by Philipp [92] 30 years ago. The thin films studied by him were characterized by optical absorption spectroscopy and visible, near ultraviolet, and UV reflectance spectroscopy. He proposed the Random-Bonding Model (RBM) of the nc-Si/a-SiO_2 nanocomposites in which the local bonding was statistical in nature, and was characterized by five different local bonding environments, Si–$Si_{4-n}O_n$, where $n=0$, 1, 2, 3, 4 that varied systematically as the oxide composition changed. However, it is well known that Si suboxides have a possibility to be phase-separated into Si and SiO_2. This concept was discussed in more detail based on a theoretical approach by Temkin [93], who proposed a Random-Mixture Model (RMM), in which the microstructure of the Si-rich oxide are mixtures of Si and SiO_2 phases.

Many researchers have confirmed these two models experimentally. It seems that the microstructures of the SiO_x films synthesized with different methods can be quite different. The SiO_x films fabricated with CVD and evaporation approaches tend to fall in the RBM regime. Temkin had theoretically calculated the probability of the nth tetrahedral configuration, C_n, for $n=1$ through 5, versus concentration x, in SiO_x, for the random model under the following assumptions [93]:

1. Each Si atom is bonded to four other atoms
2. Each O atom is bonded to two other atoms

3. Atoms do not bond with one another
4. The average S–O bond length is approximately 1.62 Å
5. The average Si–Si bond length is approximately 2.35 Å
6. The average bond angle at an Si vertex is the tetrahedral value, 109.5°
7. The average bond angle at an O vertex is approximately 144°

The probability of the *n*th tetrahedral configuration in the SiO_x amorphous films is

$$C_n(x) = \frac{4!}{(4-n)!n!}\left(\frac{1}{2}x\right)^n\left[\frac{1}{2}(2-x)\right]^{4-n} \quad (5.3)$$

The distribution probabilities of the substitution tetrahedral $Si\text{–}Si_{4-n}O_n$, with $n=0, \ldots, 4$ of the random bonding model according to Temkin's equation is shown in Figure 5.9.

Besides theoretical works, solid experimental results have also been reported to support the RBM. Figure 5.10 shows the XPS 2p core level of the as-deposited SiO_x films with different Si concentrations by thermal evaporation [95]. The fitting of the XPS 2p spectrum shows that the SiO_x films containing all Si chemical states and the fitted concentration of the Si−O components is consistent with the random model. The study in Ref. [96] reinforced the conclusions of the earlier research of Philipp, which regarded the nature of the local atomic structure of the as-deposited SiO_x thin films to be a random bonding structure, and provided additional quantitative information.

The RMM assumes in small domains silicon is bonded to silicon, or in the other types of domains, only to oxygen. This corresponds to a two-phase mixture. However, physical and chemical properties of the material are quite different from the properties of a macroscopic mixture of phase, and therefore the size of the separated regions was assumed to be ~10 Å [93]. Only a thin boundary layer between the domains of silicon and SiO_2 was postulated. Dupree et al. [97] successfully performed magic-angle spinning investigations on silicon monoxide and estimated that the domains are even larger than 20 Å. Very recently, on the basis of transmission electron microscopy, Schulmeister et al. used a series of electron spectroscopic images to investigate the configuration of the amorphous silicon monoxide powders as shown in Figure 5.11 [98]. The resulting Si (oxygen) map shown in Figure 5.11 clearly exhibits regions where silicon (oxygen) is enriched relative to the regions with lower silicon (oxygen) content. The regions which are silicon (oxygen)-rich appear well separated in the thin part of the specimen and are between 3 and 4 nm in diameter. In intensity profiles of

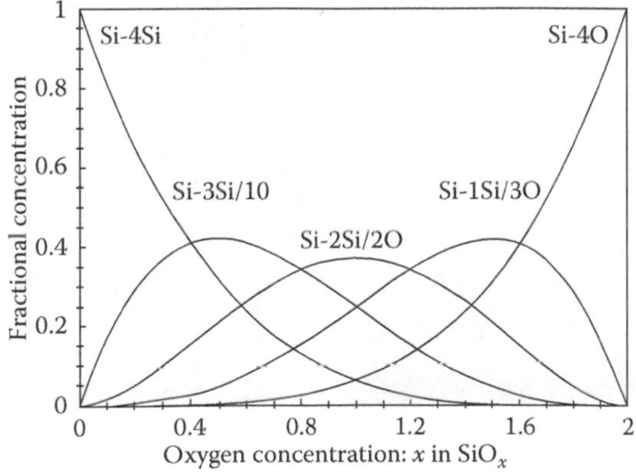

FIGURE 5.9 Distribution probabilities of the substitution tetrahedral $Si\text{—}Si_{4-n}O_n$, with $n=0,\ldots,4$ of the random bonding model in comparison with the Si/SiO_2 mixture model. (Reprinted from Lucovsky, G., *J. Non-Cryst. Solids*, 227–230, 1, 1998. With permission.)

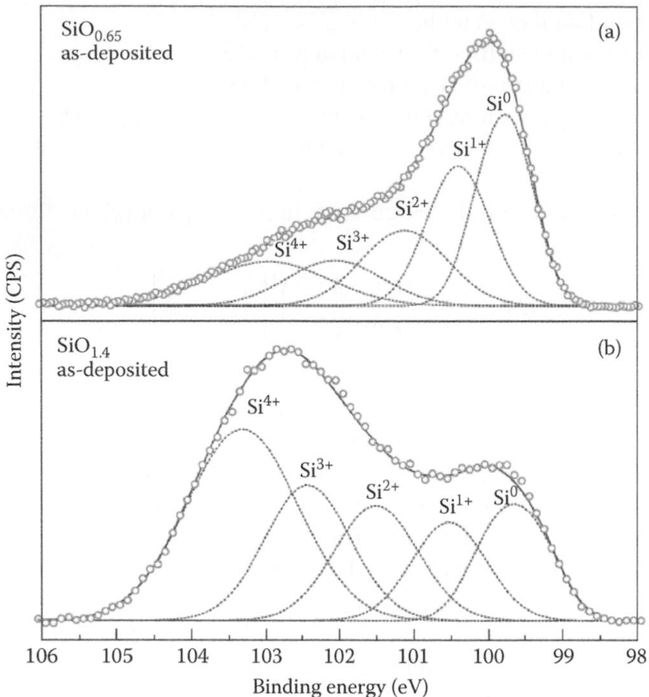

FIGURE 5.10 Si 2p core level and the fitted Si—O components for SiO_x samples with different oxygen concentrations x. (Reprinted from Zhang, W. et al., *J. Cryst. Growth*, 311, 1296, 2009. With permission.)

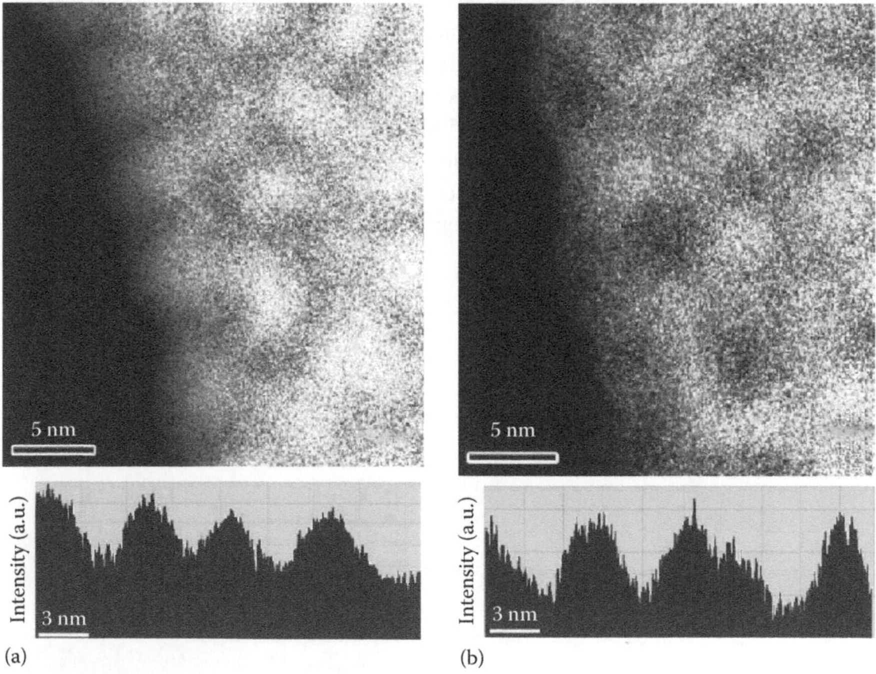

FIGURE 5.11 (a) Distribution map of silicon in silicon monoxide, brightness correlates with Si content and profile of silicon distribution. (b) Distribution map of oxygen in silicon monoxide, brightness correlates with oxygen content and profile of oxygen distribution. (Reprinted from Schulmeister, K. and Mader, W., *J. Non-Cryst. Solids*, 320, 143, 2003.)

the silicon (oxygen) map, the variation in the chemical composition and the size of the separated regions is displayed. The distribution maps of the elements yield direct proof for the existence of a chemically inhomogeneous microstructure on the nanometer scale. There are oxygen-rich regions representing the SiO_2 phase and others containing elemental silicon.

The RBM and RMM are idealistic atomic structures of the SiO_x films. More realistic models are the nanometer-scale mixture models with different cluster sizes. A cluster mixture model suggesting that the microstructure of SiO_x contains nanoscale amorphous clusters of Si and SiO_2 embedded in the a-SiO_x matrix has been proposed [99].

5.2.3 Thermal Decomposition of Si Suboxides

The as-deposited amorphous SiO_x films synthesized by most of the methods contain significant Si suboxides. However, these Si suboxides have a tendency to decompose into Si and SiO_2 [58].

$$\mathrm{SiO}_x \rightarrow \left(1-\frac{x}{2}\right)\mathrm{Si}+\frac{x}{2}\mathrm{SiO}_2 \tag{5.4}$$

The detail of the decomposition process is interpreted by assuming that the SiO_x films are formed by tetrahedral Si−(O_n,Si_{4-n}) (n=0,1,2,3,4) units, where n is the oxidation state of the central Si atoms [100]. Each oxygen atom is bound to two silicon atoms, forming a Si−O−Si bond. The dependence of the stability on Si oxidation numbers was obtained from molecular orbital calculations for (O_n,Si_{3-n})−Si−Si−(O_m,Si_{3-m}) and (O_n,Si_{3-n})−Si−O−Si−(O_m,Si_{3-m}) (n,m=0,1,2,3) clusters. The stability of a Si−O−Si bond is measured by the energy of the following oxidation reaction, in which the oxygen atom is inserted into a Si−Si bond:

$$(\mathrm{O}_n,\mathrm{Si}_{3-n})\text{−Si−Si−}(\mathrm{O}_m,\mathrm{Si}_{3-m})+\frac{1}{2}\mathrm{O}_2 \rightarrow (\mathrm{O}_n,\mathrm{Si}_{3-n})\text{−Si−O−Si−}(\mathrm{O}_m,\mathrm{Si}_{3-m}) \tag{5.5}$$

Table 5.1 shows the calculated oxidation energies, $E(n,m)$ [100]. It shows that the more exothermic is the reaction, the higher is the initial oxidation states, n and m, of the central Si atoms. The energies in Table 5.1 allow us to calculate the energies of each decomposition reaction. The first reaction is the partial decomposition

$$\mathrm{Si}^{2+}+\mathrm{Si}^{2+} \rightarrow \mathrm{Si}^{1+}+\mathrm{Si}^{3+} \tag{5.6}$$

TABLE 5.1
Calculated Energies $E(n,m)$ in kJ/mol for the Reaction (O_n,Si_{3-n})−Si−Si−(O_m,Si_{3-m}) + 1/2O_2 → (O_n,Si_{3-n})−Si−O−Si−(O_m,Si_{3-m})

n/m	0	1	2	3
0	−371			
1	−401	−409		
2	−411	−436	−444	
3	−425	−445	−461	−470

Source: Reprinted from Barranco, A. et al., *Vacuum*, 2002, 67, 491, 2002. With permission.

The energy of this process can be estimated as the energy of removal of a central oxygen in a $(O,Si_2)-Si-O-Si-(O,Si_2)$ (i.e., $-E(1,1)$) cluster that results in a $Si^{2+}-Si^{2+}$ pair, plus the energy of insertion of oxygen in a $(O_2,Si_1)-Si-Si-(O_2,Si_1)$ (i.e., $E(2,2)$) cluster that results in a $Si^{3+}-Si^{3+}$ pair. The net reaction is

$$(O,Si_2)-Si-O-Si-(O,Si_2)+(O_2,Si)-Si-Si-(O_2,Si) \rightarrow (O,Si_2)-Si-Si-(O,Si_2) + (O_2,Si)-Si-O-Si-(O_2,Si) \quad (5.7)$$

The energy of this process is $E(2,2)-E(1,1)=-35$ kJ/mol. Therefore, the decomposition into the $Si^{1+}+Si^{3+}$ mixture is exothermic. The second decomposition reaction is

$$Si^{1+} + Si^{3+} \rightarrow Si^{0} + Si^{4+} \quad (5.8)$$

This process can be modeled as the oxygen exchange between $(Si_3)-Si-O-Si-(Si_3)$ and $(O_3)-Si-Si-(O_3)$ clusters:

$$(Si_3)-Si-O-Si-(Si_3)+(O_3)-Si-Si-(O_3) \rightarrow (Si_3)-Si-Si-(Si_3)+(O_3)-Si-O-Si-(O_3) \quad (5.9)$$

The energy of this reaction is $E(3,3)-E(0,0)=-99$ kJ/mol, showing that, the Si−O−Si bonds in the SiO_x would aggregate to form separate SiO_2 and Si phases.

Both decomposition reactions can be viewed as oxygen exchanges between the silicon atoms as shown in Figure 5.12. The lowest energy pathways for the two oxygen exchange processes can be calculated through the kinetics of both reactions. The first decomposition formation of $Si^{3+}+Si^{1+}$

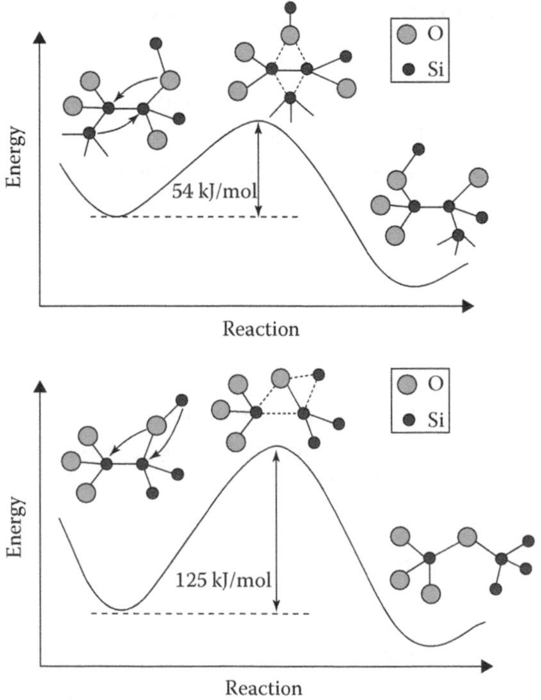

FIGURE 5.12 Calculated activation energy and structures for reactants, products, and transition state. (Reprinted from Barranco, A. et al., *Vacuum*, 67, 491, 2002. With permission.)

can be modeled as the following exchange reaction: $(O_2,Si)-Si-Si-(O_2,Si) \rightarrow (O_3)-Si-Si-(O,Si_2)$. The lowest energy pathway consists of a concerted ligand exchange between the atoms in the central Si–Si unit. An oxygen atom is transferred from left to right while, at the same time, a Si atom is transferred from right to left. The energy barrier for this mechanism is 54 kJ/mol. For the second decomposition reaction, the formation of a Si^{4+} can be modeled by inserting an oxygen atom into the central Si–Si bond, the reaction being $(O_3)-Si-Si-(O,Si_2) \rightarrow (O_3)-Si-Si-O-Si-(Si_3)$. The activation energy for this mechanism is 125 kJ/mol, which is considerably higher than that for the ligand exchange mechanism.

The above two-step decomposition reactions of the Si suboxides occur simultaneously. The first exothermic decomposition of the $Si^{2+}+Si^{2+}$ into $Si^{1+}+Si^{3+}$ occurs rapidly at relatively low thermal annealing temperatures. The second decomposition although also exothermic, however, would be much slower and need higher annealing temperature due to the high activation energy.

Figure 5.13 demonstrated Si 2p XPS spectra measured on SiO_x films after annealing at 400°C, 700°C, and 1000°C, respectively [96]. By varying the annealing temperatures, the concentrations of the various Si chemical states have been changed. The amount of Si suboxides decreases, while the intensity of the Si^0 and Si^{4+} peaks increases when the annealing temperature increases. This is because the Si suboxides were unstable and tended to thermally decompose into more stable Si and stoichiometric SiO_2 during the annealing [100].

Figure 5.14 shows the changes in the concentration of the five oxidation states in the $SiO_{1.4}$ sample with annealing temperature. Upon annealing, the concentration of the Si^{2+} decreases, while both Si^0 and Si^{4+} species increases, whereas the concentration of the Si^{1+} and Si^{3+} first slightly increases, and then decreases. The decomposition process of the Si suboxides can be described by the following two-step reactions [96]:

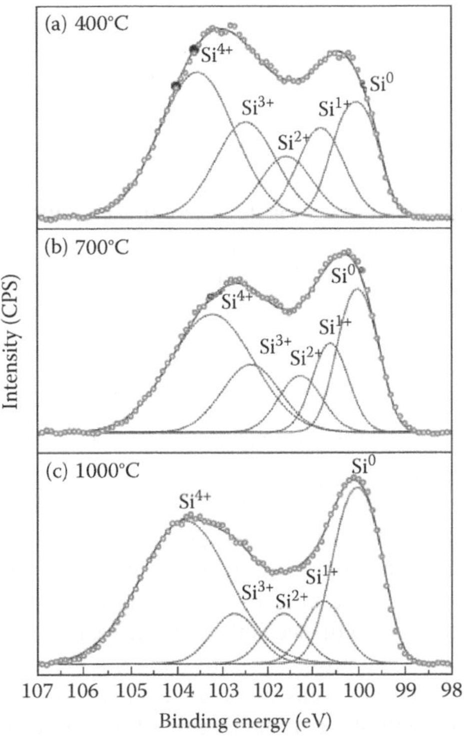

FIGURE 5.13 Thermal decomposition of SiO_x films after annealing at different temperatures by rapid thermal annealing for 3 min. (Reprinted from Zhang, W. et al., *J. Cryst. Growth*, 311, 1296, 2009. With permission.)

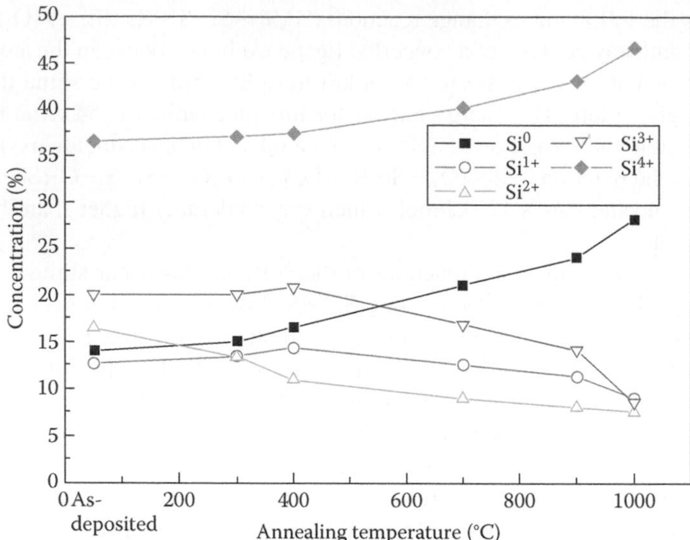

FIGURE 5.14 Changes of components concentration in SiO_x films after annealing versus annealing temperatures. (Reprinted from Zhang, W. et al., *J. Cryst. Growth*, 311, 1296, 2009. With permission.)

$$(1)\ Si^{2+} + Si^{2+} \rightarrow Si^{1+} + Si^{3+}$$

$$(2)\ Si^{1+} + Si^{3+} \rightarrow Si^{0} + Si^{4+}$$

These two reactions take place simultaneously but in different manners. Barranco et al. [100] suggested that the reaction (1) takes place by the ligand exchanging of oxygen atom within the $Si-(O_2,Si_2)$ tetrahedron with a energy barrier of at least 54 kJ/mol. Reaction (2) takes place by inserting oxygen into a $Si^{3+}-Si^{1+}$ bond with an activation energy of 125 kJ/mol. Since the kinetic energy barrier of the reaction (2) is much higher than that of the reaction (1), thus reaction (2) needs high annealing temperatures. Reaction (1) should take place easier than that of reaction (2), especially, at low-temperature annealing. As a result, reaction (1) should dominate at low-temperature annealing, and reaction (2) is more pronounced at higher-temperature annealing. This is consistent with the XPS results (Figure 5.14), in which Si^{1+} and Si^{3+} first increase because of reaction (1) and then decrease because of reaction (2) with increasing annealing temperature.

5.2.4 The Growth Mechanism of nc-Si in a-SiO_x Matrix

Si nanocrystals generally form during high-temperature annealing. The formation of nc-Si should include nucleation, growth, and ripening processes. The factors affecting the formation process can be divided into two categories: internal and external. The internal factors include the Si concentration and the local atomic microstructure of the as-deposited SiO_x films. The external factors include the annealing temperature, time, and annealing ambient. The internal factors determine the growth mechanism, and the external factors determine the size, density, and passivation of the formed nc-Si. Two models were proposed for the growth mechanism: the diffusion-controlled growth and the phase segregation growth. The diffusion-controlled growth theory believes that the nc-Si were grown by the thermal diffusion of the excess Si atoms in the amorphous SiO_x matrix; the phase segregation growth theory considers the growth of the nc-Si because of the segregation of the Si suboxides.

The diffusion-controlled growth mechanism was proposed based on the observations that the average size of the nc-Si was dictated by the initial concentration of the Si in the SiO_x for a given

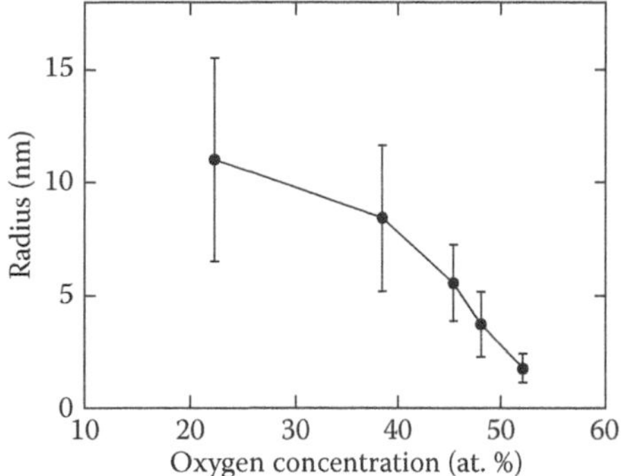

FIGURE 5.15 Average crystalline Si dot radius as a function of the oxygen concentration. (Reprinted from Lombardo, S. et al., *Mater. Sci. Eng. B: Solid State Mater. Adv. Technol.*, 69, 295, 2000. With permission.)

annealing time and temperature: the higher the Si concentration, the larger the average size of the resulted nc-Si [46,75]. Figure 5.15 shows the average nc-Si radius as a function of the oxygen concentration in the SiO_x films annealed at 1100°C for 1 h [45]. The nc-Si size decreases with increasing oxygen concentration. This shows that the nc-Si are formed by the consumption of the excess Si due to the diffusion of the elemental Si atoms toward the nucleation center.

According to the classical nucleation theory [101], the driving force for the precipitation is the reduction of the volume free energy, $\Delta G_v(X_{si}, T_A)$. This driving force increases as the Si content goes above the stoichiometric level and as the annealing temperature T_A increases. Opposing the free energy reduction is the energy required to create the interface between the Si cluster and the matrix, $A\gamma_{cm}$, where A is the surface area and γ_{cm} is the cluster-matrix interfacial energy. The total free energy passes through a maximum as a function of cluster radius where the maximum is associated with the free energy to form a critical nucleus that is given by the expression $\Delta G^* = (16\pi\gamma_{cm}^3)/3\Delta G_v^2$ for the spherical clusters with a critical cluster radius of $r^* = 2/\gamma_{cm}\Delta G_v$. In homogeneous nucleation, all the energy required to form the cluster–matrix interface must be supplied by the chemical driving force. However, in most real systems the nucleation occurs through heterogeneous nucleation at preexisting defect sites, where the threshold for the nucleation is reduced by the energy released through the annihilation of the defect. In either case, once nucleation occurs with the formation of the critically sized nuclei, cluster growth is governed by the process of coarsening that minimizes the overall surface energy of the system by promoting the growth of the large clusters at the expense of the smaller clusters. As such, Si content, annealing temperature, and time can be used to guide the fabrication of the nanocrystal assembly with the desired size, size distribution, and density.

The diffusion-controlled growth mechanism of the nc-Si was detailed by Nesbit based on TEM studies of the diffusion behaviors of SiO_x films synthesized by CVD under thermal annealing [102]. Assuming a spherical silicon cluster radius r, the silicon cluster growth rate in the SiO_x matrix at a given annealing temperature T can be expressed as

$$\left(\frac{dr}{dt}\right)_{r_i} = \frac{D}{C^b - C^a}\left(\frac{\partial C}{\partial r}\right)_{r_i} \tag{5.10}$$

where

r_i is the initial Si cluster radius
D is the diffusion coefficient of silicon in SiO_2
C^a is the composition of stoichiometric SiO_2 that is the assumed composition of the oxide matrix at the silicon cluster–oxide interface
C^b is the composition of the silicon cluster that is assumed 100% silicon

The composition in the oxide matrix near the silicon cluster, C, is assumed to be a linear function of distance into the matrix from the silicon cluster surface. The $(\partial C/\partial r)$ can be approximated by the following expression:

$$\left(\frac{\partial C}{\partial r}\right)_{r_1} = \frac{(C^{\mathrm{m}} - C^{\mathrm{a}})}{r_1} \tag{5.11}$$

where C^{m} is the average composition of the entire film. By integrating the above equations with respect to radius r and time t, the silicon diffusion coefficient as a function of temperature, T, can be expressed as

$$D(T) = \frac{r_2^2 - r_1^2}{2t}\left(\frac{C^{\mathrm{m}} - C^{\mathrm{a}}}{C^{\mathrm{b}} - C^{\mathrm{a}}}\right) \tag{5.12}$$

where

r_1 is the radius of the as-deposited silicon cluster
r_2 is the radius after annealing for a time t at temperature T

The calculated values of the diffusion coefficients are shown in the third column of Table 5.2. It is shown that the diffusion coefficient of Si in the SiO_x films is independent of composition, with an average of $1.1 \times 10^{-16}\,cm^2/s$ at 1100°C. The diffusion coefficient values of Si in the SiO_x films at different temperatures are shown in the fourth column in Table 5.3. With linear regression analysis of the data in Table 5.3, the diffusion coefficient can be expressed as

$$D(T) = D_0 \exp(-Q/RT) \tag{5.13}$$

where

Q is the activation energy
R is the universal gas constant

TABLE 5.2
Silicon Cluster Diameters and Calculated Diffusion Coefficients as a Function of Composition

O/Si Ratio	Silicon Cluster Diameter (nm)	Calculated Diffusion Coefficient ($10^{-18}\,cm^2/s$)
1.4	2.5	110
1.3	4.0	130
0.95	4.5	100
0.72	5.0	92

Source: Reprinted from Nesbit, L.A., *Appl. Phys. Lett.*, 46, 38, 1985. With permission.

Note: Annealing conditions: 1100°C, 15 min in nitrogen.

TABLE 5.3
CVD Si Cluster Diameters and Calculated Diffusion Coefficients as a Function of Annealing Time and Temperature for O/Si = 0.82

Annealing Conditions		Si Clusters	Calculated Diffusion
Temperature (°C)	Time (h)	Diameter (nm)	Coefficient (10^{-18} cm²/s)
1060	1.0	9.0	86.0
950	2.7	5.5	15.0
800	18.0	4.5	1.1
700	72.0	3.5	0.18

Source: Reprinted from Nesbit, L.A., *Appl. Phys. Lett.*, 46, 38, 1985. With permission.

The activation energy is 180 kJ/mole or 1.9 eV/atom, and D_0 is equal to 1.2×10^{-9} cm²/s. The diffusion coefficient of Si at the low temperatures (700°C and 800°C) is quite low, thus the Si nanocluster size is still very small after annealing for up to 72 h at 700°C or for 18 h at 800°C. Nesbit indicated that temperatures are required to be >1150°C for the complete crystallization of the excess silicon [102]. Nesbit also found that SiO_x films with O/Si < 1 formed crystalline Si particles when annealed at 850°C. However, for films with 1 < O/Si < 1.9, an annealing temperature of 1050°C was required to crystallize the excess Si [102].

The diffusion control growth mechanism has been verified experimentally. Researchers have observed the increase of the nc-Si size with an increase in the annealing time and annealing temperature. The mean radius of the Si nanoclusters as a function of annealing temperature synthesized by plasma-enhanced chemical vapor deposition (PECVD) is shown in Figure 5.16 [103]. The data reported in Figure 5.16 show that the nanocluster size increases with the temperature, and is in qualitative agreement with the model of diffusion-controlled mechanism.

The transition of the Si nanoparticles embedded in the SiO_2 matrix from the amorphous to the crystalline during annealing was detailed by Kachurin et al. [25] and Wang et al. [104]. The microstructure evolution from the amorphous to the crystalline of the SiO film is divided into three stages

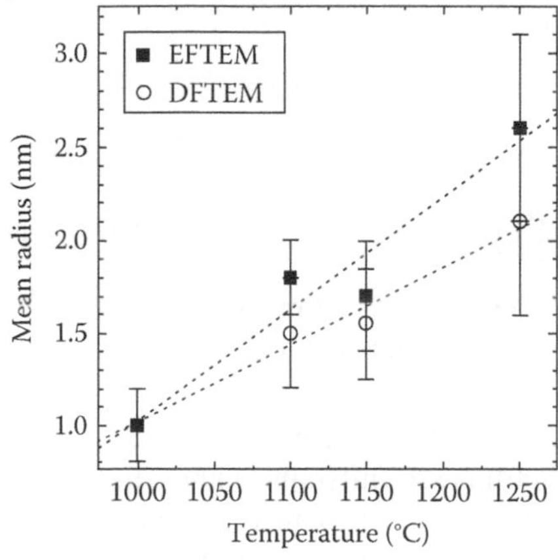

FIGURE 5.16 Si nanocluster mean radius as a function of annealing temperature. (Reprinted from Iacona, F. et al., *J. Appl. Phys.*, 95, 3723, 2004. With permission.)

FIGURE 5.17 EFTEM and HRTEM images and diffraction data from the as-deposited film and annealed films at 300°C—Stage I. (Reprinted from Wang, J. et al., *Philosophical Magazine*, 87, 11, 2007. With permission.)

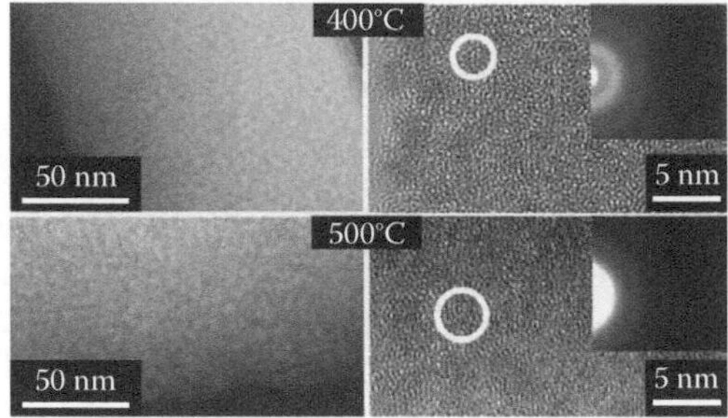

FIGURE 5.18 EFTEM and HRTEM images and diffraction patterns from the as-deposited film and annealed films at 400°C and 500°C—Stage II. (Reprinted from Wang, J. et al., *Phil. Mag.*, 87, 11, 2007. With permission.)

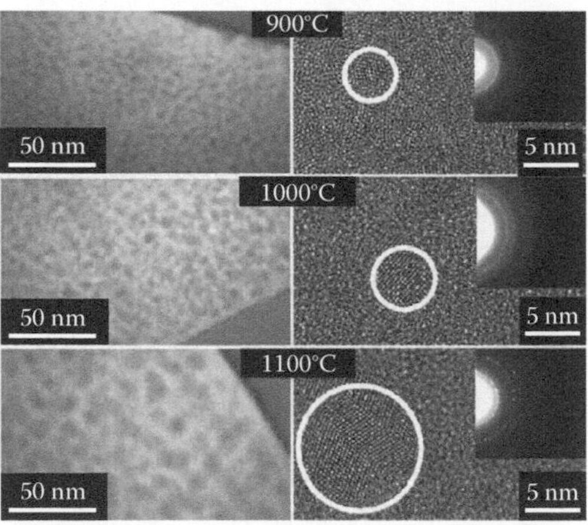

FIGURE 5.19 EFTEM and HRTEM images and diffraction patterns from as-deposited and annealed films at 900°C–1100°C—Stage III. (Reprinted from Wang, J. et al., *Phil. Mag.*, 87, 11, 2007. With permission.)

as shown in Figures 5.17 through 5.19, respectively. Stage I occurs at annealing temperatures of 300°C or lower, where the SiO_x films are fully amorphous and are structurally homogeneous. The local atomic structure of the SiO_x films during annealing at these temperatures has no changes, and there is no evidence of a cluster-matrix configuration. The thermal decomposition of the Si suboxides takes place very slowly at low temperatures.

Stage II occurs at annealing temperatures between 400°C and 800°C. In this stage, Si and O inhomogeneously distribute in the films. The amorphous silicon-rich nanoclusters are formed, and these clusters grow larger as the annealing temperature increases from 400°C to 800°C. The diameter of the amorphous Si nanoclusters in the specimen annealed at 400°C is ~2.4 nm and increases to 2.8 nm after annealing at 500°C. The clusters remain amorphous up to 800°C.

Stage III occurs at an annealing temperature of 900°C or above. In this stage, the amorphous silicon-rich nanoclusters change into nanocrystals. There are several independent processes taking place in the silicon-rich oxide films. First, the continuous decomposition of SiO_x to Si and SiO_2 that indicates that the host matrix for the Si clusters is a mixture of SiO_2 and SiO_x for the whole temperature range instead of pure SiO_2; secondly, continuous growth of the Si clusters with increasing annealing temperatures that is caused by the diffusion of the excess silicon. The diffusivity of Si in SiO_2 at these annealing temperatures is of the order of 10^{-16} cm^2/s. This is sufficient for Si nanoclusters to grow due to the diffusional sink of the excess silicon to their boundaries. The coarsening of Si inclusions helps them to be crystalline after cooling.

However, even at annealing temperatures high enough for crystallization, not all excess Si is necessarily crystallized. Some Si remains amorphous in the annealed samples. For the samples with coexisting amorphous and crystalline Si, Iacona et al. [103] have determined the crystalline fraction present at a given temperature. The ratio of the number of nc-Si over the total number of clusters has been evaluated at different annealing temperatures by energy-filtered transmission electron microscopy (EFTEM). The crystalline ratio of the clusters is shown in Figure 5.20a [103]. A clear trend of progressive increase of the crystalline fraction with temperature can be observed. It is shown that complete crystallization is achieved at 1250°C, and the nc-Si fractions at lower temperatures are about 30% at 1100°C and 60% at 1150°C. This indicates that the temperature plays a role in promoting not only the cluster growth, but also the amorphous to crystalline transition. The Si nanocluster density as a function of the annealing temperature is shown in Figure 5.20b [103]. The nanocluster density is almost constant in the range of 1000°C–1150°C, while it decreases when the annealing temperature is increased up to 1250°C. This decrease in nanocluster density results from the continuous growth of the nc-Si with annealing temperature, suggesting that nanocluster growth is not simply due to the inclusion of Si atoms diffusing from the oxide matrix, but also as a result of the Ostwald ripening, which consumes small clusters. The number of clustered Si atoms per unit volume at different temperatures is shown in Figure 5.20c [103]. It is demonstrated that the number of detectable clusters remarkably increases with the annealing temperature. It appears that there are new clusters formed at a higher temperature. These newly formed clusters may be due to

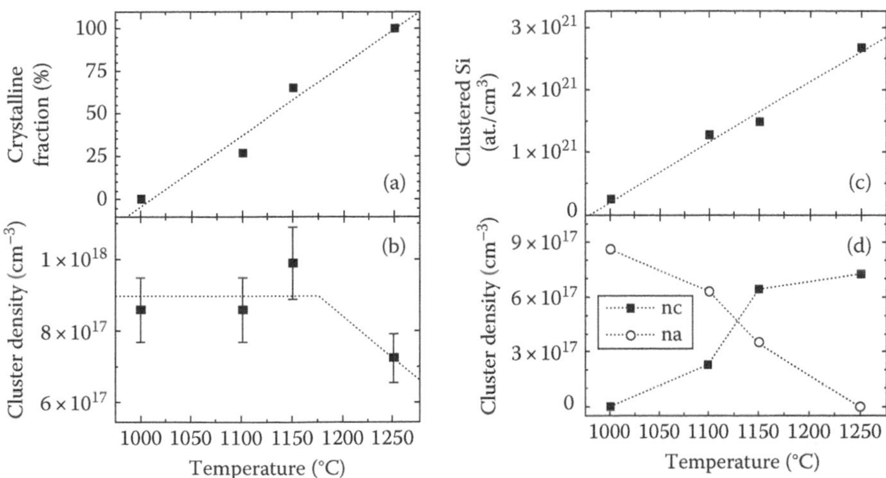

FIGURE 5.20 (a) Crystalline fraction, (b) density of Si nanoclusters, (c) concentration of clustered Si atoms, (d) density of amorphous (na) and crystalline (nc) nanoclusters, as a function of annealing temperature. (Reprinted from Iacona, F. et al., *J. Appl. Phys.*, 95, 3723, 2004. With permission.)

the growth of the smaller ones, or to the nucleation of the new ones. The concentration of the nc-Si and amorphous Si nanoclusters as a function of annealing temperature is shown in Figure 5.20d [103]. With increasing annealing temperature, the amorphous Si nanoclusters continually transform into nc-Si, which leads to the increase of the nc-Si concentration, while the concentration of the Si amorphous nanoclusters decreases.

The segregation mechanism of Si oxide is proposed based on the observation of the rapid growth of the nc-Si during pulse annealing or rapid thermal annealing. Kachurin et al. [26] observed the formation of nc-Si in the ion-implanted SiO_x films by pulse annealing. The implanted SiO_x samples were subjected to either rapid thermal annealing at 900°C–1200°C for 1 s or flash-lamp annealing at 1050°C–1350°C for 20 ms. The formation of nc-Si could not be explained by the diffusion-limited growth or solid-phase crystallization of amorphous Si phase inclusions due to the short annealing time and the low diffusivity of Si in a-SiO_x matrix. They proposed a new model considering the Si nanocrystal formation through the segregation of the Si atoms from SiO_x, rapid percolation-like formation of Si chains, or fractals and the final transformation to Si phase inclusions and nanocrystals. The a-SiO_x system is unstable and tends to segregate into Si and SiO_2 even at low temperatures. The segregation proceeds as a percolation via "weak points" in the form of "silicon cracks" or "silicon breakdowns" in a-SiO_2. It does not need long-range diffusion of Si atoms and the process could be fast. The increase in the thermal budget simultaneously leads to enhanced segregation and progressive transformation of the initial Si clusters into larger agglomerates, namely Si chains and fractals. Such inclusions have no phase boundaries and therefore could not be considered as inclusions of a Si phase in a-SiO_2. For the solid-phase crystallization to start, these chains and fractals should convolute and collapse to Si phase inclusions. These processes take some time and limit the formation of the nc-Si at a given annealing temperature. It is well known that the melting point of the metal and semiconductor particles with a mean size below 2–3 nm becomes very low. Thus, during annealing no crystals will be formed until their mean size reaches a value sufficient for them to survive the annealing. In Ref. [25] the minimum amount of excess Si in a-SiO_2 and minimum annealing temperature required to form the nc-Si is reported. The minimum crystal diameter should be about 2.5 nm. It is well known that the stability of the small crystals depends on the balance between their volume and the surface free energies. In 2–3 nm crystallites, about 50% of the atoms sit on surface. Being influenced by the surface, the nanocrystals are quite unstable and their melting point may be much lower than that of their bulk counterparts. For instance, 2–3 nm Si and Ge nanocrystals are unstable even at room temperature [96].

5.2.5 Defects in nc-Si/a-SiO_2 Nanocomposites

It is reported that the light emission and charge trapping of the nc-Si/a-SiO_2 nanocomposites is not only due to the nc-Si, but also due to the defects in the oxide. Torchynska et al. reported that the size of the nc-Si did not affect the PL spectrum for the nc-Si/a-SiO_2 nanocomposites fabricated by magnetron sputtering [59,105]. They attributed the light emission from the SiO_x films to the defect related E′-center and non-bridging oxygen hole center (NBOHC). The argument that the deep trapping center plays an important role in the charge trapping and retention in the nc-Si was proposed based on the long retention time of charge trapping, which is contrary to the concept that the conduction band of the nc-Si cannot store electrons for long time. The types of defects in the SiO_x films include the NBOHC, the oxygen vacancy related E′-center, and the P′-center as detailed below.

5.2.5.1 Non-Bridging Oxygen Hole Center

The NBOHC is an under-coordinated O atom bonded to a single Si atom with an unpaired electron in its 2p orbital, and is denoted by ≡Si−Ȯ, where (≡) stands for bonds with three oxygen and (·) indicates unpaired electron as shown in Figure 5.21. It is well accepted that in silica glasses this defect occurs as an isolated localized defect with an energy level around the midgap of the oxide, giving

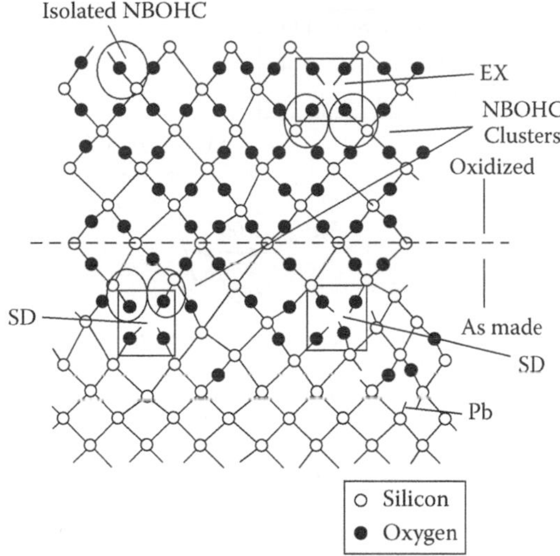

FIGURE 5.21 Structure model of the Si–SiO_2 interfacial region, showing NBOHC clusters as well as an isolated NBOHC. (Reprinted from Prokes, S.M. and Carlos, W.E., *J. Appl. Phys.*, 78, 2671, 1995. With permission.)

rise to a weak absorption band at 2.0 and 4.8 eV regions, and a PL band at 1.8–1.9 eV. Further, the NBOHCs are very stable even at high temperatures (as high as 800°C–900°C).

5.2.5.2 E′ Center

Researchers assign the emission bands centered at 620 nm to the E′ center [107]. The E′ center is an oxygen–deficiency center in amorphous SiO_2 films [108]. This defect has been widely studied and the most accepted model is a puckered positively charged oxygen vacancy: ≡Si−O · O≡Si + (where ≡ represents the bonds to three oxygen atoms, · represents an unpaired electron, and + is a trapped hole) [109]. In this model, it is supposed that, following ionization of the vacancy, the positively charged Si atom moves backward through the plane of its basal oxygen in the puckered configuration.

5.3 ELECTRICAL CHARACTERISTICS OF nc-Si/a-SiO_2 NANOCOMPOSITES

5.3.1 Coulomb Blockade Effect in Quantum Dots

In nanostructures, when the mean-free path of the electrons exceeds the dimensions of the device structure, the quantum characteristics may dictate the physical properties of the devices. In quantum devices, the charge transport properties are governed by tunneling. Due to the discrete tunneling of the electrical charge, current flow through a quantum dot is a series of events in which exactly one electron passes through the quantum dot. The quantum dot is charged with one elementary charge by the tunneling electron, causing a self-charging energy E_c given by [110]

$$E_c = \frac{e^2}{2C_{dot}} \tag{5.14}$$

where

e is the elementary charge of 1.6×10^{-19} Coulomb

C_{dot} is the self-capacitance of the quantum dot given by

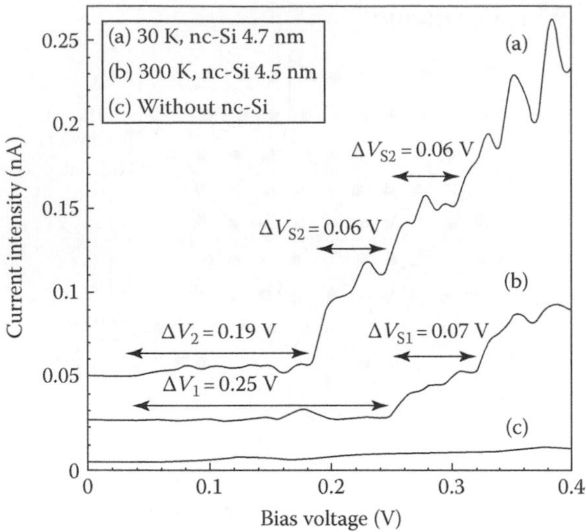

FIGURE 5.22 Coulomb staircases in current-voltage characteristics (a) 4.7 nm nc-Si in diameter at 30 K, (b) 4.5 nm nc-Si at 300 K, and (c) 1.2 nm tunneling oxide without nc-Si. (Reprinted from Baron, T. et al., *Appl. Phys. Lett.*, 79, 1175, 2001.)

$$C_{dot} = 4\pi\varepsilon_0\varepsilon_{SiO_2} r \quad (5.15)$$

where

r is the radius of nc-Si
ε_0 is the vacuum permittivity
ε_{SiO_2} is the dielectric constant of SiO_2

The self-charging energy of the quantum dot causes a voltage buildup, $U = e/C_{dot}$, which induces an electric field preventing the tunneling-in of a second electron. To inject another electron into the quantum dot, the electrical field induced by the previous injected electron needs to be overcome. Thus, the current-voltage characteristics of the quantum dot embedded dielectric films usually show a staircase behavior.

The typical *I-V* characteristic of a-SiO_x films showing Coulomb staircases is shown in Figure 5.22 [23]. Curve (c) is taken from a pure SiO_2 film. The curve shows an exponential dependence of the current on the bias voltage as expected for a single tunneling barrier. Remarkable features can be observed from the nc-Si/a-SiO_2 nanocomposites: curve (a), measured at 30 K and curve (b), measured at 300 K, exhibit threshold voltages and a staircase on the current. Thresholds of 0.19 and 0.25 V and staircases of 0.06 and 0.07 V are observed, respectively, in curve (a) and (b).

5.3.2 Charge Transport in nc-Si/a-SiO_2 Nanocomposites

In recent years, charge transport behaviors in the nc-Si/a-SiO_2 nanocomposites have attracted increasing interests along with the remarkable progress of the nc-Si nanostructures. The nanoscale nc-Si leads to various unique electronic and photonic properties associated with quantum confinement and single-electron tunneling. These unique properties have been exploited for fabricating single-electron transistors and memories, ballistic electron emitters, and silicon light emission devices. Electron transportation in the nc-Si/a-SiO_2 is a complex process of macroscopic percolation transport and nanoscopic transport, such as resonant tunneling and coulomb blockade. Investigations on electron transport in the nc-Si are still at the early stage. Most of the discussions are based on the current-voltage measurement.

5.3.2.1 Single-Electron Device

Highly sophisticated nanotechnology enables us to fabricate nanometer scale metal-oxide-semiconductor field-effect transistors. The size reduction in the device increases both the speed and the integration density, and consequently improves the performance. However, the power dissipation in an IC chip increases continuously and the cooling becomes an issue. To address the heat problem, the method used to build large-scale integrated circuits (LSIs) needs to be changed, from either the system architecture or device level, or both. In general, lowering the power operation of the LSIs requires reducing the local capacitance of the circuits and the operation voltage, which means the reduction of the number of electrons participating in the operation of the devices. Single-electron devices (SEDs), which literally have the potential to manipulate electrons at the level of their elementary charge, are thus considered the devices that allow significant reduction of power consumption.

SEDs can be made from the nc-Si/a-SiO_2 nanocomposites films by using the Coulomb blockade effect of the nc-Si. The benefit of using Si as a base material is that the existing advanced CMOS fabrication technologies can be made use of. The basic principle with respect to manipulating a single electron involves Coulomb blockade and single-electron tunneling that requires small islands connected via a tunneling barrier or a tunneling capacitor. The small island can be made of nc-Si. When the nc-Si is sufficiently small, the charging energy E_c becomes large. In this case, no additional electron can enter the island through the tunneling capacitor unless the potential of the island is reduced by an external bias.

The most fundamental three-terminal SED, which is the single-electron transistor (SET), is illustrated in Figure 5.23. It consists of two tunneling junctions sharing one common electrode with a low self-capacitance, known as the island. The electrical potential of the island can be tuned by a third electrode, that is, the gate that is capacitively coupled to the island. In the blocking state, no accessible energy levels are within the tunneling range of the electron on the source contact. All energy levels on the island electrode with lower energies are occupied. When a positive voltage is applied to the gate electrode, the energy barriers of the island electrode are lowered. The electron can tunnel into the island, occupying a vacant energy level. From there it can tunnel into the drain electrode.

The equivalent circuit of a single-electron transistor is shown in Figure 5.24. The Coulomb blockade of the SET is determined by two tunneling capacitances, C_s and C_d, and the gate capacitance, C_g. The "Coulomb diamond" shows the conduction when the source terminal is grounded. In the rhombic regions, electron transfer is prevented because the number of electrons in the island is a fixed integer. Outside the Coulomb diamonds, the number of electrons in the island fluctuates between certain numbers. If the drain voltage is low enough, the current flows by means of one-by-one electron tunneling, that is, the electron number changes only between two adjacent integers. The maximum drain voltage for a coulomb blockade state is given by e/C_Σ, where $C_\Sigma = C_g + C_s + C_d$ is the total capacitance of the island.

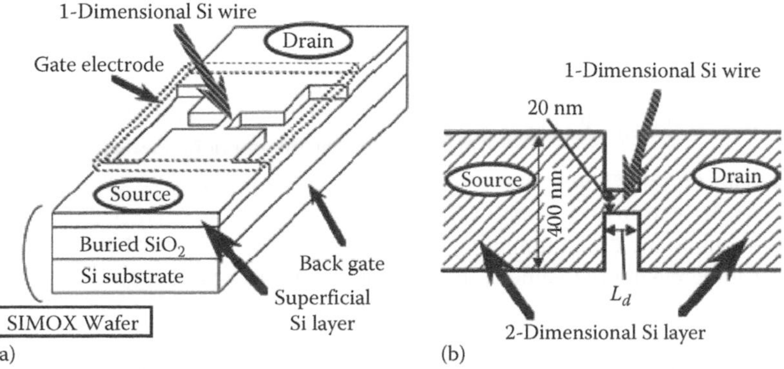

FIGURE 5.23 (a) Schematic diagram, (b) top view of single-electron transistor. (Reprinted from Takahashi, Y. et al., *J. Phys.: Condens. Matter*, 14, R995, 2002. With permission.)

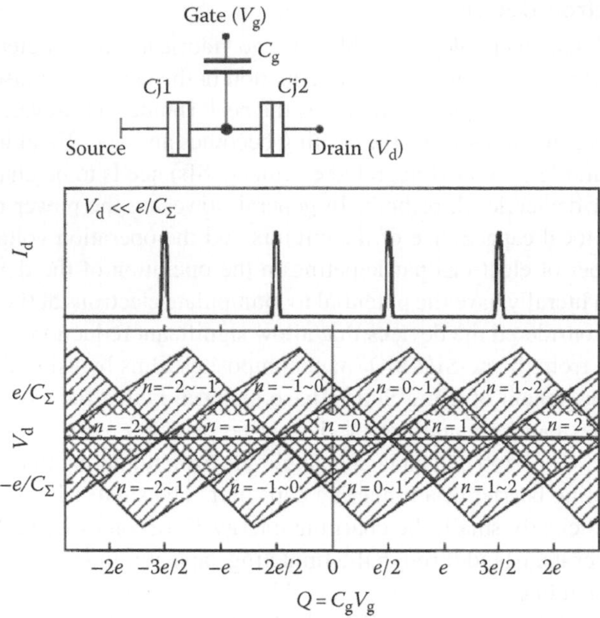

FIGURE 5.24 Equivalent circuit, drain current characteristics, and stability diagram of a SET. I_d and V_d are the drain current and drain voltage. Q and V_g are the external charge and voltage to the gate with a capacitance C_g. (Reprinted from Takahashi, Y. et al., *J. Phys.: Condens. Matter*, 14, R995, 2002. With permission.)

If the gate voltage is swept with a fixed small source-drain voltage, the Coulomb blockade states and the single-electron tunneling states appear alternately and the drain current-gate voltage characteristics exhibit a multi-peak behavior. This is called Coulomb blockade oscillation. As can be understood from the principle mentioned above, the drain current is determined by the value of C_gV_g/e that corresponds to the number of excess electrons in the gate electrode. The drain current has a minimum when C_gV_g/e is an integer because the Coulomb blockade occurs. Conversely, when C_gV_g/e is a half-integer, the current flows because the Coulomb blockade is lifted.

5.3.2.2 Current Conduction Mechanism

The electrical properties of a-SiO_2 containing nc-Si are quite different from that of pure a-SiO_2. The charge transport, charge trapping, and charge retention mechanism of the a-SiO_2 films containing nc-Si have been studied extensively. Current-voltage (*I-V*) and capacitance-voltage (*C-V*) measurements are usually employed to characterize the electrical properties.

Figure 5.25 shows the *I-V* characteristics of the nc-Si/a-SiO_2 nanocomposites with different Si concentrations. With increasing Si concentration in the a-SiO_x layer, a clear increase in the current density is observed. The increase in Si concentration of the a-SiO_2 films results in the increase in the density of the nc-Si that subsequently increases the current conduction. This is because more nc-Si present in a-SiO_2 films can form more tunneling sites for the carriers.

The mechanism of carrier injection through the device may involve two contributions: direct tunneling and Fowler–Nordheim tunneling. At low voltage, the current is dependent on the Si concentration that is due to the direct tunneling between the nanocrystals that are separated from each other by SiO_2. By increasing the voltage, the current density increases with the typical exponential behavior of the Fowler–Nordheim tunneling. A clearer picture about the tunneling mechanism is shown in Figure 5.26 [111]. The current starts at a low-threshold voltage for a regime of direct tunneling conduction of carriers, and at higher voltages it exhibits the Fowler–Nordheim regime.

Chakraborty et al. have studied the tunneling mechanism of the nc-Si/a-SiO_2 nanocomposites in detail [112]. According to their study, the direct tunneling dominates when the applied voltage

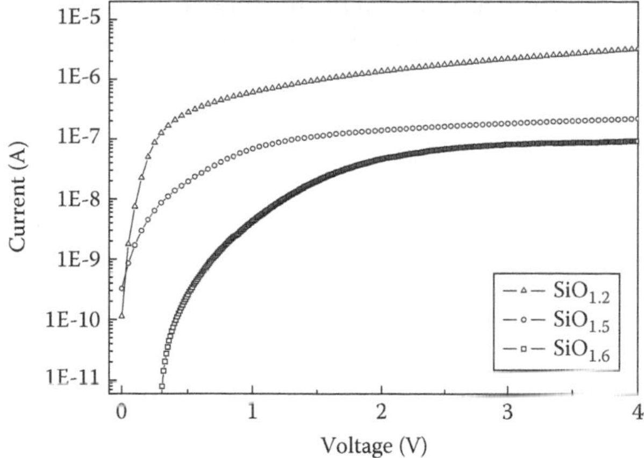

FIGURE 5.25 Current density as a function of voltage for the nc-Si devices.

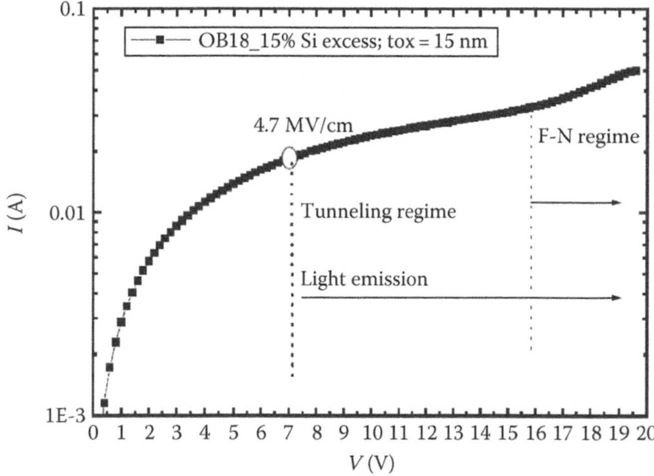

FIGURE 5.26 Current-voltage characteristics of the emitting device with SiO_x films by ion implantation. (Reprinted from Garrido, B. et al., *Nucl. Instrum. Methods Phys. Res. Sect. B: Beam Interact. Mater. Atoms*, 216, 213, 2004. With permission.)

is less than the barrier height $(\phi_b - E_0)/q$, where ϕ_b is the potential barrier height at the nanocrystal–insulator interface; E_0 is the ground state energy, where the zero energy is taken at the minimum of the Si conduction band at the interface between the Si substrate and the dielectric layer. The tunneling current density through a barrier in the direct tunneling regime is given by

$$J_D = \frac{[2m_{eff}(\phi_b - E_o)]^{1/2}\alpha q^2 V}{\hbar^2 d} \exp\left[\frac{2\alpha\sqrt{2m_{eff}(\phi_b - E_o)}}{\hbar} d\right] \tag{5.16}$$

where

m_{eff} is the effective mass of an electron
d is the barrier width
ϕ_b is the barrier height
V is the applied bias
α is a parameter that accounts for an effective mass
$\hbar$ is the reduced Plank constant

Fowler–Nordheim (FN) injection in an MOS structure is the high-field induced emission of electrons. This tunneling occurs when the applied voltage is larger than $(\phi_b - E_o)/q$. The FN current density can be estimated by [113]

$$J = \frac{A}{4\phi_B} E^2 \exp\left(\frac{-2B\phi_B{}^{3/2}}{E}\right) \tag{5.17}$$

where
E is the electrical field
A, B, and C are constant parameters

5.3.3 Memory Effect of nc-Si/a-SiO$_2$ Nanocomposites

The nanocomposite nc-Si/a-SiO$_2$ has a great potential for use in quantum dot memory. In this application, the device operation is realized by charge and discharge of the nc-Si embedded in the gate oxide. The nc-Si are normally confined in a narrow layer in the gate dielectric near the substrate. Such a structure can be realized by ion implantation of the Si ions into a thermal oxide layer or the deposition of the Si-rich oxide sandwiched with a-SiO$_2$ by low-pressure chemical vapor deposition (LPCVD), sputtering, evaporation, etc. The charging and discharging of the nc-Si are achieved by electron tunneling between the confined nc-Si layer and the Si substrate through a very thin tunneling oxide, causing a flat band voltage shift of the *C-V* characteristics.

C-V measurements of MOS capacitors provide a lot of information about the devices. To understand *C-V* measurements one must first be familiar with the changes of charges in the MOS structure under different gate bias voltage. An MOS capacitor with *p*-Si substrate is illustrated in Figure 5.27.

To understand the different bias modes of an MOS capacitor, three different bias voltages should be considered, one below the flat band voltage (V_{FB}), a second between the flat band voltage and the threshold voltage (V_T), and finally one larger than the threshold voltage. These bias regimes are called the accumulation, depletion, and inversion modes of operation, respectively. These three modes as well as the charge distribution associated with each of them are shown in Figure 5.28.

Accumulation occurs typically when one applies negative gate voltages ($V_G < V_{FB}$). The negative charge on the gate attracts holes from the substrate toward the oxide–semiconductor interface. As a positive voltage ($V_G > V_{FB}$) is applied, the mobile holes are pushed away from the interface into

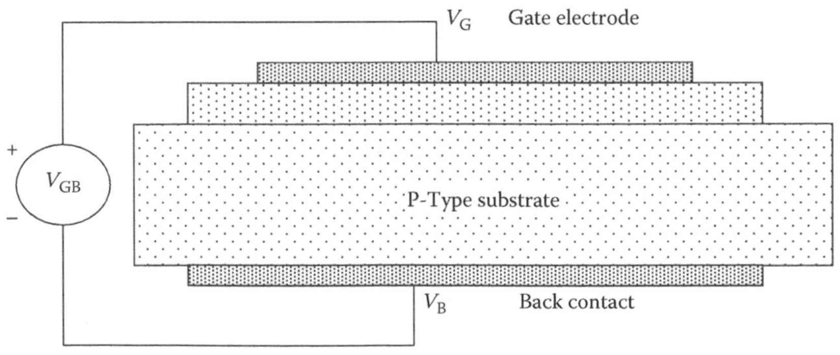

FIGURE 5.27 Schematic of an MOS structure.

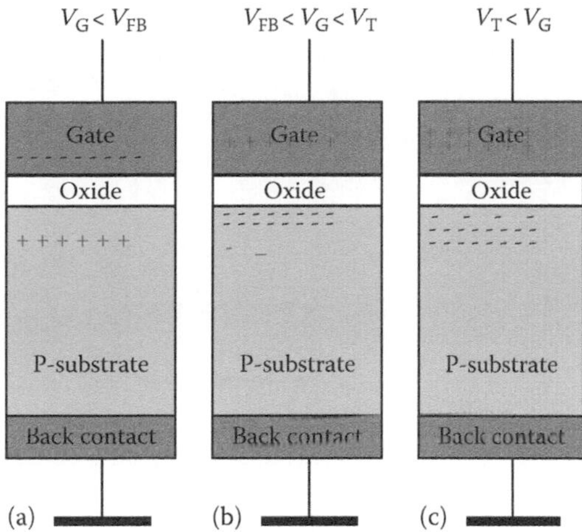

FIGURE 5.28 Changes in MOS structure under accumulation, depletion, and inversion conditions, respectively.

the substrate. Therefore, the semiconductor is depleted of mobile carriers at the interface and negative charges due to the ionized acceptor ions that are left in the space charge region. When the gate voltage is beyond the threshold voltage, the negatively charged minority carriers can be attracted to the interface and their number at the surface becomes larger than that of holes, the surface thus becomes inverted and this is the inversion case.

The memory effect of the MOS structure with the nc-Si/a-SiO_2 nanocomposites as the active layer can be observed from the hysteresis in dual-sweeping *C-V* measurements (i.e., sweeping the gate voltage from positive to negative first and then back to positive, and vice versa). Typical dual-sweeping *C-V* characteristics of an MOS device with nc-Si/a-SiO_2 nanocomposites are shown in Figure 5.29 [114]. A flat band voltage shift of ~1.2 V is observed with dual-sweeping between +7 V and −7 V from Figure 5.29, indicating significant charge trapping in the nc-Si/a-SiO_2 nanocomposites.

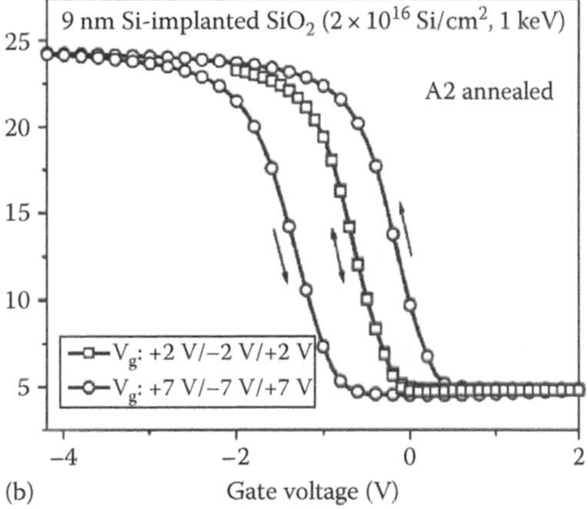

FIGURE 5.29 Typical dual-sweeping *C-V* characteristics for nc-Si/a-SiO_2 nanocomposites. (Reprinted from Normand, P. et al., Effects of annealing conditions on charge storage of Si nanocrystal memory devices obtained by low-energy ion beam synthesis, in *28th International Conference on Micro- and Nano-Engineering*, Lugano, Switzerland, September 16–19, 2003, pp. 629–634. With permission.)

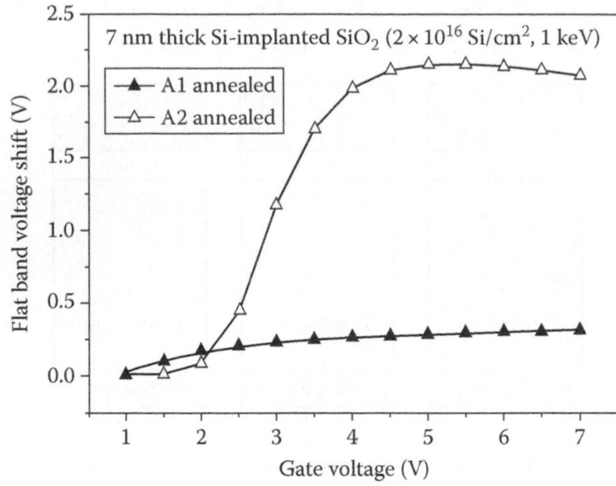

FIGURE 5.30 Flat band shift as a function of the gate voltage round sweep for the nc-Si/a-SiO_2 nanocomposites. (Reprinted from Normand, P. et al., Effects of annealing conditions on charge storage of Si nanocrystal memory devices obtained by low-energy ion beam synthesis, in *28th International Conference on Micro- and Nano-Engineering*, Lugano, Switzerland, September 16–19, 2003, pp. 629–634. With permission.)

The charging-induced shift of the *C-V* characteristics is an indication of the memory effect. The memory windows are affected by the sweeping range of the gate voltages. The larger the sweeping range, the larger the memory window, as shown in Figure 5.30 [114].

The flat band voltage shift of nc-Si/a-SiO_2 nanocomposites is determined by several factors, including the charge-trapping density, the control oxide thickness, and the dimension of the nc-Si. Under the assumption that each nanocrystal can store only one electron or hole, the flat band voltage shift can be approximated by

$$\Delta V_{\mathrm{FB}} = \frac{Q_{\mathrm{charge}}}{\varepsilon_{\mathrm{ox}}}\left(T_{\mathrm{control}} + \frac{1}{2}\frac{\varepsilon_{\mathrm{ox}}}{\varepsilon_{\mathrm{Si}}}d_{\mathrm{nc\text{-}Si}}\right) \tag{5.18}$$

where

ΔV_{FB} is the flat band voltage shift
T_{control} is the thickness of the control oxide
$d_{\mathrm{nc\text{-}Si}}$ is the diameter of the nc-Si
$\varepsilon_{\mathrm{ox}}$ and $\varepsilon_{\mathrm{Si}}$ are the permittivity of SiO_2 and Si, respectively
Q_{charge} is the charge density trapped in the nc-Si

The first term in the bracket is approximately an order larger than the second term. Thus, within this model, the flat band voltage shift can be approximated to be

$$\Delta V_{\mathrm{FB}} = \frac{Q_{\mathrm{charge}}T_{\mathrm{control}}}{\varepsilon_{\mathrm{ox}}} \tag{5.19}$$

The flat band voltage shift strongly depends on the magnitude of the gate voltage bias and the application time. Figure 5.31 shows the flat band shift as a function of gate voltage bias applied for different durations. The flat band voltage shift increases with the gate voltage until the nc-Si are fully charged or discharged. An increase in the time of voltage application will cause more charge trapping or de-trapping depending on the voltage polarity.

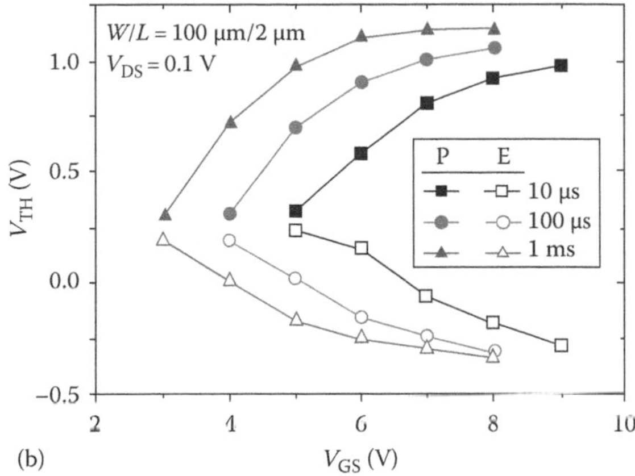

FIGURE 5.31 Program and erasing characteristics after symmetric gate pulses for the nc-Si/a-SiO_2 nanocomposites. (Reprinted from Tsoukalas, D. et al., *Mater. Sci. Eng. B-Solid State Mater. Adv. Technol.*, 124, 93, 2005. With permission.)

5.3.3.1 Influence of nc-Si Density on the Memory Effect

When using nc-Si for charge storage devices, the size distribution and density of the nanocrystals will influence the properties of the devices. Smaller Si nanocrystals will offer a better control on the number of stored charges but would also result in an increased spacing energy and coulomb charging energy.

The parameter Q_{charge} in the simplified flat band voltage shift equation is the charge density that can be replaced by qn, where q is the element charge and n is the nc-Si density. So Equation 5.19 can be rewritten as

$$\Delta V_{\text{FB}} = \frac{qnT_{\text{control}}}{\varepsilon_{\text{ox}}} \tag{5.20}$$

This equation shows that the maximum flat band voltage shift is directly proportional to the density of the nc-Si. This means that to achieve a large memory window the nc-Si should be made as small as possible with a minimum separation. The density of the nc-Si, n, can be simply expressed as [115]

$$n = \frac{1}{(s+d)^2} \tag{5.21}$$

where s is the average spacing between the nc-Si. It seems that the best memory effect can be achieved by making s and d as small as possible. However, there are two negative consequences associated with such a strategy. First, there is a limit on s to prevent the charge migration from one nc-Si to a neighboring nc-Si. To sufficiently eliminate this lateral charge migration, the nc-Si must be separated by a distance of approximately 5 nm [115]. Second, the size reduction of the nc-Si results in enhanced coulomb blockade effects due to the reduction in the nc-Si capacitance and increased confinement effects on the single-particle energy levels of the nc-Si. These effects reduce the maximum number of electrons that an nc-Si can store. The changes in the energy, ΔE, of the single-particle levels of the nc-Si while going from $N-1$ electrons to N electrons can be calculated by the energy required to bring an electron to the nc-Si containing $N-1$ electrons. The changes in the nc-Si energy levels can be calculated by [115]

$$\Delta E = \frac{q^2(N-1)}{C} \tag{5.22}$$

where

q is the electron charge
N is the number of electrons stored in the nc-Si
C is the capacitance of the nc-Si

By solving this equation for the maximum number of electrons one nc-Si can store, and combining it with the expression for flat band voltage shift, the memory window can be given approximately as in Ref. [115]

$$\Delta V_{\mathrm{FB}} = A\frac{d}{(S+d)^2} \tag{5.23}$$

where A is a constant introduced to simplify the expression. Differentiating this expression with respect to the nc-Si diameter, the maximum memory window occurs when the nc-Si diameter is equal to the spacing between the nc-Si. Combining this with the minimum requirement of nc-Si spacing, the nc-Si structure with a diameter of 5 nm and a spacing of 5 nm should yield optimum memory performance [115].

For conventional fabrication methods, such as CVD, sputtering, evaporation, etc., the reported size distributions are normally several nanometers wide. The control of size and size distribution, in single and multilayers, was demonstrated previously, including the post annealing of the films in the dilute oxygen ambient to shrink the size of the nc-Si [116], nc-Si/a-SiO_2 multilayer, and recently developed SiO_x/a-SiO_2 multilayer ($0<x<2$). So far, the SiO_x/a-SiO_2 multilayer has been considered as the best approach for controlling the density, separation, and spatial distribution of the nc-Si.

Figure 5.8 shows the nc-Si density control with SiO_x/a-SiO_2 structure by Lu et al. [11]. There are three layers of SiO_x for both samples and the thickness of the SiO_x layers are identical, but the Si concentrations in the SiO_x layers is $x=0.9$ for sample A and $x=1.63$ for sample B. The calculated average nc-Si density is $2.79\times10^{12}/\mathrm{cm}^2$ and $0.93\times10^{12}/\mathrm{cm}^2$ for samples A and B, respectively. The electrical properties have been studied. A series of *C-V* curves are shown in Figure 5.32. *C-V* hysteresis are also observed, which are caused by the trapping/de-trapping of charges in the nc-Si.

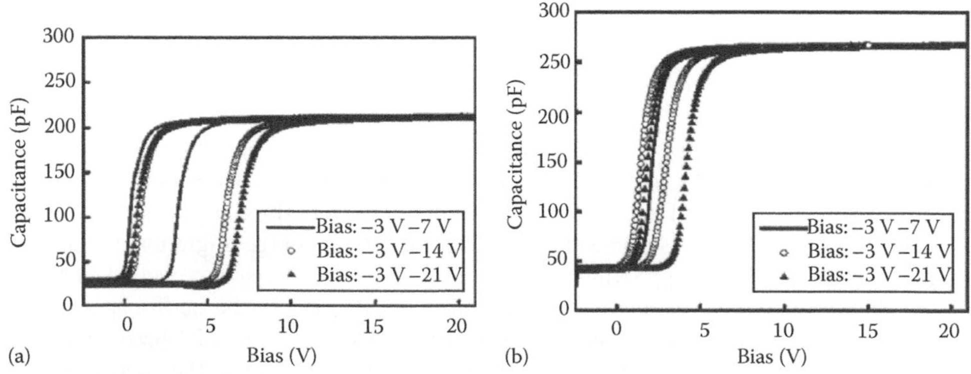

FIGURE 5.32 (a) Typical high frequency *C-V* characteristics of sample A and (b) typical high frequency *C-V* characteristics of sample B. The *C-V* hysteresis are proportional to the charging density in the nc-Si. (Reprinted from Lu, T.Z. et al., *J. Appl. Phys.*, 100, 5, 2006. With permission.)

A higher bias at the gate results in a larger hysteresis. The narrowest *C-V* loop corresponds to the charging of the first layers only, the middle one to the charging of the first and second layers, and the widest one to the charging of all three layers. The memory window is larger for the sample with a higher nc-Si density under the same voltage bias.

The theoretical calculation of the flat band shift of multilayer charges can be estimated as follows [11]:

$$\Delta V_{FB} = \frac{en_{dot}}{\varepsilon_{ox}}\left(t_{upper1} + \frac{\varepsilon_{ox}}{2\varepsilon_{Si}} t_{dot} \times N + t_{upper2} + \cdots + t_{upperN}\right) \quad (5.24)$$

where

ΔV_{FB} is the flat band voltage shift
n_{dot} is the density of the nc-Si
t_{upper} is the upper control gate oxide thickness
t_{dot} is the silicon nanocrystal diameter
ε_{ox} and ε_{Si} are the permittivity of the oxide and silicon, respectively
t_{upperN} is the upper control gate oxide thickness of the *N*th layer of nc-Si

5.3.3.2 Charge Trapping in nc-Si

For the application of a nonvolatile memory device, a long charge retention time is very critical and necessary. Usually, retention of 10 years is required for commercial applications. To achieve the long retention time of quantum dot memory, the charge storage behavior during the charge retention mode should be well understood. It is reported that for nc-Si/a-SiO_2 nanocomposites charges can be stored inside the nanocrystals (deep traps center) [117], the defect of the oxide matrix, or the interface of the nc-Si and the dielectric matrix [118,119].

For the charge trapping mechanism in the nc-Si/a-SiO_2 nanocomposites, it is widely accepted that the electrons inside the nanocrystal cannot be stored in the conduction band of the nc-Si [117]. First, the conduction band edge inside the nanocrystal is higher than that of the substrate because of the charging effect and quantum confinement effect that allows electrons to tunnel back to the channel very easily. This is not consistent with the long retention observed in the published data [118,119]. Second, the experimental retention time measurement shows heavy temperature dependence. If the electrons are stored in the nanocrystal conduction band, the retention time should only show mild change between room temperature and <100°C. Third, several researchers showed that the memory phenomenon disappears if the nc-Si memory device is annealed in hydrogen (low defects density due to the hydrogen passivation). This suggests that many deep trap states such as P_b center and E' center exist in the nc-Si/a-SiO_2 nanocomposites.

It is well known that the deep trapping centers play an important role in the charge storage of silicon nitride films. Actually, regions with a high density of interface defects can act as effective charge trap centers because of the large surface-to-volume ratios, high surface roughness, and compositional disorders of the nanocrystals. Tiwari et al. [117] suggested that the charges may be stored in the deep trapping center after injecting into the nanocrystals. Such a charging mechanism has been confirmed by Shi et al. [118,119], who studied the effect of traps on the long-term charge storage characteristics in the MOS memory based on nc-Si. To produce different defects and trap density, after high-temperature annealing to induce the nc-Si, they annealed the samples in H_2 ambient at 430°C and in vacuum at 700°C. The annealing in H_2 ambient could effectively decrease interface traps by H-passivation, and the annealing in vacuum resulted in high density of interface traps, especially P_b center. The *C-V* measurement was performed in the samples having different conditions of traps.

The *C-V* hysteresis of various annealed MOS diodes is shown in Figure 5.33 [118]. The flat band voltage shift is attributed to the injection of hole or electrons. The maximum shift in the

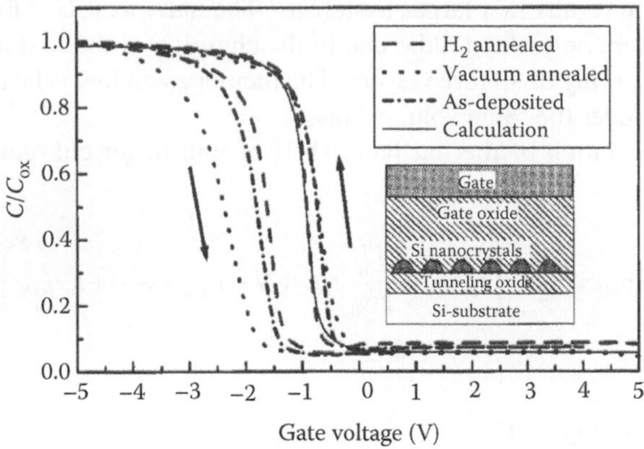

FIGURE 5.33 *C-V* hysteresis loops in various annealed MOS diodes. (Reprinted from Shi, Y. et al., *J. Appl. Phys.*, 84, 2358, 1998. With permission.)

C-V measurement is observed in the vacuum-annealed sample having the highest trap density, the minimum shift is observed in the H_2-annealed diode having the lowest trap density, and the middle is in the as-deposited sample. This indicates that more charges are stored in the vacuum-annealed nanocrystals than in the H_2-annealed ones. The long-term charge storage characteristics are illustrated in Figure 5.34 [118]. The changes in the net charge in nanocrystals as a function of waiting time are linearly proportional to the shifts of the V_{FB}. A striking difference in the charge-loss rates for various annealed diodes is observed. The vacuum-annealed sample has the fastest charge-loss rate, 0.078 V/decade, and the sample annealed in H_2 has the lowest charge-loss rate 0.014 V/decade. The as-deposited sample still showed a middle charge-loss rate, 0.064 V/decade. The positive pulse voltage bias leads to the injection of electron into nanocrystals, shifting the V_{FB} to positive direction. The electrons subsequently leak back to the substrate with logarithmic time dependence in the charge retention mode. A negative pulse voltage bias leads to the hole tunneling into the nanocrystals, shifting the V_{FB} to the negative direction.

The charging and retention behaviors of the samples with different annealing conditions can be explained by the energy band diagram of the MOS diode at flat band in retention mode. Applying a positive voltage bias, the electrons are injected from the conduction band of silicon substrate to the conduction band of the nc-Si. However, the quantum confinement of nc-Si leads to the band

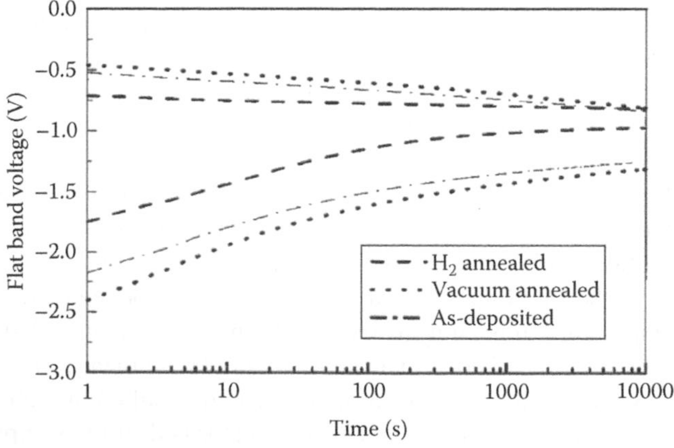

FIGURE 5.34 Long-term charge retention characteristics of various annealed MOS diodes, where the applied pulse voltages are ±5 V; measured using the constant capacitance method at flat band point. (Reprinted from Shi, Y. et al., *J. Appl. Phys.*, 84, 2358, 1998. With permission.)

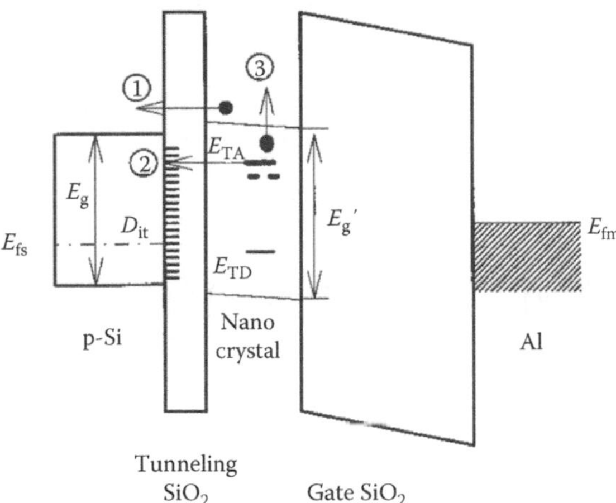

FIGURE 5.35 Energy band diagram of the MOS memory structure at a flat band point for electron tunneling processes in the retention mode after injecting electrons. (Reprinted from Shi, Y. et al., *J. Appl. Phys.*, 84, 2358, 1998. With permission.)

edge of the nanocrystals slightly higher than that of substrate, thus the electron tunneling into the conduction band of nc-Si will easily tunnel back to the substrate. Therefore, the assumption that the charges are mainly stored in the conduction band of nc-Si contradicts the long-term retention behaviors of nc-Si memory. One possible explanation is that the charges may be stored at deep trapping centers, as shown in Figure 5.35 [118].

In the tunneling process under a positive voltage, the charges injected into an nc-Si will first fill the empty states, where the trap level is deeper, and then fill the states where the level is shallower. A Coulomb charging energy E_q, proportional to the number of electron already stored in the trap, should be overcome in order to inject an additional electron into the trap. Thus, the total trap energy E_T is raised by the E_q. The more the electron injection, the higher the trap energy becomes at the deep trapping center. If the total trap energy is higher than the conduction band edge of the silicon substrate, the trapped electrons will possibly tunnel back to the substrate. That is to say that the number of trapped electrons is closely related to the number of deep trapping centers. Thus, a larger flat band shift is generally observed in the samples with more deep traps, which are those annealed in vacuum.

Two possible processes were deduced for the long-term loss of trapped electrons. One is the direct tunneling from the traps to the interface states at the SiO_2/Si substrate, process 2 as shown in Figure 5.35. Another one is an indirect process involving thermal de-trapping of the electrons to the conduction band and then tunneling back to the conduction band of the substrate, processes 3 and 1 as shown in Figure 5.35. This indirect process is possibly more difficult to occur for the charges in the deeper trap level in the nanocrystals. Thus, possibly what determines the long-term charge retention is the charges tunneling from the traps to the interface states at the SiO_2/Si substrate, process 2.

5.3.4 Conduction Modulation Caused by Charging and Discharging of nc-Si

Charging/discharging of the nc-Si can also lead to the modulations in the total capacitance and current conduction of the nc-Si/a-SiO_2 nanocomposite films. This can be explained in terms of the breaking of the nc-Si tunneling paths due to the charge trapping in the nc-Si.

The effect of charging/discharging of the nc-Si on the capacitance and tunneling current of the nc-Si/a-SiO_2 nanocomposites has been reported in detail in Refs. [120–122]. Figure 5.36 shows the effect of charging/discharging of the nc-Si on the conductance and capacitance of the films. A virgin device was stressed at a constant voltage of +4 V for 320 s. After the constant-voltage stressing, the *C-V* and *I-V* measurements were conducted again. The capacitance after the stressing decreases

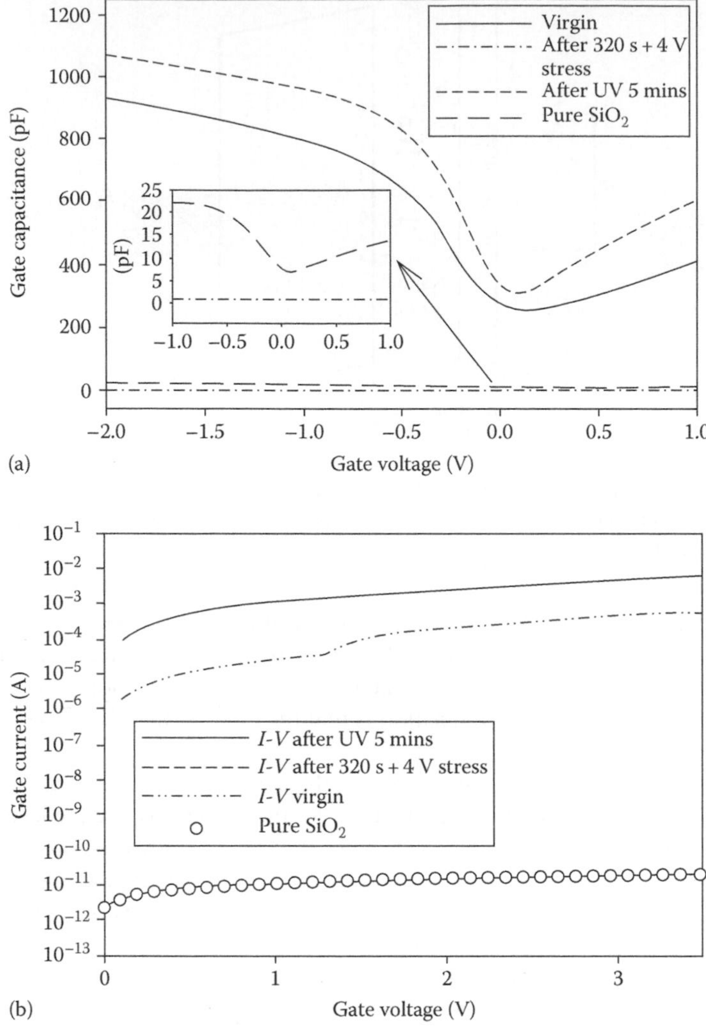

FIGURE 5.36 (a) Reduction and recovery of the gate capacitance after the electrical stressing and UV illumination and (b) reduction and recovery of the gate current after the electrical stressing and UV illumination. (Reprinted from Ng, C.Y. et al., Charging/discharging of silicon nanocrystals embedded in a SiO_2 matrix inducing reduction/recovery in the total capacitance and tunneling current, in *SPIE International Symposium on Microlectronics, MEMS and Nanotechnology*, Perth, Australia, December, 2003, pp. S43–S46.)

to a very low level. The conductance of the device also drops dramatically to a level similar to that of pure SiO_2 films. The reduction of the overall capacitance and conductance is mainly due to the charging of the nc-Si embedded in the SiO_2 matrix.

The total capacitance (C_{total}) of the MOS structure with the nc-Si/a-SiO_2 nanocomposites consist of several components, including the nc-Si capacitance ($C_{nc\text{-}Si}$), the remaining pure SiO_2 capacitance (C_{SiO_2}), and the capacitance of the silicon surface region (C_D). The reduction in the capacitance can be understood with the equivalent circuit shown in Figure 5.37. The total capacitance of the MOS capacitor can be calculated by [121]

$$C_{total} = \frac{C_D(C_{nc\text{-}Si} + C_{SiO_2})}{C_D + (C_{nc\text{-}Si} + C_{SiO_2})} \tag{5.25}$$

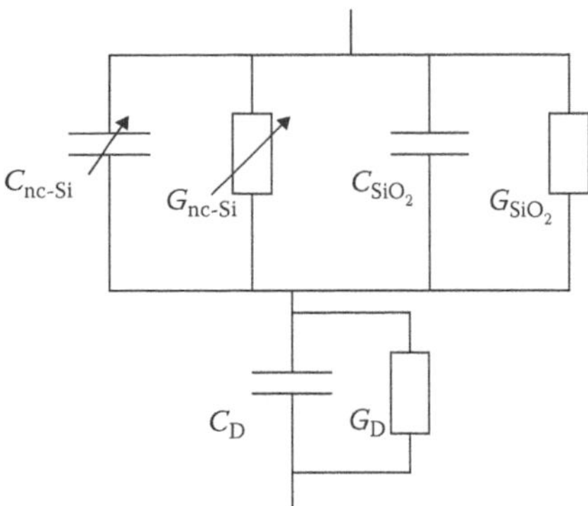

FIGURE 5.37 Equivalent circuit of the MOS structure with the nc-Si embedded in the gate oxide. (Reprinted from Ng, C.Y. et al., Charging/discharging of silicon nanocrystals embedded in a SiO_2 matrix inducing reduction/recovery in the total capacitance and tunneling current, in *SPIE International Symposium on Microlectronics, MEMS and Nanotechnology*, Perth, Australia, December, 2003, pp. S43–S46.)

As the C_D is very large [122], the overall capacitance can be roughly estimated to be $C_{total} \approx C_{nc\text{-}Si} + C_{SiO_2}$. As the remaining gate oxide capacitance is very small [122], the total capacitance is mainly determined by the nc-Si capacitance. The capacitance of the nc-Si is large if it is unoccupied (uncharged), and becomes very small when occupied (charged). The application of the positive gate bias will inject the electrons into the nc-Si, leading to the decrease of the capacitance of the nc-Si that reduces the total capacitance.

The dramatic reduction in the gate current after the positive electrical bias can be explained in terms of the breaking of the tunneling paths formed through the nc-Si due to the charging in some nc-Si as shown in Figure 5.38 [121]. In the virgin sample, most of the nc-Si are not charged and can form many conductive percolation paths, as shown in Figure 5.38a, and thus the gate current is relatively large. The positive electrical bias causes some of the nc-Si to be charged up, which blocks the electron tunneling through that particular nc-Si because of the Coulomb blockade effect. In this case, the tunneling current will drop. If most of the tunneling paths are broken, the gate current will drop significantly.

The trapped charges in the nc-Si can be released under UV illumination and the discharged nc-Si can form conductive tunneling paths again, leading to an increase in the current, as shown in Figure 5.38c. The discharge caused by UV illumination is evident from the recovery of the *C-V* and *I-V* characteristics after 5 min of UV exposure, as shown in Figure 5.36 [121].

Charging/discharging of nc-Si can be also employed to explain the current switching during the *I-V* measurement as shown in Figure 5.39 [123]. The *I-V* characteristics exhibit fluctuations and switching behaviors. This behavior is interpreted by considering a limited number of the conductive paths for the tunneling current. The nc-Si and the other traps in the oxide, located close to this path, act as trapping centers for the electrons. When one electron is trapped, the local electrostatic potential increases, and thus the distribution of the electric field in this area is modified. As a consequence, the current abruptly decreases. By increasing the gate voltage, the current level again increases until another electron charges the new nc-Si. During the voltage sweeping the applied electric field can cause charging or discharging of some nc-Si, leading to the breaking or formation of some conductive paths and thus the decrease or increase of the current, respectively.

Maeda et al. [124] have investigated the electron conduction behavior in the dynamic and static *I-V* characteristics using the diode structure. The dynamic *I-V* measurement of their films shows novel features such as the *N*-shaped tunneling current versus gate voltage characteristics and hysteresis. The time-dependent current behaviors were also observed. They attributed the electrical

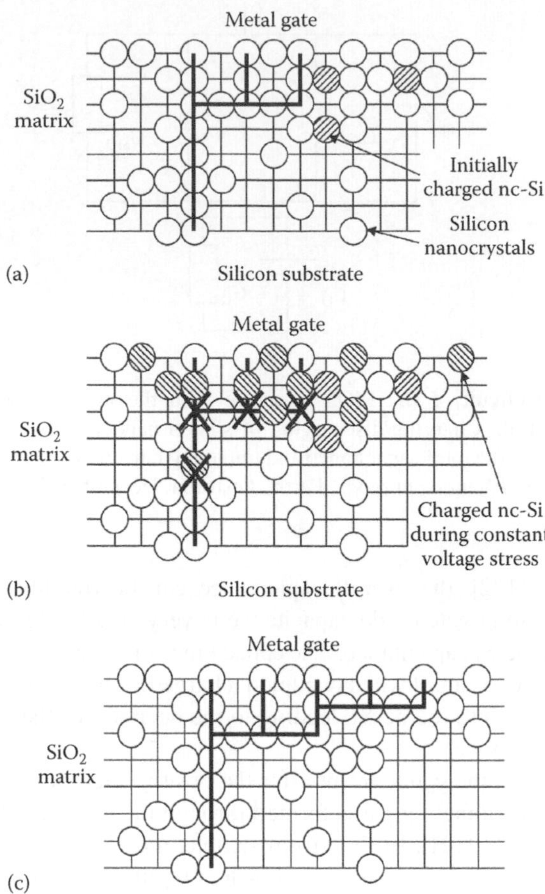

FIGURE 5.38 Schematic of the effect of charging and discharging of the nc-Si on the current conduction in the gate oxide. (a) Virgin sample, (b) after constant voltage stress, (c) after UV annealing. (Reprinted from Ng, C.Y. et al., Charging/discharging of silicon nanocrystals embedded in a SiO_2 matrix inducing reduction/recovery in the total capacitance and tunneling current, in *SPIE International Symposium on Microelectronics, MEMS and Nanotechnology*, Perth, Australia, December, 2003, pp. S43–S46.)

phenomenon to the charging effect of a fixed amount of electrons at the nc-Si and the consequent screening effect on the tunneling current.

For the *I-V* characteristics, because the carrier charging will lead to the changes in the potential or electrical field in the MOS structure, usually a time-dependent current is observed in the *I-V* measurement. To obtain the static characteristic, it is necessary to wait for a time until the current becomes stable. The waiting time until the current measurement after applying each voltage can be defined as the delay time. Figure 5.40 shows the round sweep *I-V* characteristics measured under the delay time 0 s for the sample with 10 nm Si layer sandwiched by the SiO_2 after rapid thermal annealing for 30 s and 10 min, respectively. A novel feature of the current switching at the gate voltage from 2.5 to 4.2 V in the forward sweep is observed for the sample after annealing for 10 min. In contrast, no current switching was observed in the reverse sweeping, corresponding to the static behavior, but a hysteresis loop can be observed. Moreover, there was neither current switching in the forward *I-V* measurement for the 10 min RTO sample, where the delay time is 60 s, nor the hysteresis as shown in Figure 5.41 [124]. Therefore, the current switching in their measurement should not be attributed to the resonant tunneling.

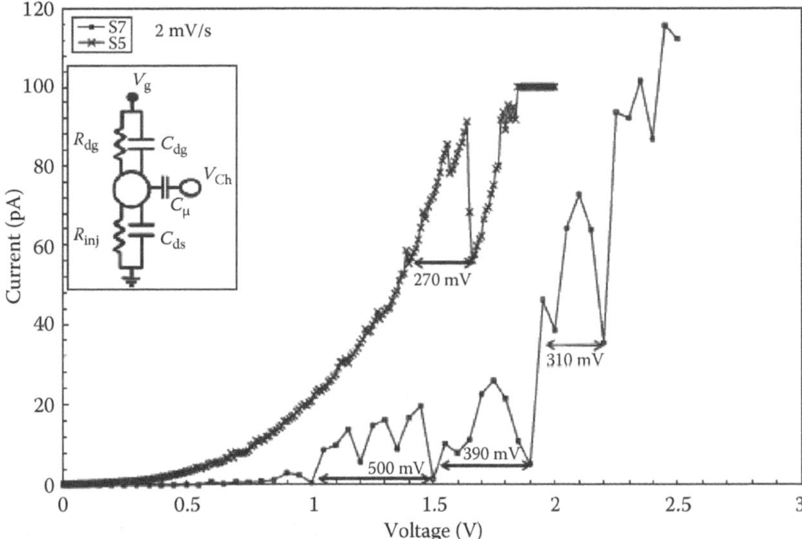

FIGURE 5.39 *I-V* characteristics of the nc-Si/a-SiO_2 nanocomposites. The inset shows the equivalent electrical model. (Reprinted from Dumas, C. et al., Influence of the thickness of the tunnel layer on the charging characteristics of Si nanocrystals embedded in an ultrathin SiO_2 layer, in *Symposium on Silicon Nanocrystals for Electronic and Sensing Applications held at the 2006 E-MRS Spring Meeting*, Nice, France, May 29–June 02, 2006, pp. 80–84.)

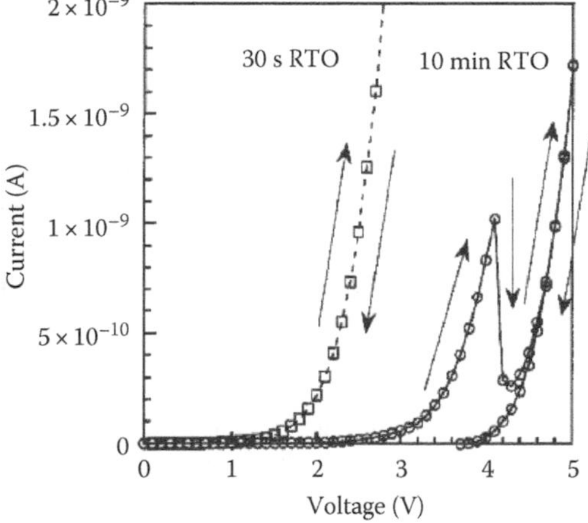

FIGURE 5.40 Typical dynamic *I-V* characteristics of a 10 min (solid curve) and 30 s (dashed curve) RTO diode sample measured at room temperature, where the measurement delay time is 0. (Reprinted from Maeda, T. et al., *Nanotechnology*, 10, 127, 1999.)

The time dependence of *I-V* characteristics is useful to investigate the above-mentioned phenomenon. Figure 5.42 shows the time dependence of the tunneling current for the 10 min RTO diode sample, where the gate bias voltage is changed from 2 to 5.5 V. A clear current reduction can be observed in Figure 5.42. The reduction in the tunneling current accounts for two orders of magnitude. However, when the gate voltage is higher than 5 V, no tunneling current reduction is observed in the measurement timescale. The time-dependent tunneling current indicates an electron charging effect and the consequent screening effect on the tunneling current flowing through the

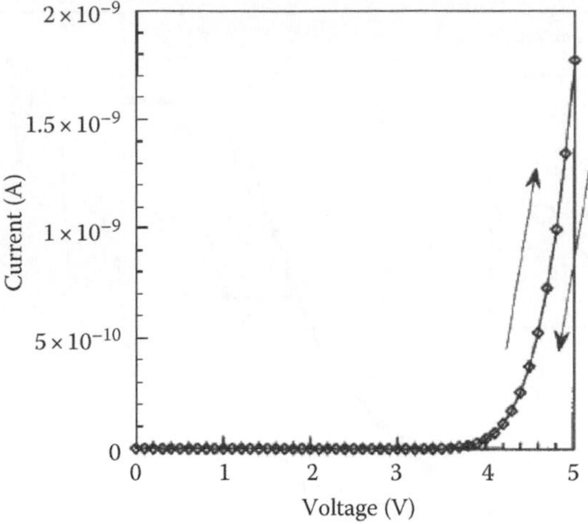

FIGURE 5.41 Typical static *I-V* characteristics of a 10 min RTO diode sample measured at room temperature, where the measurement delay time is 60 s. (Reprinted from Maeda, T. et al., *Nanotechnology*, 10, 127, 1999.)

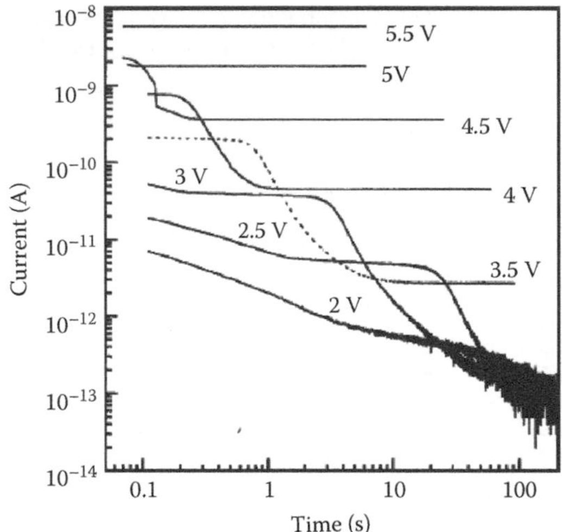

FIGURE 5.42 Time dependence of the tunneling current (100 ms to 100 s) for the 10 min RTO diode sample with the gate voltage as a parameter. (Reprinted from Maeda, T. et al., *Nanotechnology*, 10, 127, 1999.)

sample structure. The current peak and valley and the hysteresis strongly suggest that the electron charging effect is related to the nc-Si. Noting that the incubation time until the start of the current reduction exists even in the case of the relatively high initial current means that it takes some time for the electrons in the conduction band of the nc-Si to be stabilized in the trapping sites. Thus, the conduction band of the nc-Si should be out of consideration for the trapping sites. The trapping sites are expected to locate at the nc-Si surface adjacent to the amorphous SiO_2.

The current peak and valley can be explained as follows. Once charges are trapped at the nc-Si and fixed, the electronic potential of the nc-Si itself rises up and simultaneously forces up the surrounding electric potential. Current peak and valley are due to the narrowing of the tunneling current paths because of the current screening effect. Noting that the tunneling current paths of the nc-Si are formed with a wide size distribution, the current peak valley occurs when the tunneling paths are broken due to the screening effect of the smaller nc-Si once they are charged. Due to the small

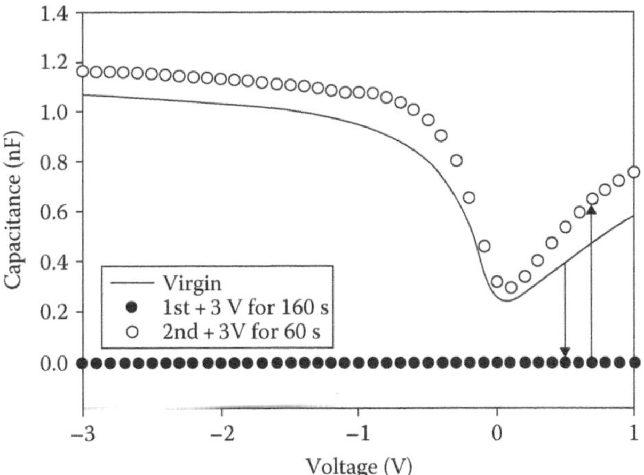

FIGURE 5.43 Changes in the MOS capacitance after two consecutive applications of + 3 V for different durations. (Reprinted from Liu, Y. et al., *Nanotechnology*, 16, 1119, 2005.)

charge energy of the large nc-Si, their contribution to the current screening effect can be ignored by increasing the gate voltage further, the tunneling current rises up again, keeping the charged state in the nc-Si. The second threshold voltage refers to the start voltage of the screened tunneling current. The reverse sweep current consists of the fully screened tunneling current and coincides with the static tunnel current curve as shown in Figure 5.41.

Liu et al. have demonstrated that charging and discharging of the nc-Si could occur at the same voltage [122]. Figure 5.43 shows the changes in the MOS capacitor with the nc-Si/a-SiO_2 nanocomposites as the active layer after two applications of +3 V for different durations. In the virgin case, the MOS structure shows a capacitance of ~1 nF under accumulation. The application of +3 V for 160 s reduces the capacitance to nearly zero. The reduction in the capacitance is due to the charging up in the nc-Si caused by the voltage application. The charged nc-Si does not respond to the small AC signal. Therefore, the capacitance of the nc-Si decreases, and thus the MOS capacitance is reduced. On the other hand, the second +3 V for 60 s will release the trapped charge in the nc-Si, leading to the recovery of the capacitance. The charging and discharging of the nc-Si could occur at the same voltage magnitude suggesting that the capacitance modulation could be observed in the time–domain capacitance measurement under a constant voltage. Figure 5.44 shows the capacitance–time characteristic measured at −6 and

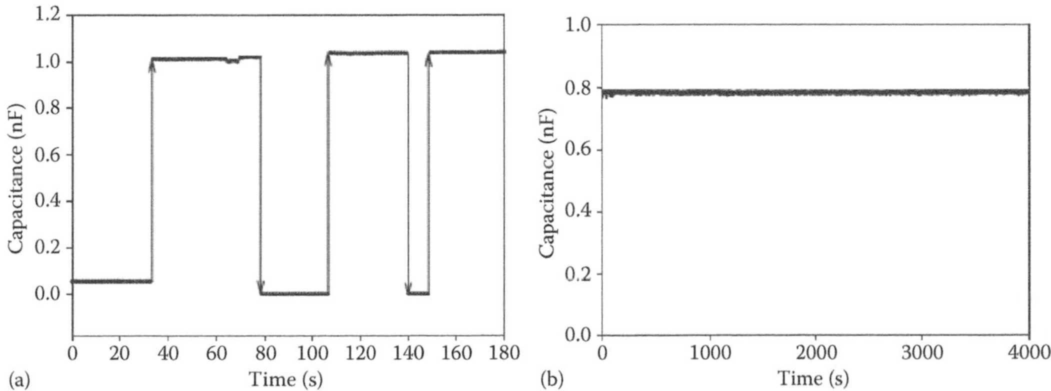

FIGURE 5.44 Capacitance modulations in the capacitance–time measurement at −6 V (a) and −0.5 V (b). (Reprinted from Liu, Y. et al., *Nanotechnology*, 16, 1119, 2005.)

−0.5 V, respectively. The capacitance shows a significant increase and decrease during the time-domain capacitance measurement. The changes of the capacitance are quite random and unpredictable. As the charging and discharging in the nc-Si depends on the applied voltage, the capacitance modulation also depends on the voltage. This is confirmed by Figure 5.44b. The capacitance modulation at −0.5 V is insignificant, indicating that the charging and discharging are insignificant at low gate voltage.

5.4 SUMMARY

The Si nanocrystals (nc-Si) imbedded in amorphous SiO_2 (i.e., nc-Si/a-SiO_2) are promising candidates for the next generation quantum dot nonvolatile flash memory. There are mainly three types of nc-Si/a-SiO_2 nanocomposites synthesized, including single-layer Si-rich SiO_x ($0<x<2$) films, Si/SiO_2 multilayers, and SiO_x/SiO_2 multilayers structures. The later two structures are usually employed to control the spatial and size distribution of the nc-Si. However, the nc-Si produced by the Si/SiO_2 multilayer are brick-shaped, containing high density of grain boundaries, while SiO_x/SiO_2 multilayer can generate spherical and well-passivated-nc-Si. The as-produced Si-rich SiO_x films are amorphous. The local atomic structures are depicted by the RBM, where local bonding contains five different chemical structures, $Si{-}Si_{4-n}O_n$ ($n=0$, 1, 2, 3, 4) or RMM where the local bonding is due to mixtures of Si and SiO_2 nanosized domains. Correspondingly, two nc-Si growth mechanisms were proposed, including the segregation-controlled mechanism of the Si suboxides or the diffusion-controlled mechanism of the excess Si during annealing. The decomposition of the Si suboxides takes place by two consequent reactions, $Si^{2+}+Si^{2+} \rightarrow Si^{1+}+Si^{3+}$ (1) and $Si^{1+}+Si^{3+} \rightarrow Si^0+Si^{4+}$ (2). The memory effect of the nc-Si/a-SiO_2 nanocomposite is realized by the charging and discharging of the nc-Si that usually can lead to a flat band voltage shift in the capacitance–voltage characteristics. The Coulomb blockade effect and quantum confinement effect of the nc-Si enable the nc-Si/a-SiO_2 nanocomposite to be one of the promising candidates for single-electron devices. The presence of the nc-Si in the SiO_2 strongly enhances its conductance due to the tunneling paths formed by the nc-Si. The current conduction mechanisms are direct tunneling at low electrical voltages and Fowler–Nordheim tunneling at high electrical voltages. The charges in the nc-Si/a-SiO_2 nanocomposites are trapped in the deep trapping centers of the nc-Si or the defects in oxides rather than the conduction band of the nc-Si. Besides memory effect, the capacitance and conductance of the nc-Si/a-SiO_2 nanocomposites also can be modulated by charging and discharging of the nc-Si that can be potentially used in two-terminal memory devices.

ABBREVIATIONS

a-SiO_2	Amorphous SiO_2
CMOS	Complementary metal-oxide-semiconductor
C-V	Capacitance-voltage
CVD	Chemical vapor deposition
EFTEM	Energy-filtered transmission electron microscopy
HRTEM	High-resolution transmission electron microscopy
I-V	Current-voltage
LED	Light-emitting diode
LPCVD	Low-pressure chemical vapor deposition
LSIs	Large-scale integrated circuits
MOS	Metal-oxide-semiconductor
NBOHC	Non-bridging oxygen hole center
nc-Si	Si nanocrystals
PECVD	Plasma-enhanced chemical vapor deposition
PL	Photoluminescence
RBM	Random-bonding model

RMM	Random-mixture model
SEDs	Single-electron devices
SET	Single-electron transistor
TEM	Transmission electron microscopy
XPS	X-ray photoelectron spectroscopy

SYMBOLS

A	Surface area
C^a	Composition of stoichiometric SiO_2
C^b	Composition of the silicon cluster
C_d	Control oxide capacitances
C_{dot}	Self-capacitance of the quantum dot
C_g	Gate capacitance
C^m	Average composition of the entire film
C_s	Tunneling oxide capacitances
C_{SiO_2}	SiO_2 capacitance
C_{total}	Total capacitance
d	Diameter of the nanocrystal
d_m	Maximum size of nanocrystals
D	Diffusion coefficient of silicon in SiO_2
D_0	Mean size of nanocrystals
e	Elementary charge
E	Electrical field
E_0	Ground state energy
E_c	Coulomb charging energy
E_g	Band gap of nc-Si
E_g^0	Band gap of bulk silicon
$\hbar$	Reduced Plank constant
k	Wave vector
m_{eff}	Effective mass of an electron
n_{dot}	Density of nc-Si
Q	Activation energy
Q_{charge}	Charge density trapped in the nc-Si
r	Radius of nc-Si
r_i	Initial Si cluster radius
R	Universal gas constant
R^*	Critical cluster radius
S	Spacing between nc-Si
T_A	Annealing temperature
V_{FB}	Flat band voltage
V_G	Applied gate voltage
V_T	Threshold voltage
X_{si}	Si content goes above the stoichiometric level
α	Parameter that accounts for an effective mass
γ_{cm}	Cluster-matrix interfacial energy
ΔV_{FB}	Flat band voltage shift
ΔG_v	Gibbs free energy
ε_0	Vacuum permittivity
ε_{SiO_2}	Dielectric constant of SiO_2
ϕ_b	Potential barrier height at the nanocrystals–insulator interface

REFERENCES

1. De Salvo B, Gerardi C, Lombardo S, Baron T, Perniola L, Mariolle D, Mur P, Toffoli A, Gely M, Semeria MN et al., How far will silicon nanocrystals push the scaling limits of NVMs technologies? In *Electron Devices Meeting, 2003 IEDM '03 Technical Digest IEEE International*. Washington, DC, 2003, pp. 26.21.21–26.21.24.
2. Compagnoni CM, Ielmini D, Spinelli AS, Lacaita AL, Previtali C, Gerardi C, Study of data retention for nanocrystal Flash memories. In *Reliability Physics Symposium Proceedings, 2003 41st Annual 2003 IEEE International*. Dallas, TX, 2003, pp. 506–512.
3. Compagnoni CM, Gusmeroli R, Ielmini D, Spinelli AS, Lacaita AL, Silicon nanocrystal memories: A status update. *Journal of Nanoscience and Nanotechnology* 2007, **7:**193–205.
4. Lombardo S, De Salvo B, Gerardi C, Baron T, Silicon nanocrystal memories. *Microelectronic Engineering* 2004, **72:**388–394.
5. Tiwari S, Rana F, Hanafi H, Hartstein A, Crabbe EF, Chan K, A silicon nanocrystals based memory. *Applied Physics Letters* 1996, **68:**1377–1379.
6. Takahashi Y, Ono Y, Fujiwara A, Inokawa H, Silicon single-electron devices. *Journal of Physics-Condensed Matter* 2002, **14:**R995–R1033.
7. Heitmann J, Muller F, Zacharias M, Gosele U, Silicon nanocrystals: Size matters. *Advanced Materials* 2005, **17:**795–803.
8. Oda S, Huang SY, Salem MA, Hippo D, Tanaka A, Tsuchiya Y, Mizuta H, Nanocrystalline silicon quantum dot devices. In *8th International Conference on Solid-State and Integrated Circuit Technology 2006*, Shanghai, China, 2006, pp. 1045–1048.
9. Dimitrakis P, Kapetanakis E, Tsoukalas D, Skarlatos D, Bonafos C, Ben Asssayag G, Claverie A, Perego M, Fanciulli M, Soncini V et al., Silicon nanocrystal memory devices obtained by ultra-low-energy ion-beam synthesis. *Solid-State Electronics* 2004, **48:**1511–1517.
10. Tsoukalas D, Dimitrakis P, Kolliopoulou S, Normand P, Recent advances in nanoparticle memories. *Materials Science and Engineering B-Solid State Materials for Advanced Technology* 2005, **124:**93–101.
11. Lu TZ, Alexe M, Scholz R, Talalaev V, Zhang RJ, Zacharias M, Si nanocrystal based memories: Effect of the nanocrystal density. *Journal of Applied Physics* 2006, **100:**5.
12. Zeghbroeck BV, *Principles of Semiconductor Devices*. Boulder, CO: University of Colorado at Boulder, 2004.
13. Chelikowsky JR, Cohen ML, Electronic structure of silicon. *Physical Review B* 1974, **10:**5095.
14. Ranjan V, Kapoor M, Singh VA, The band gap in silicon nanocrystallites. *Journal of Physics-Condensed Matter* 2002, **14:**6647–6655.
15. Delerue C, Allan G, Lannoo M, Theoretical aspects of the luminescence of porous silicon. *Physical Review B* 1993, **48:**11024–11036.
16. Kobayashi H, Mori T, Namba K, Nakato Y, New method for determination of energy distribution of surface states in the semiconductor band-gap: XPS measurements under biases. *Solid State Communications* 1994, **92:**249–254.
17. Delley B, Steigmeier EF, Size dependence of band-gaps in silicon nanostructures. *Applied Physics Letters* 1995, **67:**2370–2372.
18. Hill NA, Whaley KB, A theoretical study of light emission from nanoscale silicon. *Journal of Electronic Materials* 1996, **25:**269–285.
19. Agrawal BK, Agrawal S, First-principles study of one-dimensional quantum-confined H-passivated ultrathin Si films. *Applied Physics Letters* 2000, **77:**3039–3041.
20. Franceschetti A, First-principles calculations of the temperature dependence of the band gap of Si nanocrystals. *Physical Review B* 2007, **76:**4.
21. Ledoux G, Gong J, Huisken F, Guillois O, Reynaud C, Photoluminescence of size-separated silicon nanocrystals: Confirmation of quantum confinement. *Applied Physics Letters* 2002, **80:**4834–4836.
22. Tiwari S, Rana F, Chan K, Shi L, Hanafi H, Single charge and confinement effects in nano-crystal memories. *Applied Physics Letters* 1996, **69:**1232–1234.
23. Baron T, Gentile P, Magnea N, Mur P, Single-electron charging effect in individual Si nanocrystals. *Applied Physics Letters* 2001, **79:**1175–1177.
24. Lu TZ, Alexe M, Scholz R, Talelaev V, Zacharias M, Multilevel charge storage in silicon nanocrystal multilayers. *Applied Physics Letters* 2005, **87:**3.

25. Kachurin GA, Zhuravlev KS, Pazdnikov NA, Leier AF, Tyschenko IE, Volodin VA, Skorupa W, Yankov RA, Annealing effects in light-emitting Si nanostructures formed in SiO_2 by ion implantation and transient preheating. *Nuclear Instruments and Methods in Physics Research Section B: Beam Interactions with Materials and Atoms* 1997, **127–128:**583–586.
26. Kachurin GA, Tyschenko IE, Zhuravlev KS, Pazdnikov NA, Volodin VA, Gutakovsky AK, Leier AF, Skorupa W, Yankov RA, Visible and near-infrared luminescence from silicon nanostructures formed by ion implantation and pulse annealing. *Nuclear Instruments and Methods in Physics Research Section B: Beam Interactions with Materials and Atoms* 1997, **122:**571–574.
27. Kim HB, Kim TG, Son JH, Whang CN, Chae KH, Lee WS, Im S, Song JH, Effects of Si-dose on defect-related photoluminescence in Si-implanted SiO_2 layers. *Journal of Applied Physics* 2000, **88:**1851–1854.
28. Dimitrakis P, Kapetanakis E, Normand P, Skarlatos D, Tsoukalas D, Beltsios K, Claverie A, Benassayag G, Bonafos C, Chassaing D et al., MOS memory structures by very-low-energy-implanted Si in thin SiO_2. *Materials Science and Engineering B-Solid State Materials for Advanced Technology* 2003, **101:**14–18.
29. Bonafos C, Carrada M, Cherkashin N, Coffin H, Chassaing D, Assayag GB, Claverie A, Muller T, Heinig KH, Perego M et al., Manipulation of two-dimensional arrays of Si nanocrystals embedded in thin SiO_2 layers by low energy ion implantation. *Journal of Applied Physics* 2004, **95:**5696–5702.
30. Meldrum A, Buchanan KS, Hryciw A, White W, Micropixelated luminescent nanocrystal arrays synthesized by ion implantation. *Advanced Materials* 2004, **16:**31–34.
31. Sias US, Moreira EC, Ribeiro E, Boudinov H, Amaral L, Behar M, The influence of the implantation temperature on the photoluminescence characteristics of Si nanocrystals embedded into SiO_2 matrix. *Nuclear Instruments and Methods in Physics Research Section B: Beam Interactions with Materials and Atoms* 2004, **218:**405–409.
32. Prabakaran R, Kesavamoorthy R, Amirthapandian S, Ramanand A, Raman scattering and photoluminescence studies on O^+ implanted porous silicon. *Materials Letters* 2004, **58:**3745–3750.
33. Ross GG, Smirani R, Levitcharsky V, Wang YQ, Veilleux G, Saint-Jacques RG, Effect of Si implantation on the microstructure of silicon nanocrystals and surrounding SiO_2 layer. *Nuclear Instruments and Methods in Physics Research Section B: Beam Interactions with Materials and Atoms* 2005, **230:**198–202.
34. Giri PK, Kesavamoorthy R, Bhattacharya S, Panigrahi BK, Nair KGM, Simultaneous formation of Si and Ge nanocrystals in SiO_2 by one step ion implantation. *Materials Science and Engineering: B* 2006, **128:**201–204.
35. Giri PK, Kesavamoorthy R, Panigrahi BK, Nair KGM, Studies on the formation of Si nanocrystals in SiO_2 by Ge ion implantation. *Nuclear Instruments and Methods in Physics Research Section B: Beam Interactions with Materials and Atoms* 2006, **244:**56–59.
36. Zhang CS, Sun JZ, Wang X, Zhang F, Influence of thermal treatment on photoluminescence of Er-doped Si-rich SiO_2 prepared by ion implantation. *Nuclear Instruments and Methods in Physics Research Section B: Beam Interactions with Materials and Atoms* 2006, **242:**279–281.
37. Serincan U, Kulakci M, Turan R, Foss S, Finstad TG, Variation of photoluminescence from Si nanostructures in SiO_2 matrix with Si^+ post implantation. *Nuclear Instruments and Methods in Physics Research Section B: Beam Interactions with Materials and Atoms* 2007, **254:**87–92.
38. Podhorodecki A, Zatryb G, Misiewicz J, Wojcik J, Mascher P, Influence of the annealing temperature and silicon concentration on the absorption and emission properties of Si nanocrystals. *Journal of Applied Physics* 2007, **102:**5.
39. Ding L, Chen TP, Liu Y, Yang M, Wong JI, Liu YC, Trigg AD, Zhu FR, Tan MC, Fung S, Influence of nanocrystal size on optical properties of Si nanocrystals embedded in SiO_2 synthesized by Si ion implantation. *Journal of Applied Physics* 2007, **101:**6.
40. Sias US, Behar M, Moreira EC, Passivation effect and photoluminescence decay lifetime of Si nanocrystals produced by hot implantation. *Nuclear Instruments and Methods in Physics Research Section B: Beam Interactions with Materials and Atoms* 2008, **266:**3125–3128.
41. Lucovsky G, Lin SY, Richard PD, Chao SS, Takagi Y, Pai P, Keem JE, Tyler JE, Local atomic structure of silicon suboxides (SiO_x, $x < 2$). *Journal of Non-Crystalline Solids* 1985, **75:**429–434.
42. Dehan E, Temple-Boyer P, Henda R, Pedroviejo JJ, Scheid E, Optical and structural properties of SiO_x and SiN_x materials. *Thin Solid Films* 1995, **266:**14–19.
43. Spiga S, Tallarida G, Borghesi A, Sassella A, De Santi G, Annealing effects on silicon-rich oxide films studied by spectroscopic ellipsometry. *Thin Solid Films* 1998, **325:**36–41.

44. Iacona F, Franzo G, Spinella C, Correlation between luminescence and structural properties of Si nanocrystals. *Journal of Applied Physics* 2000, **87:**1295–1303.
45. Lombardo S, Coffa S, Bongiorno C, Spinella C, Castagna E, Sciuto A, Gerardi C, Ferrari F, Fazio B, Privitera S, Correlation of dot size distribution with luminescence and electrical transport of Si quantum dots embedded in SiO_2. *Materials Science and Engineering B-Solid State Materials for Advanced Technology* 2000, **69:**295–298.
46. Franzo G, Iacona F, Spinella C, Cammarata S, Grazia Grimaldi M, Size dependence of the luminescence properties in Si nanocrystals. *Materials Science and Engineering: B* 2000, **69–70:**454–458.
47. Fazio B, Vulpio M, Gerardi C, Liao Y, Crupi I, Lombardo S, Trusso S, Neri F, Residual crystalline silicon phase in silicon-rich-oxide films subjected to high temperature annealing. *Journal of the Electrochemical Society* 2002, **149:**G376–G378.
48. Chen WD, Wang YQ, Chen CY, Diao HW, Liao XB, Kong GL, Hsu CC, Strong red light emission from silicon nanocrystals embedded in SiO/sub 2/ matrix. In *Conference on Optoelectronic and Microelectronic Materials and Devices*, Beijing, China, 2002, pp. 267–270.
49. Wu LC, Dal M, Huang XF, Zhang YJ, LI W, Xu J, Chen KJ, Room temperature electron tunneling and storage in a nanocrystalline silicon floating gate structure. *Journal of Non-Crystalline Solids* 2003, **338–340:**318–321.
50. Naito S, Satake M, Kondo H, Sakashita M, Sakai A, Zaima S, Yasuda Y, Growth of silicon nanocrystal dots with high number density by ultra-high-vacuum chemical vapor deposition. *Japanese Journal of Applied Physics Part 1-Regular Papers Short Notes & Review Papers* 2004, **43:**3779–3783.
51. Peralvarez M, Garcia C, Lopez M, Garrido B, Barreto J, Dominguez C, Rodriguez JA, Field effect luminescence from Si nanocrystals obtained by plasma-enhanced chemical vapor deposition. *Applied Physics Letters* 2006, **89:**051112.
52. Chen XY, Lu YF, Wu YH, Cho BJ, Tang LJ, Lu D, Dong JR, Correlation between optical properties and Si nanocrystal formation of Si-rich Si oxide films prepared by plasma-enhanced chemical vapor deposition. *Applied Surface Science* 2006, **253:**2718–2726.
53. Ivanda M, Gebavi H, Ristic D, Furic K, Music S, Ristic M, Zonja S, Biljanovic P, Gamulin O, Balarin M et al., Silicon nanocrystals by thermal annealing of Si-rich silicon oxide prepared by the LPCVD method. *Journal of Molecular Structure* 2007, **834–836:**461–464.
54. Hernandez S, Pellegrino P, Martinez A, Lebour Y, Garrido B, Spano R, Cazzanelli M, Daldosso N, Pavesi L, Jordana E et al., Linear and nonlinear optical properties of Si nanocrystals in SiO_2 deposited by plasma-enhanced chemical-vapor deposition. *Journal of Applied Physics* 2008, **103:**6.
55. Sullivan BT, Lockwood DJ, Labbe HJ, Lu ZH, Photoluminescence in amorphous Si/SiO_2 superlattices fabricated by magnetron sputtering. *Applied Physics Letters* 1996, **69:**3149–3151.
56. Kanzawa Y, Hayashi S, Yamamoto K, Raman spectroscopy of Si-rich SiO_2 films: Possibility of Si cluster formation. *Journal of Physics-Condensed Matter* 1996, **8:**4823–4835.
57. Seifarth H, Grotzschel R, Markwitz A, Matz W, Nitzsche P, Rebohle L, Preparation of SiO_2 films with embedded Si nanocrystals by reactive rf magnetron sputtering. *Thin Solid Films* 1998, **330:**202–205.
58. Yoshida S, Hanada T, Tanabe S, Soga N, Annealing characteristics of Si doped amorphous silica films by rf sputtering. *Journal of Materials Science* 1999, **34:**267–271.
59. Torchynska T, Espinoza FGB, Goldstein Y, Savir E, Jedrzejewski J, Khomenkova L, Korsunska N, Yukhimchuk V, Nature of visible luminescence of co-sputtered Si-SiO_x systems. *Physica B-Condensed Matter* 2003, **340:**1119–1123.
60. Kim W, Lee J, Lee J, Ko EK, Lyou JH, Structural changes in nanocrystalline silicon deposited by rf-magnetron sputtering. *Applied Physics A-Materials Science & Processing* 2004, **79:**1813–1817.
61. He Y, Bi L, Feng JY, Wu QL, Properties of Si-rich SiO_2 films by RF magnetron sputtering. *Journal of Crystal Growth* 2005, **280:**352–356.
62. Boyd E, Blaikie RJ, Development of Si/SiO_2 super-lattices deposited by RF reactive sputtering. *Current Applied Physics* 2006, **6:**491–494.
63. Lu YW, Du XW, Hu SL, Han X, Li H, Formation and luminescent properties of face-centered-cubic Si nanocrystals in silica matrix by magnetron sputtering with substrate bias. *Applied Physics Letters* 2007, **90:**241910.
64. Hidetoshi, M, Nanocrystalline silicon embedded in SiO films by rf magnetron sputtering. *Japanese Journal of Applied Physics* 2008, **46:**3766–3768.
65. Shabalov AL, Feldman MS, Optical and dielectric properties of thin SiO_x films of variable composition. *Thin Solid Films* 1983, **110:**215–224.

66. Shabalov AL, Feldman MS, Optical properties and structure of thin SiO_x films. *Thin Solid Films* 1987, **151:**317–323.
67. Yi LX, Heitmann J, Scholz R, Zacharias M, Si rings, Si clusters, and Si nanocrystals-different states of ultrathin SiO_x layers. *Applied Physics Letters* 2002, **81:**4248–4250.
68. Heitmann J, Scholz R, Schmidt M, Zacharias M, Size controlled nc-Si synthesis by SiO/SiO_2 superlattices. *Journal of Non-Crystalline Solids* 2002, **299:**1075–1078.
69. Fang YC, Li WQ, Qi LJ, Li LY, Zhao YY, Zhang ZJ, Lu M, Photoluminescence from SiO_x thin films: Effects of film thickness and annealing temperature. *Nanotechnology* 2004, **15:**494–500.
70. Barranco A, Yubero F, Espinos JP, Groening P, Gonzalez-Elipe AR, Electronic state characterization of SiO_x thin films prepared by evaporation. *Journal of Applied Physics* 2005, **97:**113714.
71. Wilson WL, Szajowski PF, Brus LE, Quantum confinement in size-selected, surface-oxidized silicon nanocrystals. *Science* 1993, **262:**1242–1244.
72. Hill NA, Whaley KB, Size dependence of excitons in silicon nanocrystals. *Physical Review Letters* 1995, **75:**1130–1133.
73. Ogut S, Chelikowsky JR, Louie SG, Quantum confinement and optical gaps in Si nanocrystals. *Physical Review Letters* 1997, **79:**1770–1773.
74. Soni RK, Fonseca LF, Resto O, Buzaianu M, Weisz SZ, Size-dependent optical properties of silicon nanocrystals. *Journal of Luminescence* 1999, **83–84:**187–191.
75. Takeoka S, Fujii M, Hayashi S, Size-dependent photoluminescence from surface-oxidized Si nanocrystals in a weak confinement regime. *Physical Review B* 2000, **62:**16820–16825.
76. Muller T, Heinig KH, Moller W, Size and location control of Si nanocrystals at ion beam synthesis in thin SiO_2 films. *Applied Physics Letters* 2002, **81:**3049–3051.
77. Liu M, Lu GH, Chen JH, Synthesis, assembly, and characterization of Si nanocrystals and Si nanocrystal carbon nanotube hybrid structures. *Nanotechnology* 2008, **19:**5.
78. Grisolia J, Shalchian M, BenAssayag G, Coffin H, Bonafos C, Schamm S, Atarodi SM, Claverie A, Oxidation effects on transport characteristics of nanoscale MOS capacitors with an embedded layer of silicon nanocrystals obtained by low energy ion implantation. *Materials Science and Engineering: B* 2005, **124–125:**494–498.
79. Coffin H, Bonafos C, Schamm S, Cherkashin N, Assayag GB, Claverie A, Respaud M, Dimitrakis P, Normand P, Oxidation of Si nanocrystals fabricated by ultralow-energy ion implantation in thin SiO_2 layers. *Journal of Applied Physics* 2006, **99:**044302.
80. Yang MD, Chu AHM, Shen JL, Huang YH, Yang TN, Chen MC, Chiang CC, Lan SM, Chou WC, Lee YC, Improvement of luminescence from Si nanocrystals with thermal annealing in CO_2. *Journal of Crystal Growth* 2008, **310:**313–317.
81. Lu ZH, Lockwood DJ, Baribeau JM, Visible light emitting Si/SiO_2 superlattices. *Solid-State Electronics* 1996, **40:**197–201.
82. Schmidt M, Zacharias M, Richter S, Fischer P, Veit P, Blasing J, Breeger B, Er doping of nanocrystalline-Si/SiO_2 superlattices. *Thin Solid Films* 2001, **397:**211–215.
83. Khomenkova L, Korsunska N, Torchynska T, Yukhimchuk V, Jumayev B, Many A, Goldstein Y, Savir E, Jedrzejewski J, Defect-related luminescence of Si/SiO_2 layers. *Journal of Physics-Condensed Matter* 2002, **14:**13217–13221.
84. Naciri AE, Mansour M, Johann L, Grob JJ, Eckert C, Optical study of Si nanocrystals in Si/SiO_2 layers by spectroscopic ellipsometry. *Nuclear Instruments & Methods in Physics Research Section B-Beam Interactions with Materials and Atoms* 2004, **216:**167–172.
85. Modreanu M, Aperathitis E, Androulidaki M, Audier M, Chaix-Pluchery O, Characterisation of room temperature blue emitting Si/SiO_2 multilayers. *Optical Materials* 2005, **27:**1020–1025.
86. Du XW, Li H, Lu YW, Sun J, Preferential growth of Si nanocrystals in $SiO_2/Si/SiO_2$ sandwich structure. *Journal of Crystal Growth* 2007, **305:**59–62.
87. Heitmann J, Schmidt M, Zacharias M, Timoshenko VY, Lisachenko MG, Kashkarov PK, Fabrication and photoluminescence properties of erbium doped size-controlled silicon nanocrystals. *Materials Science and Engineering B-Solid State Materials for Advanced Technology* 2003, **105:**214–220.
88. Zacharias M, Heitmann J, Scholz R, Kahler U, Schmidt M, Blasing J, Size-controlled highly luminescent silicon nanocrystals: A SiO/SiO_2 superlattice approach. *Applied Physics Letters* 2002, **80:**661–663.
89. Sato S, Yamamoto N, Yao H, Kimura K, Synthesis of three-dimensional silicon cluster superlattices. *Chemical Physics Letters* 2002, **365:**421–426.
90. Yi LX, Heitmann J, Scholz R, Zacharias M, Phase separation of thin SiO layers in amorphous SiO/SiO_2 superlattices during annealing. *Journal of Physics-Condensed Matter* 2003, **15:**S2887–S2895.

91. Lau WM, Wu XW, Measurements of interface state density by X-ray photoelectron spectroscopy. *Surface Science* 1991, **245:**345–352.
92. Philipp HR, Optical and bonding model for non-crystalline SiO_x and SiO_xNy materials. *Journal of Non-Crystalline Solids* 1972, **8–10:**627–632.
93. Temkin RJ, An analysis of the radial distribution function of SIO_x. *Journal of Non-Crystalline Solids* 1975, **17:**215–230.
94. Lucovsky G, Atomic structure and thermal stability of silicon suboxides in bulk thin films and in transition regions at Si-SiO_2 interfaces. *Journal of Non-Crystalline Solids* 1998, **227–230:**1–14.
95. Alfonsetti R, Desimone G, Lozzi L, Passacantando M, Picozzi P, Santucci S, SiO_x Surface stoichiometry by XPS: A comparison of various methods. *Surface and Interface Analysis* 1994, **22:**89–92.
96. Zhang W, Zhang S, Liu Y, Chen T, Evolution of Si suboxides into Si nanocrystals during rapid thermal annealing as revealed by XPS and Raman studies. *Journal of Crystal Growth* 2009, **311:**1296–1301.
97. Dupree R, Holland D, Williams DS, An assessment of the structural models for amorphous SiO using MAS NMR. *Philosophical Magazine B-Physics of Condensed Matter Statistical Mechanics Electronic Optical and Magnetic Properties* 1984, **50:**L13–L18.
98. Schulmeister K, Mader W, TEM investigation on the structure of amorphous silicon monoxide. *Journal of Non-Crystalline Solids* 2003, **320:**143–150.
99. Rochet F, Dufour G, Roulet H, Pelloie B, Perriere J, Fogarassy E, Slaoui A, Froment M, Modification of SiO through room-temperature plasma treatments, rapid thermal annealings, and laser irradiation in a nonoxidizing atmosphere. *Physical Review B* 1988, **37:**6468–6477.
100. Barranco A, Yubero F, Espinos JP, Holgado JP, Caballero A, Gonzalez-Elipe AR, Mejias JA, Structure and chemistry of SiO_x $(x<2)$ systems. *Vacuum* 2002, **67:**491–499.
101. Porter, AD, *Phase Transformations in Metals and Alloys*, 2nd edn. London, U.K.: Chapman & Hall; 1992.
102. Nesbit LA, Annealing characteristics of Si-rich SiO_2-films. *Applied Physics Letters* 1985, **46:**38–40.
103. Iacona F, Bongiorno C, Spinella C, Boninelli S, Priolo F, Formation and evolution of luminescent Si nanoclusters produced by thermal annealing of SiO_x films. *Journal of Applied Physics* 2004, **95:**3723–3732.
104. Wang J, Wang XF, Li Q, Hryciw A, Meldrum A, The microstructure of SiO thin films: From nanoclusters to nanocrystals. *Philosophical Magazine* 2007, **87:**11–27.
105. Khomenkova L, Korsunska N, Yukhimchuk V, Jumayev B, Torchynska T, Hernandez AV, Many A, Goldstein Y, Savir E, Jedrzejewski J, Nature of visible luminescence and its excitation in Si-SiO_x systems. *Journal of Luminescence* 2003, **102:**705–711.
106. Prokes SM, Carlos WE, Oxygen defect center red room-temperature photoluminescence from freshly etched and oxidized porous silicon. *Journal of Applied Physics* 1995, **78:**2671–2674.
107. Oliver A, Cheang-Wong JC, Crespo A, Rodriguez-Fernandez L, Hernandez JM, Munoz E, Espejel-Morales R, E′ and B2 center production in amorphous quartz by MeV Si and Au ion implantation. *Materials Science and Engineering: B* 2000, **78:**32–38.
108. Garapon J, Favaro L, Poumellec B, Computed paramagnetic properties for an E′ center produced from the twofold coordinated Si or Ge in silica. *Journal of Non-Crystalline Solids* 2007, **353:**605–609.
109. Buscarino G, Agnello S, Gelardi FM, Investigation on the microscopic structure of E'_δ center in amorphous silicon dioxide by electron paramagnetic resonance spectroscopy. *Modern Physics Letters B* 2006, **20:**451–474.
110. Lu MP, Chen MJ, Oxide-trap-enhanced Coulomb energy in a metal-oxide-semiconductor system. *Physical Review B* 2005, **72:**5.
111. Garrido B, Lopez M, Perez-Rodriguez A, Garcia C, Pellegrino P, Ferre R, Moreno JA, Morante JR, Bonafos C, Carrada M et al., Optical and electrical properties of Si-nanocrystals ion beam synthesized in SiO_2. *Nuclear Instruments and Methods in Physics Research Section B: Beam Interactions with Materials and Atoms* 2004, **216:**213–221.
112. Chakraborty G, Chattopadhyay S, Sarkar CK, Pramanik C, Tunneling current at the interface of silicont and silicon dioxide partly embedded with silicon nanocrystals in metal oxide semiconductor structures. *Journal of Applied Physics* 2007, **101:**6.
113. Lu TZ, Shen J, Mereu B, Alexe M, Scholz R, Talalaev V, Zacharias M, Electrical behavior of size-controlled Si nanocrystals arranged as single layers. *Applied Physics A-Materials Science & Processing* 2005, **80:**1631–1634.
114. Normand P, Kapetanakis E, Dimitrakis P, Skarlatos D, Tsoukalas D, Beltsios K, Claverie A, Benassayag G, Bonafos C, Carrada M et al., Effects of annealing conditions on charge storage of Si nanocrystal memory devices obtained by low-energy ion beam synthesis. In *28th International Conference on Micro- and Nano-Engineering*, Lugano, Switzerland, Vol. 67–68, pp. 629–634, September 16–19, 2003.

115. Koshida N, *Device Applications of Silicon Nanocrystals and Nanostructures*. Ottawa, Canada: Springer Verlag; 2009.
116. Bonafos C, Coffin H, Schamm S, Cherkashin N, Ben Assayag G, Dimitrakis P, Normand P, Carrada M, Paillard V, Claverie A, Si nanocrystals by ultra-low-energy ion beam-synthesis for non-volatile memory applications. *Solid-State Electronics* 2005, **49:**1734–1744.
117. Tiwari S, Wahl JA, Silva H, Rana F, Welser JJ, Small silicon memories: Confinement, single-electron, and interface state considerations. *Applied Physics A-Materials Science & Processing* 2000, **71:**403–414.
118. Shi Y, Saito K, Ishikuro H, Hiramoto T, Effects of traps on charge storage characteristics in metal-oxide-semiconductor memory structures based on silicon nanocrystals. *Journal of Applied Physics* 1998, **84:**2358–2360.
119. Shi Y, Gu SL, Yuan XL, Zheng YD, Saito K, Ishikuro H, Hiramoto T, Silicon nano-crystals based MOS memory and effects of traps on charge storage characteristics. In *Proceedings of the Fifth International Conference on Solid-State and Integrated Circuit Technology*, Beijing, China, 1998, pp. 838–841.
120. Ng CY, Chen TP, Liu Y, Tse MS, Gui D, Modulation of capacitance magnitude by charging/discharging in silicon nanocrystals distributed throughout the gate oxide in MOS structures. *Electrochemical and Solid State Letters* 2005, **8:**G8–G10.
121. Ng CY, Liu Y, Chen TP, Tse MS, Charging/discharging of silicon nanocrystals embedded in an SiO_2 matrix inducing reduction/recovery in the total capacitance and tunneling current. In *SPIE International Symposium on Microlectronics, MEMS and Nanotechnology*, Perth, Australia, December 2003, pp. S43–S46.
122. Liu Y, Chen TP, Ng CY, Tse MS, Zhao P, Fu YQ, Zhang S, Fung S, Random capacitance modulation due to charging/discharging in Si nanocrystals embedded in gate dielectric. *Nanotechnology* 2005, **16:**1119–1122.
123. Dumas C, Grisolia J, BenAssayag G, Bonafos C, Schamm S, Claverie A, Arbouet A, Carrada M, Paillard V, Shalchian M, Influence of the thickness of the tunnel layer on the charging characteristics of Si nano-crystals embedded in an ultra-thin SiO_2 layer. In *Symposium on Silicon Nanocrystals for Electronic and Sensing Applications Held at the 2006 E-MRS Spring Meeting*, Nice, France, 2006, pp. 80–84.
124. Maeda T, Suzuki E, Sakata I, Yamanaka M, Ishii K, Electrical properties of Si nanocrystals embedded in an ultrathin oxide. *Nanotechnology* 1999, **10:**127–131.

6 Properties and Applications of Sol-Gel-Derived Nanostructured Thin Films: Optical Aspects

Ana C. Marques and Pierre Chevalier

CONTENTS

6.1 INTRODUCTION

The ability to modify the structure of thin films and coatings at the nanoscale brings through quantum size effect the possibility to achieve peculiar characteristics that fit specific needs. The thin film nanostructurization approach has been widely applied as a potential solution, when the aim is going beyond in the development of materials with specific optical properties to enable desired applications.

The aim of this chapter is to review the recent progress in the area of sol-gel-derived nanostructured thin films, with special emphasis on their optical properties and applications. Clearly, there is no pretension to be exhaustive in such a rapidly growing field with advances occurring on a daily basis and, therefore, omissions and oversights can easily happen. Another issue is the fact that although some authors deal with nanostructured thin films in their publications, they do not state it, which makes the bibliographic search on this area quite challenging.

This chapter consists of a brief introduction on the sol-gel method and on the materials obtained from it in the form of nanostructured thin films. An analysis of the evolution of sol-gel technology toward nanostructured thin films will also be described, as well as an overview of the synthesis aspects and the optical properties and applications of the nanostructured thin films.

6.1.1 The Sol-Gel Method

It is not the aim of this section to describe extensively the sol-gel method, but to give a brief introduction, in a thin film point of view, on this method, as well as clarify its advantages and disadvantages compared to other techniques.

The sol-gel method is a chemical route, based on wet chemistry processing, used to synthesize glassy or ceramic materials, as well as composites, at relatively low temperatures. It has been used to produce a wide range of compositions (mostly oxides) in various forms, including coatings and thin films, monoliths, composites, porous membranes, powders, and fibers. Thin films and coatings represent by far the most important use of sol-gel-derived products, where the advantages offered by this method are relatively easy to achieve and its disadvantages can be minimized.

This technique involves the preparation of a sol, a homogeneous dispersion, or colloidal suspension, of nanometer-sized entities in a liquid,[1] mostly by a bottom-up route that starts from molecular precursors like metal alkoxides and/or organometallic compounds (or metal salts or oxides), followed by the gelation of the sol and the removal of the liquid existing in fine interconnected channels within the gel. Hydrolysis and condensation reactions occur throughout this process, where the by-product from hydrolysis is an aliphatic alcohol, which will be removed by volatilization, and (poly)condensation leads to the establishment of metal (M)–oxygen–metal bond, which in the case of silicon is Si–O–Si. The following equations describe the hydrolysis (a) and condensation (b–d) reactions:

(a) $M(OR)_n + H_2O \rightarrow M(OH)(OR)_{n-1} + ROH$
(b) $2M(OH)(OR)_{n-1} \rightarrow (RO)_{n-1}M\text{–}O\text{–}M(OR)_{n-1} + H_2O$
(c) $2M(OH)_n \rightarrow (OH)_{n-1}M\text{–}O\text{–}M(OH)_{n-1} + H_2O$
(d) $M(OR)_n + M(OH)(OR)_{n-1} \rightarrow (RO)_{n-1}M\text{–}O\text{–}M(OR)_{n-1} + ROH$

Most metal alkoxides are partially or completely hydrolyzed to the corresponding hydroxides, and different approaches can be carried out to accelerate or slow down their hydrolysis, such as the use of catalysts (selected acids or bases) or the use of chelating agents (e.g., acetyl acetone), respectively.

As condensation proceeds, the increasingly larger molecules, in the form of colloids, or polymer chains, or clusters dissolved in the sol, become so large that the material stiffens, by strong increase in the viscosity, forming a gel.[2–4] In the case of thin film or coating preparation, there is a fast drying of the solvent, which, together with the small amount of material involved, forces gelation to occur much more rapidly than for bulk materials. The gel, at this stage, consists of a weak amorphous solid structure and an interconnected network of very fine pores filled with a substantial amount of liquid, which will be dried and densified or sintered by heat treatment, photo-based processes (e.g., ultraviolet (UV) irradiation), etc.

Nonoxides, such as nitrides, sulfides, carbides, and fluorides can also be prepared by sol-gel. In the particular case of sulfides (e.g., chalcogenide glasses), there are several routes for their preparation: direct solvation of sulfide glasses using appropriate organic solvents, which leads to a sulfide sol and therefore allows the preparation of thin films[5]; thiolysis*-condensation reaction of organometallic precursors and H_2S; synthesis from inorganic precursors such as nitrates and chlorides; and vulcanization of oxide gels in a reactive atmosphere using anhydrous hydrogen sulfide gas.

In the aqueous sol-gel chemistry, hydrolysis, condensation, and aggregation occur almost simultaneously, enhancing the complexity of the process, which is mainly derived from the high reactivity of the metal oxide precursors and the double role of water as ligand and solvent. The nonaqueous (or nonhydrolytic) sol-gel process occurs in organic solvents under the exclusion of water,[6] avoiding, therefore, some issues of aqueous sol-gel. Moreover, it can be preferable for applications where the presence of water in the final material structure is not desirable. The organic components (organic solvents) in the reaction mixture act as the oxygen-supplying agent; however, the moderate reactivity of the C–O bond and the stabilizing effect of the organic species lead to slow reaction rates. The condensation reactions leading to M–O–M bonds in the nonaqueous process are well described in Ref. [7], where the typical phenomena occurring are exhibited, such as alkyl halide elimination, ether elimination, ester elimination, and aldol†-like condensation.

One of the most important parameters in the sol-gel process is the pH of the starting solution. The isoelectric point of silica, at which the electron mobility and surface charge are zero, occurs at ca. pH = 2. This pH value sets the boundary between acid and base catalysis of the polymerization process. In silicates, the cross-linking between the polymer chains is much higher at high pH and high H_2O:SiO_2 ratio (R), so that highly branched clusters are formed under these conditions, whereas more weakly branched clusters, with microporous structure with pore size <2 nm, are formed at low pH.[8] Therefore, for acid-catalyzed gels, the polymer chains are weakly cross-linked and the structure can be highly compacted before it is sufficiently cross-linked to produce residual porosity.

The microstructure of sol-gel-derived materials is dependent on a large number of processing parameters. Apart from pH and R values, mentioned above, other important parameters, such as the precursor type, solvent concentration, nature of catalyst, ageing time and temperature, and drying/heating time and temperature also affect the structure.

The main advantages of the sol-gel process applied to the preparation of thin films and coatings are listed in Table 6.1.

The sol-gel method, although apparently simple, is quite complex in what concerns parameters that can be changed to tailor specific properties of the final product, such as pore volume, size, etc.[9] This method is, therefore, extremely versatile in terms of processing parameters, compositions, and final product properties and shapes.

In particular, the sol-gel process has been extensively compared to other techniques for the preparation of antireflective (AR) coatings, such as vacuum-based techniques (e.g., E-beam evaporation and sputtering), and it has been proven to be an advantaged process, namely, in terms of the capability of coating large substrate sizes (by, e.g., dip-coating), process control, and the possibility to produce low-cost AR coatings.[10]

The main drawbacks of sol-gel processing are listed in Table 6.2.

One of the ways to control the high reactivity of some alkoxides (of Ti, Zr, etc.) and, therefore, to avoid preferential hydrolysis, is to chemically modify their coordination state by adding a chelating agent, such as acetylacetone, or to partially hydrolyze the less reactive alkoxide first and then add the sensitive alkoxide, to facilitate the condensation reaction with the partially hydrolyzed alkoxide.

* Reaction with a thiol (R-SH) that cleaves one compound into two. This reaction is similar to hydrolysis, but involves a thiol, instead of water.

† Abbreviation of aldehyde and alcohol.

TABLE 6.1
Advantages of the Sol-Gel Process

Versatility	Selection of processing parameters for tailoring and optimizing the composition, structure, texture and porosity, and shape of the coating material
	Possibility of incorporating nanoparticles, noble metal elements and RE ions, organic molecules, biomolecules, microorganisms, and liquid crystal droplets into sol-gel-derived materials
	Large-area and bent-shape substrates
Wet chemistry method with mixing at the molecular level	Better homogeneity and high purity
	Uniform doping, at the molecular scale
	Extended composition ranges: fabrication of any oxide composition, but also some non-oxides, as well as new hybrid organic–inorganic materials and nanocomposites
Lower processing temperatures	More variety of substrates
	Less energy consumption
Relatively cheap	No need for special or expensive equipment (such as vacuum coating or sputtering), or a controlled atmosphere

TABLE 6.2
Disadvantages of the Sol-Gel Process

Cracking[a]	Stress by shrinkage of the gel upon drying
	Thick coatings (above ca. 2 μm) are only achieved by adding an organic component, or by controlling the pore size
Preferential hydrolysis	Different reactivity of the alkoxide precursors may lead to preferential precipitation of a particular oxide during sol formation (in multicomponent materials)
Sensitivity to process conditions	Limited lifetime of the sol

[a] If coating thickness is above the critical thickness.

6.1.2 Inorganic, Hybrid, and Organic Materials by Sol-Gel

The main goals of studying nano–hybrid organic–inorganic materials are to overcome the mechanical and thermal disadvantages of organic materials and the mechanical drawbacks, such as poor flexibility of inorganic materials, along with the realization of unique properties that either material cannot provide alone. Hybrid organic–inorganic materials, also called ORMOCER®s, combine the inorganic (glass/ceramic) and organic polymer structures, which means they combine mechanical strength, chemical resistance, thermal stability, and optical transparency from glasses and ceramics; the toughness, flexibility, lightweight, good processability, and functionalization from organic polymers; and elasticity, low surface energy, and high gas permeability from silicones.[11] They can be seen as either inorganically modified carbon (C)-based polymers or organically modified inorganic polymers. In this sense, they can be divided either into (1) materials, which involve precursors with hydrolytically stable/passive organic bonds (e.g., Si-CH_3 units); or (2) inorganic–organic polymers, which involve organic polymers with C–atoms in the main chain and inorganic or inorganic–organic polymers such as polysiloxanes; or (3) composites, when their organic phase acts as the polymeric matrix and the inorganic phase as a filler, and its particle size can vary from nano- to micrometers.

The use of inorganic fillers in the nanoscale range is particularly attractive since it allows the control of the degree of interaction between the polymer and the nanofillers. Previous studies have shown that the in situ generation of inorganic nanoparticles through a sol-gel process leads to significantly better control of their dispersion within the organic matrix, than the incorporation of preformed nanoparticles.[12] The in situ generation of nanoparticles results in the absence of macroscopic phase separation and homogeneous, transparent coatings. This (in situ) approach for preparing nanoparticles uses sol-gel precursors, and involves hydrolysis and condensation reactions of the metal alkoxide as precursor for the inorganic domain formation. A coupling agent may be added to promote bonding between the inorganic and organic phases and, therefore, preventing macroscopic phase separation. As a matter of example, the preparation of nanostructured TiO_2/epoxy polymeric films has been prepared following these two approaches: dispersion of preformed, commercially available TiO_2 nanoparticles; and their in situ generation through a sol-gel dual-cure process.[12] A dicycloaliphatic epoxy resin, 3,4-epoxycyclohexylmethyl-3′,4′-epoxycyclohexanecarboxylate (CE), was employed in this study as a photocurable system. TiO_2 nanoparticles were generated in situ from the sol-gel precursor titanium-tetraisopropoxide (TIP) and 3-glycidoxypropyltrimethoxysilane (GPTMS) was used as the coupling agent. GPTMS can act as the coupling agent, due to the occurrence of copolymerization with the organic matrix through its epoxy group, and condensation with the TiOH groups through its SiOH groups generated by the hydrolysis of $SiOCH_3$ groups. Figure 6.1 shows a transmission electron microscopy (TEM) image of a nanostructured TiO_2/epoxy polymeric film, with in situ generated TiO_2 nanoparticles, where well-dispersed particles can be observed.

TiO_2 imparts to the polymer matrix a high refractive index, an increase of surface hydrophilicity behavior, through the formation of TiOH species and, when in the crystalline form of anatase, leads to strong photocatalytic properties. Dynamic mechanical thermal analysis (DMTA) showed a decrease in Tg values when increasing TIP content in the photocurable formulations, because the competitive absorption of UV light between the TiO_2 nanoparticles and the cationic photoinitiator leads to less epoxy group conversion. However, there was an increase in storage modulus above Tg in the rubber region, which was attributed by the authors to a strong and extensive interfacial interaction between the organic and inorganic phase.[12] The advancing contact angle with water was found to decrease from 75° to 55° by increasing TIP content. Besides TiO_2, SiO_2 nanoparticles have

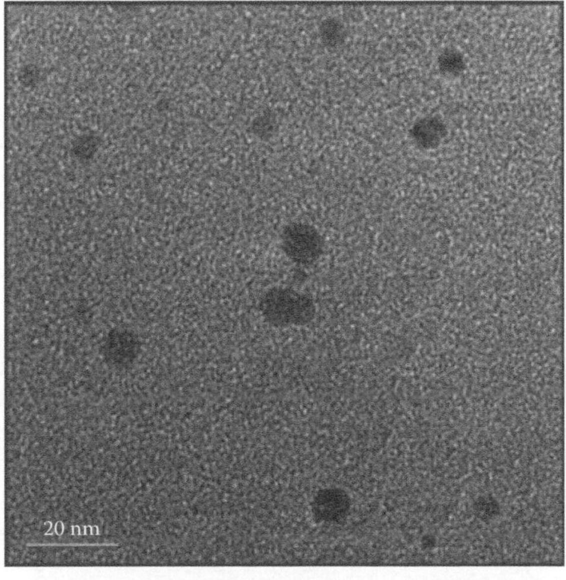

FIGURE 6.1 Bright-field TEM micrograph of a film containing 5 wt% of in situ generated TiO_2 nanoparticles, well dispersed in the epoxy network. (From Sangermano, M., Malucelli, G., Amerio, E., Bongiovanni, R., Priola, A., Di Gianni, A., Voit, B., Rizza, G. 2006, *Macromol. Mater. Eng.* 291: 517. Copyright Wiley-VCH Verlag GmbH & Co. KGaA. Reproduced with permission.)

also been generated in situ via sol-gel process.[13] Also, silica nanoparticles have been employed to obtain organic–inorganic hybrid coatings. Colloidal silica sols have been dispersed in methacryl oligosiloxanes nano-hybrid resins synthesized by sol-gel reaction of methacryloxypropylmethoxysilane and diphenylsilanediol, leading to enhanced mechanical and thermal characteristics.[14] Colloidal silica has been also added to epoxy systems, which were subsequently cured by UV. An increase on kinetics and epoxy conversion was observed by increasing the silica content, as well as an increase on modulus and a decrease on water uptake, which makes these nanocomposites a good choice for gas barrier coating applications.[15]

Sol-gel processing allows inorganic–organic hybrid materials (ORMOCERs) to be produced, which can be functionalized in order to tailor their physical and chemical properties such as refractive index, optical loss, the water or oil repellency or affinity, etc. This can be done by molecular design of the silane precursors by incorporation of other chemical elements, as well as by selection of the sol-gel-processing conditions.

The presence of organic components in hybrid materials influences the surface morphology of the thin films. In particular, it leads to a decrease in surface roughness and porosity. However, by blending two different polymerized silanes, such as a methyl silane (e.g., methyltriethoxysilane, MTES) and tetraethoxysilane (TEOS), phase separation between polymer and solvent-rich phases is promoted with the development of roughness of the coating. This will be described in more detail later in this chapter.

6.1.3 Evolution of Sol-Gel Technology toward Nanostructured Thin Films

Since the accidental preparation of a glass by sol-gel reaction from TEOS, reported by Ebelman[16] in 1846, the Sol-Gel Science has emerged at first very progressively and recently quite dynamically. Curiously enough, the sol-gel method was initially developed and used for practical purposes: it emerged as an application-driven technology and, therefore, it was connected to the material science community rather than to the fundamental chemistry community. In 1939, Geffcken and Berger patented a process to fabricate thin oxide films produced from alkoxides,[17] based on silica, alumina, titania, and other oxides. These films were claimed to be crystalline, but optically highly transparent, which suggests that they would have to be nanostructured. However, this fact was not stated clearly at that time, probably because of the lack of awareness about the nanostructured materials concept. This simple consideration has been recently published by Schmidt.[18]

Especially in the last decade, the connection of sol-gel materials to nanostructured materials has become closer, with sol-gel as an enabling technique for the fabrication of nanostructured thin films. Numerous publications have indeed arisen on this topic confirming this evolution. The yearly totals of the numbers of publications (Figure 6.2) clearly show this remarkable trend.

From the 1926 references found, related to nanostructured thin films and coatings prepared by sol-gel processing, 1663 of them are about inorganic substances, while 915 are about organic substances, which means there are some with inorganic and organic materials in common. In terms of properties and applications, 482 references were found on optics and photonics. Most of the published works, including patents and journal articles, regard silica-based materials (719 references), but materials such as TiO_2 (543 references), ZnO (106 references), ZrO_2 (78 references), and transparent conductive oxides (TCO), such as indium tin oxide (ITO), are also often reported.

Whether by in-house product development or by technology transfer from universities to the industry, sol-gel-derived nanostructured thin films have been introduced in the market (or prototyped) for varied applications, ranging from automotive, architectural, ophthalmic to optical, and electronic devices.[19] Examples are IR-absorbing coatings with uniformly dispersed ITO nanoparticles, for automotive sidelights; NiO crystalline nanoparticles (bunsenite structure) embedded in an amorphous TiO_2 network, for brown coloring electrochromic devices; coatings composed by nanoscale TiO_2 with a special crystalline phase with both super-hydrophilic and super-oleophilic properties under ultraviolet irradiation, for self-cleaning glass; nanocrystalline MgF_2-SiO_2 thin

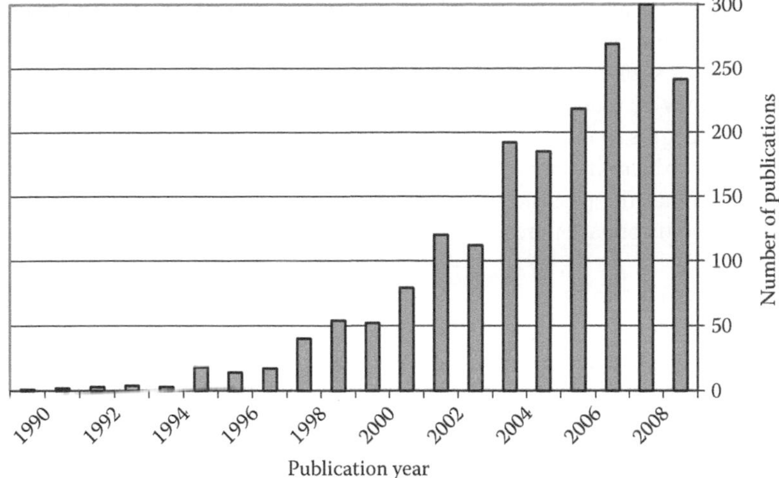

FIGURE 6.2 Plot of the yearly total publications on nanostructured thin films and coatings prepared by sol-gel. (Data obtained using the SciFinder software, September 2008.)

films with low refractive index, for optical lenses; coatings consisting of nanoparticles modified with polysiloxane components, for universal lacquer for ophthalmic lenses with superhigh refractive index; and coatings consisting of γ-Al_2O_3 nanoparticles with controllable size and shape, for phosphor lamps protection to degradation, among others. These are some examples of industrial or prototype sol-gel-derived nanostructured thin films. Japanese companies are, according to the overview on industrial utilization of sol-gel-techniques, given by Sakka,[20] the main user of these technologies, probably due to the high number of electronic and optical companies in this country. While research institutes and universities have accounted to numerous advances on materials technology enabled by sol-gel processing, there is, however, a gap between the outstanding results from research and their use in industry or their offer on the market. The reason behind this, already presented by Schmidt,[18] is that the material producers (e.g., chemical industry for ceramics or polymers) operate preferentially on a first step of the value-added chain, while the component system and device makers generate much higher value-added fractions than the material producers. Therefore, specific coatings are difficult to fit in any of these market approaches, which deal with large volume market segments. This is, however, not the case for most of the specific coatings, more directed to niche markets.

On the other hand, niche markets can also lead to significant revenues. An assessment of industrial applications for nanoscale coatings in general, not only the sol-gel-derived ones, has been recently published in a Research and Markets' new report, "Market Assessment on Nanocoatings."[21] In this report, nanomaterials are presented as a technology to provide revolutionary properties that are currently incorporated into coatings to provide evolutionary functionalities such as increased scratch resistance, thermal insulation, and antimicrobialism. These properties and corresponding functionalities already represent a significant niche market with global revenue exceeding $600 million in 2008. Self-cleaning and self-healing coatings are forecasted to be the next wave of disruptive technologies that are beginning to impact the market. By 2013 it is predicted by nanoposts that the market will exceed $5 billion.

6.1.4 Types of Nanostructured Thin Films by Sol-Gel

Nanostructured coatings are coatings with structures/features defined at nanometric scale, either ordered or not, such as pores, roughness, particles, grains, arrays, patterns, etc. They exhibit characteristic properties resulting from reducing microstructural features by factors of 100 to 1000 times compared to current engineering materials and therefore they offer the potential, through quantum size effect, for significant improvements in engineering properties. As a matter of example, refractive index may be reduced because of the presence of pores/voids in the coating, absorption at selected wavelengths in the visible region may be achieved because of the presence of noble metal nanoparticles, hardness in metals may be enhanced because of the reduced grain size, etc.

It should be noted that the presence of nanoparticles, nanograins, or nanocrystallites is often associated with nanovoids or nanopores.

6.1.4.1 Nanoporous Thin Films

During the thin film formation, at the gel point, the solid phase is extended throughout the entire volume of the film, and viscosity increases sharply. Aging and drying phenomena occur and the resulting material will exhibit porosity depending on the reactions and evaporation conditions. The effect, on pore structure evolution in silica gel, of temporal and thermal aging, pore fluids, and their pH and surface tension, has been described in several references, such as Refs. [22,23]. Inorganic materials prepared by sol-gel present intrinsic porosity (micropores, i.e., pores less than 2 nm in size, according to IUPAC definition); however, this is not straightforward for thin films, because during film formation, there is fast solvent evaporation that causes the pores to collapse, forming a fully cross-linked, denser material upon heat treatment.[24]

One strategy to induce film porosity is to reduce cross-linking by introducing organic terminal groups into the inorganic network, i.e., replace some of the alkoxide groups by unhydrolyzable organic groups, which will be directly attached to Si (or other metal) atoms and, therefore not only reduce the number of Si–O–Si linkages, but also act as a spacer in the inorganic network (Figure 6.3). A porous phase will be developed around the organic groups (e.g., $-CH_3$), since they are hydrophobic, which can be enhanced by decomposition of the organics posttreatment (e.g., heat treatment).

Figure 6.4 is a micrograph of the surface morphology, obtained by TEM, of a thin film prepared by mixing tetramethoxysilane (TMOS) with methyltrimethoxysilane (MTMS), at 75% MTMS/(MTMS + TMOS) composition, where micropores smaller than 1 nm can be observed. In a related study, an inorganic silica film was shown to be dense and, therefore, to not exhibit these nanofeatures present on the hybrid film.[24]

The porosity and refractive index of sol-gel thin films consisting of blends of silanes, namely, TEOS and MTES, have been determined by spectroscopic ellipsometry, with certain accuracy.[25]

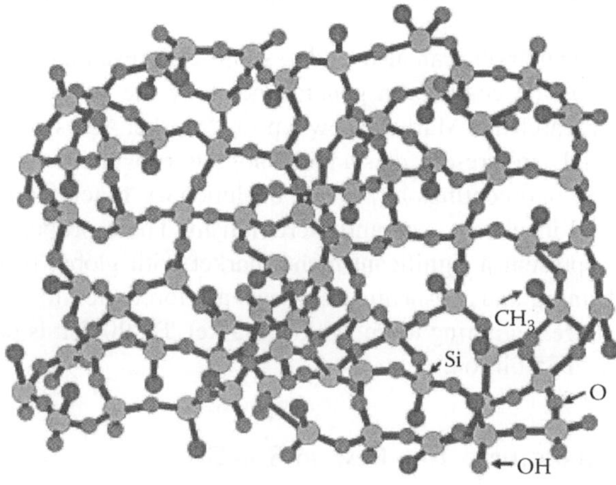

FIGURE 6.3 Organic groups ($-CH_3$) acting as a spacer in the inorganic network. (From Yan, Y. et al., *Chem. Mater.*, 9, 2583, 1997. With permission.)

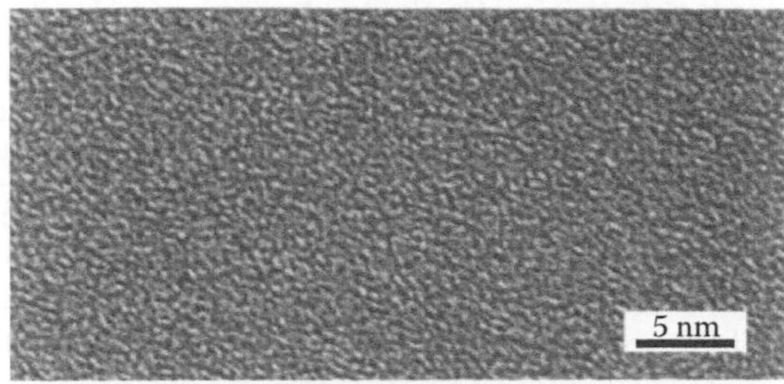

FIGURE 6.4 TEM micrograph of a 75% MTMS/(MTMS+TMOS) hybrid thin film deposited on a silicon wafer cured at 200°C. (From Yan, Y. et al., *Chem. Mater.*, 9, 2583, 1997. With permission.)

Films with increasing TEOS-to-MTES molar ratio of 1:0, 1:0.5, and 1:1 were found to exhibit increased porosity values of 3.5%, 5.5%, and 7.6%, and increased thickness values of 95, 191, and 219 nm, respectively.

As a process to prepare highly porous materials (aerogels), supercritical drying is well known,[26] but it needs high temperature and high pressure conditions, which can be obtained using an autoclave.

The micropores, in silica-derived films, can be enlarged by solvent exchange procedures using, e.g., an ammonia aqueous solution. Silica is known to exhibit appreciable solubility at high pH, so that when the wet gels are soaked into ammonia aqueous solution (pH ~ 10), the pore structure evolution is thought to be driven by the coarsening of particles in the gel, through dissolution-reprecipitation of the fractally rough surface of silica gel network,[27] which leads to the enlargement of the nanopore size. In general, the solvent exchange replaces the original solvent residing within the pores of gel by low-surface-tension solvent to reduce the capillary pressure during drying and minimizing the collapse of the pores. Ref. [28] regards dielectric layer preparation with a tailored porosity profile, for use as semiconductor insulators, and it discloses variations of the process. One example is the immersion of the as-prepared wet/gel film either in a solvent (e.g., butanol) or in a mixture of solvents with different surface tension (e.g., mixture of butanol with water, etc.) or in a saturated atmosphere at all times prior to the drying stage.[28] Silica thin films with refractive indices in the range of 1.006–1.036 (equivalent porosity 98.5%–91%) have been prepared at ambient pressure, using organosiloxane polymers and optimized processing parameters, such as selected dilution, aging, organic modification, heat treatment, and dip-coating conditions.[29]

Another approach to the preparation of nanoporous thin films is based on the phase separation between the metal oligomer adsorbed by a polymer, e.g., poly(ethylene glycol), poly(ethylene oxide), or polyacrylic acid, and the solvent mixture. Then, during drying and heat treatment stage the solvent-rich phase is evaporated leaving voids in the structure, which can be isolated (spherical pores), or interconnected pores, depending if the phase separation occurred by nucleation growth or spinodal decomposition, respectively. The pores can be further enlarged if the heat treatment temperature is above that of the polymer decomposition and burning out. First, polymer self-assembly, then, phase separation between the metal oligomer–polymer complex and solvent in the system result in a nanoporous material. The amount of polymer and solvent, the amount and type of catalyst, and the atmosphere and evaporation rate at the film formation stage strongly affect the occurrence of phase separation and the final morphology. Titania films have been prepared by a sol-gel dip-coating method from a system containing poly(ethylene glycol), PEG, together with water, ethanol, TIP, and nitric acid.[30] In this case, phase separation occurs between TiO_2 oligomer adsorbed by PEG and the solvent mixture, depending on water adsorption from the ambient atmosphere. The pore size of the TiO_2 thin film increases with an increase in the PEG content or molecular weight. However, excess addition of PEG suppresses the formation of the porous structure due to the difficulty in PEG self-assembly owing to its large steric hindrance.

Sol-gel processes combined with the templated growth of inorganic or hybrid networks, involving self-assembly with surfactant nanophases, lead to the formation of nanopores, as well as ordered metal oxide nanopatterns. This is also called template-assisted chemical solution deposition or evaporation-induced self-assembly (EISA). This method involves nanotexturization by liquid deposition of condensable inorganic precursors in the presence of micelle-forming surfactants. Amphiphilic block copolymers, which contain at least two blocks with strong hydrophilicity contrast, are generally used as templating surfactants.[31] Poly(ethylene oxide), PEO, is usually chosen as the hydrophilic chain, while poly(propylene oxide), PPO, may be chosen as the hydrophobic chain, for instance. Examples of surfactant-templating agents, also called structure-directing agents, are nonionic surfactant molecules such as amphiphilic triblock copolymers based on PEO and PPO blocks, e.g., Pluronic F127 or Pluronic P123, C_nH_{2n+1}PEO diblocks, e.g., Brij 56 ($C_{16}H_{33}(OCH_2CH_2)_{10}OH$), Triton X-100, poly-L-lysine peptide, as well as poly(styrene)-b-poly(ethylene oxide) (PS-PEO) and poly(butadiene)-b-poly(ethylene oxide) (PB-PEO) diblock copolymers,[32] oligo(p-phenylene vinylene) (OPV), and cetyltrimethylammonium bromide[33] (CTAB, $CH_3(CH_2)15N^+$ $(CH_3)_3Br^-$).

The preparation of thin films by this approach involves the preparation of a sol containing the metal alkoxide and the block copolymer as main constituents, followed by deposition of the sol on the substrate surface by coating techniques. The inorganic compounds preferentially migrate to PEO moieties, for instance, and swell the copolymer into different morphologies, depending on the amount of sol-gel precursors added, for instance. Through heat treatment the organic phase can subsequently be removed, leaving an inorganic porous nanostructured coating. Depending on the dilution and deposition rate (withdrawal rate, in the case of dip-coating), either one single layer of block copolymer micelles surrounded by inorganic precursors may be formed, or a non-organized assembling of polymers and precursors may occur instead. The former case gives rise to an organized nanostruture with ordered nanopores, while the latter case gives rise to disordered pores.

The microstructure of the thin films, including the size and shape of the pores and grains, can be controlled by changing the species and concentration of surfactants, solvent polarity, dilution, temperature, and substrate surface energy.[32,34] The templating of pores, including their alignment, is important for many applications, and the scope of applications can be further enlarged by incorporating nanoparticles in regularly arrayed pores or channels, such as photo- and opto-functional materials for the creation of new optical elements.[35] Cosolvent organic molecules, such as trimethylbenzene (TMB), have been used to expand the nanopore size, but the resulting products have been observed to possess less-resolved x-ray diffraction (XRD) patterns due to some loose in structure organization.

A sol-gel system consisting of tetrabutylorthotitanate ($Ti(OBu)_4$), diethanolamine, ethanol, decane, and CTAB has been utilized to prepare TiO_2 porous thin films.[36] Diethanolamine is used as a complexing agent for $Ti(OBu)_4$ and decane as a swelling agent. The reactions presumed to occur in this sol-gel system are displayed below[36]:

$$H\!-\!N\begin{matrix} C_2H_4OH \\ C_2H_4OH \end{matrix} + Ti(OC_4H_9)_4 \longrightarrow HN\begin{matrix} C_2H_4O \\ C_2H_4O \end{matrix}Ti(OC_4H_9)_2 + 2C_4H_9OH \tag{6.1}$$

$$HN\begin{matrix} C_2H_4O \\ C_2H_4O \end{matrix}Ti(OC_4H_9)_2 + H_2O \longrightarrow HN\begin{matrix} C_2H_4O \\ C_2H_4O \end{matrix}Ti\begin{matrix} OH \\ OC_4H_9 \end{matrix} + C_4H_9OH \tag{6.2}$$

$$HN\begin{matrix} C_2H_4O \\ C_2H_4O \end{matrix}Ti\begin{matrix} OH \\ OC_4H_9 \end{matrix} + \begin{matrix} HO \\ H_9C_4O \end{matrix}Ti\begin{matrix} OH_4C_2 \\ OH_4C_2 \end{matrix}NH \longrightarrow HN\begin{matrix} C_2H_4O \\ C_2H_4O \end{matrix}Ti\begin{matrix} \quad \\ OC_4H_9 \end{matrix}\!-\!O\!-\!\begin{matrix} \quad \\ H_9C_4O \end{matrix}Ti\begin{matrix} OH_4C_2 \\ OH_4C_2 \end{matrix}NH + H_2O \tag{6.3}$$

or

$$\mathrm{HN}\begin{smallmatrix}\diagup \mathrm{C_2H_4O} \diagdown \\ \diagdown \mathrm{C_2H_4O} \diagup\end{smallmatrix}\mathrm{Ti}\begin{smallmatrix}\mathrm{OH} \\ \mathrm{OC_4H_9}\end{smallmatrix} + \begin{smallmatrix}\mathrm{H_9C_4O} \\ \mathrm{HO}\end{smallmatrix}\mathrm{Ti}\begin{smallmatrix}\diagup \mathrm{O\,H_4C_2} \diagdown \\ \diagdown \mathrm{O\,H_4C_2} \diagup\end{smallmatrix}\mathrm{NH} \longrightarrow \mathrm{HN}\begin{smallmatrix}\diagup \mathrm{C_2H_4O} \diagdown \\ \diagdown \mathrm{C_2H_4O} \diagup\end{smallmatrix}\mathrm{Ti}\begin{smallmatrix}\mathrm{O} \\ \mathrm{O}\end{smallmatrix}\mathrm{Ti}\begin{smallmatrix}\diagup \mathrm{O\,H_4C_2} \diagdown \\ \diagdown \mathrm{O\,H_4C_2} \diagup\end{smallmatrix}\mathrm{NH} + 2\mathrm{C_4H_9OH} \qquad (6.4)$$

Reaction (6.1) represents the complexation of $Ti(OBu)_4$ by diethanolamine, which restrains the hydrolysis and condensation reactions (denoted in (6.2) through (6.4)), by increasing the coordination number of Ti and the space-steric hindrance. The porosity was found to be determined by the formation of CTAB micelle aggregates and phase separation, which was promoted by the presence of *n*-butanol, a hydrolysis product from $Ti(OBu)_4$. Then, after a heat treatment (drying and firing) there was formation of the pores, as exemplified in Figure 6.5.

The pore sizes of the TiO_2 films, prepared by this approach, ranged from 30 to 400 nm, tending to be larger with increasing amounts of CTAB, as shown in Figure 6.6, and more uniform and well distributed for lower H_2O amounts, as shown in Figure 6.7. The low H_2O amount leads indeed to slow hydrolysis and polycondensation of $Ti(OBu)_4$, which renders high uniformity of inorganic component in the sol promoting aggregation and phase separation of CTAB micelles.[36]

The addition of decane was also found to affect the pore size distribution: porosity was increased, the pore size was moderately expanded, and the pores became more interconnected, as can be observed in Figure 6.8.

Another templating agent that has been employed for the mesophase formation is CTAB, namely, in the preparation of benzene-bridged periodic mesoporous organosilica thin films, which have been synthesized by spin-coating a solution of 1,4-bis(triethoxy-silyl)benzene, water, ethanol, HCl, and CTAB on silicon substrates.[37] These films may be applied as low-κ dielectric materials and new optical materials through fluorescent bridging groups. The pores were formed by solvent extraction, i.e., the solvent was removed by placing the dried thin films in a solution of ethanol and HCl in a sealed vial at 70°C for 8 h. Some regions of these thin films have been found to contain pore channels with vertical orientation relative to the substrate surface, as shown in Figure 6.9. The silanol groups present in the interior pore wall were reported by the authors to be accessible and able to be converted into $Si–O–Si(CH_3)_3$ groups.

Pluronic P123 is a triblock copolymer ($EO_{20}PO_{70}EO_{20}$, where EO represents ethylene oxide monomer segments, and PO the propylene oxide segments), a nonionic and amphiphilic substance, which has been often used as a surfactant, or structure-directing agent to incorporate porosity and to induce nanostructured patterns on thin films made of SiO_2, TiO_2, etc.

As a matter of example, mesostructured silica thin films have been synthesized from network-forming silica precursors (e.g., TEOS) and Pluronic P123, in tetrahydrofuran (THF), by

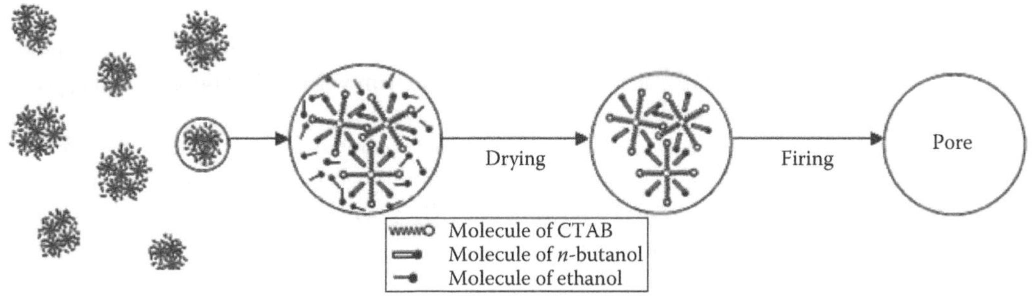

FIGURE 6.5 Pathway for pore formation. (Adapted from Liu, G.Q. et al., *J. Sol-Gel Sci. Technol.*, 41, 49, 2007.)

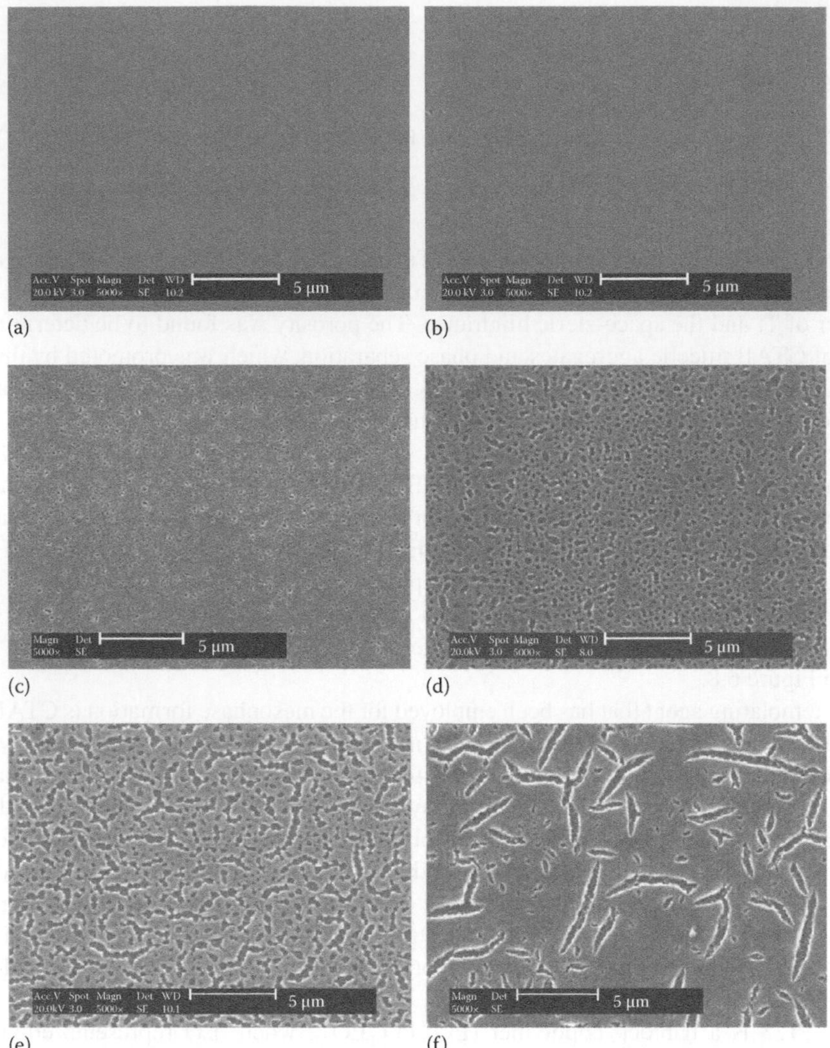

FIGURE 6.6 SEM images of porous TiO_2 thin films with $[H_2O]/[Ti] = 1$, showing the evolution of porosity for increasing CTAB amount, namely, [CTAB]/[Ti] of (a) 0, (b) 0.1, (c) 0.2, (d) 0.3, (e) 0.4, and (f) 0.5. (With kind permission from Springer Science+Business Media: Liu, G.Q. et al., *J. Sol-Gel Sci. Techn.*, 41, 49, 2007.)

dip-coating.[38] The water molecules released (in situ) during condensation reactions associate preferentially with the hydrophilic PEO blocks of P123, leading to self-assembly of the copolymer micelles. Then, silica cross-linking occurs and a composite film with highly ordered mesophases, lamellar or hexagonal, was formed, as can be seen in Figure 6.10.

Conjugated polymer guest species, such as a blue-emitting polymer (poly(9,9′-dioctylfluorene), PFO), have been confined in the silica mesochannels without occurrence of macroscopic phase separation.[38]

Another approach that has been pursued to control the porosity and morphology of thin films involves the preparation of organic–inorganic hybrid material starting from alkoxides (e.g., TEOS) and functionalized silanes, such as alkylalkoxides (e.g., MTES). In these materials, TEOS or TMOS precursors function as building blocks to construct the framework, while the organosiloxanes with non-hydrolyzable organic groups contribute with both framework silicate units and

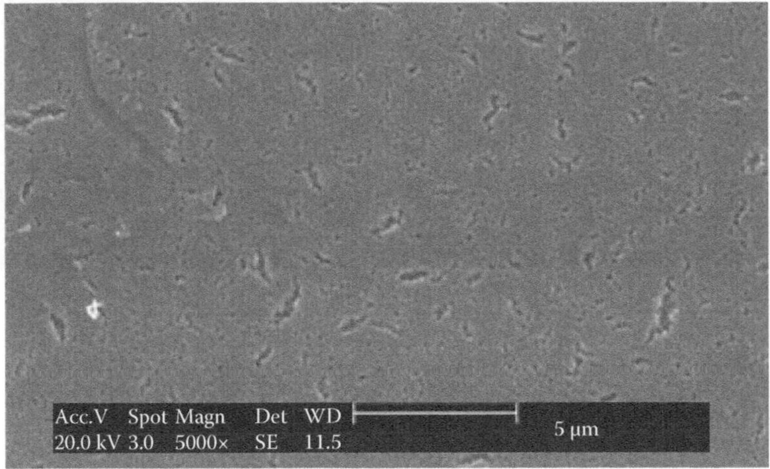

FIGURE 6.7 SEM image of a porous TiO_2 thin film with [CTAB]/[Ti] = 0.2 and $[H_2O]/[Ti] = 2$. (With kind permission from Springer Science + Business Media: Liu, G.Q. et al., *J. Sol-Gel Sci. Techn.*, 41, 49, 2007.)

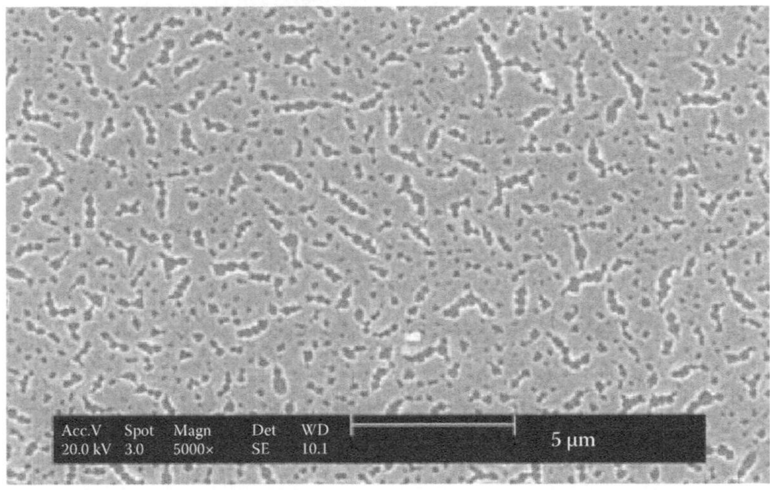

FIGURE 6.8 SEM image of a porous TiO_2 thin film with [CTAB]/[Ti] = 0.2, $[H_2O]/[Ti] = 1$ and decane, at molar ratio of $[C_{10}H_{22}]$/[CTAB] = 2. (With kind permission from Springer Science + Business Media: Liu, G.Q. et al., *J. Sol-Gel Sci. Techn.*, 41, 49, 2007.)

the organic surface functional groups.[39] Besides hydrophobic functionality, coatings with AR property can also be achieved, due to the decrease in refractive index derived from porosity generation.

Blends of MTES and TEOS have been used at selected molar ratio and have been shown to affect the roughness at the nano- and microscale. The addition of an organically modified precursor, MTES, containing non-hydrolyzable Si-CH_3 groups, to the TEOS sol results in the incorporation of hydrophobic groups (–CH_3) into the coating solution. The presence of hydrophobic groups promotes nano- and microphase separation into polymer-rich and solvent-rich phases. The phase separation induces various types of surface morphologies, depending on the MTES/TEOS molar ratio selected, such as pores for MTES/TEOS = 10/90, or micro-protuberances for MTES/TEOS = 75/25, or small spikes for MTES/TEOS = 100/0 (Figure 6.11).

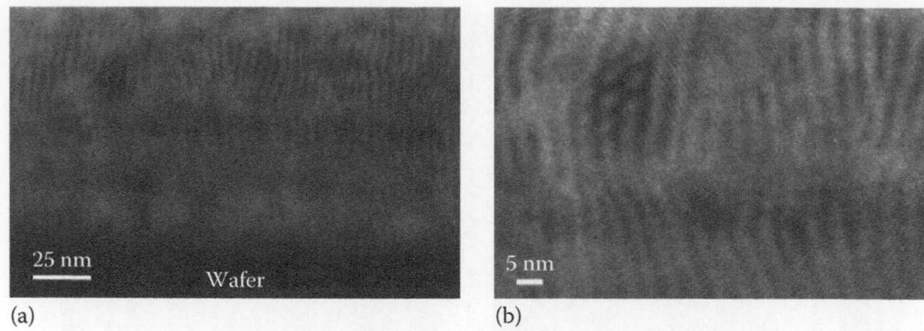

FIGURE 6.9 (a) Cross-sectional TEM image of a periodic mesoporous organosilica thin film. (b) TEM image showing the magnification of the upper left corner in (a), where two sets of lattice fringes with different directions are revealed. (From Wu, S.-Y. et al., *Chem. Mater.*, 19, 5986, 2007. With permission.)

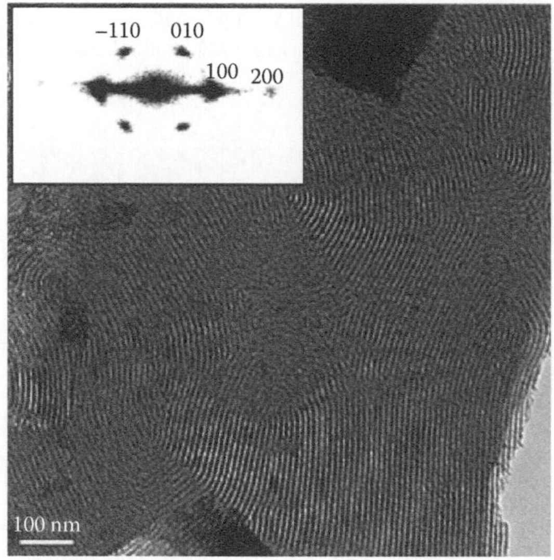

FIGURE 6.10 TEM image and glancing-incidence 2D-SAXS patterns (inset) of silica films synthesized from TEOS, THF, and P123, with TEOS:P123 molar ratio of 1:0.017. (From Kirmayer, S. et al., *Chem. Mater.*, 20, 3745, 2008. With permission.)

Besides the relative amount of Si precursors, there are other sol-gel-processing parameters, such as the order at which the different silanes are added, or the type of solvent used,[39,40] which also may contribute to varied film surface morphologies, such as those observed in Figure 6.12.

6.1.4.2 Nanograin and Nanoparticulate Thin Films

Nanograin sol-gel thin films, in most of the cases, derive from the crystallization of amorphous as-deposited thin films, through heat treatment or photo-assisted techniques. It should be noted that voids are situated at the boundaries of nanocrystallites and can be quantitatively assessed by using spectroscopic ellipsometry and the effective medium approximation (EMA), for instance.

Figures 6.13 through 6.15 show the atomic force microscopy (AFM) micrographs of TiO_2, HfO_2, and SiO_2 thin films, with AFM roughness Rms of 8.9, 2.7, and 0.7 nm, respectively (Table 6.3). TiO_2 sol was prepared from titanium isopropoxide in ethanol with diluted acetic acid, while the HfO_2

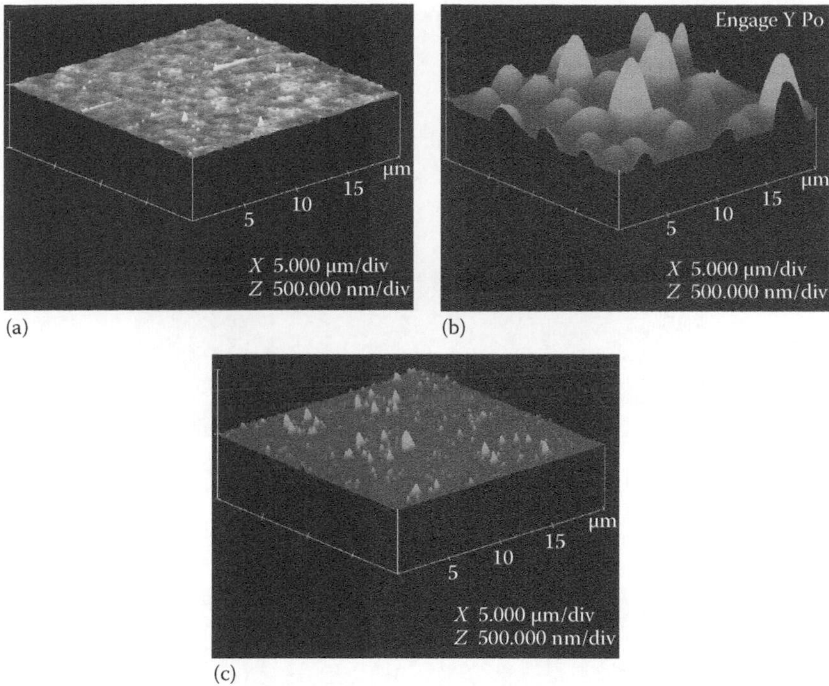

FIGURE 6.11 AFM micrographs (1 µm × 1 µm) of a SiO_2 thin film with MTES/TEOS equal to (a) 10/90, (b) 75/25, and (c) 100/0 (*Z* axis: 500 nm/div).

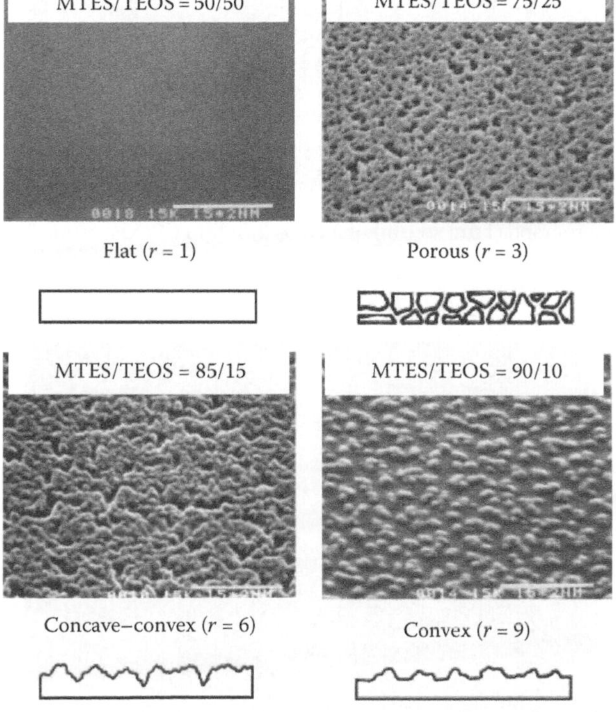

FIGURE 6.12 SEM images exhibiting the film morphology evolution with the MTES/TEOS molar ratio. (With kind permission from Springer Science + Business Media: Makita, K. et al., *J. Sol-Gel Sci. Techn.*, 14, 174, 1999.)

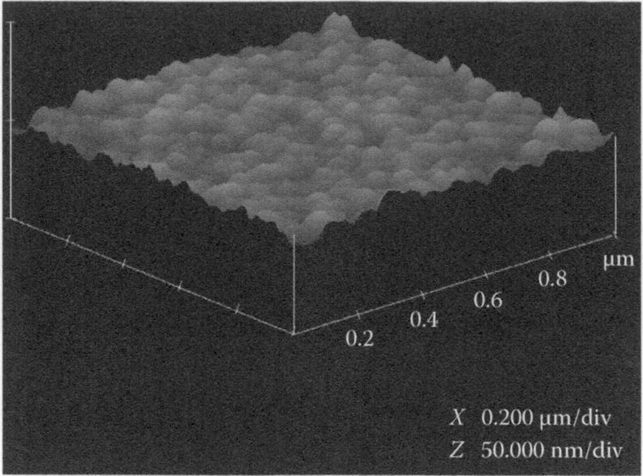

FIGURE 6.13 AFM micrograph (1 μm × 1 μm) of a TiO_2 thin film (Z-axis: 50 nm/div).

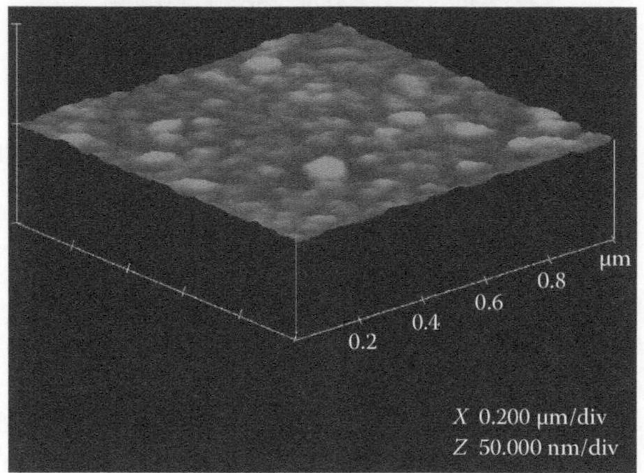

FIGURE 6.14 AFM micrograph (1 μm × 1 μm) of a HfO_2 thin film (Z-axis: 50 nm/div).

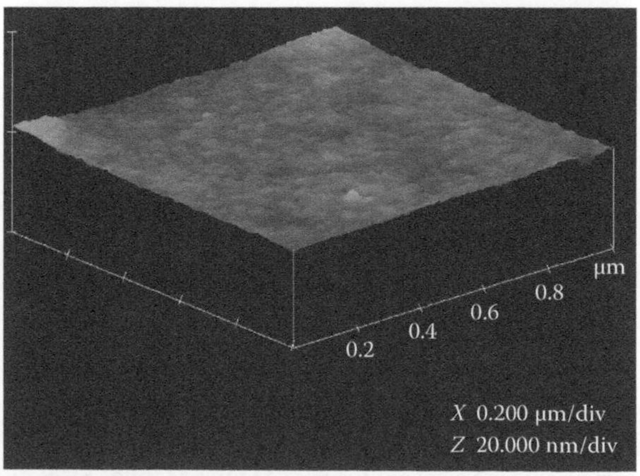

FIGURE 6.15 AFM micrograph (1 μm × 1 μm) of a SiO_2 thin film (Z-axis: 50 nm/div).

TABLE 6.3
Roughness Values (AFM Rms and Ra) for Thin Films

	Rms (nm)		Ra (nm)	
Sample	**Mean**	**St. Dev**	**Mean**	**St. Dev**
TiO_2	8.9	7.2	6.3	4.4
HfO_2	2.8	0.8	2.3	0.6
SiO_2	0.7	0.4	0.4	0.3

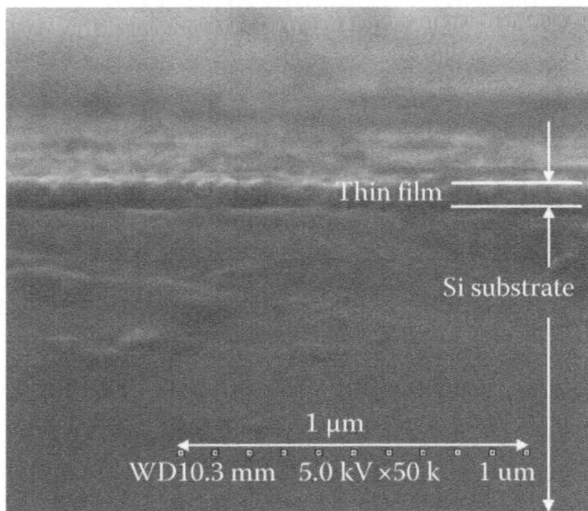

FIGURE 6.16 FE-SEM micrograph of a TiO_2 thin film on silicon substrate, heat treated at 800°C for 4 h. (From Marques, A.C. et al., Physical characterization of sol-gel derived titania thin films, Poster presented at *14th International Workshop on Sol-Gel Science and Technology*, Montpellier, France, 2007.)

FIGURE 6.17 FE-SEM micrograph of the same film shown in Figure 6.16, at higher magnification. (From Marques, A.C. et al., Physical characterization of sol-gel derived titania thin films, Poster presented at *14th International Workshop on Sol-Gel Science and Technology*, Montpellier, France, 2007.)

solution was derived from the $HfOCl_2$ precursor. Inorganic SiO_2 sol was prepared using TEOS, water, HNO_3, and ethanol. The inorganic thin films derived from these sols were obtained by heat treating at 930°C for 5 min.

The TiO_2 thin film exhibits protruding grain tops at the surface, also observed by field-emission gun SEM (Figures 6.16 and 6.17), for films with similar heat treatment. These grains do not exceed ca. 70 nm in diameter and have heights of just a few nanometers. The HfO_2 thin film is found to have less and not-so-protruding grains at the surface than TiO_2, with smaller heights. The roughness observed, at the nanoscale, for both TiO_2 and HfO_2 films, is mainly due to the crystallization typical of these compositions when heat treated at 900°C. In particular, it consists of the presence of TiO_2 (anatase) and HfO_2 (tetragonal) nanocrystals identified by x-ray diffraction (graphics not shown in this chapter). HfO_2 has been found to have less tendency to crystallize than TiO_2 when mixed with SiO_2, which might explain the lower roughness values obtained by AFM, in the present case. On the other hand, SiO_2 films, which have been found to be amorphous at such treatment temperature,[41] reveal a very smooth surface. Therefore, by selecting the composition and crystallization rate, and the heat treatment conditions, the derived roughness or morphology of the coatings can be controlled. There are other approaches to obtain sol-gel-derived nanostructured TiO_2 thin films, composed of nanograins or nanocrystallites, based in the exposure to microwave (2.45 GHz) radiation,[42] or to Excimer (KrF)-pulsed laser irradiation,[43] for instance.

6.1.4.3 Nanoparticles

Nanoparticles (e.g., Ag, Au, Cu, CdS, CdTe, and PbS) can be either directly incorporated into the sol-gel solution,[45] or in situ generated by spontaneous reduction process of the oxide species to the corresponding metallic ones by thermal,[46–48] chemical,[49] electrochemical,[50] or photochemical[51] precipitation (i.e., nanocrystallization).

Among the various techniques used for the preparation of nanocomposite films consisting of metal nanoparticles embedded in glassy matrices, sol-gel is one of the most versatile and cheapest methods (see Section 6.1.1). The sol-gel technique offers great possibilities, given the ease of introduction of a wide variety of dopants into the glass composition and the possibility of controlling the molecular structure of the material through chemistry. The different species are mixed at the molecular level, such that high homogeneity of the structure and easy incorporation or in situ generation of nanoparticles can be achieved. The heat treatment temperature and time, as well as the nanoparticle material content, both influence the size of the particles that are obtained by thermal reduction of, e.g., ionic silver into the metallic form.

As a matter of example, silver colloids (metallic Ag° nanoparticles), obtained by spontaneous reduction process of Ag^+ ions to Ag° through UV radiation, have been generated into the channels of a sol-gel thin film with a 2D hexagonal nanostructure, previously prepared by the dip-coating method using a nonionic diblock copolymer Brij58 (surfactant) to produce channels into the film.[52] In this case, a surfactant was employed to template the precipitation sites of the metallic silver nanoparticles. Two directions were combined in the nanostructurization effort of these films: nanoporosity through surfactant incorporation and metallic nanoparticles through reduction of the corresponding oxide species using UV radiation.

Colloidal silica particles may be incorporated into sol-gel solutions with the aim of improving the mechanical properties and thermal stability, as well as decrease the coefficient of thermal expansion (CTE) of the resulting thin films. Silica nanoparticles can also be incorporated to render roughness and texture to the thin film, or to prepare antiglare coatings, for instance. They have been utilized to prepare a hydrophilic film,[19] by mixing with TEOS, $ZrOCl_2.8H_2O$, ethanol, methyl propylene glycol, and deionized water. The films heat treated at 150°C for 20 min had a convex morphology due to the presence of 40–50 nm colloidal particles, which, besides increasing the surface area of the films, renders hydrophilicity due to surface SiOH groups. The SiO_2 and ZrO_2 promote wear resistance as well as chemical durability.

MgF_2 nanoparticles have been synthesized from magnesium acetate and hydrofluoric acid in methanol,[53] followed by the preparation of nanocrystalline MgF_2 thin films. The nanoparticles-containing sol was heated in an autoclave at 120°C–180°C to increase its crystallinity and improve the hardness of its surface. The resulting films had high porosity due to the presence of voids between the particles and, therefore, presented low refractive index ($n = 1.20$ at 550 nm). In order to improve their mechanical resistance, SiO_2 was added as a binder, connecting the MgF_2 particles. The resulting film is shown in Figure 6.18.

The mean crystallite size of MgF_2 films was found to increase with increasing temperature as the average pore diameter increased.[54] This behavior can be followed in Table 6.4 and Figure 6.19. The films exhibited high surface area (e.g., 821 m^2/g at 150°C), high pore volume (1.00 cm^3/g), and small particle size (e.g., 5–7 nm).

Transparent composite silica-based coatings with AlOOH boehmite nanorods in amounts up to 40 wt% have been obtained by spin-coating the sol-gel solution on a glass substrate,[55] in which glycidoxypropyltrimethoxysilane (GPTS) was used as Si precursor. The nanorods were found to be aligned parallel to the glass substrate surface, as observed in Figure 6.20, and crack toughness of the resulting composite was improved.

The presence of the boehmite nanorod-filled coating was found to impact the transparency of the glass in the visible region of the electromagnetic spectrum, through a decrease of 7% in

FIGURE 6.18 SEM image of the SiO_2 binder reinforced MgF_2 film. (From Ishizawa, H. et al., *Appl. Opt.*, 47, C200, 2008. With permission.)

TABLE 6.4
Evolution of MgF_2 Film Texture with the Heat Treatment Temperature

T (°C)	*S* (m^2/g)	Pore Volume (cm^3/g)	Av. Pore Diameter (nm)
150	821	1.00	3.9
350	212	0.84	6.7
500	178	0.84	7.2
600	78	0.60	13.8
900	<10	0.02	24.8

Source: Sermon, P.A. and Badheka, R., *J. Sol-Gel Sci. Technol.*, 32, 149, 2004. With permission.

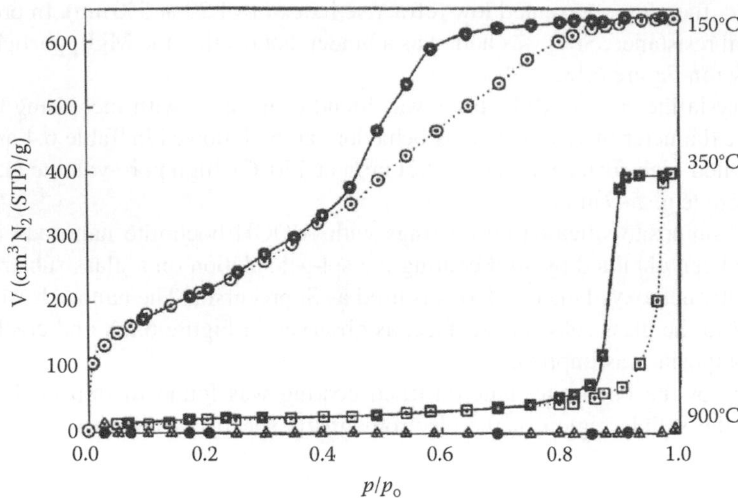

FIGURE 6.19 Evolution of N_2 adsorption isotherms on MgF_2 with the heat treatment temperature. (From Sermon, P.A. and Badheka, R., *J. Sol-Gel Sci. Technol.*, 32, 149, 2004. With permission.)

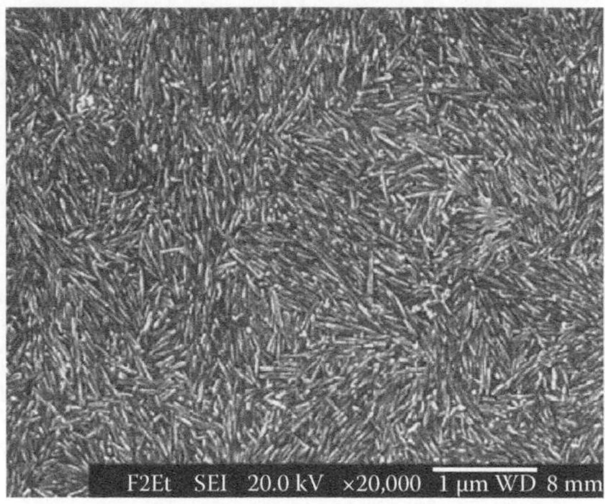

FIGURE 6.20 SEM image of a 40% boehmite nanorod filled coating, showing the nanorods aligned parallel to the surface. (From Chen, Q. et al., *J. Sol-Gel Sci. Technol.*, 44, 125, 2007. With permission.)

transmission. Such a transmission difference was not high enough to be discerned by the human eye,[55] as the authors reported.

6.1.4.4 Multi-Nanolayered Thin Films

This type of thin films consists of multilayers of a few tens of nanometers deposited on a substrate. The multilayers exhibit different properties, such as different composition and therefore different refractive index, same composition but different porosity, etc., which can be used in the fabrication of, e.g., interference filters, Bragg mirrors, or "graded" refractive index coatings.

In particular, a multilayer coating, produced by forming individual layers using a sol-gel-coating process, has been patented[56] for the production of optical components (e.g., x-ray optics, vacuum UV optics, multilayered gratings for x-ray fluorescence analysis, and dielectric filters). The multilayer was produced by structuring successive layers of the same material with the smallest possible layer thickness.

6.2 PREPARATION OF NANOSTRUCTURED THIN FILMS BY SOL-GEL

6.2.1 Precursors and Synthesis

The link between inorganic and organic network-forming sites is chemically stable and leads to the functionalization of both networks, and that is why hybrid coating materials are so interesting to pursue. Namely, the spacer length between inorganic and organic network sites may be used to control mechanical properties, such as high stiffness for short spacer length and high flexibility for larger spacer lengths.

The inorganic precursors lead to inorganic highly cross-linked oxidic networks by hydrolysis and polycondensation reactions. Examples of inorganic precursors are Si- or other metal alkoxides or M-OH-containing compounds (type I precursors). Besides Si, various other metal alkoxides of Ti, Hf, Zr, Al, Sn, etc. may be used, sometimes modified with complexing or chelating agents in order to slow down their hydrolysis and condensation reactions that are fast compared to Si-alkoxides. In the case of silanes, the Si–OR bonds (with OR being an alkoxy group, e.g., methoxy, ethoxy, etc.) hydrolyze readily with water, even if only with moisture adsorbed on the surface, to form silanol SiOH groups. These silanol groups can then condense with each other to form polymeric structures with very stable siloxane Si–O–Si bonds. They can also condense with other metal hydroxyl groups (present in the sol or on the surface of the glass or metal substrate) to form stable Si–O–M bonds (M = Ti, Fe, Al, Si, Zr, Hf…).

Examples of hybrid precursors are organically modified Si-alkoxides, also called organosilanes, which contain at least one silicon–carbon bond (e.g., Si–CH_3). The silicon–carbon bond is very stable and nonpolar, and in the presence of an alkyl group (e.g., methyl, ethyl, etc.) it gives rise to low surface energy and hydrophobic effects. They can be either with no reactive organic groups for the modification of the inorganic network (type II precursors), or with reactive organic groups (type III precursors).

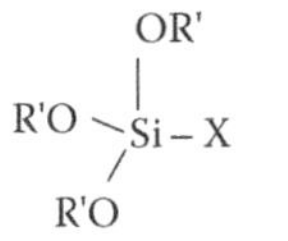

OR'
R'O–Si–X
R'O

OR' = methoxy, ethoxy, or acetoxy

X = alkyl, aryl, or organofunctional group

FIGURE 6.21 Basic organosilane structure—asymmetrical.

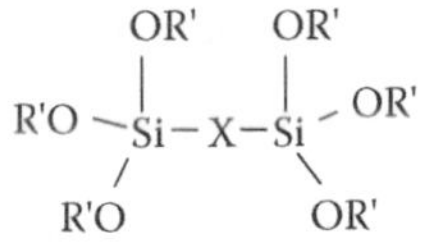

OR' OR'
R'O–Si–X–Si–OR'
R'O OR'

OR' = methoxy, ethoxy, or acetoxy

X = alkyl, aryl, or organofunctional group

FIGURE 6.22 Alternative "bis" organosilane structure—symmetrical.

Type II precursors contain stable Si–C bonds (R group), which are not modified during hydrolysis and polycondensation reactions, decreasing the inorganic network density from fourfold to threefold or twofold Si–O–Si connectivity. The organic group R is selected depending on the organic functionalization desired, such as polarity (hydrophobic modification with methyl groups, as well as phenyl groups or hydrophilic modification with amino groups), refractive index (phenyl groups to increase the refractive index and, e.g., trifluoropropyl groups to decrease the refractive index), conductivity, etc.

Type III precursors can lead to the formation of an additional organic network or to the cross-linking between inorganic-oxidic network clusters. Examples of organic reactive groups are glycidoxypropyl, methacryloxypropyl, vinyl, mercaptopropyl, isocyanatopropyl (for promotion of the reaction with OH functionalities), etc. Methacryl-based silane systems are more reactive than vinylsilane systems in terms of organic polymerization/cross-linking. Epoxy groups are very suitable to be incorporated into sol-gel systems, because of the epoxy ring opening with water. Aminosilanes are often used for cross-linking with epoxy- or isocyanate groups.

Type III precursors can also be called of organofunctional silanes (either asymmetrical or symmetrical structure, as shown in Figures 6.21 and 6.22). They are molecules carrying two different reactive groups on their silicon atom: hydrolyzable groups such as methoxy, ethoxy, or acetoxy; and organofunctional groups, such as epoxy, amino, methacryloxy, or sulfido (examples of these molecules are displayed in Figure 6.23). These molecules can react and couple with very different

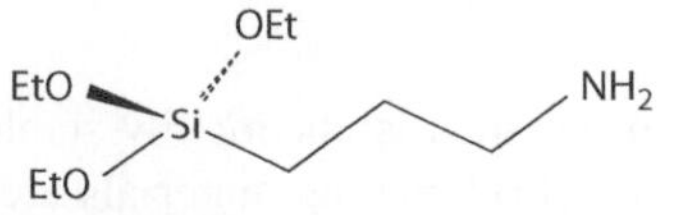

Amino-silane
γ-Aminopropyltriethoxysilane

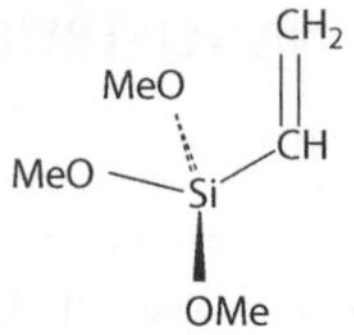

Vinyl-silane
Vinyltrimethoxysilane

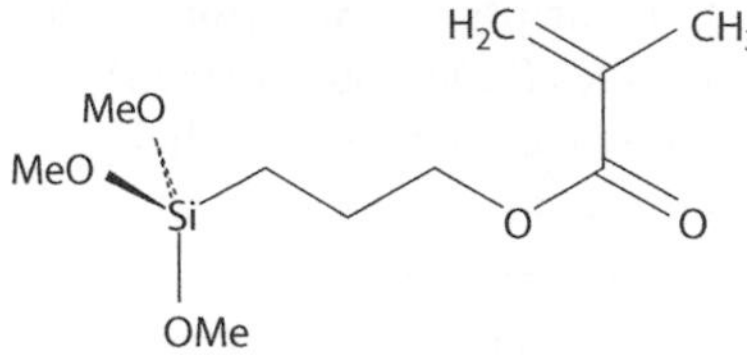

Methacryloxy-silane
γ-Methacryloxypropyltrimethoxysilane

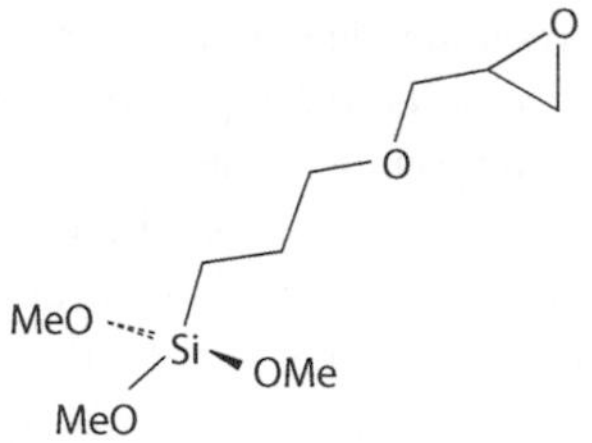

Epoxy-silane
γ-Glycidoxypropyltrimethoxysilane

FIGURE 6.23 Examples of organofunctional silanes showing two different functionalities: hydrolyzable alkoxy groups and organic-functional groups. Dow Corning is one of the companies manufacturing and supplying these silanes, of commercial name Dow Corning® Z-6011, Z-6300, Z-6030, and Z-6040, respectively.

materials (e.g., inorganic surfaces and organic resins via covalent bonds and often via a polymeric "transition" layer between these different materials). The value of organofunctional silanes as coupling agents was discovered in the 1940s during the development of fiberglass-reinforced composites.[57]

In the case of resins, matching the organofunctional group on silicon with the resin polymer type to be bonded will dictate which silane-coupling agent should be used in a particular application. The properties of the organofunctional silane should match the reactivity of the resin with appropriate groups on the silane to react with the resin (e.g., epoxy or amino groups to react with epoxy resins, amino groups to react with phenolic resins, or a methacrylate group to react with styrene in unsaturated polyester resins).

The organic precursors (type IV) are non-Si-based organic monomers that increase the organic nature of the hybrid materials. Epoxies, acrylics, and diols are examples of organic monomers that react with organic reactive alkoxides (type III precursors).

6.2.2 Thin Film Deposition Techniques and Posttreatments for Nanostructured Thin Film Generation

There are varied deposition techniques and posttreatments to dictate the thin film roughness, texture, and morphology.

Prior to gelation, the sol obtained by hydrolysis and condensation reactions can be used for preparing thin films by common methods such as spinning, dipping, and spraying. Besides these most common techniques, meniscus-coating,[10] roll-coating, and flow-coating have also been used, among other not-so-common techniques. In the spin-coating technique, the solution is dropped over the substrate, which is then made to spin at a selected (high) velocity,[3,58,59] dip-coating involves the immersion of the substrate into the sol and withdrawal of the substrate at a selected rate,[60] spray-coating consists of pulverization of the sol, but it is not the most suitable technique for the production of thin uniform films with optical quality on large area substrates. The thickness of the resulting films is affected by the viscosity of the sols, spinning/withdrawal rate, pulverization output, etc., depending on the coating technique. All of these coating techniques are well described in the literature, namely, in Ref. [61].

Nanocrystallization or nanopore formation occurs during the drying and sinterization process, either by heat treatment, or photo-assisted treatments.

Top-down and bottom-up strategies, whether in combination with sol-gel-coating deposition techniques or not, may be also applied to generate nanostructured thin films, with varying degrees of speed, cost, and structural quality. Top-down techniques involve the use of various forms of conventional lithographic techniques to pattern nanostructures (e.g., onto planar substrates), whereas bottom-up methods exploit the interactions of atoms and molecules together with process kinetics, to "assemble" nanostructures, either in solution or on the substrates.

6.2.2.1 Top-Down Approaches

Top-down approaches consist of either using macroscopic tools to directly "write" materials on a substrate, or using macroscopic tools to transfer a computer-generated pattern onto the thin film material, and then "sculpting" a nanostructure by physically removing material (e.g., by etching techniques). It should be noted that either the solution/material to be used in the printing techniques, or the films to be "sculpted" may be prepared via sol-gel.

The former case includes scanning tunneling microscopy (STM), scanning probe lithography (SPL), nanoimprint lithography (NIL), and microcontact printing (μCP). STM allows the position of individual atoms to pattern structures with subnanometer precision, and SPL leads to features as small as 15 nm. μCP allows conformal contact between the stamp and the substrate for a range of topologies, including curved substrates/inner surfaces, due to the mechanical flexibility of the rubbery stamp. The imprinting device used in the NIL approach can be reused numerous times, leading to cost-effective, sub-100 nm lithographic replication; however, the initial fabrication of the ultrahigh-resolution master molds remains very difficult to attain. It should be noted that these techniques, which allow the patterning of structures with nanometer precision, are neither cost- nor time-effective.[62]

The most common techniques used in the latter case ("sculpting" a nanostructure) are based on photolithography, which is unfortunately limited by optical diffraction effects to 200–500 nm, and nanolitography (electron- and ion-based methods), which already allow the preparation of ordered nanostructured arrays with high resolution, but have low throughput, since they consist of a line-by-line pattern.

6.2.2.2 Bottom-Up Approaches

They are based on physicochemical interactions for the hierarchical synthesis of ordered (or not) nanoscale structures on the coatings. Chemically or topologically patterned surfaces can be used as a templating substrate, but traditional top-down techniques are required for the preparation of these surfaces; inorganic mesoporous structures can also be used as the templating substrate, as well as organic supramolecular complexes, such as block copolymers, which can be incorporated into sol-gel solutions. And more recently, but not so well established, is the use of biological scaffolds for the fabrication of novel types of nanostructures.[62] Among these techniques, the use of block copolymers, through the self-assembly of basic building blocks, is one of the most employed for the nanostructurization of coatings. They can be used as surfactants, nanopore directing agents, since they involve nano- or microphase separation of domains, with periodicity typically in the range of 10–200 nm.

For instance, the direct-write technique applied to sol-gel-derived xerogel thin films has been employed for the fabrication of planar ridge and hollow waveguides;[63] procedures to provide surface roughness, such as phase separation have been applied to sol-gel systems;[64] a two-photon polymerization process of photosensitive inorganic–organic hybrid polymers (ORMOCERs) prepared by sol-gel has been claimed to be a promising approach for the fabrication of complicated three-dimensional micro- and nanostructures;[65] a nano-replication process has been used for the fabrication of diffractive optical elements in sol-gel-derived glasses, where a polydimethylsiloxane (PDMS) replica is obtained from a Si or PMMA/PI master structure, followed by applying a layer of sol-gel material on the PDMS replica to get micro- and nanostructure in glass after drying and annealing.[66]

6.3 PROPERTIES AND APPLICATIONS: OPTICAL AND PHOTONIC ASPECTS

As a result of the molecular mixing occurring in the sol-gel method, the obtained materials are often transparent and with good mechanical properties. Their advantages, listed previously in the chapter, as well as their versatility to be nanostructured, can be very useful in the field of optics. The thin films can be spontaneously nanostructured indeed, when, e.g., the material is not completely densified resulting in nanoporosity, or the densification treatment is such that it leads to crystallization at the nanoscale in the thin films. The optical (and photonic) studies carried out on sol-gel-derived nanostructured thin films cover a wide range of objectives; some of them are listed below:

- Investigation of the optical/spectroscopic properties of dye molecules (e.g., Rhodamine G6) isolated in nanopatterned or nanoporous sol-gel materials[67–69]
- Investigation of the photoluminescence (PL) of, e.g., Eu^{3+}-doped nanocrystalline thin films, and the influence of the rare-earth (RE) ion environment on its PL[70]
- Preparation of AR coatings, in the form of nanoporous or nanoparticulate coatings, made of SiO_2, MgF_2[54,71], TiO_xN_y-ATO[72], SiO_2, and ITO nanolayers[73]
- Optical monitoring of nanoparticles or nanocrystallites generated and grown in situ[74,75]
- Investigation of PL and photoelectrochemical properties of nanoparticulate thin films, and correlation with quantum size effect; intrinsic fluorescence properties and RE (lanthanide) fluorescence in nanoparticulate films, e.g., nano-ZnO films[76–80]

One of the most-studied optical aspects of nanostructured thin films is the tuning of refractive index by controlling the porosity of the material. The presence of nanopores or oriented nanochannels can be also advantageously applied to provide new opportunities for fabrication of new optical elements by incorporating photo- and opto-functional materials into the pores.

On the other hand, topics such as glass-based photonic systems, where light is confined or where RE ions are hosted by nanostructured thin film matrices, have been also extensively studied due to the possibility they offer of enhancing and controlling the optical and spectroscopic properties in thin films.[81]

In optical science and technology, the refractive index of a material is a fundamental quantity that dictates the characteristics and performance of optical components. The refractive index is related to the phase and group velocity of light in the material and, therefore, it determines the effects of refraction and reflection, being in this sense a critical parameter for optical phenomena, such as refraction, Fresnel reflection, wave propagation, and wave guiding.

The presence of voids or empty spaces, either ordered or not, in thin films leads to a lower refractive index (composite dense material—air) than that of the corresponding dense material. Highly porous optical coatings are attractive materials that can be used in optics and microelectronics, sensing, etc., as a result of their low dielectric constant, tunable refractive index, and high surface area. It is also important to note that the pore or feature sizes in the thin film structure are typically below 100 nm, in order to limit the effect of Rayleigh scattering rendering specular films of high optical quality.

Therefore, the refractive index needs to be controlled and tuned as required by many target applications, such as AR coatings, where the aim might be getting light efficiently into a material (e.g., glass cover in photovoltaic solar cells), or out of a material, such as in light-emitting diodes (LEDs). Another interesting application for low refractive index materials is on the preparation of distributed Bragg reflectors (DBRs) using a single optical material, but alternated layers of porous and dense material. This approach would avoid the drawbacks of TiO_2, normally used as high refractive index material, in terms of absorption in the UV, or the lack of conductivity presented by SiO_2 or TiO_2.

In thin films, such as planar waveguides for optical amplification purposes, the host material determines the atomic scale structure around the RE ions (e.g., erbium ions), influencing their spectroscopic properties. So, the PL of the RE species is expected to change depending on the crystallization degree of the matrix where the RE species (e.g., Er^{3+}) are embedded, for instance. Nanograin or nanocrystalline thin films play a role here, in the sense that the shape of the PL emission spectrum from Er^{3+} ions is directly related to the surrounding environment of these ions[82] (disordered/amorphous or ordered/crystalline). Disordered/amorphous Er^{3+} hosts are known to result in larger inhomogeneous broadening of the Er^{3+} optical transitions, allowing the simultaneous operation of several wavelength channels, at multigigabit per second (Gbps) data transmission rates, in a single fiber-optical link.[82]

Next, the optical and photonic aspects of nanostructured thin films will be divided by the type of material: silicon-based materials, e.g., SiO_2, non-silicon-based oxide materials, and non-oxide materials. Some case studies will be described for each type of material.

6.3.1 Silicon-Based Oxide Materials

Silicon dioxide (SiO_2)-based materials have been often used as a matrix in many kinds of thin films and coatings, including planar waveguides, sensors, and filters. They have been the preferred materials for use as hosts for many RE elements in optical amplifiers, or for noble metal or semiconductor nanoparticles, due to their superior optical, mechanical, and chemical properties, as well as their inherent compatibility with the passive transport fiber in photonics (integrated optics, IO) applications.

Amorphous and transparent SiO_2-based coatings with low optical loss at data and telecom wavelengths (810, 1310, and 1550 nm, respectively) have been prepared by a nonhydrolytic process, where Si–O–Si network is established solely by polycondensation reaction without releasing water.[83] Then, in order to structurize the thin film, for the production of (planar and stacked) waveguides and microlenses, the nanometer-sized inorganic-oxidic units may be organically functionalized by methacrylic groups, which enable the material to be patterned either by conventional UV lithography or two-photon absorption (2PA) processes. An example of ORMOCER, which has been developed for waveguide applications, results from the mixture of 3-methacryloxypropyltrimethoxysilane and disphenysilanediol, type III precursors. The refractive index can be adjusted, in this case, by mixing the appropriate inorganic–organic hybrid polymer. Kwon et al. have reported another silane combination for refractive index tuning consisting of the mixture of (3-acryloxypropyl)trimethoxysilane (ATMS) and 4,4′-(Hexafluoroisopropylidene)diphenol (FBPA), which leads to a decrease in refractive index for increasing FBPA amount.[84] This is mainly due to the increase in the number of fluoromethyl units. The refractive index could be further reduced by adding ethylene glycol dimethacrylate (EDMA) to the sol at a FBPA:ATMS:EDMA mixing ratio of 0.2:1:0.2. This process occurred under nonhydrolytic conditions and allowed the synthesis of a flexible optical waveguide by soft-lithography technique. There was no additional bending optical loss above 2 mm diameter curvature. The transmission performance of each waveguide channel was tested using a 10 Gbps data stream and the electrical output signal from a photodetector, connected to a wide-band oscilloscope, displayed a clear 10 Gbps eye pattern.

Another example of patternable sol-gel material is reported in Ref. [67] by Della Giustina et al., namely, an epoxy-based hybrid sol-gel material to be applied as negative resist for electron beam lithography (EBL), in the fabrication of lasers and resonators. It derives from GPTMS, as Si precursor (type III) and germanium tetraethoxide (TEOG), a type I precursor, which contributes to the inorganic network formation and is known for not opening the epoxy ring of GPTMS, for molar content around 20%. 2-Methoxyethanol (MeEtOH) was used as solvent. EBL was performed on the previously dried thin films prepared by spin-coating. The electron beam induces polymerization of the organic part and mechanical stiffness of the film. The feature size ranges from 2 μm to 50 nm and their resolution depends on the electron dose, which is required to be high in order to achieve small features. A laser structure has been patterned and demonstrated using a Rhodamine 6G-doped hybrid film.[67]

The next example shows the synergy between mesoporosity generation and organic functionalization in a hybrid coating, which allows the fabrication of receptors for incorporating inorganic ions and biomolecules such as enzymes, DNA, and proteins. Organic functionalization is indeed an approach that can be used to promote the guest loading into the mesoporous films, either by physical forces (i.e., van der Waals, hydrogen bonding, hydrophobic, and electrostatic force), or by chemical bonding with the help of a cross-linker.[85] In particular, the distribution of amino groups in the mesoporous surface enhances the adsorption of cytochrome c, the guest species employed in the present example.[68] For this purpose, amino-functionalized ordered mesoporous hybrid silica thin films have been synthesized by sol-gel dip-coating on glass substrates using TEOS, 3-aminopropyltriethoxysilane (APTES), and the triblock copolymer Pluronic P123 as the silica source, amino-precursor, and structure-directing agent, respectively. Figure 6.24 shows the TEM images (along and perpendicular to the pore channel), where a well-ordered hexagonal mesostructure can be observed. N_2 adsorption–desorption isotherms of the same film exhibit isotherms of type IV (and type H_1 hysteresis loop), and the average BJH pore size was found to be 8.3 nm. The thin film presented a nitrogen BET surface area of 680 m^2/g and a pore volume of 1.06 cm^3/g.

The adsorption of cytochrome c protein to the mesoporous material was monitored by UV-Vis spectroscopy, since it absorbs in the UV-Vis range, exhibiting a maximum at 409 nm and two weak peaks at 365 and 528 nm. These bands indeed were found to be more intense for amino-functionalized mesoporous silica thin films, than for the non-functionalized ones.

Nanostructured thin films have been also applied in optical sensors, such as optical oxygen sensors. In this case the approach reported here consisted of affecting the morphology at the nano- and submicron scale of a film surface, by using blends of TEOS with MTES or with octyltriethoxysilane (OTES), at 3 and 1 molar ratio, respectively.[69] The sols were doped with Ru(II)-tris(4,7-diphenyl-1,10-phenanthroline) dye, before spin- and dip-coatings.[69] The presence of MTES and OTES (type II precursors) significantly increased the surface roughness of the hybrid films, while the TEOS-derived films (type I precursor) were found to be smooth. The structured morphology strongly depended on the hydrocarbon chain length (CH_3 *versus* CH_3-$(CH_2)_7$). The films produced by TEOS plus OTES precursor mixture exhibited an intermediate

FIGURE 6.24 TEM images (and corresponding FFT images) of amino-functionalized mesoporous silica thin film after surfactant extraction with hot ethanol under reflux for 12 h; (a) along and (b) perpendicular to the pore channel. (From Zhang, X. et al., *J. Sol-Gel Sci. Technol.*, 43, 305, 2007. With permission.)

surface morphology between that obtained for the individual precursors. Hollows with a diameter of 35–65 nm were observed on the surface of the TEOS/OTES-based films with a depth twice of that for pure TEOS-based films. The thickness as well as the deposition parameters and the presence of the photosensitive dye were found to be responsible for the film's roughness. The morphology generated through silane blending and the resulting increase in surface area, associated to a homogeneous distribution of the dye, gave rise to a better-performing optical oxygen sensor, by improving the PL response to oxygen.

Nanoporous silica films with tuned refractive index (namely, low refractive index values) have been obtained by a base–acid two-step catalytic sol-gel process,[86] where an acid-catalyzed sol is mixed with a base-catalyzed sol. These films have a combined character of the acid-catalyzed films, which are dense with a smooth surface, and of the base-catalyzed films, which are composed of silica particles and are rough, as shown in Figure 6.25.

Figure 6.26 displays the wavelength dependence (in the visible region) of the refractive index of the films prepared from the three different silica sols, and Figure 6.27 shows the corresponding reflectivity spectra.

Finally, another way of decreasing the refractive index of a thin film, in order to be applied as a low reflectivity coating for instance, consists of blending different silanes, namely, MTES and TEOS,[87] at selected molar ratios. The refractive index of silica films with a MTES/TEOS molar ratio between 2.5 and 5 was found to decrease to 1.23 due to the porosity generated at the film. The effect of solvent type and the relative humidity of the coating atmosphere, as well as the heat treatment temperature (Figure 6.28), was discussed by the authors in this paper. Coating a glass panel with these silica films led to a reflectance in the visible region lower than 0.2%, as can be observed in Figure 6.29.

Sol-gel-derived SiO_2-TiO_2 thin films (as-deposited by dip-coating) have been treated in hot water at 90°C under vibration, with the aim of affecting their surface morphology and still maintain their transparency (monitored by UV-Vis spectroscopy), at the same time that other interesting properties are achieved. Figure 6.30 exhibits the precipitation of titania nanosheet features at the surface of the coatings prepared on glass substrates. Matsuda et al. have identified this titania nanosheet as being an hydrated titania layer having a spacing of about 0.6 nm with a lepidocrocite-type structure.[88]

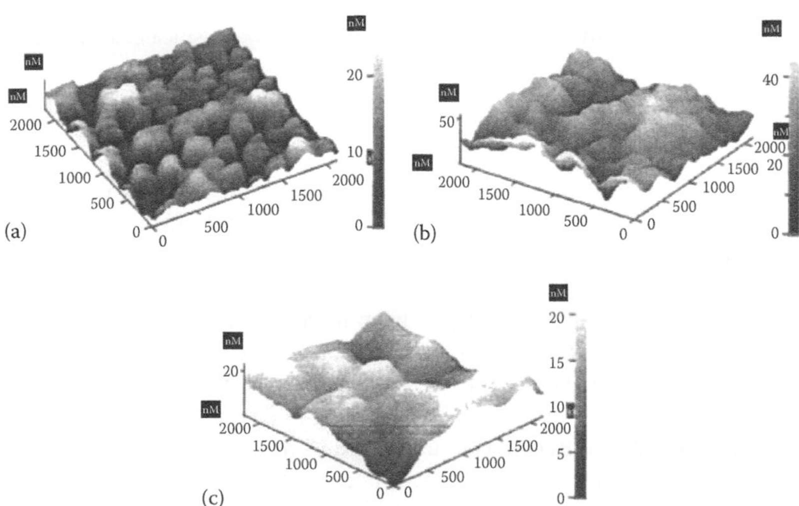

FIGURE 6.25 AFM images of the silica films prepared by (a) base catalysis, (b) two-step catalysis, and (c) acid catalysis. (From Wu, G. et al., *Mater. Sci. Eng.*, B78, 135, 2000. With permission.)

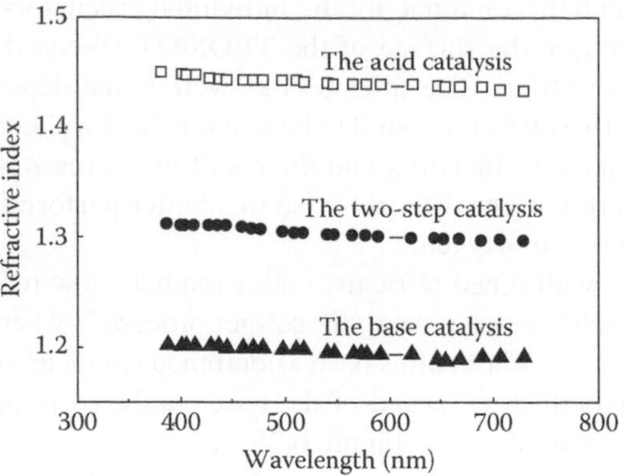

FIGURE 6.26 Refractive index spectra of the silica films resulting from the different catalyzed sols. (From Wu, G. et al., *Mater. Sci. Eng.*, B78, 135, 2000. With permission.)

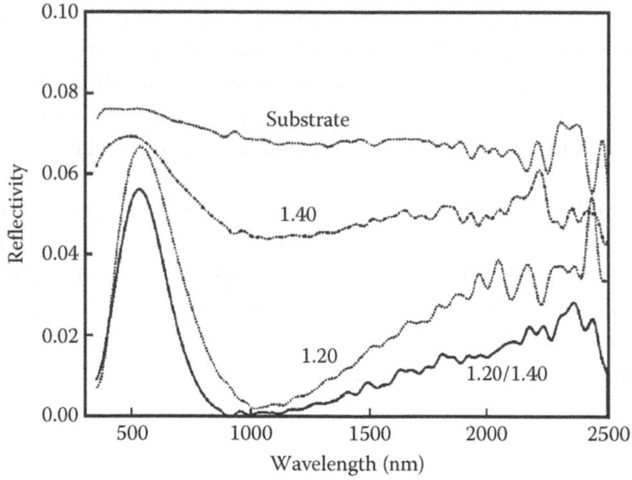

FIGURE 6.27 Reflectivity spectra of an uncoated glass (substrate) and silica-coated glasses. The values displayed at the spectra are the corresponding refractive index values. (From Wu, G. et al., *Mater. Sci. Eng.*, B78, 135, 2000. With permission.)

The role played by the nanostructure of a silicate thin film in the optical and photonic properties of the material, when doped with RE ions, namely, Er^{3+}, will be described below, through two different examples: one dedicated to the effect of nanoporosity and the other describing the effect of a nanocrystallized Er^{3+} environment on the Er^{3+} PL behavior.

The occurrence of residual nanoporosity has been found to be at the origin of a reversible PL quenching behavior at 1.5 µm, in Er^{3+}-doped silica–titania (20% mol TiO_2) (S20TEr) sol-gel planar waveguides,[89,90] which is exhibited in Figures 6.31 and 6.32. Basically, upon exposure to the laboratory atmosphere, there was a decrease in the Er^{3+} PL peak intensity, which was found to be reversible, as long as a subsequent heat treatment at temperatures as low as 200°C was performed on the same samples. Figure 6.31 shows the fluorescence spectra, corresponding to the $^4I_{13/2} \rightarrow ^4I_{15/2}$ transition of Er^{3+} ions, of the S20TEr waveguide, as a function of different final heat treatments, exposure time to laboratory atmosphere, and the subsequent heat treatment, while a graphical representation of the 1.5 µm PL intensity evolution with all the treatments performed on this sample is shown in Figure 6.32. A more pronounced decrease was observed when those samples were kept in an

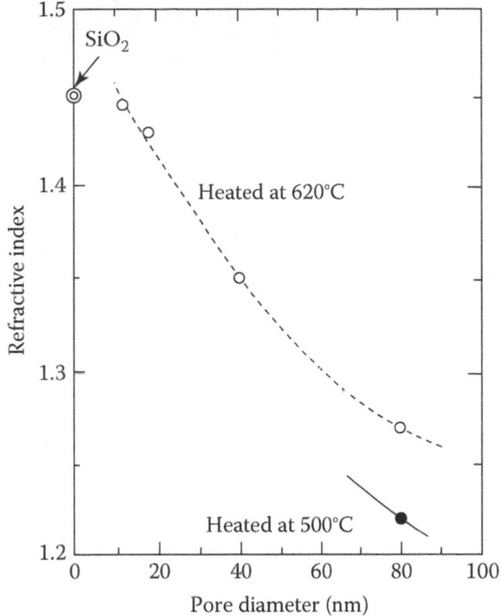

FIGURE 6.28 Refractive index dependence on the heat-treatment temperature and pore diameter. (With kind permission from Springer Science + Business Media: Makita, K. et al., *J. Sol-Gel Sci. Techn.*, 14, 174, 1999.)

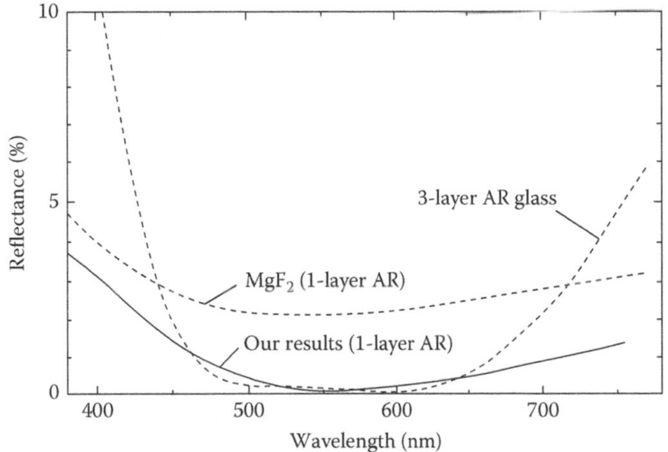

FIGURE 6.29 Comparison of the reflectance spectra between the coating obtained by the approach described here, a glass coated with MgF_2, and a three-layer AR coating on glass. (With kind permission from Springer Science + Business Media: Makita, K. et al., *J. Sol-Gel Sci. Techn.*, 14, 174, 1999.)

ultrasonic water bath. On the contrary, a significant inhibition of the PL degradation was observed when the samples were kept under a controlled atmosphere with low moisture content, such as liquid nitrogen, or a controlled atmosphere dry box. These results are presented in Table 6.5, which lists the percentage of PL quenching, as a function of different times of exposure to various atmospheres, following selected heat treatments.

On the other hand, together with the aging of the waveguides upon exposure to the laboratory atmosphere, there was a simultaneous increase in the OH-related IR absorption, as can be seen in Figure 6.33.

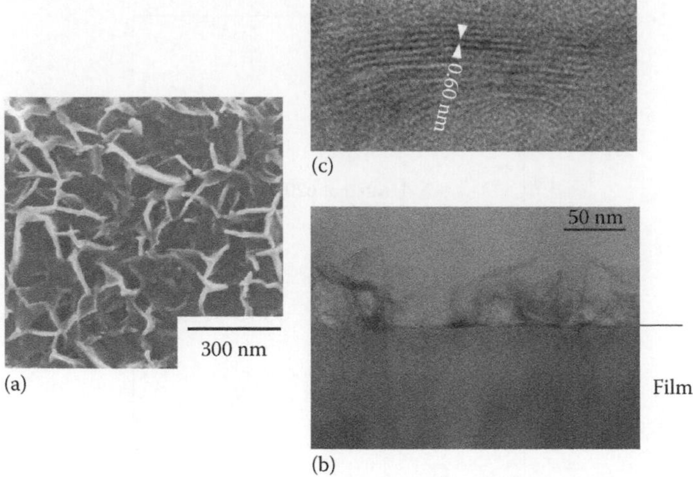

FIGURE 6.30 (a) SEM image of the surface of a SiO_2-TiO_2 coating treated in hot water at 90°C under vibration, (b) TEM image of the cross section of the coating, and (c) high-resolution TEM image of the selected area of the nanosheet in (b). (With kind permission from Springer Science+Business Media: Matsuda, A. et al., *J. Sol-Gel Sci. Techn.*, 31, 229, 2004.)

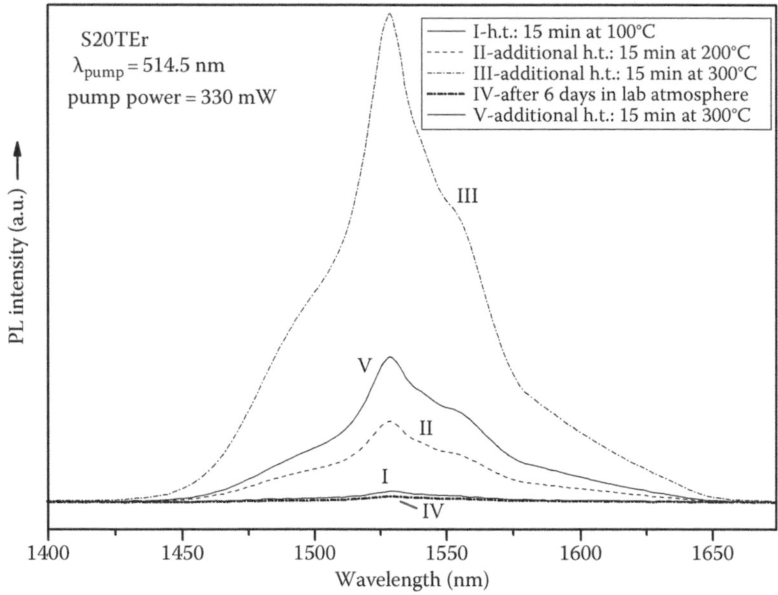

FIGURE 6.31 PL spectra of the S20TEr (0.5% mol $ErO_{1.5}$) waveguide, for different heat treatments in air and subsequent exposure times to laboratory atmosphere. (λ_{pump}=514.5 nm, pump power=330 mW). (From Matusda, A. et al., *J. Sol-Gel Sci. Technol.*, 31, 229, 2004. With permission.)

The samples whose PL decreased most and whose OH-related IR peak increased most were, indeed, those that were not yet fully densified, i.e., those that were still porous.

Also, the Er^{3+} fluorescence lifetime was found to decrease, in a reversible way, for porous waveguides when exposed to laboratory atmosphere or to other moisture-containing atmospheres.

Table 6.6 lists the corresponding 1/e fluorescence lifetimes, calculated from the PL decay as the 1/e decay time of the fluorescence intensity, after different heat treatments in air and selected

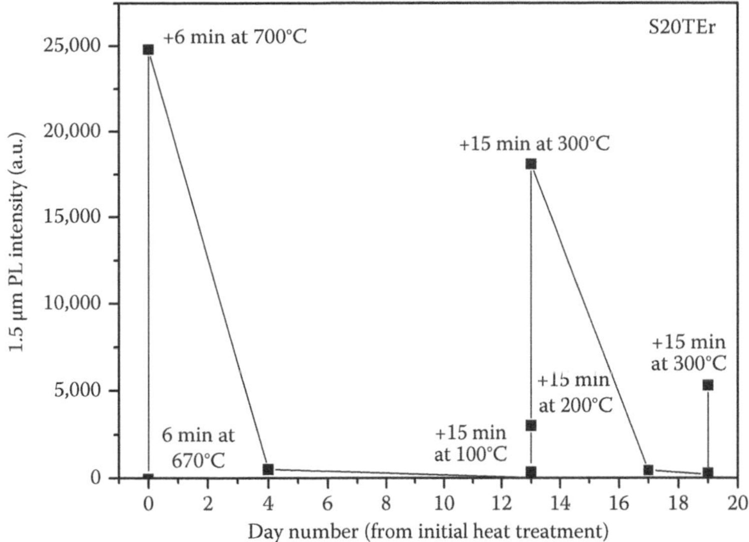

FIGURE 6.32 Evolution of the 1.5 μm PL intensity with the time of exposure to laboratory atmosphere, for different heat treatments, for sample S20TEr. (From Marques, A.C., Erbium/silver glass matrix nanocomposites by sol-gel processing, for integrated optics, PhD thesis, Technical University of Lisbon, IST, Lisbon, Portugal, 2005.)

TABLE 6.5
PL Quenching (% decrease = $[(I_o^{1.5\mu m} - I_f^{1.5\mu m})/I_o^{1.5\mu m}]$), as a Function of Different AFM Rms Times of Exposure to Various Atmospheres, Following Different Heat Treatments, for S20TEr Waveguide Samples

Heat Treatment Performed on the Samples	Time of Exposure to Selected Atmospheres	% PL Quenching
6 min at 670°C/in air	6 days in lab atmosphere (air)	72
	6 days in a desiccator	28
	6 days in a sealed container in liquid N_2	5
2 min at 700°C/in air	5 days in a glove box	17
7 min at 700°C/in air	3 h in an ultra sonic water bath	97
10 min 900°C/in air	7 days in lab atmosphere (air)	76
10 min 900°C/in vacuum	7 days in lab atmosphere (air)	34

Source: Marques, A.C. et al., *J. Non-Cryst. Solids*, 322, 272, 2003. With permission.

exposures to laboratory atmosphere. The slow decay component (SDC) lifetime, assuming a double exponential decay, is also shown. The changes observed in the fluorescence lifetime were also reversible, as long as a heat treatment, even at a low temperature ~200°C, was performed on the samples.

The mechanism proposed in this work was the quenching of the Er^{3+} luminescence by nonradiative de-excitation involving OH groups originating from atmospheric humidity, which are apparently able to interact with the Er^{3+} ions present near the inner surface of the residual pores of the sol-gel waveguides.[89] And, the surface area, therefore in contact with atmospheric humidity,

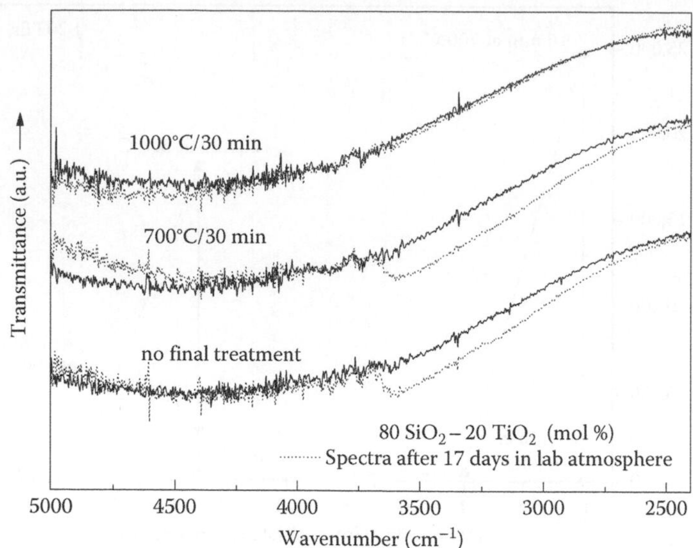

FIGURE 6.33 FTIR absorption spectra of S20T films, following different final heat treatments, before and after 17 days of exposure to laboratory atmosphere. (From Marques, A.C. et al., *J. Non-Cryst. Solids.*, 322, 272, 2003. With permission.)

TABLE 6.6
Er^{3+} Fluorescence Lifetimes of S20TEr Waveguide, for Different Heat Treatments in Air and Times of Exposure to Laboratory Atmosphere (λ_{pump} = 514.5 nm, pump power = 380 mW)

80 SiO_2 – 20 TiO_2 – 0.5 Er	$\tau_{1/e}$ (±0.5 ms)	τ_{SDC} (±0.5 ms)
40 days after a h.t. of 25 min at 750°C	1.6 ms	7.6 ms
63 days after a h.t. of 25 min at 750°C	<1 ms	4.3 ms
Additional h.t.: 30 min at 200°C	<1 ms	4.6 ms
Additional h.t.: 30 min at 400°C	3.5 ms	6.6 ms
Additional h.t.: 15 min at 900°C	6.6 ms	7.2 ms

Source: Marques, A.C. et al., *J. Non-Cryst. Solids*, 322, 272, 2003. With permission.
Note: h.t., heat treatment.

is as large as the volume fraction of nanoporosity. This case shows the effect of the presence of nanoporosity spontaneously generated in the planar waveguides, or not completely eliminated through heat treatment, which plays a negative role if such waveguides are to be used in optical amplification or a positive role if the waveguides are to be used as humidity sensors.

The example described next exemplifies how a crystallized Er^{3+} environment can affect the Er^{3+} PL behavior. Figure 6.34 shows the PL spectra of Er^{3+}-doped planar waveguides of composition 75 SiO_2 – x HfO_2 – (25 – x) TiO_2, with x = 0%, 15%, 20%, and 25%, whose acronyms are S25TEr, S15H10TEr, S20H5TEr, and S25HEr, respectively.[91] All the spectra consist of an emission peak at 1.53 μm, with shoulders near 1.48, 1.56, and 1.61 μm. The silica–titania and silica–hafnia binary waveguides show

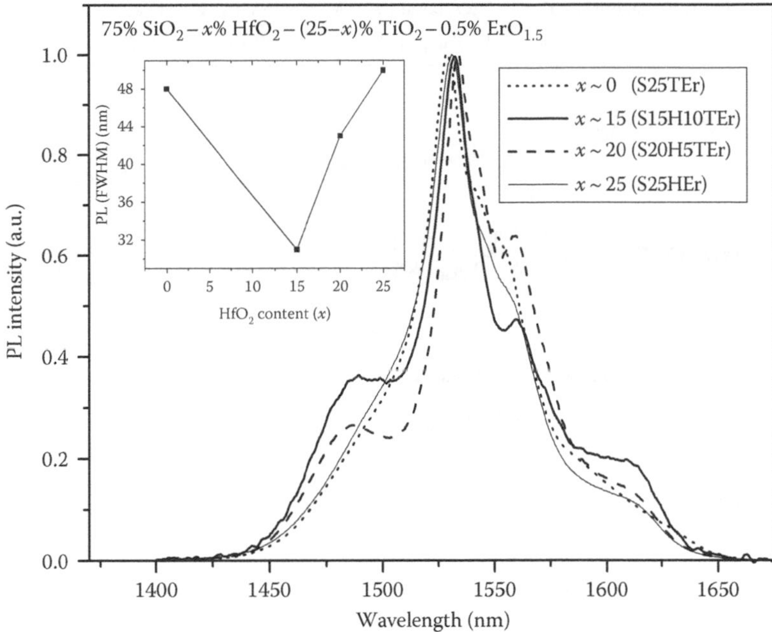

FIGURE 6.34 PL spectra of different Er^{3+}-doped (0.5% mol $ErO_{1.5}$) waveguides with base composition $75SiO_2-xHfO_2-(25-x)TiO_2$, with $x=0$, 15, 20, 25. On each waveguide, a heat treatment of 2 min at 900°C per layer + 15 min at 900°C at the end was performed. The inset shows the FWHM of the PL peak as a function of the HfO_2 content ($\lambda_{pump}=514.5$ nm). (From Almeida, R.M. et al., *J. Sol-Gel Sci. Technol.*, 31, 317, 2004. With permission.)

the broadest Er^{3+} emission peaks (in this case, the full width at half maximum (FWHM) was 48 and 50 nm, respectively). On the other hand, the intermediate, ternary compositions have a smaller FWHM, and their PL peak presents well-resolved Stark components at ~1.48 and ~1.56 μm. The inset of Figure 6.35 shows the FWHM variation between 50 and 31 nm, depending on the HfO_2/TiO_2 ratio.

The FWHM of the PL peak reaches a minimum for the intermediate composition 74.6 SiO_2 – 14.9 HfO_2 – 10 TiO_2 – 0.5 $ErO_{1.5}$ ($x=15$, sample S15H10TEr), which suggests a different local environment of the Er^{3+} ions[92] for this particular composition, including a possible increase in the degree of order around the Er^{3+} species. The x-ray diffraction patterns (Figure 6.35) exhibit indeed broad peaks near ~30° for the ternary systems, which indicate the formation of mixed oxide crystals, containing Ti and Hf ($HfTiO_4$, HT and $HfTi_2O$, HT2), as well as tetragonal hafnia to a small extent. However, the S5T20H waveguide, heat treated for just 10 min at 900°C, instead of 20 min, presents only very incipient crystallization.

Moreover, the patterns of the S90HEr waveguide (10% SiO2 – 90% HfO2 – 0.5% $ErO_{1.5}$) suggest a significantly crystallized structure, where the presence of tetragonal HfO_2 dominates. The PL spectrum for this latter waveguide where significant amount of nanocrystallites is present is shown in Figure 6.36 and was found to be much narrow than that of the S25HEr waveguide.

Concerning the waveguide Raman spectroscopy results, in Figure 6.37, the S25H spectrum appears typical of an amorphous material, with broad bands characteristic of the silica network (whose spectrum is also included in Figure 6.37 for comparison). The ternary compositions exhibit a few additional broad peaks, but the S25T Raman spectrum contains several sharp peaks, typical of anatase nanocrystals,[93] however, not detected through x-ray diffraction.

The corresponding PL spectrum, however, did not indicate the presence of anatase nanocrystals, since it remained (inhomogeneously) broadened, with no well-resolved Stark components. In fact, previous work[94] reported that when Er^{3+} ions were incorporated in a TiO_2 matrix, their environment

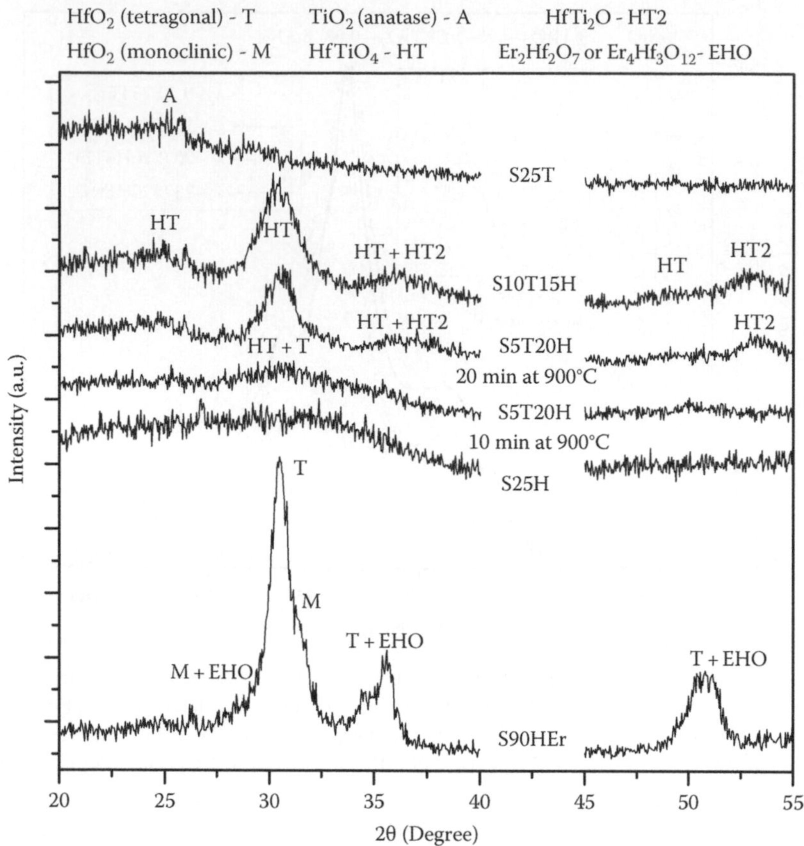

FIGURE 6.35 Grazing incidence XRD patterns for $75SiO_2-xHfO_2-(25-x)TiO_2$, $x=0$, 10, 15, 25, and 10 $SiO_2-90\ HfO_2-0.5\ ErO_{1.5}$ (S90HEr) planar waveguides. (From Almeida, R.M. and Marques, A.C., *J. Non-Cryst. Solids.*, 352, 475, 2006. With permission.)

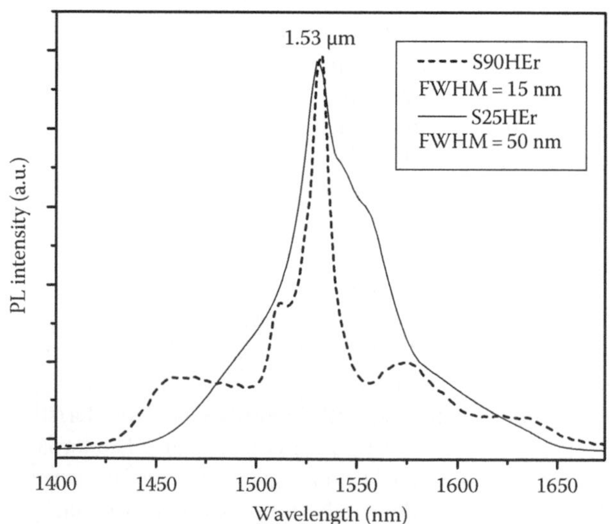

FIGURE 6.36 PL spectra of Er^{3+}-doped planar waveguides with different HfO_2 contents: 25 mol% and 90 mol% ($\lambda_{pump}=514.5$ nm). (From Marques, A.C., Erbium/silver glass matrix nanocomposites by sol-gel processing, for integrated optics, PhD thesis, Technical University of Lisbon, IST, Lisbon, Portugal, 2005.)

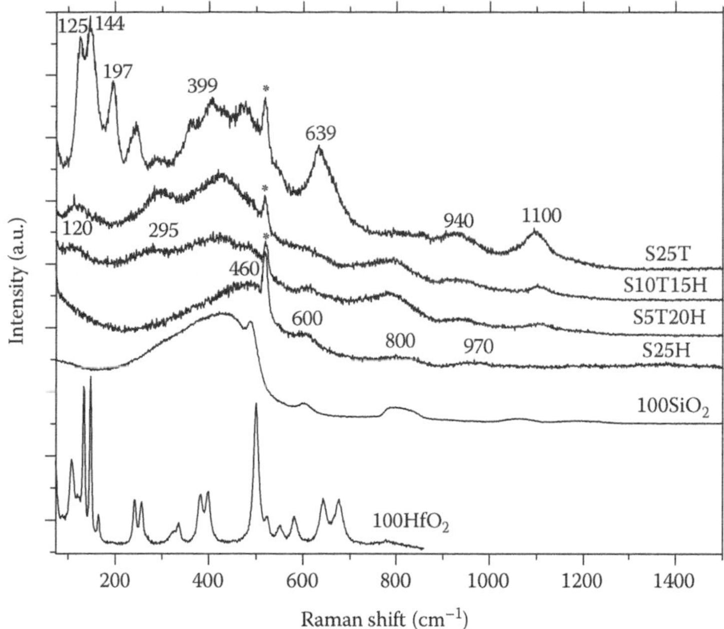

FIGURE 6.37 Raman spectra of silica-based planar waveguides. The peaks marked with (*) was due to the Si substrate. (From Almeida, R.M. and Marques, A.C., *J. Non-Cryst. Solids.*, 352, 475, 2006. With permission.)

appeared to be amorphous, by means of EXAFS analyses, even when anatase nanocrystals were found in the matrix structure, revealing that Er^{3+} ions, in these samples do not have appreciable tendency to be located near anatase (TiO_2) crystals. On the other hand, they are likely to be located near mixed hafnia–titania nanocrystals, i.e., they tend to form Er–O–Hf bridges upon annealing at 900°C, as shown in a recent EXAFS study.[95]

The presence of noble metal nanoparticles or Si nanoparticles (Si-nc)[96–98] embedded in an amorphous silica glass matrix has been shown to affect (in this case to enhance) the Er^{3+} or Eu^{3+} PL spectrum intensity. As a matter of example, the optical properties have been reported to depend on the Si-nc size, e.g., Si-nc red-IR emission, with lifetime ranging from a hundred of microsecond to some millisecond.

Metal nanoparticles are known to exhibit interesting optical properties, which result from the collective electronic excitations at the interface between the metal and the dielectric, i.e., surface plasmon resonance (SPR). Glass/metal nanoparticle composites have been in use for centuries, formerly as colored glasses in church windows and other works of art. These composites absorb light in a particular wavelength region, depending on the metal, via the SPR phenomenon. This behavior was first described by Mie.[99] The SPR of metal nanoparticles is known by providing not only a great enhancement of the optical third-order nonlinear susceptibility of the dielectric matrix, but also an increase in the local electromagnetic field strength around the nanometer-sized metal particle.[100] The frequency and intensity of the SPR are known to depend on the concentration, shape, and size of the nanoparticles; the dielectric function of the surrounding medium; the near-field interactions between the nanoparticles; and the presence of adsorbed species at the surface of those particles.[101]

Indeed, SPR effects may be adequately employed, leading to the evolving of new and better materials for photonic applications. This may lead to another type of mechanism responsible for PL enhancement effects, based on a local field enhancement around the RE^{3+} ions, due to the SPR of metal nanoparticles embedded in the glassy matrix. What has been described to happen in this case is when the electric field (E) of the incident light interacts with a spherical particle of diameter a, with $\lambda_{pump} > 10a$, it will polarize electrons to one side of the particle, and this polarization will track the incident field. Provided the metal has high conductivity at the optical frequency, its electrons will oscillate along with the field, to create a dipolar surface plasmon (SP). This localized electron polarization in the particle produces a local electric field (much larger than the optical field) and the enhanced (electric) field depends on the optical conductivity of the metal (a function of the wavelength, λ), and on the size and shape of the particle. In this case (local electric field intensification), the PL enhancement occurs only when the excitation wavelength is absorbed by the RE ions, whereas the energy transfer-based mechanisms for PL intensification occur both under resonant and nonresonant excitation, with the excitation spectrum having a shape similar to the absorption spectra of the nanoparticle-doped materials.

Two different procedures for silver and erbium ion incorporation into the glassy planar waveguides have been investigated[102]: the direct incorporation of a metal salt ($AgNO_3$ and/or $Er(NO_3)_3$) into the sol-gel solution; as well as a modified sol-gel process, based on pore-doping of a precursor gel with $AgNO_3$ and/or $Er(NO_3)_3$ solutions. The presence of Ag° nanoparticles in the Er-doped silica–titania planar waveguides, having average sizes ranging from ~2 to 15 nm,[102] was revealed by visible absorption spectroscopy (with an SPR observed at ~415 nm, shown in Figure 6.38), XRD, XPS, and TEM, shown in Figure 6.39.

The presence of Ag° nanoparticles in Er^{3+}-activated silica-based waveguides was shown to cause a significant enhancement of the Er^{3+} PL at 1.5 μm, irrespective of the method used for Ag incorporation, but the PL intensification was particularly remarkable, up to ~60 times, in samples prepared by sol doping with both Er^{3+} and Ag^+ ions (followed by Ag° precipitation), as can be seen in Figure 6.40.

The dominant mechanism responsible for the PL intensification in Ag-containing samples, which occurred especially for resonant excitation, is proposed to be the local electric field enhancement around the Er^{3+} ions, due to SPR of the Ag° nanoparticles. Such a process only changes the absorption probability at wavelengths where there is a nonzero absorption cross section, as opposed to the possible energy transfer from Ag° particles to Er^{3+} ions. Another possible effect that might have also contributed to the PL enhancement was the stronger densification of the sol-gel materials in the presence of silver particles.

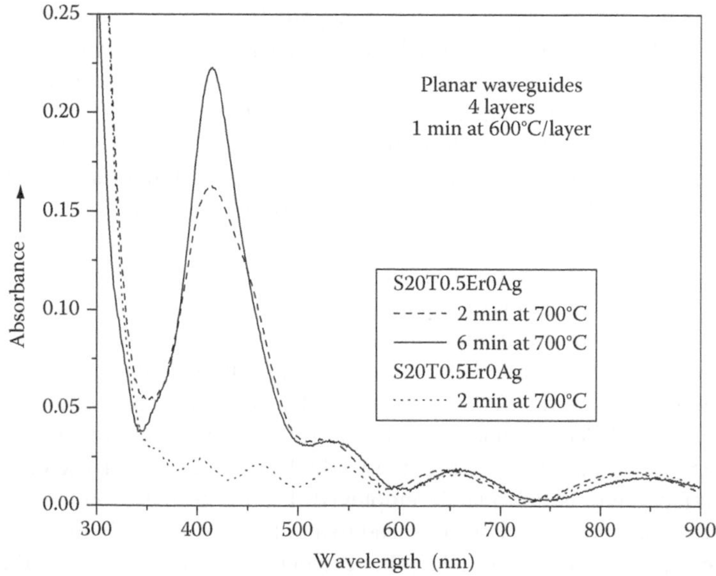

FIGURE 6.38 Optical absorption spectra of 78.8 SiO_2 – 19.7 TiO_2 – 0.5 $ErO_{1.5}$ – 1 Ag (molar ratio) planar waveguides with a final heat treatment at 700°C for 2 and 6 min. The samples were prepared by sol doping with Er^{3+} and Ag^+. (From Marques, A.C. and Almeida, R.M., *J. Non-Cryst. Solids.*, 353, 2613, 2007. With permission.)

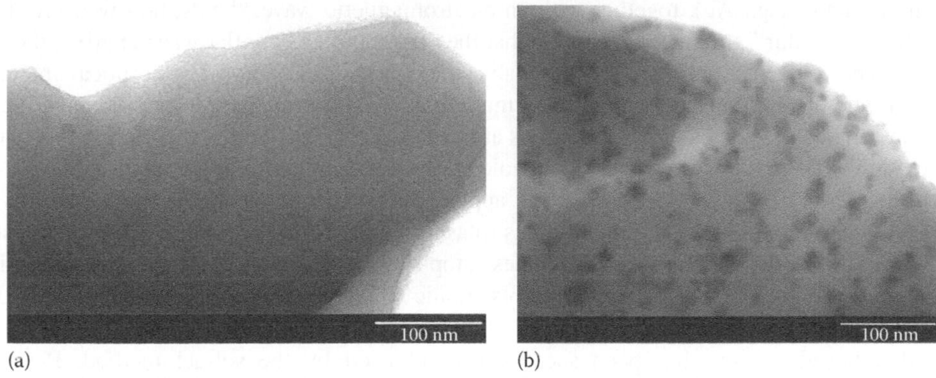

FIGURE 6.39 Dark-field transmission electron micrographs of bulk silica glass samples treated for 5h at 900°C: (a) sample E; (b) sample EA2. (From Marques, A.C. and Almeida, R.M., *J. Non-Cryst. Solids.*, 353, 2613, 2007. With permission.)

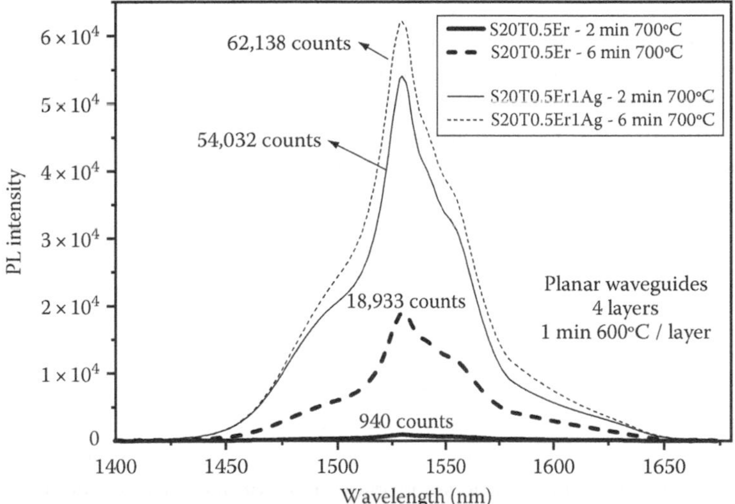

FIGURE 6.40 PL spectra of planar waveguides S20T0.5Er and S20T0.5Er1Ag after a final heat treatment at 700°C, for 2 and 6min. (From Marques, A.C. and Almeida, R.M., *J. Non-Cryst. Solids.*, 353, 2613, 2007. With permission.)

The SPR phenomenon derived from certain nanoparticles embedded in an amorphous matrix has been generally applied in colored coatings, e.g., by Nippon Sheet Glass Co.,[103] for automotive windshields. In this case, chlorauric acid tetrahydrate was used as a starting material for Au fine particles dispersed in a silica-based matrix, with high visible transmission. In the case of Au-particles-dispersed SiO_2 the transmitted color is pink. The plasmon absorption peak by Au particles has been found to shift to longer wavelengths (from ca. 550 to 650nm) with the increase of the refractive index of the surrounding matrix (from 1.46 to 2.20). The matrix consists of a multicomponent oxide made up of SiO_2 and TiO_2, and the refractive index is controlled by the ratio of these components, in the sense that it increases with the increase of TiO_2 amount. Such a shift of the plasmon absorption, which can even lead to a change in color of the resulting nanocomposite coating, has also occurred for other metal nanoparticles (e.g., Ag) and matrices.

Since 2001, there has been a large growth of scientific interest in the role of surface plasmons, which are longitudinal waves that propagate along the interface between a dielectric (e.g., silica glass) and a metal (e.g., Ag), together with an electromagnetic wave.[104] SPs, however, are distinguished from "regular" photons in the sense that they have a much smaller wavelength at the same frequency. The "plasmon regime" should enable the fabrication of nanoscale optical integrated circuits, in which light can be localized and manipulated in appropriately designed metallic and metallodielectric nanoparticle array structures at the subwavelength scale. The term "plasmonics," therefore, describes the new science and technology of optics/photonics at the nanoscale.

There are several approaches that can be applied for synthesizing metal nanostructures that support propagating and/or localized plasmons (plasmonic nanostructures) and exhibit interesting/useful optical properties. Solution-phase synthesis, top-down lithography, and other unconventional lithographic techniques for the fabrication of plasmonic nanostructures are described in Ref. [105].

The fabrication of metallic nanostructures within a ceramic matrix, as is the case of the Ag°-doped silica-based glasses, has been successfully achieved by the sol-gel method. Plasmonic waveguides may therefore be prepared, in which light is confined to metal-dielectric interfaces, as opposed to the dielectric waveguides studied above, where light is confined in the high refractive index core region. In these plasmonic devices, light can in principle be localized in thin metallic (e.g., Ag) nanolayers, or in periodic chain arrays of metal (Ag°) nanoparticles, with particle spacing much smaller than the wavelength of light. In both cases, the guided modes will be in the subwavelength scale.

As a matter of example, silver colloids (metallic Ag° nanoparticles), obtained by the spontaneous reduction process of Ag^+ ions to Ag° through UV radiation, have been generated into the channels of a sol-gel thin film with a 2D hexagonal nanostructure, previously prepared by the dip-coating method using a nonionic diblock copolymer Brij58 (surfactant) to produce channels into the film.[52] An absorption band located at 438 nm was detected on these films, corresponding to the SPR of Ag species. TEM analysis showed the presence of core-shell structures of metallic silver-silver oxide species, which have higher refractive index than the host matrix. In this type of metallic nanostructures within an amorphous (silica) matrix, light may be guided through the metallic silver nanoparticles, i.e., by plasmonic effects.

6.3.2 Non-Si-Based Oxide Materials

Indium oxide is a wide bandgap material (3.3 eV), with high optical transparency in the visible wavelength range and a high conductivity due to the oxygen vacancies, acting as donor states.[106] If mixed with tin, in order to obtain a solid solution of ITO, conductivity is increased and this material can find applications in optoelectronics, as antistatic coatings, and as infrared (IR) reflectors,[107] for instance. Nanostructured (nanograin) ITO thin films, consisting of In_2O_3:Sn nanocrystallites and nanopores, have been achieved by sol-gel, exhibiting porosity up to 50%.[108] These films were prepared by multilayer deposition of single layers with approximately 9 nm in thickness. A very high density of voids, which appear at the limit of the ITO nanograins is exhibited in Figure 6.41. Moreover, each 9 nm thick sol-gel single layer was found to be basically formed by a monolayer of ITO nanograins.

Figure 6.42 shows the void or pore evolution throughout successive depositions. The single layer has in this case a thickness of about 50 nm, in order to be easy to distinguish it in the SEM image. A decrease in the pore size is observed to occur for the previously deposited layers and therefore the authors could suppose that the successive depositions were filling partially the pores in the already deposited layers.

Also by spectroscopic ellipsometry it was found that the void density can reach a value of 50% and decreases with the number of layers, which is in agreement with Figure 6.42. As stated before, the refractive index is affected by the density and size of voids. The refractive index, in this particular work was found to vary from 1.85 to 1.55, when the void density varied from 10% to 45%.

Tin oxide thin films doped with other elements have also been prepared by sol-gel, using the dip-coating method, namely, fluorine doped (FTO) or antimony doped (ATO). FTO thin films[74]

FIGURE 6.41 TEM image of a 9 nm thick ITO single layer. Pores are located at the boundaries of the ITO nanocrystallites. (From Stoica, T.F. et al., *J. Optoelectron. Adv. Mater.*, 7, 2353, 2005. With permission.)

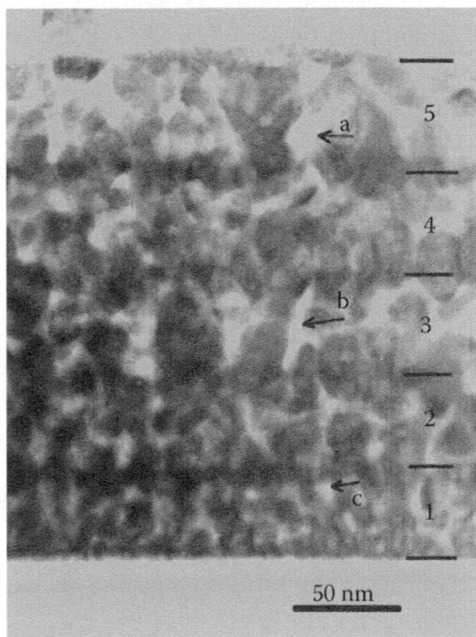

FIGURE 6.42 Cross-section TEM image of a five-layer ITO film: larger voids are identified in the last deposited layer (label a); needle-like voids are identified in the intermediate layer (label b); small voids are visible in the first deposited layer (label c). (From Stoica, T.F. et al., *J. Optoelectron. Adv. Mater.*, 7, 2353, 2005. With permission.)

have been derived from stannous chloride ($SnCl_2.2H_2O$) and hydrogen fluoride (HF) mixed with isopropyl alcohol. A heat treatment of 300°C–325°C, for 30 min, was employed to form the nanocrystalline thin films, with particles of ~6 nm in diameter. Spectral dependence of the absorption coefficient indicated a direct bandgap of ~3.34 eV, calculated from the extrapolation of the linear portion of the graph (Figure 6.43) to the $h\nu$ axis. This optical bandgap value is similar to that of SnO_2 (3.4 eV). Finally, the optimum percentage of fluorine doping into the SnO_2 matrix, which gave higher conductivity (minimum resistance), was found to be 10.8% (F/Sn).

Sb-doped tin oxide films can be applied as transparent conducting and heat reflecting films; however, aggregation and precipitation of nanometer-sized ATO colloidal particles in the sol before deposition limit the transparency of the films in the visible light region. The use of oxalic acid as a dispersion agent of such particles in the synthesis of ATO sol was reported by X. C. Chen,[109] as

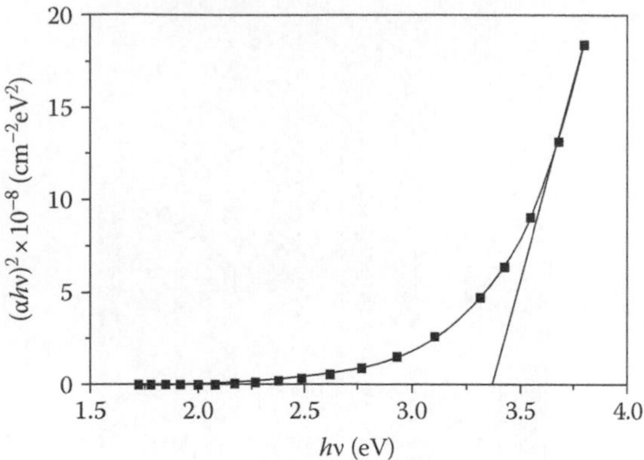

FIGURE 6.43 Direct bandgap transition for FTO thin films on glass. (From Banerjee, A.N. et al., *J. Sol-Gel Sci. Technol.*, 28, 105, 2003. With permission.)

being due to chemisorption of the oxalic acid on the ATO nanoparticle surface with formation and ionization reaction of the surface complex Sn-O-$(CO)_2$-OH. These films presented high transmission in the visible and low transmission in the near IR (Table 6.8).

Withdrawal speed, for thin films deposited by dip-coating, has been found to play a role in the thickness, porosity, and finally refractive index. Varied TiO_2 thin films were prepared at different withdrawal rates, from 3.6 to 22 cm/min.[61] As the annealing temperature increased, the as-deposited amorphous TiO_2 film crystallized to anatase, as expected, and, as the withdrawal rate increased, the thickness increased and the porosity was found to decrease. This fact, associated with the number of crystallites per unit area on the film surface, which was found to be larger, has led to the increase in refractive index.

Alumina (Al_2O_3) thin films, with roughness up to 100 nm, have been prepared from aluminum tri-*sec*-butoxide (Al(O*sec*-Bu)$_3$) and ethylacetoacetate (EAcAc), for low-reflectance purposes. The thin film deposition, by dip-coating, was followed by a heat treatment at 400°C for 30 min, the immersion into hot water at 100°C for 30 min, and again a heat treatment at 400°C.[110] The resulting films, of 250 nm in thickness, were previously flat and smooth and after the hot water treatment became porous, with a "flowerlike" structure, as shown in Figure 6.44, and roughness up to 100 nm. The "flowerlike" structure was reported in this work to be formed by dissolution of the porous Al_2O_3

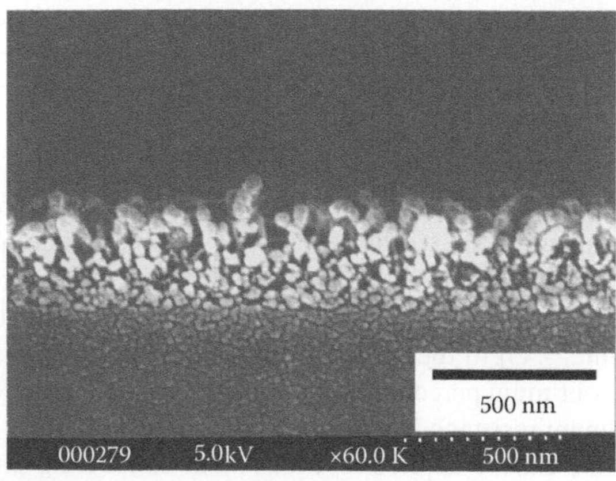

FIGURE 6.44 SEM image of the cross section of porous Al_2O_3 thin film deposited on soda-lime glass substrate after immersion in hot water at 100°C. (From Yamaguchi, N. et al., *J. Sol-Gel Sci. Technol.*, 33, 117, 2005. With permission.)

film and the associated precipitation of pseudo-boehmite. There is a gradient in porosity, and therefore in refractive index, from the surface to the substrate.

The 1-side reflectance from quartz glass substrate coated with this "flowerlike" Al_2O_3 film was found to be less than ~0.5% from 300 to 720 nm.

Nanocrystallization in non-silicon-based materials and its effect on RE PL is described, e.g., in Ref. [70], where nanograin or nanocrystalline Eu^{3+}:$BaTiO_3$ thin films, with 21 nm size crystallites, were formed upon a heat treatment at 750°C. They were found to exhibit a strong inhomogeneous broadening of the PL spectrum due to the distortion of the Eu^{3+} sites. The nanocrystallites have grown less in the thin films than in the powder form samples (21 nm *versus* 32 nm), due to higher space constraints.

ZnO nanostructured thin films are currently being extensively explored. ZnO possesses a wide bandgap of 3.37 eV at room temperature, as well as other properties such as piezoelectricity, chemical stability, biocompatibility, optical transparency in the visible region, high voltage-current nonlinearity, etc. The electric, optoelectronic, and photochemical properties exhibited by undoped ZnO have resulted in the use for solar cells, blue/UV light-emitting devices, transparent electrodes, etc.[80] Moreover, ZnO is attracting considerable attention as a promising candidate for UV LEDs and laser diodes.[77] ZnO exhibits generally two kinds of luminescence: one in the UV region (~380 nm); and the other in the visible, in the green or yellow region (~450–730 nm). The visible emission is related to intrinsic defects and dopants in ZnO crystal and, therefore, depends on the processing conditions.[79] As observed before in other systems, ZnO nanoparticles show surface effects and quantum size effects in the optical properties. For instance, the presence of OH groups or organic groups adsorbed at the particle surface significantly affect the emission wavelength and intensity.

The energy-level structure tends to vary with the size of ZnO particles: in the case of nanoparticles (or nanoclusters), there will be exciton absorption besides the absorption due to the transitions between valence band and conduction band; e.g., if the nanoparticle size is below 10 nm, absorption due to energy splitting will be observed, with evidence of a fine structure in the absorption spectrum,[77] as shown in Figure 6.45, for a ZnO sol made from $Zn(OAc)_2$ and from LiOH-doped $Zn(OAc)_2$ solution. This fine structure was reported to be also observed in the excitation spectra.

Thin films have been prepared by dip-coating using this ZnO sol, and the particle size was found to significantly increase with the annealing treatment (Figure 6.46). The non-annealed film (with ZnO nanoparticles of ~10 nm in size) exhibits blue emission, as can be observed by the emission

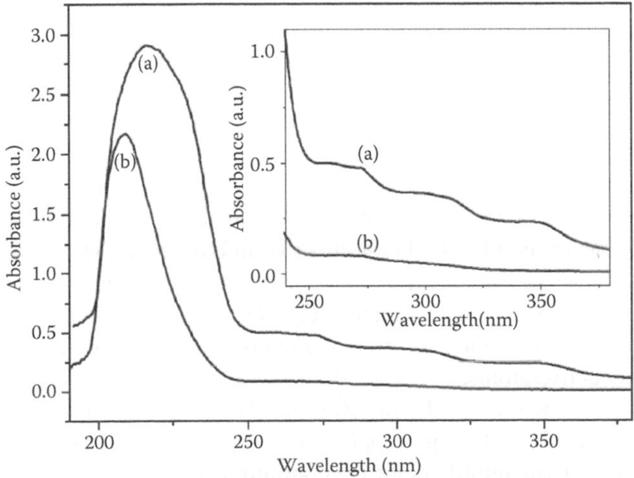

FIGURE 6.45 Absorption spectra of ZnO sol after (a) and before (b) LiOH was added in the $Zn(OAc)_2$ solution. (From Zhang, J. et al., *J. Sol-Gel Sci. Technol.*, 39, 37, 2006. With permission.)

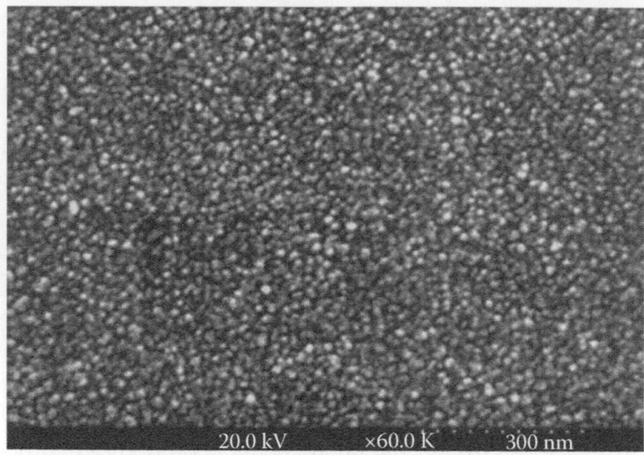

FIGURE 6.46 Thin film heat treated at 200°C. (From Zhang, J. et al., *J. Sol-Gel Sci. Technol.*, 39, 37, 2006. With permission.)

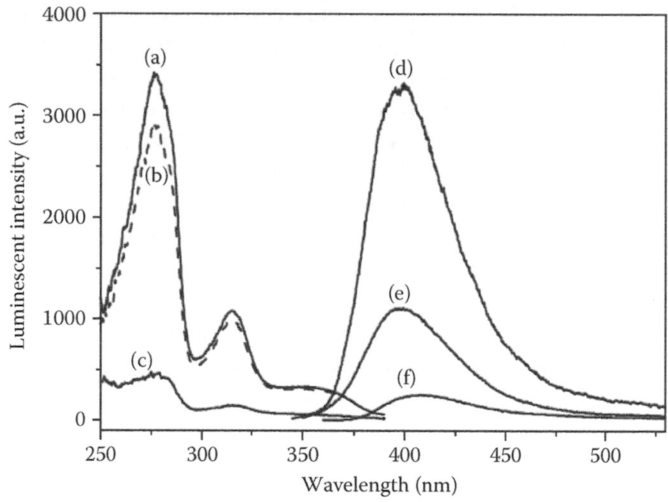

FIGURE 6.47 Excitation and emission spectra of a ZnO thin film. The excitation spectra indicated in (a), (b), and (c) are obtained by monitoring emission at 398, 407, and 469 nm, respectively. If the film is excited at 276, 315, and 360 nm, the emission spectra are shown in (d), (e), and (f), respectively. (From Zhang, J. et al., *J. Sol-Gel Sci. Technol.*, 39, 37, 2006. With permission.)

bands (d, e, and f) peaked at ~400 nm, in Figure 6.47. However, in the annealed film, depicted in Figure 6.46, the emission of the film gradually turns from blue to yellow.

ZnO nanoparticles have been doped and co-doped with other elements, such as transition metal cations (e.g., Fe^{3+}, Co^{2+}, and Ni^{2+}) or RE elements (e.g., Tb^{3+} and Er^{3+} ions). This does not only allow the control of the particle shape, but also offers many possibilities to use ZnO as nanohost of optically and magnetically active atoms.

The emission of Er^{3+} functionalized nano-ZnO has been reported for nanocolloidal solutions, sintered films, and chemically etched micropatterns. This research has been motivated by the need for new, cheap, planar 1.54 μm amplifiers for telecommunication technologies.[76]

Co, Fe, Mn, and V-doped ZnO thin films, prepared from a hydrated zinc acetate precursor and dimethyl formamide (DMF) solvent, were reported by Farley et al.[80] The ZnO crystallites generated

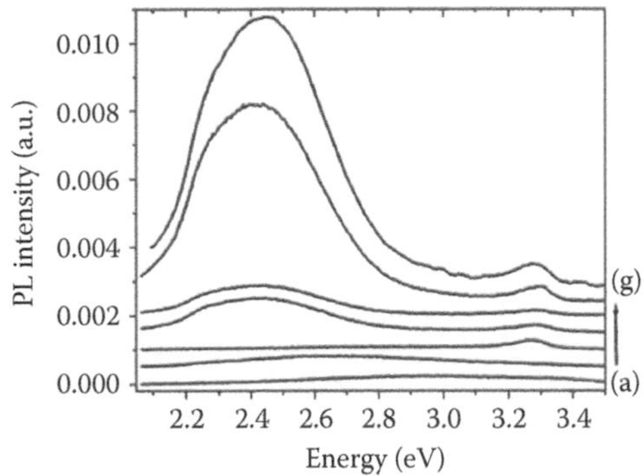

FIGURE 6.48 PL spectra of (a) sapphire substrate, (b) Co^{2+}, (c) V^{3+}, (d) Mn^{3+}, (e) Mn^{2+}, (f) Fe^{3+}, and (g) undoped ZnO films. The excitation wavelength is 325 nm. Scans are arbitrarily offset for clarity. (Reproduced with permission of The Royal Society of Chemistry from Farley, N.R.S. et al., *J. Mater. Chem.*, 14, 1087, 2004.)

in the film were found to be almost purely uniaxial and with a high degree of long-range order. The structure of the ZnO crystals depended on the dopant element: cobalt doping resulted in the formation of columnar ZnO nanocrystals, while vanadium-doped films formed the spinel oxide $ZnAl_2O_4$, resulting from the reaction between the substrate (sapphire) and ZnO.

Figure 6.48 shows the PL spectra of the doped and undoped ZnO films compared to that of the substrate (no PL peaks, as expected). The authors have suggested that, due to the low volume of the Co^{2+}-doped sample, the PL intensity hardly exceeded the background level (spectrum 6.48b). The undoped film exhibited the typical green mid-band (2.4 eV) and UV band-edge (3.3 eV) emissions, as well as the Fe^{3+}, Mn^{2+}, and Mn^{3+} films (spectra 6.48g through d, respectively). The broad deep-level emission band originates from structural defects, which form electron-hole recombination centers. The band-edge luminescence is due to free exciton emission or exciton transitions with shallow donors or acceptors. Mid-band emission was suggested by the authors to be linked with the presence of $ZnAl_2O_4$, which was found to lower the ZnO structural quality.

Cadmium has been incorporated into ZnO in order to narrow the bandgap of ZnO (3.32 eV), since CdO bandgap is much lower (~2.3 eV). It also keeps the crystalline structure and lattice parameter close to ZnO. Maiti et al. have reported on the preparation of 6% Cd-doped ZnO thin films,[78] which exhibited, by x-ray diffraction, the hexagonal wurtzite structure of ZnO and no secondary phases corresponding to oxides of cadmium, or metallic cadmium, suggesting the incorporation of Cd into the ZnO lattice. The films, reported in this work, consisted of densely packed nano/submicron ZnO rods.

As the film is crystalline in nature, the relation between the absorption coefficients (α) and the incident photon energy ($h\nu$) can be written as $(\alpha h\nu)^2 = A(h\nu - E_g)$, where A is a constant and E_g is the bandgap of the material. The $(\alpha h\nu)_2$ vs. $h\nu$ plot is shown in Figure 6.49b. Extrapolating the linear portion of the graph to the $h\nu$ axis, the direct bandgap may be extrapolated from the intercept. With the increase of Cd doping, the bandgap was found to decrease, as shown in Figure 6.49c, and the Urbach parameter increased. Note that the Urbach parameter gives the width of the localized states in the bandgap. All the films were found to exhibit more than 80% transmittance in the visible wavelength region, and a red shift in the UV absorption edge occurred for increasing the Cd content. Due to the large variation of the bandgap with small Cd-doping percentage, these films are candidates for the preparation of, e.g., quantum wells and superlattices.[78]

6.3.3 Non-Oxide Materials

In what concerns nanostructured non-oxide materials prepared by sol-gel, one of the most studied is the nanocrystalline MgF_2 thin films, for protective and AR purposes. They have shown to be very porous,

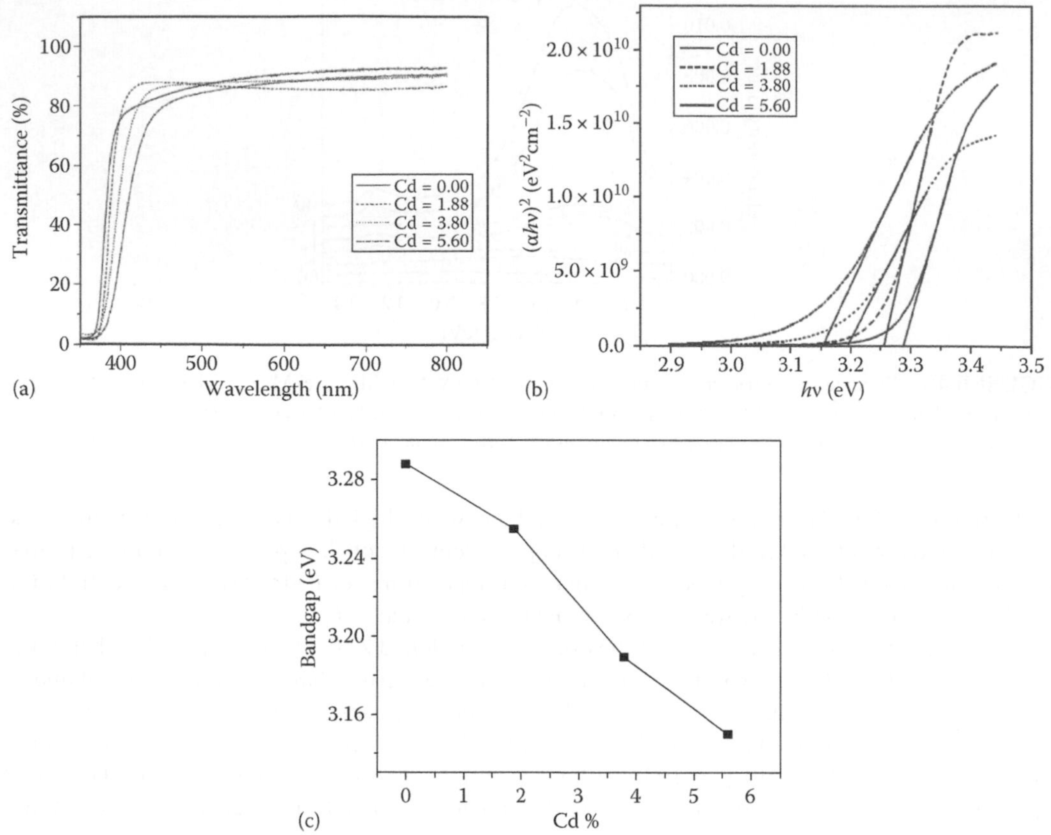

FIGURE 6.49 (a) Transmittance spectra of Cd-doped ZnO thin films for doping levels ranging from 0% to 5.6%; (b) plot to determine the direct bandgap for different doping percentage; and (c) the variation of bandgap energy with cadmium doping. (From Maiti, U.N. et al., *J. Sol-Gel Sci. Technol.*, 41, 87, 2007. With permission.)

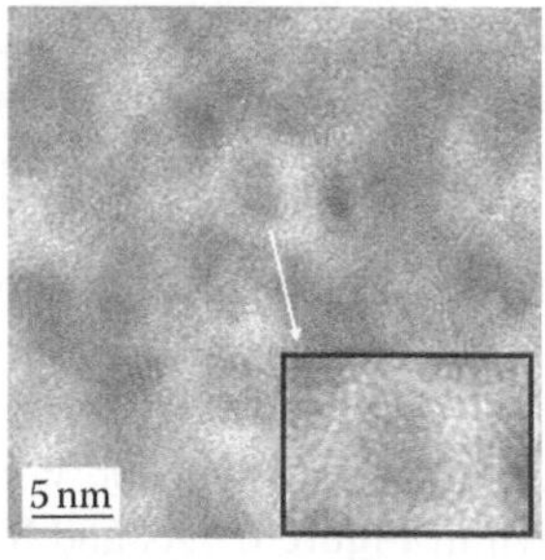

FIGURE 6.50 High-resolution TEM image of a ZnO-MgF_2 film with 40% ZnO, which was heat treated at 300°C for 10 min with a single coating procedure. Dark spots indicate the ZnO nanoparticles. (Reprinted from Fujihara, S. et al., *Thin Solid Films*, 389, 227, Copyright (1997), with permission from Elsevier.)

which could be an advantage for application in sensors. MgF_2 has been extensively used in optical coatings, due to its transparency from 132 nm through the IR, its low refractive index, and poor miscibility with oxides in optical multilayers.[54] MgF_2 nanoparticles have been synthesized from magnesium acetate and hydrofluoric acid in methanol,[53] followed by the preparation of nanocrystalline MgF_2 thin films. The resulting films have high porosity involving pores between the particles and, therefore, low refractive index ($n = 1.20$ at 550 nm). Their mechanical resistance could be improved by adding SiO_2 as a binder, connecting the particles, with a resulting slight increase in refractive index to 1.26 at 550 nm,[19] which corresponds to a lower volume fraction of porosity.

ZnO/MgF_2 nanocomposite thin films consisting of ZnO nanoparticles, of 6–7 nm in size, dispersed in a MgF_2 matrix, have been prepared in order to combine the green PL of ZnO nanoparticles with the transparency of the MgF_2 matrix.[79] The content of ZnO was varied by 30, 40, and 50 mol%; Z30, Z40, and Z50, respectively. Figure 6.50 exhibits the ZnO nanoparticles, as dark spots, slightly larger than 5 nm in diameter, embedded in the MgF_2 matrix.

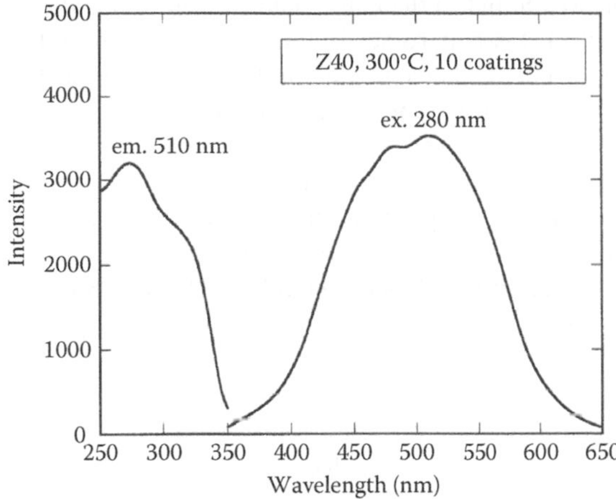

FIGURE 6.51 Excitation and PL spectra of the Z40 film, heat treated at 300°C for 8 min. (Reprinted from Fujihara, S. et al., *Thin Solid Films*, 389, 227, Copyright (1997), with permission from Elsevier.)

The films were synthesized from metal acetates, isopropanol, trifluoroacetic acid, monoethanolamine, and water as starting materials.[79] The ZnO emission intensity was found to be dependent on the heating temperature, 300°C–500°C; and time, 5–14 min; as well the transmittance of the films. By optimizing the heating conditions (300°C and 8 min), a film with high transparency and high PL intensity was obtained. Moreover, the efficient excitation at 280 nm was possible by utilizing the MgF_2 matrix. Figure 6.51 shows the excitation and emission spectra, peaked at 280 and 510 nm, respectively.

6.3.4 Multi-Nanolayer Coatings

There are performances that are enabled by building nanolayer stacks of different compositions, or of different morphology, or volume fraction of porosity, i.e., by playing a role also at the topological level. AR properties, for instance, can be obtained by preparing alternated layers of low refractive index material (e.g., SiO_2) and high refractive index material (e.g., TiO_2). As a matter of example, Chen et al.[111] have reported on an interference-type AR coating prepared on large plastic substrates for display applications, by dip-coating. It consisted of a stack of four alternated layers of SiO_2, of 42.5 nm thickness, with refractive index of 1.428; and TiO_2, of 32.5 nm thickness and refractive index of 1.965, at 632.8 nm. The average photopic (1-side) reflectance level between 400 and 700 nm, which represents the visual response of the human eye to the reflectance of the coating, was below 0.1% each side.[10,111] Another example for AR nanolayer stacks of SiO_2/TiO_2 was given in Ref. [112] with applications in solar cells. These materials were either transparent or colored with Rhodamine. AR coatings, with electromagnetic shielding as functionality, were reported on Ref. [73], for application in cathode ray tubes (CRTs). They consisted of an outer SiO_2 nanolayer formed over a porous inner ITO particle layer. The outer SiO_2 layer was found to penetrate into the inner layer and adhere to the glass substrate surface. The colloidal solution containing ITO particles was diluted with water, ethyl alcohol, isopropyl alcohol, and methyl alcohol to a solid content of 2.0 wt%. The outer layer was prepared with ethanol, TEOS, water, and aqueous nitric acid solution 61 wt% as starting materials. The resulting sol was diluted with the mixed solvent consisting of isopropyl alcohol, methyl alcohol, propylene glycol monoethyl ether, and 4-hydroxy-4-methyl-2-pentanone to a solid content of 0.8 wt%.

Different heat treatments were performed in this study: 600°C and 210°C in air and in N_2. As the curing condition of the coating became more reductive, the surface resistance of the film to electromagnetic radiation was found to decrease. The transmittance of the films was found to be affected by the presence of ITO nanoparticles in the near-IR region, in the sense that it decreased as the curing condition of the coating became more reductive and the surface resistance of the films became small. Although ITO is transparent in the visible region, it is not transparent in the near-IR region due to the plasma oscillation. And, therefore the transmittance of the films in the near-IR region depends on the plasma frequency of ITO particles.[73]

Two other types of AR coatings composed of nanoparticles, also developed for CRTs, have been reported by K. Abe et al.[72]. One of them consisted of an outer layer of SiO_2 and an inner layer of titanium oxynitride (TiO_xN_y), antimony-doped tin oxide (ATO), and SiO_2, with the latter layer being porous. The other coating described, besides AR properties, also displayed electromagnetic wave-shielding properties, and consisted of an outer layer also of SiO_2 and an electric conductive inner layer composed of silver colloids and TiO_xN_y nanoparticles.

Another example of multi-nanolayer thin films relies in the photonic bandgap (PBG) structures, or photonic crystals, for light confinement, which evolved from the original ideas of Yablonovitch[113] and John,[114] less than 20 years ago. This phenomenon is based on the existence of electronic gaps of forbidden energy in semiconductors, due to a periodic electrostatic potential with a period of the order of the interatomic spacings. Artificial structures (rather than materials) with a periodic modulation of their refractive index, or dielectric constant, have been proposed to possess PBGs, or energy gaps, where light within a certain frequency range cannot propagate (irrespective of its direction or polarization state), but where the isolated materials that form the composite structure are otherwise transparent.

PBGs structures or photonic crystals can be obtained by sol-gel processing by preparing multi-nanolayer thin films,[115] in particular stacks of alternating high and low refractive index dielectric layers of optical thickness (the physical thickness, x, times the refractive index, n) equal to $\lambda/4$. If the photonic crystal is designed with a periodicity of the same order of the optical wavelengths, then a certain range of frequencies in the visible or near-IR range, forming a stop band, will be completely reflected (due to Bragg reflection, rather than Fresnel reflection). If the periodicity is broken, for example, when a defect is introduced in the PBG structure, then a range of wavelengths within the stop band, proportional to the index contrast and the defect size, is transmitted through the periodic structure (pass band), at $\lambda = 2nx$. A one-dimensional (1D) PBG structure, having one defect only, constitutes a single Fabry–Perot microcavity.[116] When more than one defect is introduced into the PBG structure, coupled Fabry–Perot microcavities are obtained.[117–119] The electric field of the emitted light beam whose wavelengths match the cavity position is enhanced by a quantity corresponding to the quality factor (Q) of the cavity,[120] due to multiple reflections inside the defect layer. The quality factor of the microcavity is defined by the ratio of the resonance wavelength to the FWHM of the pass band. Defects in microcavities can also be doped with RE elements, like erbium and, when the emitted light is resonant with the cavity, the Er^{3+} PL intensity is expected to be enhanced by its Q factor.[117,121]

Multilayer stacks of silica and titania have been built by spin-coating in order to prepare a PBG structure with a pass band at around 1530 nm, the Er^{3+} PL peak wavelength. Enhancement of the Er^{3+} spontaneous PL emission when inserted into the cavity was achieved (by a factor of up to ~18), compared to the corresponding isolated doped films, which was similar to the Q value obtained for that structure. Angle-dependent cavity resonance (Figure 6.52) and Er^{3+} PL spectrum intensity and shape variations (Figure 6.53) were found for the cavities prepared. Several interesting features were reported when the incidence angle increases (from 20° to 70°)[122]: (1) the pass band position shifts toward shorter wavelengths (blue shift), from 1660 to 1354 nm; (2) the stop band centre also has a blue shift; and (3) the stop band narrows, with its bandwidth decreasing from ~800 to ~520 nm. Moreover, the PL spectra, shown in Figure 6.53b were found to be broader for detection angles far from 45°, in particular for lower incidence angles (that correspond to pass band wavelengths longer than 1530 nm). For example, for detection at 25°, the PL FWHM

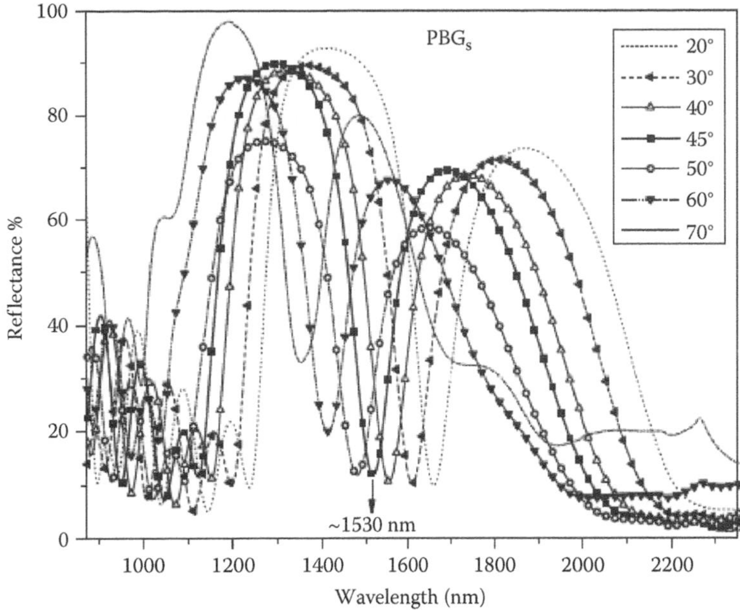

FIGURE 6.52 Angle-dependent reflectance spectra of a simple Fabry–Perot microcavity, PBGs, consisting of two 3-pair DBRs of alternating SiO_2 and TiO_2 layers, plus a sandwiched Er^{3+} -doped SiO_2 defect layer. (Reprinted from Almeida, R.M. et al., *J. Non-Cryst. Sol.*, 353, 490, Copyright (2007), with permission from Elsevier.)

was 37 nm, instead of the 25 nm obtained for 45° detection. It should be noted, therefore, that the broadening of the Er^{3+} emission band for detection angles far from 45° may provide an interesting way to obtain a wider emission band for the Er^{3+} ions, since the typical value for a SiO_2 matrix is very low (ca. 8 nm[92]).

For this particular case, where the aim is to achieve and affect the Er^{3+} PL, peaked at ~1530 nm, the TiO_2 layers may be at the nano-range, but the thickness of the cavity must be ca. 300 nm, taking into account the refractive index of SiO_2 and TiO_2 materials. However, if doping the cavity with other RE ions, such as Eu^{3+}, the thickness of layers would be at the nanoscale, e.g., TiO_2 layers would be around 65 nm in thickness and SiO_2 layers around 100 nm. The same optical and spectroscopic phenomena were achieved for Eu^{3+}-doped cavities,[123] as well, as shown in Figures 6.54 and 6.55: a blue shift occurs for the pass and stop band together with the modification of the luminescence line ($^5D_0$ state to the ^7F_J ($J = 2$, 1) levels of the Eu^{3+} ion) both in intensity and bandwidth by the cavity (12 nm compared to the typical FWHM of about 25 nm observed in sol-gel-derived bulk glasses).

6.3.5 Other Applications

Depending on the application, the optical aspects may be combined with other properties, such as low surface energy (related to hydrophobic performance), for anti-stain coatings with AR properties, or high surface energy, e.g., mirror hydrophilic coatings made of SiO_2, ZrO_2, and colloidal SiO_2.[19]

Other compositions, beyond SiO_2, display interesting features, such as TiO_2 nanostructures[19,43,69,124] due to (1) a high refractive index (of the order of 1.95–2.2), (2) excellent transparency in the visible range, and (3) photocatalytic activity under UV illumination and superhydrophilicity. The latter property is of importance for antifogging and self-cleaning properties of the optical structures.

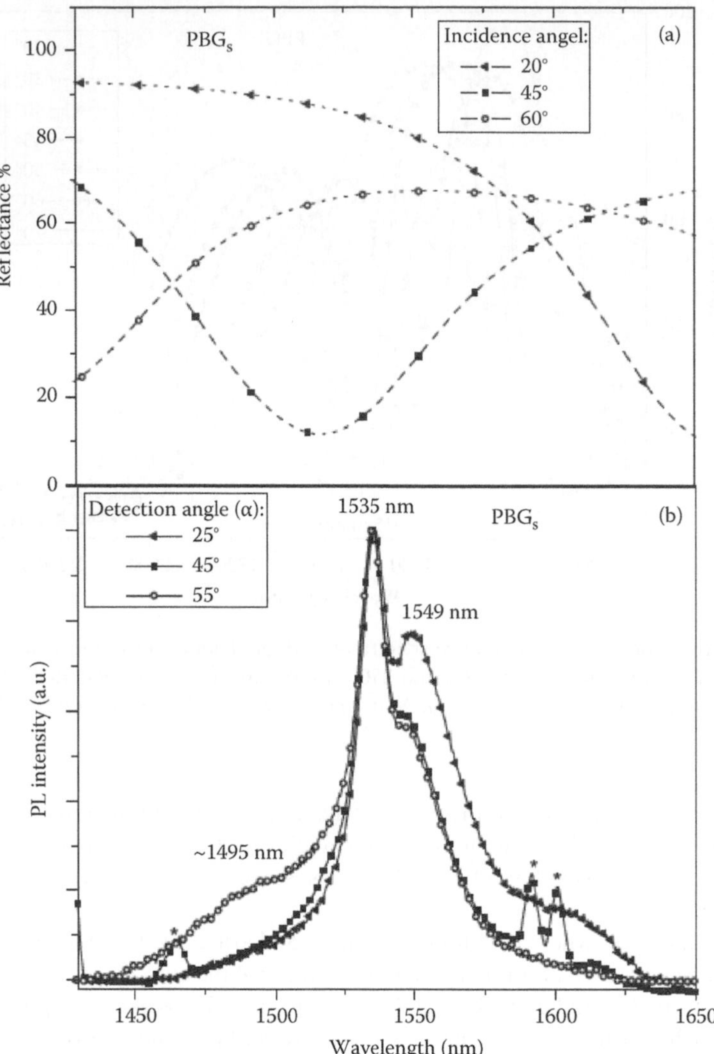

FIGURE 6.53 (a) Angle-dependent reflectance spectra of PBGs in the range of 1430–1650 nm (enlarged detail from Figure 6.52). (b) Normalized Er^{3+} PL spectra of PBGs as a function of the detection angle, for excitation at 514.5 nm, in the range of 1430–1650 nm. Peaks marked with (*) were artifacts not related to the sample PL. (Reprinted from Almeida, R.M. et al., *J. Non-Cryst. Sol.*, 353, 490, Copyright (2007), with permission from Elsevier).

The current global market for titania thin film products is forecasted to increase by at least one order of magnitude, as a result of recent advances in the ability to nanostructure titania films, especially with respect to shifting their photoactivity into the visible region of the spectrum. Their practical application will be hugely impacted by this attribute. New multifunctional properties have arisen, such as photocatalytic oxidation or reduction activity under visible light, photochromism and photomagnetic switching, sensing behavior, photoelectrochemical behavior, photoconductivity, photovoltaic response, and controlled and "switchable" wetting behavior. The exploitation of these properties is of strategic importance in the fields of health and environment protection (air and water purification), sustainable energy production (photovoltaic materials), national and personal security (advanced sensors, devices for deactivation of biological agents), and advanced materials (surface coatings).

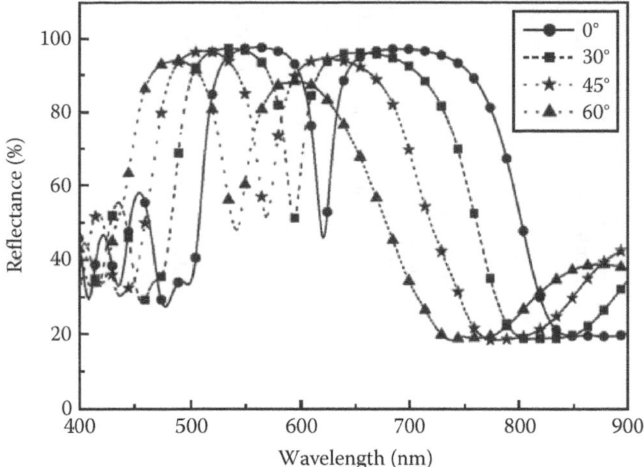

FIGURE 6.54 Angle-dependent reflectance spectra of a sol-gel fabricated Fabry–Perot microcavity, consisting of two mirrors of seven alternated quarter wave layers of TiO_2 and SiO_2, plus a 2 mol% Eu^{3+}-activated TiO_2 half-wave defect layer, with incident light at 0° (dots), 30° (squares), 45° (stars), and 60° (triangles). (Reprinted from Zampedri, L. et al., *Appl. Surf. Sci.*, 248, 3, Copyright (2005), with permission from Elsevier.)

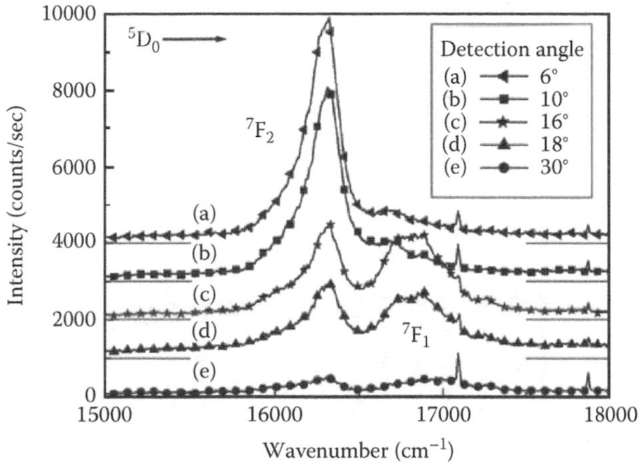

FIGURE 6.55 Angle-dependent luminescence spectra of the same Fabry–Perot microcavity analyzed in Figure 6.54, obtained exciting at 514.5 nm and detecting the luminescence at different angles. (Reprinted from Zampedri, L. et al., *Appl. Surf. Sci.*, 248, 3, Copyright (2005), with permission from Elsevier.)

The sol-gel method can be combined with other methods, for instance, to give rise to hard composite coatings of titanium nitride (TiN) and TiO_2, in particular by reacting a sol-gel-derived TiO_2 film with a nitrogenous admixture under laser irradiation.[125]

Tables 6.7 through 6.9 show some representative examples of different materials, their nanostructurization procedure, and the corresponding applications for silicon-based and non-silicon-based oxides, as well as for non-oxide materials. A few of the examples displayed in these tables have already been referred throughout the text.

TABLE 6.7
Composition, Synthesis Procedure, Structure, Properties, and Applications of Silicon-Based Oxide Nanostructured Thin Films

Thin Films

Composition	Synthesis		Structure	Properties	Applications	Refs.
	Precursors	Nanostructurization Process/Coating Process (Dip, Spin, Spray, etc.)				
SiO_2	TEOS, APTES	Triblock copolymer Pluronic P123 as surfactant for porosity achievement/dip-coating	Highly ordered hexagonal mesoporous silica thin film with pore size of 8.3 nm, specific surface area of 680 m^2/g, and a large pore volume of 1.06 cm^3/g	Aminofunctionalized thin films with improved properties for immobilization of cytochrome c	Receptors for incorporating inorganic ions and biomolecules such as enzyme, DNA, and protein	[68]
	TEOS	Amphiphilic PEO/ spin-coating (two layers)	Nano-sized SiO_2 containing coatings	High laser damage threshold	AR coatings	[126]
	TEOS	Ferrocenyl surfactant for porosity achievement/ spin-coating	Oriented silica mesostructured film with lamellar structure with (100) d-space of 36.3 Å	—	Optic and electronic devices	[127]
	TEOS	A two-step method: sol-gel with hydrolysis of TEOS and solvent-thermal process	Mesoporous structure composed by Si nanocrystals and amorphous $SiO_{1.4}$; mesh diameter ~15 nm	Stable and intensive blue emission	Integrated optical devices	[128]
	TEOS, MTES, perfluoroalkyl-silane	Underlayer made of silica and upperlayer of fluorocarbon containing silica (topological effect)	Concave–convex surface morphology with AFM roughness (Rms) of 6.3. nm	Good water-repellence with a contact angle of 113° and excellent durability	Front door and back automotive windows	[19]

Encapsulated pigments with SiO_2 shell (thin film)	TEOS	Particles stabilized with amphiphilic polyvinylpyrrolidone, poly(diallyldimethyl ammonium chloride) and poly(sodium 4-styrenesulfonate); then transferred into ethanol and ammonia for encapsulation with silica	SiO_2 nanoparticles on the surface of the coated organic pigment; nanopores of ca. 50 nm in diameter	Scattering of UV radiation (<270 nm)	Silica coating of organic pigments for improved durability, better thermal stability, and decrease of diffuse reflection (weather durability)	[129]
Noble-metal colloids-doped SiO_2-PbO	TEOS, GPTES, $Pb(Ac)_2$, $AgNO_3$, $H[Au(Cl)_4]$, $Pd(Ac)_2$	Thermal reduction (200°C–600°C) of Au^{3+}, Ag^{+}, and Pd^{2+} ions to the corresponding metallic nanoparticles embedded in 87% SiO_2–13% PbO/ spin-coating	Au^{o}, Ag^{o}, and Pd^{o} nanoparticles embedded in an amorphous SiO_2-PbO matrix	Transparent colored coatings with chemical, mechanical, thermal, and UV stability; pink, brown, green, and gray colors with optical density below 6	Eye glass lenses	[130]
SiO_2-NiO	—	Nickel oxide nanocrystals precipitated at 500°C while the film was still porous; bifunctional ligand, bearing amine groups for anchoring the metal complex moiety to the silicate matrix	Nickel oxide nanocrystals embedded in a silica matrix	Reversible change in the optical transmittance in the visible and near-IR range when exposed to carbon monoxide gas	Sensors: optical carbon monoxide gas sensitivity	[131]
Colloidal SiO_2-doped SiO_2-ZrO_2	TEOS and $ZrOCl_2$–$8H_2O$	Incorporation of colloidal SiO_2 into the sol/ roll-coating or spray-coating	Convex surface morphology	Colloidal SiO_2 renders retention of hydrophilicity, due to surface-SiOH groups and convex morphology	Hydrophilic mirror	[19]

(continued)

TABLE 6.7 (continued)
Composition, Synthesis Procedure, Structure, Properties, and Applications of Silicon-Based Oxide Nanostructured Thin Films

Composition	Synthesis		Structure	Properties	Applications	Refs.
	Precursors	Nanostructurization Process/Coating Process (Dip, Spin, Spray, etc.)				
Thin Films						
MgF_2–SiO_2	Magnesium acetate and hydrofluoric acid in methanol, TEOS	MgF_2 nanocrystals precipitated at 120°C–180°C/spin-coating	Nanograin, nanocrystalline MgF_2 thin film, with SiO_2 as a binder of the particles	Low refractive index ($n < 1.30$)	AR coatings for optical lenses such as cameras and optical microscopes	[53]
Er^{3+}-doped SiO_2-TiO_2-HfO_2	TEOS, Ti-isopropoxide, $HfOCl_2$, $Er(NO)_3$	$HfTiO_2$ nanocrystals precipitated at 900°C/spin-coating	Er^{3+} ions embedded in a silica matrix containing $HfTiO_2$ nanocrystals	Strong PL peak at 1.53 μm	Optical communication and information processing: planar waveguides, optical amplifiers	[91,102]
SnO_2:SiO_2	—	Phase separation between SnO_2-rich phase and SiO_2-rich phase/spin-coating	Nanostructured SnO_2-SiO_2 glass ceramic thin film	Optically transparent; effective energy transfer between SnO_2 nanoclusters and RE ions. Promising competitor for Si nanodots in silica	Photonics and UV-emitting devices	[132]
SiO_2-TiO_2	TEOS and titanium tetra *n*-butoxide.	Treatment of SiO_2-TiO_2 gel film with hot water at 90°C under vibration (longitudinal vibrations at ~6 Hz)/spin-coating	Titania nanosheet-precipitated coatings (each nanosheet consisting of several layers with a spacing of about 0.6 nm)	Transparent in the visible range, high photocatalytic activity, and superhydrophilicity	Photocatalysis and antifogging articles	[88]
SiO_2, TiO_xN_y, ATO, Ag	TEOS, TiO_xN_y, ATO, $Ag(NO)_3$	Incorporation of colloidal SiO_2, ATO, and TiO_xN_y nanoparticles/spin-coating	Outer layer of Ag:SiO_2 and inner layer composed of colloidal silver, TiO_xN_y, and ATO particles	Haze = 0.3%, surface resistance = 5×10^2 Ω/sq, and reflectance = 0.2%–1%	AR electromagnetic shielding coating for cathode ray tubes	[72,73]

TABLE 6.8
Composition, Synthesis Procedure, Structure, Properties, and Applications of Non-Silicon-Based Oxide Nanostructured Thin Films

Thin Films						
Composition	Synthesis		Structure	Properties	Applications	Refs.
	Precursors	Nanostructurization Process/Coating Process (Dip, Spin, Spray, etc.)				
Sb-doped tin oxide (ATO)	$SnCl_4 \cdot 5H_2O$ and $SbCl_3$; oxalic acid as surface-modifying (dispersing) agent	Annealing of the deposited film at temperatures up to 400°C in a reducing atmosphere containing the carry gas N_2 and vapor of alcohol that is used as reducing agent/dip-coating	ATO nanocrystalline particles, between 3 and 5 nm in diameter	Transmission >85% in the visible region and <25% in the near-IR region	Transparent, electrically conductive, and heat-reflecting thin films	[109]
Fluorine-doped tin oxide (FTO)	Stannous chloride ($SnCl_2 \cdot 2H_2O$) and hydrogen fluoride (HF)	Annealing at 300°C–325°C in air for half an hour/dip-coating	Fine grains in the film, of 6 nm in size	*n*-type wide bandgap transparent material: high transparency in the visible region; bandgap = 3.34 eV; room temperature resistivity ~1 Ω cm	Electronic devices, such as window layer of solar cells, electroluminescent devices, and flat panel displays	[74]
TiO_2	—	Excimer (KrF) pulsed laser irradiation of amorphous TiO_2 and anatase thin films/spin-coating	Formation of rutile and brookite TiO_2 accompanied by evolution of oxygen-deficient Ti_nO_{2n-1} phases: increase of surface roughness and pore formation	Optical absorption edge shifts to IR wavelength range; increase of photocatalytic efficiency	Photocatalysis	[45]
	$Ti(OPr_i)_4$ or $Ti(OBu_n)_4$ and acetylacetone as complexing agent	Drying at 105°C, and annealing at 300°C for 10 min/spray-coating	Spheres with the average size ~30 nm	Good hydrophilic property with a contact angle of water θ ~ 0°	Self-cleaning glasses	[19]

(*continued*)

TABLE 6.8 (continued)
Composition, Synthesis Procedure, Structure, Properties, and Applications of Non-Silicon-Based Oxide Nanostructured Thin Films

Thin Films

Composition	Synthesis		Structure	Properties	Applications	Refs.
	Precursors	Nanostructurization Process/Coating Process (Dip, Spin, Spray, etc.)				
	Titanium isopropoxide	Brij 56, Triton X-100, or Pluronic F-127 as structure-directing agents; annealing at 500°C/ dip-coating	Nanoporous TiO_2 thin films with anatase nanocrystals up to 10 nm in size	Increase of photocatalytic efficiency	Photocatalysis	[124]
TiO_2-Ga_2O_3, TiO_2-Ta_2O_5	Titanium isopropoxide and gallium (III) nitrate hydrate (or tantalum ethoxide)	Hydroxypropyl cellulose as a polymeric fugitive agent, and annealing	Nanostructured and nanoporous TiO_2-Ga_2O_3 films, rhombohedral α-Ga_2O_3, and monoclinic β-Ga_2O_3 phases, as well as anatase and rutile. Nanostructured and nanoporous TiO_2-Ta_2O_5 films: columnar-like morphology with nanosized grains	The smallest crystallite size, the smallest grain size, the highest specific surface area, and the highest roughness for Ti:Ga (or Ti:Ta) = 50:50 (at%/at%) binary oxide annealed at 500°C	From optics, electronics to gas sensors	[133,134]
ZnO	—	Annealing at 150°C and 400°C	ZnO grain thin films	Energy gap of the films annealed at 150°C is 3.33 eV; SPR investigations show reversible responses to different concentrations of methanol, ethanol, isopropanol, and hexane vapors in dry air	Optical gas sensing (by SPR)	[135]

	$(Zn(CH_3OO)_2 \cdot 2H_2O)$	Drying at 80°C/dip-coating	ZnO particles of ca. 10 nm in diameter	Bright blue-emitting coating; single wide blue emission band under excitation of UV light; quantum yield = 0.52 for excitation at 348 nm	Vacuum-fluorescent displays, field-emission displays, and electroluminescent displays	[77]
Cd^{2+}:ZnO	Zinc acetate dehydrate, monoethanolamine, dip-coating	Annealing at 673 K for 1 h in open-air furnace/ dip-coating	Densely packed randomly oriented nano/submicron ZnO:Cd rods, hexagonal wurtzite structure	An increase of Cd doping leads to the bandgap decrease and the Urbach parameter increase	Humidity sensors or quantum wells, superlattices, and other configurations that involve bandgap engineering	[78]
Al_2O_3	—	Treatment in boiling water; treatment with a fluorine-containing silane/ spray-coating	Boehmite (AlOOH) structure; coatings with smooth surfaces become rougher with 20–50 nm roughness after the hot water treatment	Superhydrophobic and transparent coatings	Water resistance on ships, submarines, etc.; antisticking effect for snow and ice	[19]
Cadmium tungstate ($CdWO_4$)	Tungsten oxychloride or tungstic acid	Fast sintering at 500°C for 20 min or slow sintering at 500°C for 1 h with a heating ramp of 8°C/min	Dense nanocrystalline films for fast sintering and porous films for slow sintering; nanocrystalline grains in $CdWO_4$ films of 15–52 nm in diameter	Higher transmittance and PL output for $CdWO_4$ dense films	Scintillation films for optics	[136]
Eu^{3+} : $BaTiO_3$	Barium acetate ($Ba(Ac)_2$), titanium butoxide, ($Ti(C_4H_9O)_4$) and $Eu(NO_3)_3$	Crystallization to tetragonal phase $BaTiO_3$ doped with Eu^{3+}/spin-coating	Tetragonal phase $BaTiO_3$ doped with Eu^{3+}	PL spectra of nanostructure Eu^{3+} : $BaTiO_3$ is dominated by $^5D_0 \rightarrow {}^7F_2$ transition, suggesting a strong distortion of the europium sites	IO, PL	[70]

TABLE 6.9
Composition, Synthesis Procedure, Structure, Properties, and Applications of Non-Oxide Nanostructured Thin Films

Thin Films						
Composition	Synthesis		Structure	Properties	Applications	Ref.
	Precursors	Nanostructurization Process/Coating Process (Dip, Spin, Spray, etc.)				
MgF_2	Magnesium acetate tetrahydrate and hydrofluoric acid	Autoclaved sol/spin-coating	MgF_2 thin films consisting of fine particles of 10–20 nm in size	Ultralow refractive indices, even in the vacuum UV region (200–150 nm)	AR films	[71]
	Magnesium methoxide ($Mg(OCH_3)_2$ and methanolic solution of HF (in a N_2 glovebox)	Annealing/dip-coating	Homogeneous but textured coating with particles of 5–7 nm in size	High surface area, high porosity, and low refractive index	AR and protective coatings, sensors	[54]
Potassium lithium niobate	—	Annealing/spin-coating	Nanostructured low surface roughness	Refractive indices close to those of their single crystal; highly transparent in the visible–near-IR spectral range; absorption edge shifted toward the violet spectral side	Waveguides	[137]

6.4 SUMMARY

As shown in the present chapter, the use of sol-gel technology enables the achievement of nanostructured thin films with specific optical properties and functionalities, allowing novel or optimized applications. The films are made of silica, other oxides, or even non-oxide materials, and have been prepared either by bottom-up or top-down approaches. Optical properties such as reflection, transmission, refractive index, absorption coefficient, optical bandgap, and Urbach parameter are discussed. Among the examples described in this chapter there are the optical sensors, low-reflection thin films, optical amplification waveguides, and transparent and colored coatings, which are presented as being enabled either by the effect of nanopores, guest loading improvement, nanocrystals, SPR, nanoparticles, or topology.

The exponential increase in the rate of publications and patent generation is an evidence of the growing importance of sol-gel processing for nanostructured thin films. This technology, being a relatively cheap and versatile approach, has evolved in the preparation effort of such nanostructures for application in optics, photonics, and more recently plasmonics.

ACKNOWLEDGMENTS

The authors would like to acknowledge all the contributors to the work reported in this chapter. Ana C. Marques is also indebted to Prof. R.M. Almeida and Dr. M. Ferrari (and to the corresponding research group) for part of the work presented in this chapter that was carried out in their lab.

ABBREVIATIONS

2PP	Two-photon absorption
Al(O*sec*-Bu)$_3$	Aluminum tri-*sec*-butoxide
APTES	3-Aminopropyltriethoxysilane
AR	Antireflective
ATMS	(3-Acryloxypropyl)trimethoxysilane
ATO	Antimony-doped tin oxide
CE	3,4-Epoxycyclohexylmethyl-3′,4′-epoxycyclohexanecarboxylate
CRTs	Cathode ray tubes
CTAB	Cetyltrimethylammonium bromide
CTE	Coefficient of thermal expansion
DMF	Dimethyl formamide
DBRs	Distributed Bragg reflectors
DMTA	Dynamic mechanical thermal analyses
EAcAc	Ethylacetoacetate
EBL	Electron beam lithography
EDMA	Ethylene glycol dimethacrylate
EISA	Evaporation induced self-assembly
EMA	Effective medium approximation
FBPA	4,4′-(Hexafluoroisopropylidene)diphenol
GPTES	Glycidoxypropyltriethoxysilane
GPTMS	Glycidoxypropyltrimethoxysilane
GPTS	3-Glycidoxypropyltrimethoxysilane
IO	Integrated optics
IR	Infrared
ITO	Indium tin oxide

LEDs	Light-emitting diodes
MeEtOH	2-Methoxyethanol
MTES	Methyltriethoxysilane
MTMS	Methyltrimethoxysilane
NIL	Nanoimprint lithography
OPV	Oligo(p-phenylene vinylene)
OTES	Octyltriethoxisilane
PB-PEO	Poly(butadiene)-b-poly(ethylene oxide)
PBG	Photonic bandgap
PDMS	Polydimethylsiloxane
PEG	Poly(ethylene glycol)
PEO	Poly(ethylene oxide)
PFO	(Poly(9,9′-dioctylfluorene)
PL	Photoluminescence
PPO	Poly(propylene oxide)
PS-PEO	Poly(styrene)-b-poly(ethylene oxide)
RE	Rare-earth
SEM	Scanning electron microscopy
SPL	Scanning probe lithography
SPR	Surface plasmon resonance
STM	Scanning tunneling microscopy
TCO	Transparent conductive oxides
TEM	Transmission electron microscopy
TEOG	Germanium tetraethoxide
TEOS	Tetraethoxysilane
$Ti(OBu)_4$	Tetrabutylorthotitanate
TiO_xN_y	Titanium oxynitride
TIP	Titanium-tetraisopropoxide
TMB	Trimethylbenzene
TMOS	Tetramethoxysilane
UV	Ultraviolet
XRD	X-ray diffraction
μCP	Microcontact printing

SYMBOLS

E	Electric field
E_g	Bandgap
h	Planck's constant
I_f	Final PL intensity
I_o	Initial PL intensity
n	Refractive index
Q	Quality factor of a microcavity
R	Water:SiO_2 ratio
S	Surface area
T	Temperature
α	Absorption coefficients
λ	Wavelength
ν	Frequency

REFERENCES

1. Brinker, C.J., Scherer, G.W. 1990. *Sol-Gel Science: The Physics and Chemistry of Sol-Gel Processing*. San Diego, CA: Academic Press.
2. Scholze, H. 1991. *Glass, Nature, Structure and Properties*. New York: Springer-Verlag.
3. Rahaman, M.N. 2003. *Ceramic Processing and Sintering*. New York: Marcel Dekker, Inc.
4. Pierre, A.C. 1998. *Introduction to Sol-Gel Processing*. Boston, MA: Academic Publishers.
5. Chern, G.C., Lauks, I. 1982. Spin-coated amorphous-chalcogenide films. *J. Appl. Phys*. 53: 6979.
6. Acosta, S., Corriu, R., Leclercq, D., Mutin, P.H., Vioux, A. 1994. Novel non-hydrolytic sol-gel route to metal oxides. *J. Sol-Gel Sci. Technol*. 2: 25.
7. Niederberger, M. 2007. Nonaqueous sol–gel routes to metal oxide nanoparticles. *Acc. Chem. Res*. 40: 793.
8. McDonagh, C., Sheridan, F., Butler, T., MacCraith, B.D. 1996. Characterization of sol-gel derived silica films. *J. Non-Cryst. Solids* 194: 72.
9. Prassas, M. 2001. Silica glass from aerogels. www.solgel.com/articles/april01/gelglassl.htm
10. Chen, D. 2001. Anti-reflection (AR) coatings made by sol-gel process: A review. *Sol. Energy Mater. Sol. Cells* 68: 313.
11. Haas, K.H. 2003. Crosslinked heteropolysiloxanes as inorganic-organic polymers: Precursors, synthesis, properties and applications. In *Handbook of Organic-Inorganic Hybrid Materials and Nanocomposites; Vol. 1: Hybrid Materials*, Ed. H.S. Nalwa, p. 207. Boca Raton, FL: CRC Press.
12. Sangermano, M., Malucelli, G., Amerio, E., Bongiovanni, R., Priola, A., Di Gianni, A., Voit, B., Rizza, G. 2006. Preparation and characterization of nanostructured TiO_2/epoxy polymeric films. *Macromol. Mater. Eng*. 291: 517.
13. Amerio, E., Sangermano, M., Malucelli, G., Priola, A., Voit, B. 2005. Preparation and characterization of hybrid nanocomposite coatings by photopolymerization and sol-gel process. *Polymer* 46: 1141.
14. Kim, J., Ko, J., Bae, B.-S. 2007. Dispersion of silica nano-particles in sol-gel hybrid resins for fabrication of multi-scale hybrid nanocomposite. *J. Sol-Gel Sci. Technol*. 41: 249.
15. Sangermano, M., Malucelli, G., Amerio, E., Priola, A., Billi, E., Rizza, G. 2005. Photopolymerization of epoxy coatings containing silica nanoparticles. *Prog. Org. Coat*. 54: 134.
16. Ebelman, M. 1846. Untersuchungen über die Verbindung der Borsäure und Kieselsäure mit Aether. *Ann. Chim. Phys*. 57: 319.
17. Geffcken, W., Berger, E. 1939. Verfahren zur änderung Reflexionsuermögens Optischer Gläser. Germany. Patent 736411.
18. Schmidt, H. 2006. Considerations about the sol-gel process: From the classical sol-gel route to advanced chemical nanotechnologies. *J. Sol-Gel Sci. Technol*. 40: 115.
19. Aegerter, M.A., Almeida, R., Soutar, A., Tadanaga, K., Yang, H., Watanabe, T. 2008. Coatings made by sol–gel and chemical nanotechnology. *J. Sol-Gel Sci. Technol*. 47: 203.
20. Sakka, S. 1995. Editorial: The task of sol-gel science and technology. *J. Sol-Gel Sci. Technol*. 4: 5.
21. "Market Assessment on Nanocoatings" in Coatings World website; 2008. http://www.coatingsworld.com/news/2008/12/11/research_and_markets_has_published_22market_assessment_on_nanocoatings%22.
22. Davis, P.J., Brinker, C.J., Smith, D.M. 1992. Pore structure evolution in silica gel during aging/drying I. Temporal and thermal aging. *J. Non-Cryst. Solids* 147: 189.
23. Deshpande, R., Hua, D., Smith, D.M., Brinker, C.J. 1992. Pore structure evolution in silica gel during aging/drying. III. Effects of surface tension. *J. Non-Cryst. Solids* 144: 32.
24. Yan, Y., Hoshino, Y., Duan, Z., Chaudhuri, S.R., Sarkar, A. 1997. Design and characterization of interconnected microporous hybrid thin films by a sol-gel process. *Chem. Mater*. 9: 2583.
25. Xie, H., Wei, J., Zhang, X. 2006. Characterization of sol-gel thin films by spectroscopic ellipsometry. *J. Phys.: Conf. Ser*. 28: 95.
26. Kistler, S.S. 1931. Coherent expanded aerogels and jellies. *Nature* 127: 741.
27. Nakanishi, K. 1997. Pore structure control of silica gels based on phase separation. *J. Porous Mater*. 4: 67.
28. Gnade, B.E., Cho, C.C., Smith, D.M. 1996. Porous composites as a low dielectric constant material for electronics applications. US Patent 5,561,318.
29. Prakash, S.S., Brinker, C.J., Hurd, A.J. 1995. Silica aerogel films at ambient pressure. *J. Non-Cryst. Solids* 190: 264.
30. Kajihara, K., Yao, T. 1998. Macroporous morphology of the titania films prepared by a sol-gel dip-coating method from the system containing poly(Ethylene Glycol). I. Effect of humidity. *J. Sol-Gel Sci. Technol*. 12: 185.

31. Stucky, G., Chmelka, B., Zhao, D., Melosh, N., Huo, Q., Feng, J., Yang, P., Pine, D., Margolese, D., Lukens, W. Jr., Fredrickson, G., Schmidt-Winkel, P. 2003. Block copolymer processing for mesostructured inorganic oxide materials. U.S. Patent 09/554259.
32. Kuemmel, M., Boissiere, C., Nicole, L., Laberty-Robert, C., Sanchez, C., Grosso, D. 2008. Highly ordered metal oxide nanopatterns prepared by template-assisted chemical solution deposition. *J. Sol-Gel Sci. Technol.* 48: 102.
33. Lu, P., Liu, J., Sayari, A. 1997. Preparation of porous hafnium oxide in the presence of a cationic surfactant. *Chem. Commun.* 6: 577.
34. Yusuf, M.M., Imai, H., Hirashima, H. 2002. Preparation of porous titania film by modified sol-gel method and its application to photocatalyst. *J. Sol-Gel Sci. Technol.* 25: 65.
35. Waki, K. 2007. Regularly arrayed nanostructured material. U.S. Patent WO2006049320/EP20050805944.
36. Liu, G.Q., Jin, Z.G., Liu, X.X., Wang, T., Liu, Z.F. 2007. Anatase TiO_2 porous thin films prepared by sol-gel method using CTAB surfactant. *J. Sol-Gel Sci. Technol.* 41: 49.
37. Wu, S.-Y., Hsueh, H.-S., Huang, M.H. 2007. Spin-coated periodic mesoporous organosilica thin films with molecular-scale order within the organosilica wall. *Chem. Mater.* 19: 5986.
38. Kirmayer, S., Dovgolevsky, E., Kalina, M., Lakin, E., Cadars, S., Epping, J.D., Fernández-Arteaga, A., Rodríguez-Abreu, C., Chmelka, B.F., Frey, G.L. 2008. Syntheses of mesostructured silica films containing conjugated polymers from tetrahydrofuran-water solutions. *Chem. Mater.* 20: 3745.
39. Yao, L., Zhu, Y., Qu, D., Du, M., Shen, J., Wang, J. 2006. Optical and hydrophobic nano-porous silica thin films with low refractive index. *Proc. SPIE, ICO20 (Optical Design and Fabrication)* 6034: 6034V.
40. Makita, K., Akamatsu, Y., Takamatsu, A., Yamazaki, S., Abe, Y. 1999. Sol-gel preparation of silica films with controlled surface morphology and their application to a low reflective glass. *J. Sol-Gel Sci. Technol.* 14: 175.
41. Marques, A.C., Almeida, R.M. 2006. Raman spectra and structure of multicomponent oxide planar waveguides prepared by sol-gel. *J. Sol-Gel Sci. Technol.* 40: 371.
42. Mechiakh, R., Meriche, F., Kremer, R., Bensaha, R., Boudine, B., Boudrioua, A. 2007. TiO_2 thin films prepared by sol–gel method for waveguiding applications: Correlation between the structural and optical properties. *Opt. Mater.* 30: 279.
43. Starbova, K., Yordanova, V., Nihtianova, D., Werner, H., Jürgen, T., Starbov, N. 2008. Excimer laser processing as a tool for photocatalytic design of sol-gel TiO_2 thin films. *Appl. Surf. Sci.* 254: 4044.
44. Marques, A.C., Santos, L.F., Almeida, R.M. 2007. Physical characterization of sol-gel derived titania thin films. Poster presented at *14th International Workshop on Sol-Gel Science and Technology*, Montpellier, France.
45. Trbojevich, R., Pellegri, N., Frattini, A., de Sanctis, O., Morais, P.J., Almeida, R.M. 2002. Preparation and isolation of gold nanoparticles coated with a stabilizer and sol-gel compatible agent. *J. Mater. Res.* 17: 1973.
46. Almeida, R.M., Marques, A.C., Ferrari, M. 2003. Optical nanocomposite planar waveguides doped with rare-earth and noble metal elements. *J. Sol-Gel Sci. Technol.* 26: 891.
47. Almeida, R.M., Morais. P.J., Marques, A.C. 2002. Planar waveguides for integrated optics prepared by sol-gel methods. *Philos. Mag. B.* 82: 707.
48. Tanahashi, I., Yoshida, M., Manabe, Y., Tohda, T. 1995. Effects of heat treatment on Ag particle growth and optical properties in Ag/SiO2 glass composite thin films. *J. Mater. Res.* 10: 362.
49. Kosuka, H., Zhao, G., Sakka, S. 1994. Control of optical properties of gel-derived oxide coating films containing fine metal particles. *J. Sol-Gel Sci. Technol.* 2: 741.
50. Rodríguez-Sánchez, L., Blanco, M.C., López-Quintela, M.A. 2000. Electrochemical synthesis of silver nanoparticles. *J. Phys. Chem. B.* 104: 9683.
51. Mennig, M., Spanhel, J., Schmidt, H.K., Betzholz, S. 1992. Photoinduced formation of silver colloids in a borosilicate sol-gel system. *J. Non-Cryst. Solids* 147&148: 362.
52. Valverde-Aguilar, G., Rentería, V., García-Macedo, J.A. 2007. Modeling of core-shell silver nanoparticles in nanostructured sol-gel thin films. *Proc. SPIE (Plasmonics: Metallic Nanostructures and Their Optical Properties)* 6641: 66411T.
53. Ishizawa, H., Niisaka, S., Murata, T., Tanaka, A. 2008. Preparation of MgF_2-SiO_2 thin films with a low refractive index by a solgel process. *Appl. Opt.* 47: C200.
54. Sermon, P.A., Badheka, R. 2004. MgF_2 xerogels. *J. Sol-Gel Sci. Technol.* 32: 149.

55. Chen, Q., Tan, J.G.H., Shen, S.C., Liu, Y.C., Ng, W.K., Zeng, X.T. 2007. Effect of boehmite nanorods on the properties of glycidoxypropyltrimethoxysilane (GPTS) hybrid coatings. *J. Sol-Gel Sci. Technol.* 44: 125.
56. Gutmann, E., Levin, A., Meyer, D., Paufler, P. 2006. Method for producing a component comprising nanometric multilayers for optical uses. WO/2006/074642.
57. Plueddemann, E.P. 1991. *Silane Coupling Agents*. New York/London, U.K.: Plenum Press.
58. Bornside, D.E., Macosko, C.W., Scriven, L.E. 1987. On the modeling of spin coating. *J. Imaging Technol.* 13: 122.
59. Scriven, L.E. 1988. Physics and applications of dip coating and spin coating. *Mater. Res. Soc. Symp. Proc.* 121: 717.
60. Brinker, C.J., Hurd, A.J., Schunk, P.R., Frye, G.C., Ashley, C.S. 1992. Review of sol-gel thin film formation. *J. Non-Cryst. Solids* 147&148: 424.
61. Attia, S.M., Wang, J., Wu, G., Shen, J., Ma, J. 2002. Review on sol-gel derived coatings: Process, techniques and optical applications. *J. Mater. Sci. Technol.* 18: 211.
62. Sotiropoulou, S., Sierra-Sastre, Y., Mark, S.S., Batt, C.A. 2008. Biotemplated nanostructured materials. *Chem. Mater.* 20: 821.
63. Hajj-Hassan, M., Gonzalez, T., Djeghelian, H., Ghafar-Zadeh, E., Therriault, D., Chodavarapu, V., Andrews, M. 2008. Direct-write patterning of nanostructured sensory waveguides for integrated optical bioimaging applications. *Proc. SPIE (Nanoscale Imaging, Sensing, and Actuation for Biomedical Applications V)* 6865: 686504.
64. Hou, H., Chen, Y. 2007. Preparation of super-hydrophobic silica films with visible light transmission using phase separation. *J Sol-Gel Sci Technol.* 43: 53.
65. Serbin, J., Chichkov, B., Houbertz, R. 2003. Three-dimensional nanostructuring of hybrid materials by two-photon polymerization. *Proc. SPIE* 5222: 171.
66. Parashar, V.K., Sayah, A., Pfeffer, M., Schoch, F., Gobrecht, J., Gijs, M.A.M. 2003. Nano-replication of diffractive optical elements in sol–gel derived glasses. *Microelectron. Eng.* 67–68: 710.
67. Della Giustina, G., Guglielmi, M., Brusatin, G., Prasciolu, M., Romanato, F. 2008. Electron beam writing of epoxy based sol–gel materials. *J. Sol-Gel Sci. Technol.* 48: 212.
68. Zhang, X., Wang, J., Wu, W., Liu, C., Qian, S. 2007. Preparation of amino-functionalized mesoporous silica thin films with highly ordered large pore structures. *J. Sol-Gel Sci. Technol.* 43: 305.
69. Anastasova, S., Milanova, M., Kashchieva, E., Funakubo, H., Kamo, T., Grozev, N., Stefanov, P., Todorovsky, D. 2008. Morphology of sol–gel produced composite films for optical oxygen sensors. *Appl. Surf. Sci.* 254: 1545.
70. Badr, Y., Salah, A., Battisha, I.K. 2005. Effect of europium ion concentrations on the photoluminescence emission of nano-crystalline $BaTiO_3$ prepared by sol–gel technique. *J. Sol-Gel Sci. Technol.* 34: 219.
71. Murata, T., Ishizawa, H., Motoyama, I., Tanaka, A. 2004. Investigations of MgF_2 optical thin films prepared from autoclaved sol. *J. Sol-Gel Sci.Technol.* 32: 161.
72. Abe, K., Sanada, Y., Morimoto, T. 2001. Anti-reflective coatings for CRTs by sol-gel process. *J. Sol-Gel Sci. Technol.* 22: 151.
73. Abe, K., Sanada, Y., Morimoto, T. 2003. Anti-reflective coatings for CRTs by sol-gel process. *J. Sol-Gel Sci. Technol.* 26: 709.
74. Banerjee, A.N., Kundoo, S., Saha, P., Chattopadhyay, K.K. 2003. Synthesis and characterization of nanocrystalline fluorine-doped tin oxide thin films by sol-gel method. *J. Sol-Gel Sci. Technol.* 28: 105.
75. Fujihara, S., Tada, M., Kimura, T. 1997. Preparation and characterization of MgF_2 thin film by a trifluoroacetic acid method. *Thin Solid Films* 304: 252.
76. Spanhel, L. 2006. Colloidal ZnO nanostructures and functional coatings: A survey. *J. Sol-Gel Sci. Technol.* 39: 7.
77. Zhang, J., Feng, H., Hao, W., Wang, T. 2006. Blue-emitting ZnO sol and film obtained by sol-gel process. *J. Sol-Gel Sci. Technol.* 39: 37.
78. Maiti, U.N., Ghosh, P.K., Ahmed, Sk.F., Mitra, M.K., Chattopadhyay, K.K. 2007. Structural, optical and photoelectron spectroscopic studies of nano/micro ZnO:Cd rods synthesized via sol-gel route. *J. Sol-Gel Sci. Technol.* 41: 87.
79. Fujihara, S., Naito, H., Kimura, T. 2001. Visible photoluminescence of ZnO nanoparticles dispersed in highly transparent MgF_2 thin-films via sol-gel process. *Thin Solid Films* 389: 227.
80. Farley, N.R.S., Staddon, C.R., Zhao, L.X., Edmonds, K.W., Gallagher, B.L., Gregory, D.H. 2004. New sol-gel synthesis of ordered nanostructured doped ZnO films. *J. Mater. Chem.* 14: 1087.

81. Almeida R.M., Marques, A.C. 2006. Rare-earth photoluminescence in sol–gel derived confined glass structures. *J. Non-Cryst. Solids* 352: 475.
82. Desurvire, E. 1994. *Erbium-Doped Fiber Amplifiers: Principles and Applications*. New York: John Wiley & Sons, Inc.
83. Houbertz, R., Frohlich, L., Popall, M., Streppel, U., Dannberg, P., Brauer, A., Serbin, J., Chichkov, B.N. 2003. Inorganic-organic hybrid polymers for information technology: From planar technology to 3D nanostructures. *Adv. Eng. Mater*. 5: 551.
84. Kwon, Y.K., Han, J.K., Lee, J.M., Ko, Y.S., Oh, J.H., Lee, H.-S., Lee, E.-H. 2008. Organic–inorganic hybrid materials for flexible optical waveguide applications. *J. Mater. Chem*. 18: 579.
85. Chong, A.S.M., Zhao, X.S. 2004. Functionalized nanoporous silicas for the immobilization of penicillin acylase. *Appl. Surf. Sci*. 237: 398.
86. Wu, G., Wang, J., Shen, J., Yang, T., Zhang, Q., Zhou, B., Deng, Z., Fan, B., Zhou, D., Zhang, F. 2000. A novel route to control refractive index of sol-gel derived nano-porous silica films used as broadband antireflective coatings. *Mater. Sci. Eng*. B78: 135.
87. Makita, K., Akamatsu, Y., Takamatsu, A., Yamazaki, S., Abe, Y. 1999. Sol-gel preparation of silica films with controlled surface morphology and their application to a low reflective glass. *J. Sol-Gel Sci. Technol.* 14: 174.
88. Matsuda, A., Matoda, T., Kogure, T., Tadanaga, K., Minami, T., Tatsumisago, M. 2004. Preparation of titania nanosheet-precipitated coatings on glass substrates by treating SiO_2-TiO_2 gel films with hot water under vibrations. *J. Sol-Gel Sci. Technol*. 31: 229.
89. Marques, A.C., Almeida, R.M., Chiasera, A., Ferrari, M. 2003. Reversible photoluminescence quenching in Er^{3+} -doped silica–titania planar waveguides prepared by sol–gel. *J. Non-Cryst. Solids* 322: 272.
90. Marques, A.C. 2005. Erbium/silver glass matrix nanocomposites by sol-gel processing, for integrated optics. PhD thesis, Technical University of Lisbon, IST: Lisbon, Portugal.
91. Almeida, R.M., Marques, A.C., Cabeca, R., Zampedri, L., Chiasera, A., Ferrari, M. 2004. Photoluminescence of erbium-doped silicate sol-gel planar waveguides. *J. Sol-Gel Sci. Technol.* 31: 317.
92. Miniscalco, W.J., Digonnet, M.J.F. 1993. *Rare Earth Doped Fiber Lasers and Amplifiers*. New York: Marcel Dekker, Inc. p. 19.
93. Chiasera, A., Montagna, M., Tosello, C., Gonçalves, R.R., Chiappini, A., Ferrari, M., Zampedri, L., Pelli, S., Righini, G.C., Monteil, A., Foglietti, V., Minotti, A., Almeida, R.M., Marques, A.C., Soares, V. 2003. Erbium/Ytterbium activated silica-titania planar and channel waveguides prepared by rf-sputtering. *Proc. SPIE* 4990: 38.
94. Mignotte, C. 2001. EXAFS studies on erbium-doped TiO_2 and ZrO_2 sol-gel thin films. *J. Non-Cryst. Solids* 291: 56.
95. D'Acapito, F., Marques, A.C., Santos, L.F., Almeida, R.M. 2008. EXAFS study of the Er^{3+} ion coordination in SiO_2–TiO_2–HfO_2 sol–gel films. *J. Non-Cryst. Solids* 354: 4940.
96. Enrichi, F., Mattei, G., Sada, C., Trave, E., Pacifici, D., Franzo, G., Priolo, F., Iacona, F., Prassas, M., Falconieri, M., Borsella, E. 2005. Study of the energy transfer mechanism in different glasses co-doped with Si nanoaggregates and Er^{3+} ions. *Opt. Mater*. 27: 904.
97. Kik, P.G., Brongersma, M.L., Polman, A. 2000. Strong exciton-erbium coupling in Si nanocrystal-doped SiO_2. *Appl. Phys. Lett*. 76: 2325.
98. Fujii, M., Yoshida, M., Kanzawa, Y., Hayashi, S., Yamamoto, K. 1997. 1.54 μm photoluminescence of Er^{3+} doped into SiO_2 films containing Si nanocrystals: Evidence for energy transfer from Si nanocrystals to Er^{3+}. *Appl. Phys. Lett*. 71: 1198.
99. Mie, G. 1908. Beiträge zur optik trüber medien, speziell kolloidaler metallösungen. *Ann. Phys*. 25: 377.
100. Yamane, M., Asahara, Y. 2000. *Glasses for Photonics*. Cambridge, U.K.: Cambridge University Press.
101. Mulvaney, P. 1996. Surface plasmon spectroscopy of nanosized metal particles. *Langmuir* 12: 788.
102. Marques, A.C., Almeida, R.M. 2007. Er photoluminescence enhancement in Ag-doped sol–gel planar waveguides. *J. Non-Cryst. Solids* 353: 2613.
103. Muromachi, T., Tsujino, T., Kamitani, K., Maeda, K. 2006. Application of functional coatings by sol-gel method. *J. Sol-Gel Sci. Technol*. 40: 267.
104. Polman, A., Atwater, H.A. 2005. Plasmonics: Optics at the nanoscale. *Mater. Today* 8: 56.
105. Stewart, M.E., Anderton, C.R., Thompson, L.B., Maria, J., Gray, S.K., Rogers, J.A., Nuzzo, R.G. 2008. Nanostructured plasmonic sensors. *Chem. Rev*. 108: 494.

106. Mohan Babu, P., Radhakhrishna, B., Venkata Rao, G., Sreedhara Reddy, P., Uthanna, S. 2004. Bias voltage dependence properties of dc reactive magnetron sputtered indium oxide films. *J. Optoelectron. Adv. Mater.* 6: 205.
107. Lewis, B.G., Paine, D.C. 2000. Applications and processing of transparent conducting oxides. *MRS Bull.* 25: 22.
108. Stoica, T.F., Gartner, M., Stoica, T., Losurdo, M., Teodorescu, V.S., Blanchin, M.G., Zaharescu, M. 2005. Properties of high porosity sol-gel derived indium-tin oxide films. *J. Optoelectron. Adv. Mater.* 7: 2353.
109. Chen, X.C. 2005. Effect of oxalic acid on stabilization and property of Sb-doped tin oxide sol. *J. Sol-Gel Sci. Technol.* 34: 267.
110. Yamaguchi, N., Tadanaga, K., Matsuda, A., Minami, T., Tatsumisago, M. 2005. Anti-reflective coatings of flowerlike alumina on various glass substrates by the sol–gel process with the hot water treatment. *J. Sol-Gel Sci. Technol.* 33: 117.
111. Chen, D., Yan, Y., Westenberg, E., Niebauer, D., Sakaitani, N., Chaudhuri, S.R., Sato, Y., Takamatsu, M. 2000. Development of antireflection coating on plastic panels for display applications. *J. Sol-Gel Sci. Technol.* 19: 77.
112. Zakharchenko, R.V., Diaz-Flores, L.L., Perez-Robles, J.F., Gonzalez-Hernandez, J., Vorobiev, Y.Y. 2005. Nanostructured porous sol-gel materials for applications in solar cells engineering. *Phys. Status Solidi* 2: 3308.
113. Yablonovitch, E. 1987. Inhibited spontaneous emission in solid-state physics and electronics. *Phys. Rev. Lett.* 58: 2059.
114. John, S. 1987. Strong localization of photons in certain disordered dielectric superlattices. *Phys. Rev. Lett.* 58: 2486.
115. Almeida, R.M., Rodrigues, A.S. 2003. Optical nanocomposite planar waveguides doped with rare-earth and noble metal elements. *J. Non-Cryst. Solids* 326 & 327: 405.
116. Righini, G., Pelli, S., Ferrari, M., Armellini, C., Zampedri, L., Tosello, C., Ronchin, S., Rolli, R., Moser, E., Montagna, M., Chiasera, A., Ribeiro, S.J.L. 2002. Er-doped silica-based waveguides prepared by different techniques: RF-sputtering, sol-gel and ion-exchange. *Opt. Quant. Electron.* 34: 1151.
117. Bellessa, J., Rabaste, S., Plenet, J.C., Dumas, J., Mugnier, J., Marty, O. 2001. Eu^{3+} doped microcavities fabricated by sol-gel process. *Appl. Phys. Lett.* 79: 2142.
118. Bayindir, M., Kural, C., Ozbay, E. 2001. Coupled optical microcavities in one-dimensional photonic bandgap structures. *J. Opt. A, Pure Appl. Opt.* 3: S184.
119. Almeida, R.M., Wang, Z. 2002. Sol-gel preparation of one-dimensional photonic bandgap structures. *Proc. SPIE* 4655: 24.
120. Inouye, H., Arakawa, M., Ye, J.Y., Hattori, T., Nakatsuka, H., Hirao, K. 2002. Optical properties of a total-reflection-type one-dimensional photonic crystal. *IEEE J. Quant. Electr.* 38: 867.
121. Rabaste, S., Bellessa, J., Brioude, A., Bovier, C., Plenet, J.C., Brenier, R., Marty, O., Mugnier, J., Dumas, J. 2002. Sol–gel fabrication of thick multilayers applied to Bragg reflectors and microcavities. *Thin Solid Films* 416: 242.
122. Almeida, R.M., Marques, A.C., Chiasera, A., Chiappini, A., Ferrari, M. 2007. Rare-earth doped photonic crystal microcavities prepared by sol–gel. *J. Non-Cryst. Solids* 353: 490.
123. Zampedri, L., Tosello, C., Portales, H., Montagna, M., Mattarelli, M., Chiappini, A., Righini, G.C., Pelli, S., Nunzi Conti, G., Martino, M., Portal, S., Marques, A.C., Almeida, R.M., Jestin, Y., Ferrari, M., Chiasera, A. 2005. Spectroscopic assessment of rare-earth activated planar waveguides and microcavities. *Appl. Surf. Sci.* 248: 3.
124. Cernigoj, U., Štangar, U.L., Trebše, P., Krašovec, U.O., Gross, S. 2006. Photocatalytically active TiO_2 thin films produced by surfactant-assisted sol–gel processing. *Thin Solid Films* 495: 327.
125. Ezz, T., Crouse, P., Li, L., Liu, Z. 2006. Laser/sol-gel synthesis: A novel method for depositing nanostructured TiN coatings in non vacuum conditions. *Appl. Phys. A, Mater. Sci. Process.* 85: 79.
126. Sun, J.-H., Fan, W.-H., Xu, Y., Wu, D., Sun, Y.-H. 1999. Nano-sized SiO_2 sol-gel for structure-controlled optical coatings. *Mol. Cryst. Liq. Cryst. Sci. Technol., Section A.* 337: 85.
127. Zhou, H.S., Honma, I. 1998. Synthesis of oriented meso-structure silica film by ferrocenyl surfactant. *Mater. Res. Soc. Symp. Proc.* 519: 77.
128. Li, X.X., Tang, Y.H., Lin, L.W., Li, J.L. 2008. Blue light emission in mesoporous SiO_x nano-structure. *Micropor. Mesopor. Mater.* 111: 591.
129. Yuan, J., Zhou, S., Gu, G., Wu, L. 2005. Encapsulation of organic pigment particles with silica via sol-gel process. *J. Sol-Gel Sci. Technol.* 36: 265.

130. Mennig, M., Endres, K., Schmitt, M., Schmidt, H. 1997. Colored coatings on eye glass lenses by noble metal colloids. *J. Non-Cryst. Solids* 218: 373.
131. Martucci, A., Pasquale, M., Guglielmi, M., Post, M., Pivin, J.C. 2003. Nanostructured silicon oxide-nickel oxide sol-gel films with enhanced optical carbon monoxide gas sensitivity. *J. Am. Ceram. Soc.* 86: 1638.
132. Chiodini, N., Paleari, A., Romagnoli, M. 2005. Nanostructured SnO_2-SiO_2 glass ceramic: A competitor for Si nanodots in silica. *Proc. SPIE* 5925: 59250J/1.
133. Mohammadi, M., Ghorbani, M., Cordero-Cabrera, M., Fray, D. 2007. Preparation and characterisation of nanostructural TiO_2-Ga_2O_3 binary oxides with high surface area derived form particulate sol-gel route. *J. Mater. Sci.* 42: 4976.
134. Mohammadi, M.R., Fray, D.J., Sadrnezhaad, S.K., Mohammadi, A. 2007. A simple particulate sol–gel route to synthesize nanostructural TiO_2–Ta_2O_5 binary oxides and their characteristics. *Mater. Sci. Eng. B.* 142: 16.
135. Fernández, C.J., Manera, M.G., Pellegrini, G., Bersani, M., Mattei, G., Rella, R., Vasanelli, L., Mazzoldi, P. 2008. Surface plasmon resonance optical gas sensing of nanostructured ZnO films. *Sens. Actuator B: Chem.* 130: 531.
136. Shang, H.M., Wang, Y., Bliss, M., Cao, G.Z. 2005. Processing and optical properties of sol-gel derived nanostructured $CdWO_4$ films. *Proc. SPIE* 6008: 60081B-1.
137. Zhang, H.X., Zhou, Y., Kam, C.H., Lam, Y.L., Ooi, B.S., Chan, Y.C. 1999. Deposition of potassium lithium niobate films by sol-gel method. *Proc. SPIE* (*Design, Fabrication, and Characterization of Photonic Devices*) 3896: 425.

7 Controllably Micro/Nanostructured Films and Devices

Guotao Duan, Weiping Cai, Yue Li, and Lichao Jia

CONTENTS

7.1 INTRODUCTION

In more and more cases, the surface structures of films have proven to be one of the vital factors to their properties in addition to material category. In the past decades, the micro/nanostructured films, i.e., the films consisting of micro-, nano-, or micro/nanocomposited structures, have attracted increasing attention due to their novel surface properties and important potential applications, including catalysis, sensors, cells, surface-enhanced Raman scattering (SERS), data storage, superhydrophobic or superhydrophilic films, photonic crystals, optoelectronics, microelectronics, optical devices, and electrochemical electrolytes [1].

The properties for micro/nanostructured films are strongly correlated with the size and the morphology of the building blocks in the films, the packing of the building block, and the supporting substrate, in addition to the intrinsic characters. Usually, the supporting substrate is selected according to device realization and compatibility with the desired materials. Building blocks should be controlled in both morphology and size for a favorable functionality. The packing of the building blocks is designed according to practical applications. In many cases, ordered packing of the building blocks is of much importance. First, such orderly packing allows a homogeneous surface characteristic in a large dimension on film, which is obviously advantageous to the designed devices and thus the subsequent stability of output functions. Second, the orderly packing of micro/nanostructured films will give enhanced output functions and some new properties due to the strong coupling effect between the building blocks. Recently, much progress in micro/nanostructured films have been directed to the orderly packing cases. Therefore, in this chapter, we mainly review the micro/nanostructured films with orderly packing. To keep things simple, we still call such films "micro/nanostructured films."

In general, micro/nanostructured films can be synthesized by photolithography [2–4], electron-beam lithography [5–8], microcontact printing [9,10], and self-assembly techniques [11,12]. Recently, using two-dimensional (2D) colloidal crystals (i.e., colloidal crystal in a limited layer) as a template has shown great promise for the fabrication of micro/nanostructured films (i.e., the orderly micro/nanostructured arrays on desired substrates). It has been found that monodispersed colloidal spheres can naturally self-assemble into an ordered hexagonally close-packed arrangement on a cleaned substrate driven by surface tension. Based on such ordered, arranged colloidal spheres and other assistant techniques, such as physical vapor deposition [13], solgel [14], solution dipping [15], and electrodeposition [16,17], one can obtain micro/nanostructured films after the removal of the colloidal spheres. This strategy, what we call colloidal lithography, is of great advantage due to the material and the substrate general, which are inexpensive and flexible in surface morphology and size.

This chapter focuses mainly on recent progress in the synthesis and devices of micro/nanostructured films by such colloidal lithography strategies. The remainder of this chapter is organized as follows. In Section 7.2, we systematically survey the synthetic approaches for micro/nanostructured films, including the evaporation deposition strategy, the solution-dipping strategy, the electrodeposition strategy, and some extensions of the above strategies based on monolayer colloidal crystal. In Sections 7.3 through 7.5, we review the structure-related properties and corresponding device-design and realization based on the micro/nanostructured films, including surface wettability, optical properties, gas sensors, and optical devices. Finally, we close with a brief remark on the advantages and the potential improvement for the micro/nanostructured films as devices.

7.2 SYNTHESIS OF MICRO/NANOSTRUCTURED FILMS

7.2.1 Monolayer Colloidal Crystal

The elementary step for colloidal lithography is to form a highly orderly 2D colloidal crystal on a substrate. Generally, monolayer colloidal crystal (or colloidal monolayer) is used. Colloidal monolayer is the ordered array with hexagonally packed lattice structure on a certain substrate, as shown in Figure 7.1. The building blocks in the monolayer are of spherical colloidal particles with uniform size, shape, composition, and surface properties. Nowadays, commonly used monodispersive colloidal particles are prepared by the inorganic oxide deposition method [18] and the emulsion polymer method [19,20], with particle diameters ranging from 50 nm to 100 μm. For instance, the monodispersive SiO_2 colloidal spheres can be prepared by using the hydrolysis of $Si(OC_2H_5)_4$, and the monodispersive colloidal polystyrene spheres (PSs) can be prepared using the polymerization of styrene. Now, there are various monodispersive colloidal goods. They may be used in scientific research and industry production. Normally, the colloidal monolayer is formed by self-assembly on a specially treated cleaned substrate. Here, we briefly introduce the formation mechanism, preparation techniques, and the transferability of the colloidal monolayer.

7.2.1.1 Self-Assembly Mechanism

Monodispersive colloidal spheres can form monolayer colloidal crystal via self-assembly on a solid substrate or even a liquid surface [21,22]. The formation process may be divided into six types [23]: dip-coating, interface-forming, electrophoretic deposition, chemical or electrochemical deposition, templating self-organization, and spin-coating (see Figure 7.2). Among them, the dip-coating and spin-coating are widely used and lead to the colloidal monolayer on a gas–solid interface. The formation process can be depicted as illustrated in Figure 7.3. When a droplet of colloidal sphere suspension is put on a substrate and dried, first a meniscus forms between the adjacent colloidal spheres on the edge of the droplet due to evaporation of the solvent. Subsequently, the colloidal spheres move together due to the capillary force or surface tension and arrange orderly, forming monolayer colloidal crystals, as shown in Figure 7.1. In the practical synthesis, the following conditions are required, i.e., slow evaporation, and a clean and flat substrate surface with good hydrophilicity. In addition, the small diameter deviation of colloidal spheres is beneficial to the formation of large-scale 2D colloidal crystals.

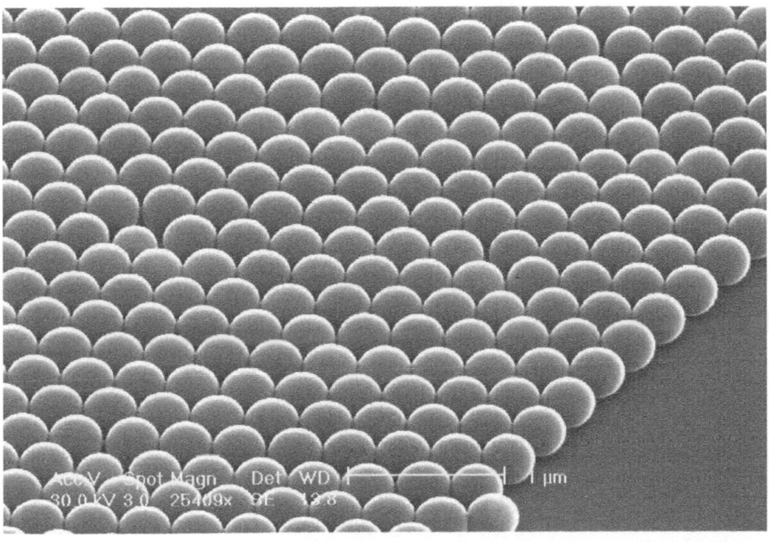

FIGURE 7.1 A SEM image of PS colloidal monolayer on glass substrate.

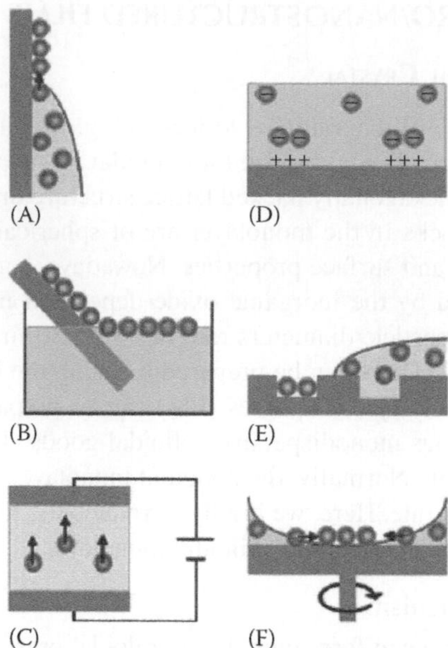

FIGURE 7.2 Diverse self-assembly strategies to create colloid monolayer: (A) dip-coating in which capillary forces and evaporation induce colloidal self-organization, (B) lifting up a colloid array from an interface using the substrate, (C) electrophoretic deposition of colloids, (D) chemical or electrochemical deposition of colloids with a patterned array, (E) physical template-guided self-organization of colloids, (F) spin-coating in which shear and capillary forces drive the colloidal self-organization. (Reproduced from Yang, S.M. et al., *Small*, 2, 458, 2006. With permission.)

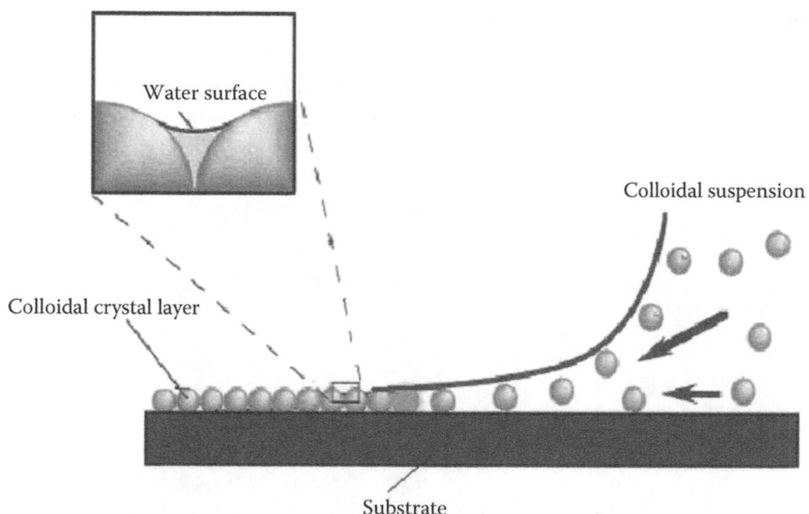

FIGURE 7.3 The formation scheme of a colloidal monolayer.

7.2.1.2 Cleaning Substrates

As mentioned above, before it is used for self-assembly of a colloidal monolayer, the flat substrate should be cleaned carefully according to a strict procedure. After cleaning, the substrate must be of good hydrophilicity with a water contact angle (CA) <30°. A bad hydrophilicity will lead to a failure of the fabrication of large-scale colloidal monolayer due to the conglomeration of the colloidal sphere, as shown in Figure 7.4.

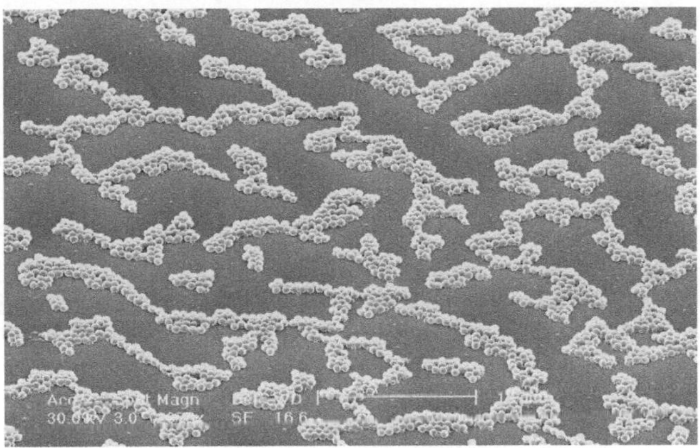

FIGURE 7.4 Failure of fabrication of PS colloidal monolayer on an uncleaned glass substrate.

Typically, for flat glass or silicon substrates, they can be cleaned according to the following procedures [24–26]: first, ultrasonic washing in acetone, ethanol, and distilled water, in turn, for 1 h, respectively; subsequently immersing in a solution of mixture of H_2SO_4:30% H_2O_2/3:1 for 1 h at 80°C, and then washing several times with distilled water; finally, keeping in solution of H_2O: NH_4OH:30% H_2O_2/5:1:1 for 1 h at 80°C, and then washing several times with distilled water. The cleaned substrates are stored in distilled water for less than two weeks before use.

7.2.1.3 Fabrication of Colloidal Monolayer

Nowadays, commercial monodispersed colloidal spheres with different diameters in the range of 50 nm–50 µm and in different materials (such as polystyrene, PMMA, silica, etc.) can be purchased easily from companies, such as Alfar Aesar, Polyscience, Soken Chemisnow, etc., as suspensions or powders. The standard deviation in the diameter of colloidal spheres should be less than 5%, because a broad-size distribution will lead to the dislocation, defect, or even a multilayer in the resulting colloidal monolayer. To fabricate the colloidal monolayer, colloidal spheres with different diameters should be prepared in aqueous suspension with suitable concentrations. With a decrease in the colloidal sphere diameter, the concentration should also be decreased. The concentrations for different diameters should be in the range of 1.0–5.0 wt%. Besides these, for colloidal suspension with small sphere diameters, the addition of surfactants, such as sodium dodecyl sulfate (SDS), Triton X-100, etc., is more propitious to the fabrication of colloidal monolayer with large area. The colloidal sphere suspension should be well dispersed ultrasonically to avoid colloidal sphere deposition owing to its long-time placement before use.

Generally, the colloidal particle number can be calculated by the following equation [27]:

$$\eta = 6\omega/(\pi Q\phi^3)\times 10^{12} \text{ particle number/mL} \quad (7.1)$$

where

η is the colloidal sphere number in a unit volume
ω is the concentration of colloidal suspension in wt%
Q is the density of the colloidal sphere
ϕ is the diameter of the colloidal sphere in µm

According to this equation, the suspension volume for different sphere diameters, required to fabricate the colloidal monolayer with a certain area, can be estimated easily. Here, as examples, we introduce two main fabrication techniques for colloidal monolayers: spin-coating and dip-coating.

7.2.1.3.1 Spin-Coating

Spin-coating is conducted on a spin-coater, as shown in Figure 7.5. The base rotating speed can be controlled in a large range. The base is heavy to keep the whole device running stably without vibrating. The sealing cover makes the suspension have a low evaporating rate, leading to the formation of large-scale colloidal monolayer. As a typical fabrication, a droplet of PS colloidal suspension is first added on the cleaned substrate fixed on the rotating plate of the spin-coater. At a suitable rotating speed, colloidal monolayer was fabricated via self-assembly on the substrate under combination with the centrifugal force and the water capillary force until the added suspension dried. The rotating speed is a key parameter. Generally, with a decrease of the colloidal sphere diameter, the rotating speed should be increased. Colloidal monolayers with square centimeter size have been successfully synthesized by spin-coating; some typical photos are showed in Figure 7.6. The colloidal monolayer with different sphere diameter assumes diverse colors due to diffraction and reflection effects. This method has an advantage of high yield. Especially for colloidal spheres with large diameters (>500 nm), it is easy to synthesize the large-scale colloidal monolayer. However, for colloidal spheres with smaller-sized diameters (<500 nm), it is difficult to obtain a large-area colloidal monolayer with high quality.

7.2.1.3.2 Dip-Coating

The dip-coater is mainly composed of a controller, an elevating device, a microscope with CCD camera, and a monitor, as illustrated in Figure 7.7. The withdraw speed is an important operating parameter that changes from zero to that, such as 25 μm/s. A cleaned hydrophilic substrate is vertically withdrawn from the colloidal suspension in a work cell, driven by a step motor. The suspension will spread out along the substrate owing to the capillary force of the solvent. With evaporation of the solvent, the nucleation occurs at the substrate top outside the suspension. With the withdrawal of substrate vertically, the colloidal crystal (monolayer or multilayer, depending on the withdrawing

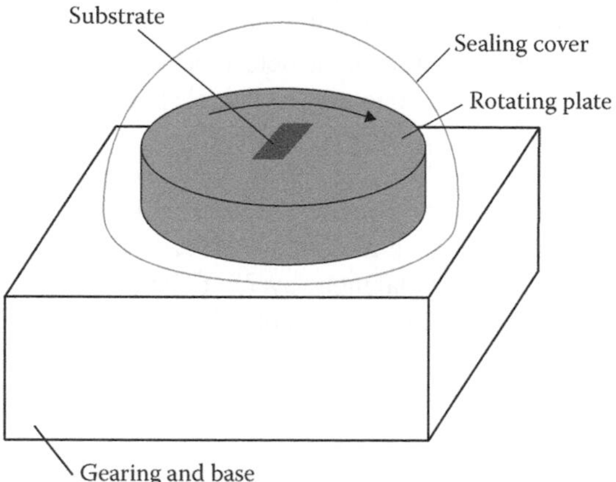

FIGURE 7.5 A scheme of spin-coater for synthesis of colloidal monolayer.

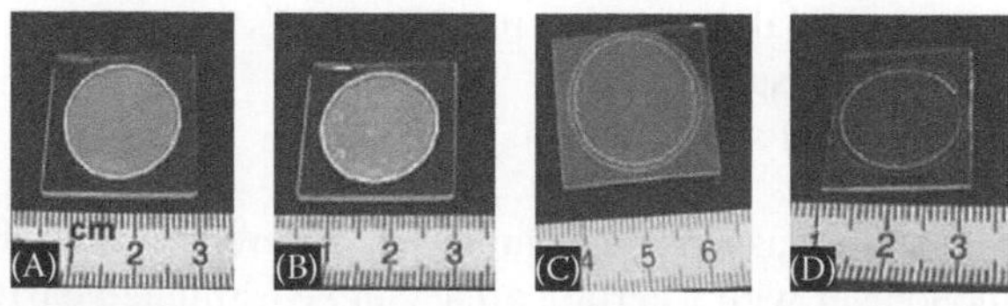

FIGURE 7.6 Square centimeter-sized PS colloidal monolayer with different sphere sizes by spin-coating. Colloidal sphere diameter: (A) 1000 nm, (B) 500 nm, (C) 350 nm, (D) 200 nm.

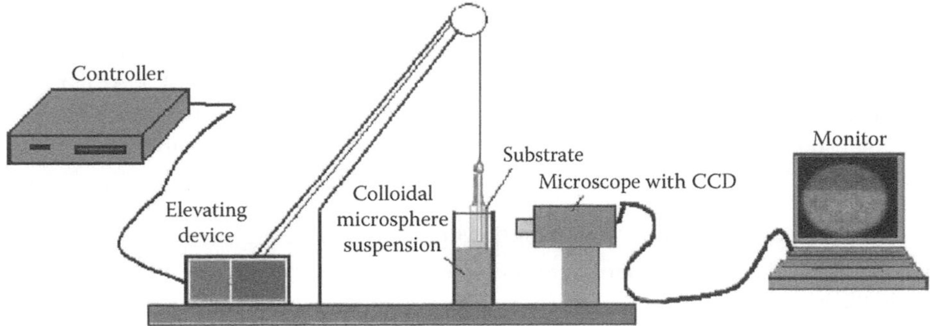

FIGURE 7.7 Scheme of a dip-coater.

speed) can be continuously fabricated on its surface due to the convective effect and capillary force of the solvent [28]. Obviously, the downward growth rate of the colloidal crystal along the substrate should be equal to the withdrawing speed. For a monolayer colloidal crystal, the withdrawing speed should be controlled in a special value. In general, the growth rate is associated with the concentration of colloidal sphere suspension, the evaporation rate, and the diameter of the colloidal spheres. The smaller the concentration of the colloidal sphere suspension, the slower the evaporation rate of the solvent, and the larger the diameter of the colloidal spheres, the smaller the growth rate of the colloidal crystal and the slower the corresponding withdrawing speed of the substrate. When the concentration, temperature, humidity, and evaporation rate of the colloidal suspension are suitably controlled, the monolayer or double-layer colloidal crystal membranes can form on the substrates via self-organization. It generally takes several hours to fabricate a centimeter-size sample due to the low withdrawing speed. Thus, this method has a disadvantage of low output. In addition, for colloidal suspension, long-time withdrawing will lead to a settle of colloidal spheres due to gravity and a bad dispersion of colloidal spheres in the suspension, especially for colloidal spheres with larger diameters. Therefore, this method is more suitable for the synthesis of the colloidal monolayer by spin-coating; samples prepared by dip-coating have a much better quality with less voids, defects, and boundaries.

7.2.1.4 Transferability of Colloidal Monolayer

It was found that the as-prepared PS colloidal monolayer has a good transferability from its substrate to other ones including silicon, silica, mica, transmission electronic microscopic (TEM) grid, and even other hydrophobic and curved surfaces. This is of great importance for templating application on any desired substrate. We can thus design and fabricate micro/nanostructured films on any substrate according to the requirements in the devices [29]. Figure 7.8 is a demonstration for such transferability. First, a PS colloidal monolayer on a glass slide is put into the distilled water in a

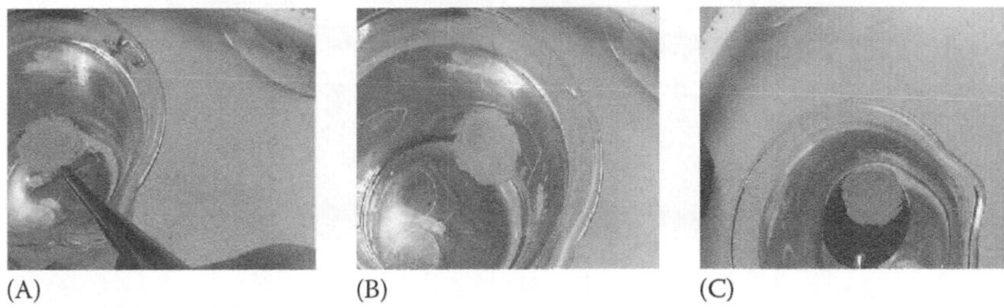

FIGURE 7.8 Photos depicting transferring a centimeter-square-sized PS colloidal monolayer from a glass substrate onto a silicon substrate. (A) The monolayer on a glass substrate, (B) liftoff onto water, (C) pickup of the monolayer with a silicon substrate.

cup slowly (see Figure 7.8A). Then the monolayer can peel out from the glass due to surface tension of water, and floats onto the water surface as integrity (Figure 7.8B). Finally, the monolayer on the water surface is picked up with a desired substrate (silicon wafer in Figure 7.8C). After drying, the monolayer is thus obtained on a new substrate. It should be mentioned that if a colloidal monolayer on the original substrate was placed for a long time (say, several weeks), it could not be transferred because of binding between the colloidal spheres and the original substrate.

7.2.1.5 Non-Close-Packed Colloidal Monolayer

The above-mentioned colloidal monolayer is of close-packing due to self-assembly drive. However, in many applications, the colloidal monolayer with non-close-packing are also desired to realize micro/nanostructured films in controllable size and morphology. Nowadays, there also exist some routes to obtain non-close-packed colloidal monolayer in controlled spacing between neighboring spheres. Jiang and McFarland [30] developed a changed spin-coating technique to embed non-close-packed silica colloidal crystals in a poly-(ethoxylated trimethylolpropane triacrylate) (PETPTA) matrix; after removal of PETPTA, the non-close-packed colloidal monolayer was thus obtained. In addition, non-close-packed colloidal monolayer can also be fabricated by insetting close-packed colloidal monolayer into an extendable PDMS film. If extending in different ways, the morphology of the non-close-packed colloidal monolayer can also be controlled.

Our group also obtained a non-close-packed PS colloidal monolayer by plasma etching the corresponding close-packed one. Figure 7.9 shows two typical samples after etching 2000 nm PS close-packed colloidal monolayer for different times. It can be seen, after etching for 30 min, PSs were reduced to 1820 nm in diameter, but still connected with necks, due to the area contact between PSs induced by heating treatment (see Figure 7.9A). If etching for 45 min, PSs were completely isolated from each other with size reduced to 1620 nm in diameter (Figure 7.9B). The tilted images show that the etched PSs are still spherical in shape (see Figure 7.9C and D), owing to an isotropic etching process. Obviously, spacing between the etched PSs (or size of the etched PSs) can be easily controlled by the etching time.

7.2.2 Evaporation Deposition Strategy

Based on the PS (or silica) colloidal monolayer, various micro/nanostructured films can be prepared by some assistant techniques. There are four categories of such assistant techniques: evaporation

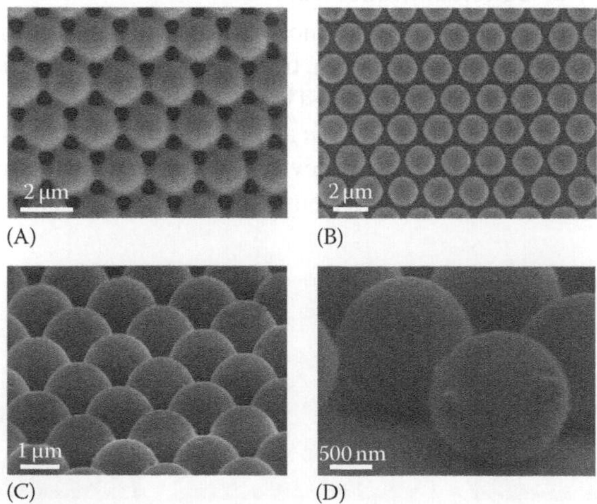

FIGURE 7.9 SEM images of PS non-close-packed colloidal monolayer on an ITO substrate prepared by plasma-etching treatment of close-packed monolayer. The etching time is (A) 30 min and (B) 45 min, respectively. (C) Typical corresponding tilted image for (B). (D) Tilted image in edge region for (B). Both samples were coated with a 10 nm gold layer for SEM investigation.

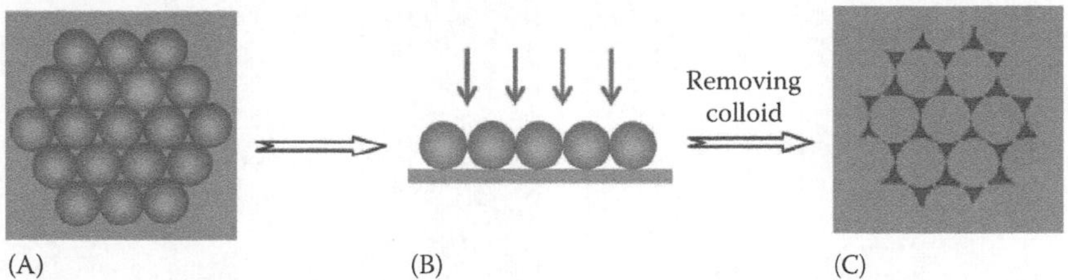

FIGURE 7.10 A schematic illustration for the evaporation deposition strategy. (A) Colloidal monolayer, (B) deposition of desired material, (C) desired film.

deposition strategy, solution-dipping strategy, electrodeposition strategy, and lithography-assistant strategy. Also, there are some extensions of the above-mentioned strategies. Generally, the former three strategies are to deposit materials on the substrates with colloidal monolayer, while the latter is to structure the substrate. Here we only introduce the former three strategies based on PS colloidal templates (silica colloidal spheres can also be used in most cases instead of PSs).

Figure 7.10 is a schematic illustration for evaporation deposition strategy based on the colloidal monolayer. After evaporating deposition of the desired materials on the monolayer and the subsequent removal of it, the desired nanoparticle array (or film) will be produced on the supporting substrate. As an example, the substrate coated with a PS colloidal monolayer was mounted in an ultrahigh vacuum chamber (under a base pressure of 10^{-6} Pa). The desired material (such as Au or Ag) was thermally evaporated and deposited vertically onto the substrate at a certain rate. A quartz crystal microbalance was used to monitor the thickness of the deposited film. After deposition to a desired thickness, the sample was immersed in methylene chloride under sonication to remove the colloidal monolayer mask, leaving a highly ordered nanoparticle array on the substrate. The resultant nanoparticles are triangular in shape due to the interstitial geometry in the monolayer. Figure 7.11 is a typical result based on 2000nm PS colloidal monolayer. Similarly, if a double-layer colloidal crystal, obtained by transfer of the monolayers, is used as a template in this strategy, another pattern of nanoparticle array films will be produced, as shown in Figure 7.12. As for the nanoparticles' size, it can be controlled by PS's size and deposition time.

Such evaporation deposition strategy was also named "nanosphere lithography" by Van Duyne and others [26]. Based on this strategy, many nanoparticle array films were prepared, including metal nanoparticle arrays such as Pt [31], Au [32–34], silver [35,36], and Ni [37,38]; semiconductor arrays, such as TiO_2 [39] and Y_2O_3 [40]; and organic solid arrays such as CoPc [41]. For the metals with low melting point, evaporation was conducted by heating the targets; while for the metals with high melting point or some metal oxide and sulfide, deposition was performed by bombarding the targets using electron beam or laser irradiation. In addition, the nanoparticle arrays with multicomponents can also be prepared by this strategy based on step-by-step (or layer-by-layer) deposition of desired materials and the final removal of PSs.

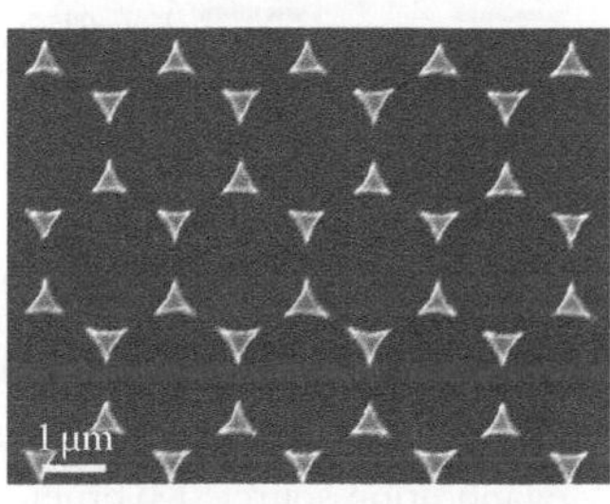

FIGURE 7.11 SEM image of Au ordered triangular nanoparticle array prepared by thermal evaporation deposition based on 2000nm PS colloidal monolayer. The film thickness (i.e., the height of nanoparticle) is controlled at 70nm.

In recent years, such nanosphere lithography technique has been further developed. Haynes et al. developed an "angle-resolved lithography" method [42]. The key to this method is to control the angle between evaporating orientation and substrate, which is usually not 90° in the case. By multistep angle-resolved lithography, one can control the nanoparticle film in material, morphology, and size. Kosiorek and coworkers presented a "shallow nanosphere lithography" method [43]. The deposition was conducted on a modified evaporation system, as is shown in Figure 7.13. In this method, the supporting substrate is also moving with a rotation

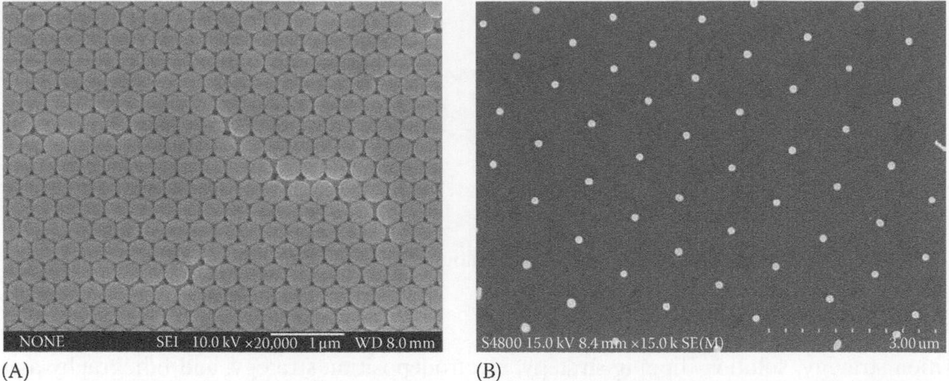

FIGURE 7.12 The FESEM images of double-layer PS colloidal crystal (A) and the corresponding Au nanoparticle pattern (B).

angle. At an appropriate evaporation angle θ, rotation angle α, and also a treatment of colloidal monolayer, micro/nanostructured films with dot, rod, ring, and other morphological building blocks can be obtained. Further, Zhang et al. fabricated heterogeneous binary arrays by stepwise angle-resolved colloidal lithography [44].

7.2.3 Solution-Dipping Strategy

Solution-dipping strategy based on colloidal monolayer is a simple route for the fabrication of large-scale morphology-controlled ordered pore films. It is universal and can be used for various metals, semiconductors, and compounds on any substrate. Importantly, one can extend this strategy for many other patterned arrays, such as, direct synthesis of micro/nanostructured porous films on any substrate with flat or even curved and rough surface based on the transferability of the colloidal monolayer, the production of hollow polymer sphere array films based on the two-step template strategy, the fabrication of nanopillar arrays based on heating-induced template deformation, etc.

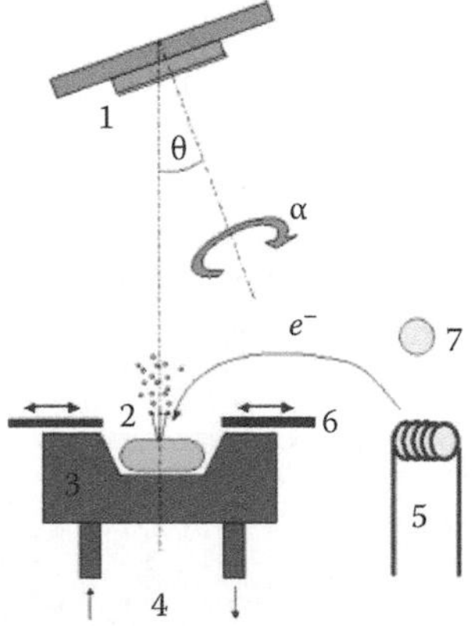

FIGURE 7.13 Schematic of the modified evaporation system. (1) Sample holder, (2) evaporation source, (3) crucible, (4) water cooling system, (5) electron beam source, (6) shutter, (7) magnetic field. θ is the evaporation angle, and α the rotation angle of the sample. (Reproduced from Kosiorek, A. et al., *Small*, 1, 439, 2005. With permission.)

7.2.3.1 Synthesis Route

Figure 7.14 shows the whole synthesis route. Briefly, a colloidal monolayer with a large area is fabricated on a substrate. First, a droplet of the precursor solution is put onto the colloidal monolayer, and the monolayer floats on the surface of the solution due to surface tension of the solution. After subsequent drying at a low temperature (say, less than the glass transition temperature of the PS colloidal monolayer) and removal of it by calcinations or ultrasonic cleanout, the ordered micro/nanostructured pore arrays can thus be obtained. The morphologies of ordered pore array films can be controlled by the solution concentration. The ordered porous films of different materials, such as metals, semiconductors, and compounds, can be fabricated based on this strategy [15].

7.2.3.2 Ordered Micro/Nanostructured Porous Films

Figure 7.15 presents typical results for α-Fe_2O_3 based on the solution-dipping strategy. Using the $Fe(NO_3)_3$ precursor solution, after drying at 80°C for 2 h and subsequent heating at 400°C for 8 h (the PSs were

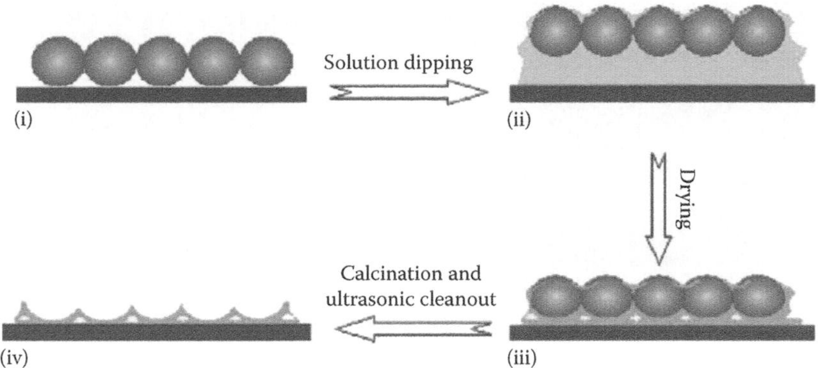

FIGURE 7.14 Schematic illustrations of the solution-dipping template strategy. (i) Colloidal monolayer template with a large area on the substrate. (ii) The colloidal monolayer floats on the surface of the precursor solution. (iii) Integrity of the solute and colloidal spheres. (iv) Ordered porous film. (Reproduced from Sun, F. et al., *Adv. Funct. Mater.*, 14, 283, 2004. With permission.)

burned away at this temperature); the α-Fe_2O_3 ordered porous film was obtained. It can be clearly seen that the morphology of the ordered pore film depends on the precursor solution concentration. When the concentration is 0.02 M or much higher, honeycomb structures are always formed, with the pore shapes and sizes at the film surface depending on the concentrations. A high concentration (0.8 M) gives rise to pores with a nearly circular upper-end opening (from top view). As the concentration decreases, the truncated shape of the pore at the film surface gradually becomes a regular hexagon (Figure 7.15B, C). When the concentration is reduced to 0.02 M, a through-pore structured film (with two open-ended pores) is formed (Figure 7.15D). However, if the concentration is decreased to a very low level (say, 0.002 M), a ring array, rather than the closely arranged pore array, is formed (Figure 7.15E). All the rings surround the positions where the PSs were originally located on the substrate, and hence take on a hexagonal arrangement. Additionally, when the periodicity of the colloidal template is changed, the ordered pore array can also be formed. For example, when a precursor solution with a concentration of 0.8 M is applied onto a template with much smaller PSs (200 nm), the morphology of the porous film is similar to Figure 7.15A, except for the pore size.

The formation of the porous films with concentration-dependent morphologies can be attributed to the deformation of PSs during solvent evaporation and the subsequent treatments. When the colloidal monolayer template is dropped with the precursor solution, it floats on top of the solution. The interstitial spaces of the closely packed PSs are then filled with the solution due to capillary force, and a meniscus is formed on the solution surface, as illustrated schematically in Figure 7.16A. During subsequent drying at 80°C, the liquid surface and the colloidal monolayer gradually decrease due to evaporation of solvent, thereby deforming the PSs. When the concentration of the solution reaches the saturation, further drying will lead to solute precipitation on the sphere's surface and the substrate. Finally, a nearly circular hole is left between two adjacent spheres and the substrate due to lack of sufficient solution compensation (Figure 7.16B through E). Obviously, the lower the solution concentration, the longer the time to reach saturation induced by evaporation at the same drying temperature, leading to a more obvious change in sphere shape.

For a high concentration (0.8 M) of precursor, the above-mentioned process is finished in a shorter time. Therefore, only a small deformation of PSs is produced before complete evaporation, and there is only a small contact area between two adjacent PSs, leading to the formation of small circular holes in the pore walls and circular openings from top view (Figure 7.16B). When the solution concentration is lower (0.08 M and 0.06 M), it takes a longer time for the solution to become

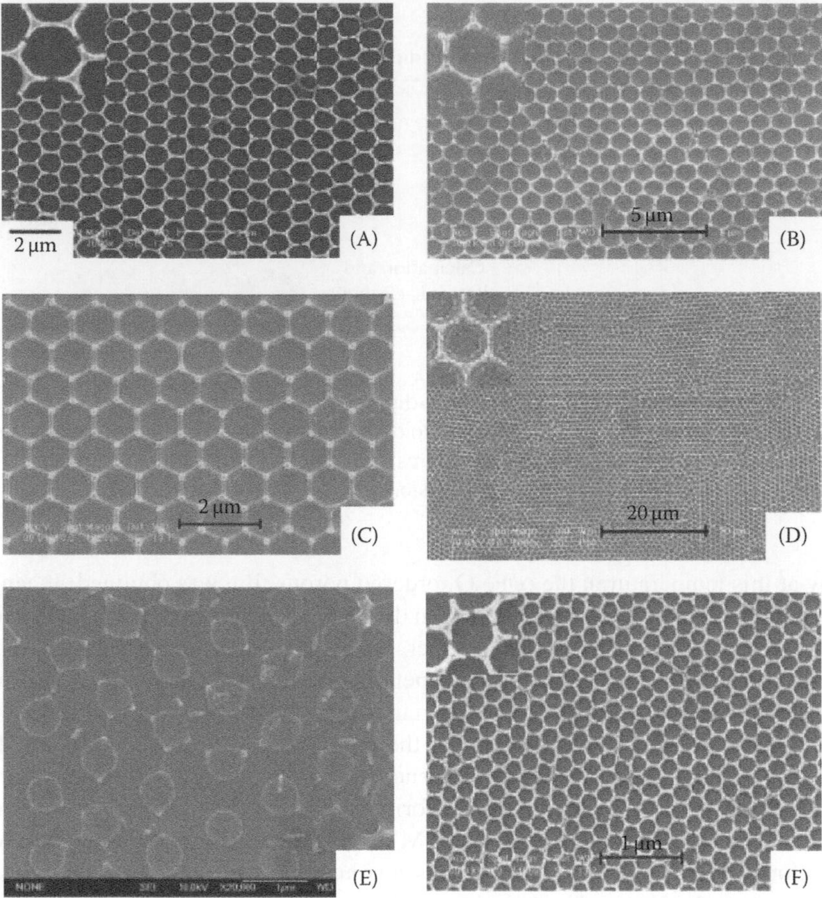

FIGURE 7.15 SEM images of α-Fe_2O_3 samples with different concentration of precursor solutions, dried at 80°C (2 h) and subsequently calcinated at 400°C (8 h), followed by ultrasonic cleanout (0.5 h). (A) 0.8 M, (B) 0.08 M, (C) 0.06 M, (D) 0.02 M, (E) 0.002 M, (F) 0.8 M, (A)–(E) correspond to the template with 1000 nm PSs, and (F) corresponds to the template with 200 nm PSs. (Reproduced from Sun, F. et al., *Adv. Funct. Mater.*, 14, 283, 2004. With permission.)

saturated, leading to a larger contact area between two adjacent spheres (Figure 7.16C) or larger holes in the pore wall and due to the larger deformation of spheres. The large contact area between two adjacent spheres originates from their deformation during drying and results in the formation of a hexagonal opening at the film surface. If the concentration of the precursor solution is further reduced (0.02 M), before solution saturation and solute precipitation, the PSs have already contacted with the substrate, and this contact area has become large enough due to their deformation during drying (see Figure 7.16D), pores with two-ended opening are formed. When the concentration of the precursor is decreased to a very low level (say, 0.002 M), only a thin solute deposition (shell) preferentially formed on the free surface of the latex spheres above and below the tangent-point plane (Figure 7.16E). Subsequent calcinations and ultrasonic vibration remove the colloidal template but leave the deposited shell, as shown in Figure 7.17A. There are two layers, and the top layer is slightly displaced with respect to the bottom layer. After ultrasonically washing for an additional 0.5 h, the top layer is removed and the bottom layer, a ring array, can be found (see Figure 7.15E). Further experiments indicate that the noncircular shape of the rings results from calcinations at 400°C and ultrasonic cleanout. Ultrasonically washing after drying at 80°C, without calcinations, shows an ordered array of circular rings (Figure 7.17B).

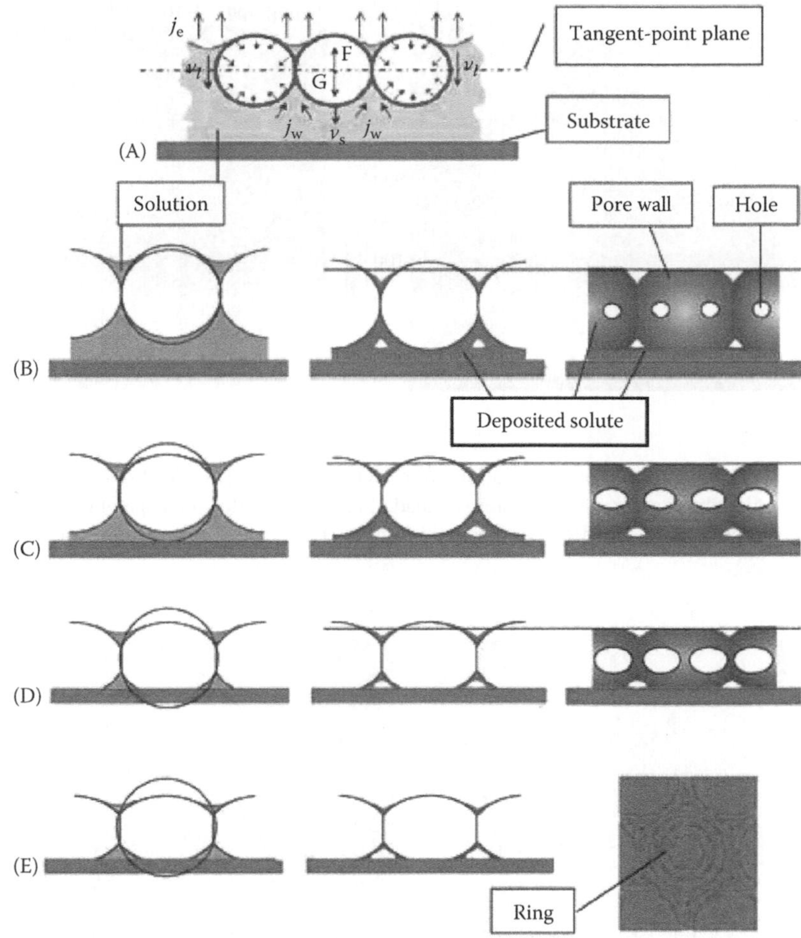

FIGURE 7.16 Schematic illustrations of the formation process of ordered pore array film for different concentrations of precursor solutions. (A) Latex spheres are floating on the solution. F and G denote the flotage and gravitation of the colloidal spheres respectively. j_e is the water evaporation flux, j_w the water influx, v_l and v_s are the descending rate of the solution surface and the colloid monolayer, respectively. (B) High concentration (0.8 M). (C) 0.08 and 0.06 M. (D) Low concentration (0.02 M). (E) Very low concentration (0.002). Left column: solutions reach saturation. Middle column: complete evaporation of solvent. Right column: after removal of the template by dissolution. The illustration for (E) in this column is a top view. (Reproduced from Sun, F. et al., *Adv. Funct. Mater.*, 14, 283, 2004. With permission.)

This solution-dipping strategy is universal and can be used for other metal or oxide ordered pore array films on any desired substrate by means of transferability of the PS colloidal monolayer from one substrate to the others. A series of other morphology-controlled ordered pore arrays, such as zinc, ZnO, Co_2O_3, CuO, CeO_2, Eu_2O_3, and Dy_2O_3, can be fabricated. Figure 7.18 shows the examples for Zn and CeO_2 pore array films on silicon substrate fabricated by this strategy.

7.2.3.3 Extensions

Using a colloidal monolayer as template, solution-dipping strategy can be well extended. For instance, using the sols to replace solutions, 2D ordered pore films of In_2O_3 [45], SiO_2 [46], TiO_2 [47], etc., can be fabricated. In addition, the porous films can be formed on any flat or even curved surface. Polymer hollow sphere arrays and ordered nanorod arrays can be also produced by an extended solution-dipping method. Here we show the extension methods mainly from our group.

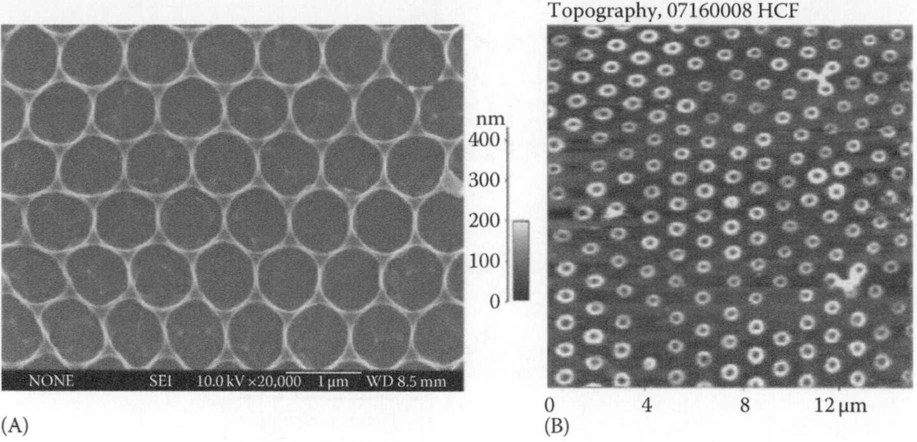

FIGURE 7.17 The sample from 0.002 M solution after drying at 80°C (2 h). (A) After subsequent heating at 400°C and ultrasonic vibration for a few minutes (instead of 0.5 h). (B) After subsequent ultrasonic washing in dichloromethane solution for 0.5 h. (Reproduced from Sun, F. et al., *Adv. Funct. Mater.*, 14, 283, 2004. With permission.)

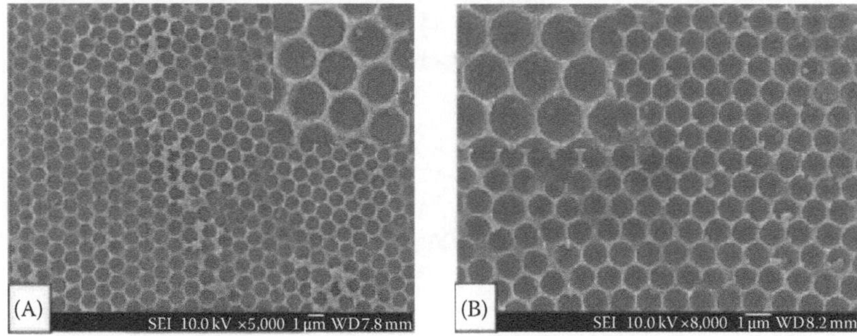

FIGURE 7.18 Pore array films of zinc (A) and CeO_2 (B) on a silicon substrate with the solution-dipping template method. Treatment conditions: (A) Precursor $Zn(OOCCH_3)_2$ (0.1 M), 80°C for 2 h and 200°C for 8 h in air, and 350°C for 4 h under H_2 followed by ultrasonic cleanout for 0.5 h. (B) Precursor-$Ce(NO_3)_3$ (0.05 M), 80°C for 2 h, and 200°C for 8 h followed by ultrasonic cleanout for 0.5 h. (Reproduced from Sun, F. et al., *Adv. Funct. Mater.*, 14, 283, 2004. With permission.)

7.2.3.3.1 Sol-Dipping Method [46]

Using a colloidal monolayer as template, dropping the preformed sol onto it, subsequently gelling, removing the template, and finally heat treating, some 2D ordered pore array films were fabricated with hierarchical pore structure.

Typically, Li et al. obtained SiO_2 ordered pore array film based on sol-dipping method [46]. SiO_2 precursor sol was prepared by mixing the tetraethyl orthosilicate, alcohol, and distilled water (molar ratio 1:4:20), and its pH value is controlled at about 1 by dropping nitric acid into it. A droplet (about 10 μL) of the precursor solution was first dropped onto the 2D monolayer colloidal crystal with 2 cm^2 using a quantitative pipette, which could infiltrate into the interstices between the substrate and the colloidal monolayer. The sample was then placed into a beaker sealed with a cover (solvent can evaporate very slowly) for one week at room temperature to form gel and dry. Subsequently, it was ultrasonically washed in CH_2Cl_2 for 2 min to dissolve the template, and then annealed in air at 500°C for 1 h. The 2D SiO_2 ordered pore array film was thus obtained, as shown in Figure 7.19. The isothermal nitrogen sorption for the skeletons of the SiO_2 ordered pore array film is illustrated

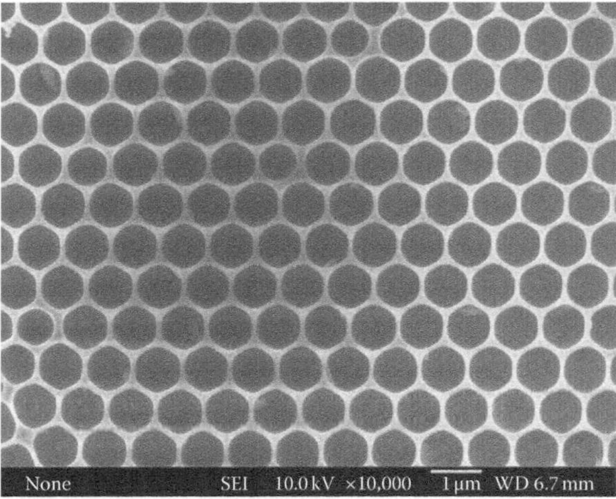

FIGURE 7.19 FESEM image of SiO_2 ordered pore array film by the sol-dipping method. (From Li, Y. et al., *Nanotechnology*, 17, 238, 2006. With permission.)

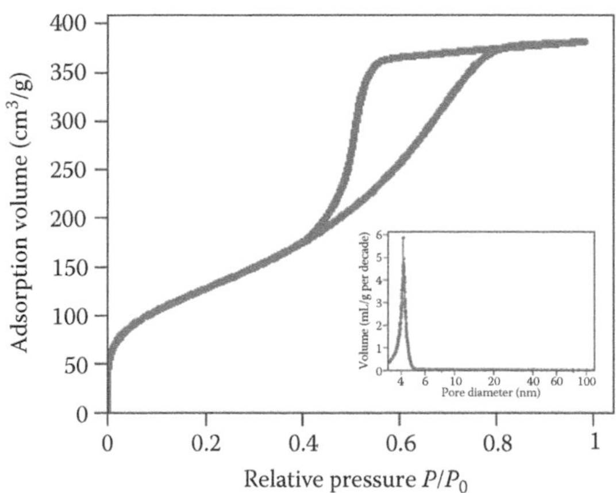

FIGURE 7.20 Isothermal nitrogen sorption curves for the skeleton of the 2D silica pore array shown in Figure 7.19. The inset is the pore size distribution. (Reproduced from Li, Y. et al., *Nanotechnology*, 17, 238, 2006. With permission.)

in Figure 7.20, which shows ink-bottle-shaped pores in the skeleton. The specific surface area of the skeletons was thus obtained to be 470.9 m^2/g. Analysis of pore diameter, by the sorption apparatus with an attached computer, shows that the pore diameter in the skeleton is sharply distributed around 4.2 nm (see the inset in Figure 7.20). The pores in the skeleton are disorderly arranged, interconnected, and open to ambient atmosphere. The porosity in the skeleton is estimated to be about 56% determined by the volume of saturated sorption [48]. The isothermal nitrogen sorption combined with the corresponding field-emission scanning electron microscope (FESEM) image indicates that the SiO_2 film is of the hierarchical structure consisting of orderly arranged macropores and disordered mesopores in its skeletons, as shown in Figure 7.21. Such hierarchical pore silica films have many potential applications in adsorbents, catalysts, chromatographic supports, microseparators, etc. [49,50].

Similarly, for In_2O_3, a wide band gap transparent n-type semiconductor, that has been widely used in the microelectronic field as selective transparent coatings, and selective gas-sensing materials, its ordered porous film can be also created by dropping the $In(OH)_3$ precursor sol on the

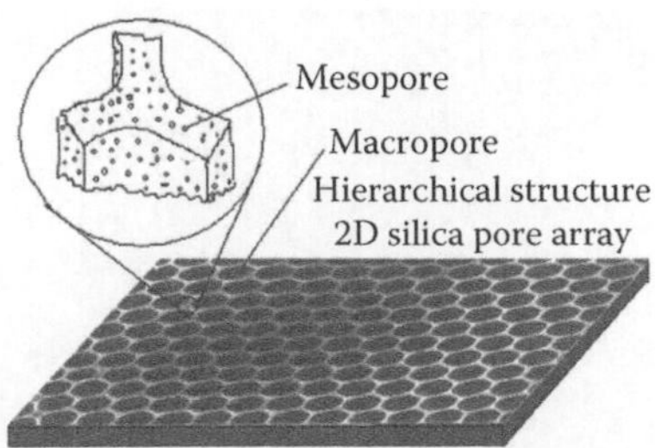

FIGURE 7.21 Schematic illustration of hierarchically structured 2D SiO_2 pore array film. (Reproduced from Li, Y. et al., *Nanotechnology*, 17, 238, 2006. With permission.)

colloidal template [45], subsequent drying at 110°C, and finally annealing at 400°C for 1 h in order to make it a decompose of In_2O_3, as shown in Figure 7.21. As we know, In_2O_3 is a good gas-sensing material. Such In_2O_3 ordered porous structure can increase the area of the sensing surface, which will promote adsorption and reaction between gases and the surface, and hence it can improve the sensitivity of this material to gas. Such properties can be useful in the design of a new generation of nano-gas sensors with high sensitivity and good selectivity.

In addition, another kind of functional material is TiO_2 that possesses good photocatalytic, optical, gas-sensing, and electronic properties, and its ordered pore array can be synthesized by the sol-dipping method. In this method, the TiO_2 precursor sol is prepared by acid hydrolysis of titanium isopropoxide. If the precursor concentration is 0.3 M, we can get the TiO_2 ordered pore array whose morphology is similar to Figure 7.21, which may find applications in photoelectric devices, photocatalytic surfaces, and gas sensors.

7.2.3.3.2 Direct Synthesis of Porous Film on Any Flat or Curved Surfaces [51]

As mentioned above, 2D large-area ordered porous films of different materials can be fabricated by the solution-dipping or the sol-dipping methods based on the PS colloidal monolayer. These porous films are generally synthesized on flat substrates, which is sufficient for certain applications. However, due to limitation of manipulation techniques, it is difficult to fabricate the porous films on the substrate with curved surfaces. In fact, for many practical applications and scientific research, the porous films need preparing on the substrate with cylindrical, spherical, or concave surfaces. The fabrication of micro/nanostructured porous films on curved surfaces will further expand the applicability of ordered porous films, especially for the construction of nanostructured devices, such as gas sensors on ceramic tubes. Although one can fabricate porous films on curved surface by some advanced lithography techniques, it has not been universal so far and is expensive. So fabrication of micro/nanostructured porous films on curved surface is a challenge.

It has been mentioned above that colloidal monolayer on a substrate can be stripped off in water and transferred onto other substrates. Recently, we have found that PS colloidal monolayer on glass substrate can also be peeled off in precursor solutions other than water, while still retaining their integrity. Based on this finding, a simple and effective strategy to directly fabricate the porous films on a curved (or non-flat) substrate by using a colloidal monolayer as template was presented [51], as shown in Figure 7.22. First, a colloidal monolayer on the flat glass substrate is slowly immersed in a precursor solution. The colloidal monolayer can be stripped off from the substrate and floated on the precursor solution. The interstitial space between the close-packed spheres is filled with the solution due to the capillary effects. Second, the floating colloidal monolayer is picked up by a desired substrate with a curved (or flat) surface. The colloidal monolayer covers the curved surface by filling the precursor solution in the interstitial space among the colloidal spheres and the space between the colloidal monolayer and the curved surface. Subsequently, the curved surface covered with the monolayer and the precursor solution is dried at a temperature slightly above the glass transition of PSs. With the solvent evaporating, the solute or hydrolyzate gradually deposits on the

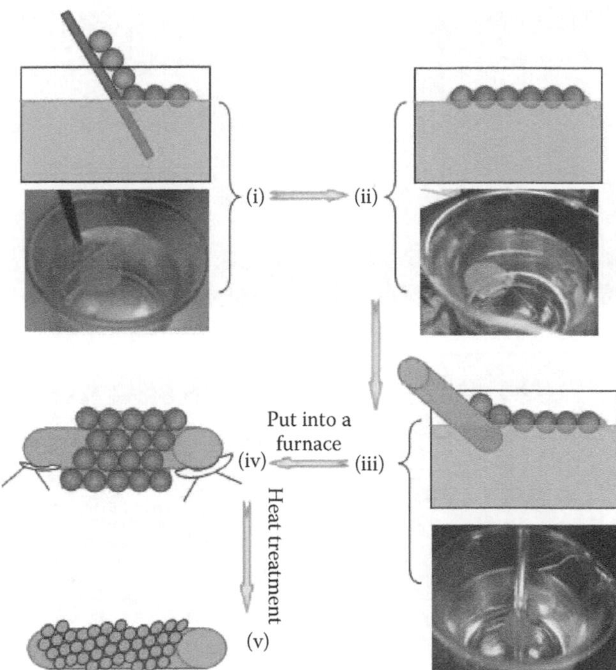

FIGURE 7.22 A route for fabricating ordered pore array films on a curved surface. (i) A flat glass substrate covered with a PS colloidal monolayer is dipped into the solution; (ii) the colloidal monolayer floats onto the precursor solution surface; (iii) the monolayer is picked up using a glass rod with a curved surface; (iv) the rod with the monolayer and the solution is heated in a furnace; (v) an ordered pore array film is formed on the curved surface after heat treatment and removal of colloidal monolayer. The lower frames in (i–iii) are photographs corresponding to the manipulations described. (Reproduced from Sun, F.Q. et al., *Adv. Mater.*, 17, 2872, 2005. With permission.)

PSs' surfaces and on the substrate. Finally, the sample is calcined to remove the colloidal template, and an ordered porous film is thus formed on the curved surface of the substrate. Since such film directly grows on the surface during the heat treatments, there will be strong adherence between the film and the substrate. By this method, we can fabricate the ordered porous film on any curved or flat surface. More importantly, the films can also directly be synthesized on the relatively rough surface of commercially supplied ceramic tubes. An important application of this strategy is direct construction of nanostructured gas sensors on such ceramic tubes, which are used for the substrates of conventional gas sensors.

Here we chose tin dioxide (SnO_2) as a model material to demonstrate the synthesis strategy above. Using 0.1 M $SnCl_4$ precursor solution and colloidal monolayers with PS diameter of 1000 nm, the ordered porous film of SnO_2 were fabricated on the curved surfaces, including the outer (Figure 7.23A) and inner (Figure 7.23B) surfaces of a glass tube, spherical steel surface (Figure 7.23C), as well as a flat surface (Figure 7.23D), respectively. Apparently, the morphology of all the ordered porous films on the curved surfaces is similar. The pore openings at the film surfaces are nearly circular. For such films, after heating at 500°C for 1 h, they adhere so strongly to the curved surfaces that they cannot be removed or destroyed, even by ultrasonic washing.

It should be pointed out that this strategy is universal and very flexible. In addition to the SnO_2 porous films, it can also be suitable for the synthesis of many other oxide ordered porous films on any desired substrate with a flat or a curved surface, for example, Fe_2O_3, TiO_2, ZnO, WO_3, and In_2O_3. Further, we can obtain the doped porous film by controlling the solution composition. More importantly, we can fabricate hetero-multilayer nanostructured porous films by repeating the procedures shown in Figure 7.23 and alternately using two or more different precursor solutions.

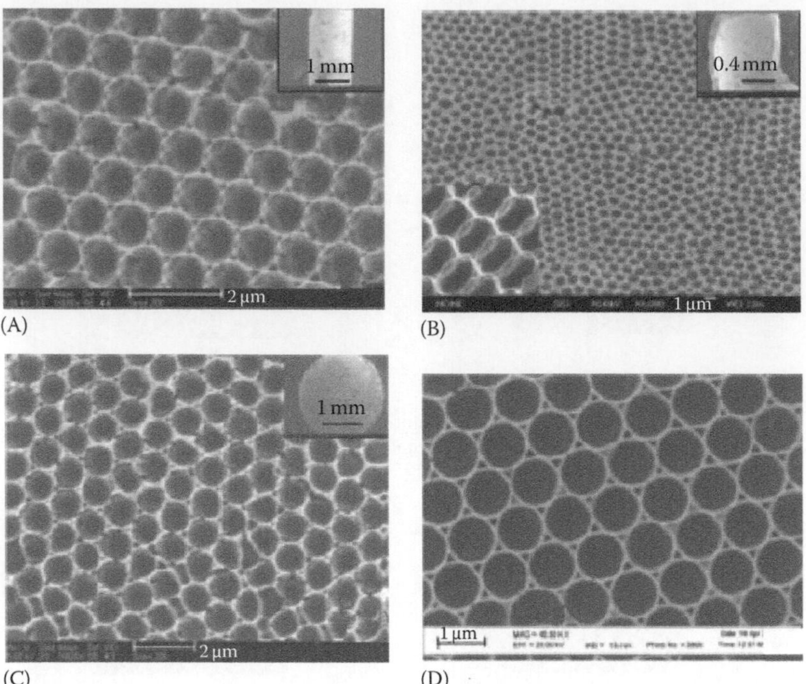

FIGURE 7.23 FESEM images of SnO_2 ordered pore array films on a curved surface. (A) On the outer surface of a glass tube (diameter: 1.2 mm), (B) on the inner surface of a glass tube (diameter: 1.0 mm), (C) on the surface of a steel sphere (diameter: 2.5 mm), and (D) on a flat surface. The insets show the corresponding low-magnification images. (Reproduced from Sun, F.Q. et al., *Adv. Mater.*, 17, 2872, 2005. With permission.)

7.2.3.3.3 Polymer Hollow Sphere Arrays Based on Two-Step Replication

Polymer hollow spheres and hollow sphere arrays with ordered arrangement have attracted much attention because of their potential applications in carriers of catalysts, enzymes, drugs, microreactor device, microacetabula, and some other devices in medicine and biotechnology [52–54]. Based on the colloidal monolayer template, we have developed a two-step replication strategy to obtain 2D hollow polymer sphere arrays with ordered arrangement on a substrate [52], as shown in Figure 7.24. Briefly, a 2D FeO(OH) ordered spherical pore array (film) is first fabricated based on our solution-dipping template strategy mentioned above, as illustrated in steps (a)–(c) of Figure 7.24A. Subsequently, the polymers polyvinyl alcohol (PVA), polyvinyl pyrrolidone (PVP), and polymethyl phenylsilane (PMPS) are dissolved into distilled water, alcohol, and cyclohexane, respectively. The polymer solutions (ranging from 20 to 50 g/L) is dropped on the 2D FeO(OH) ordered pore array until the pores are filled. Then the array full of the solution is capped with another glass substrate and reversed to locate the glass cover on the bottom. After sufficient drying at room temperature, the FeO(OH) pore array is removed by selective dissolution in 1 M oxalic acid solution. Finally, a 2D ordered polymer hollow sphere array is prepared by uncovering the top glass substrate, as shown in steps (d)–(f) of Figure 7.24A. Similarly, if the pores of the FeO(OH) are truncated, depending on $Fe(NO_3)_3$ concentration, we can obtain truncated hollow sphere arrays. Further, we can alternately drop the polymer precursor into the truncated pores of the FeO(OH) film and dry it until the pores are filled with the solid polymer. The filled arrays are covered with another glass substrate and reversed. In this way, we can obtain convex structure array, as displayed in Figure 7.24B.

Figure 7.25 shows FeO(OH) ordered spherical pore array prepared by 1.0 M $Fe(NO_3)_3$ precursor solution. We can know that the pore depth is close to the diameter of PSs. This inorganic pore array has many advantages to be used as a second template for the synthesis of other functional materials with special ordered structure, such as simple preparation, regular pore morphology with

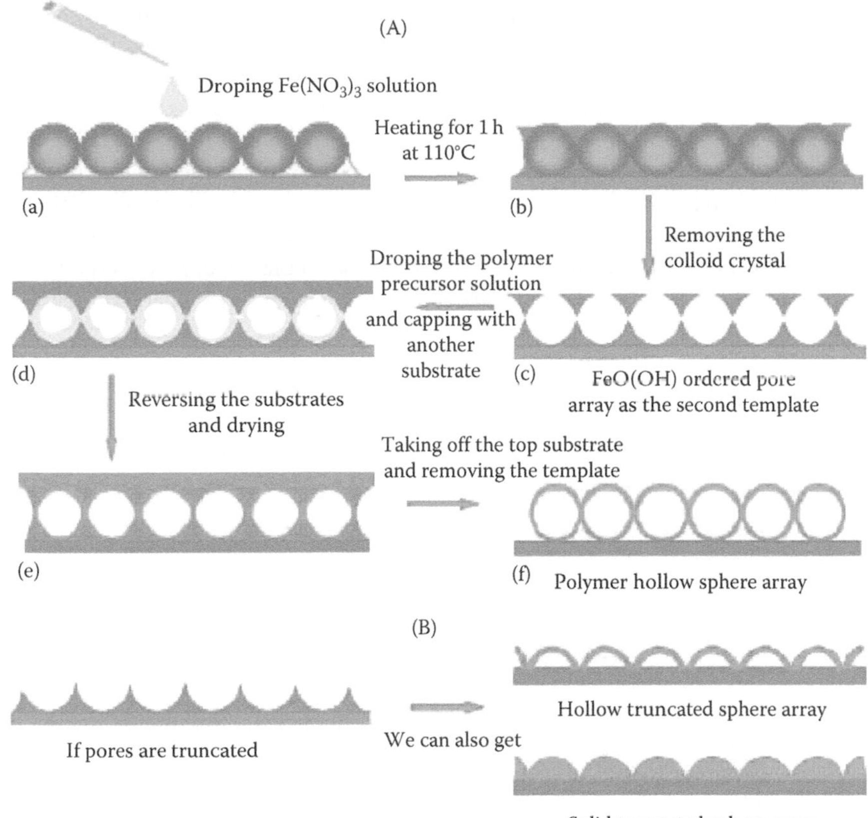

FIGURE 7.24 The fabrication strategy of the 2D polymer (A) ordered hollow sphere array and (B) hollow and solid truncated sphere arrays. (Reproduced from Li, Y. et al., *J. Mater. Res.*, 20, 338, 2005. With permission.)

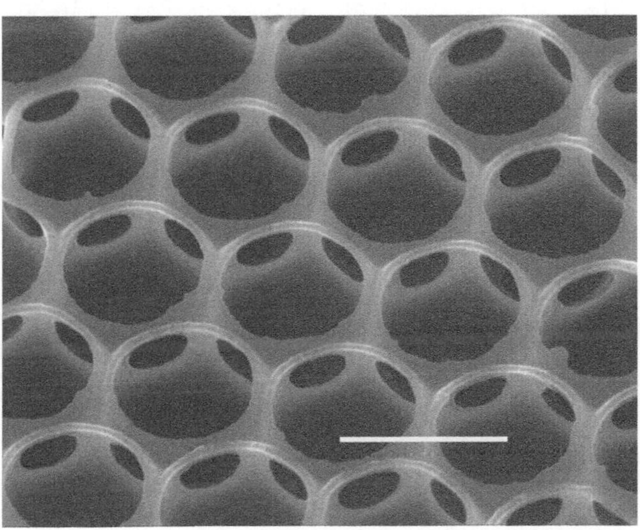

FIGURE 7.25 FESEM image of the pore array film, prepared by 1.0 M $Fe(NO_3)_3$ precursor solution and dried at 110°C for 1 h. The scale bar is 1 μm. (Reproduced from Li, Y. et al., *J. Mater. Res.*, 20, 338, 2005. With permission.)

spherical or truncated spherical shape, smooth pore wall, uneasy deformation, and low cost. What is more important is that it can be removed easily and quickly by selective dissolution in an oxalic acid solution.

Figure 7.26 shows the 2D PVA ordered array film based on the second template shown in Figure 7.25. Each unit in the film is a hollow sphere, which can be seen more clearly in the TEM image shown in Figure 7.26E and FESEM image in the edge region of the sample shown in Figure 7.26F. In the film, hollow spheres are in contact with each other and arranged orderly. When the concentration of PVA precursor solution is high enough (say, 50 g/L), hollow spheres are integrated (Figure 7.26A). With decrease of the concentration to 30 g/L, a small hole appears on some hollow sphere tops (Figure 7.26B), and if further decreased (to 20 g/L), the holes get bigger, and there exist such holes on top of the most hollow spheres in the array (Figure 7.26C), typically larger than 100 nm in size. We know that each spherical pore in the second template has six circular channels in its sidewall (Figure 7.25). However, in the corresponding hollow sphere array, no such channels are found. This can be confirmed by two neighboring hollow spheres in a rectangular frame shown in Figure 7.26F (see the sidewall indicated by an arrow) and the TEM image in Figure 7.26E. In

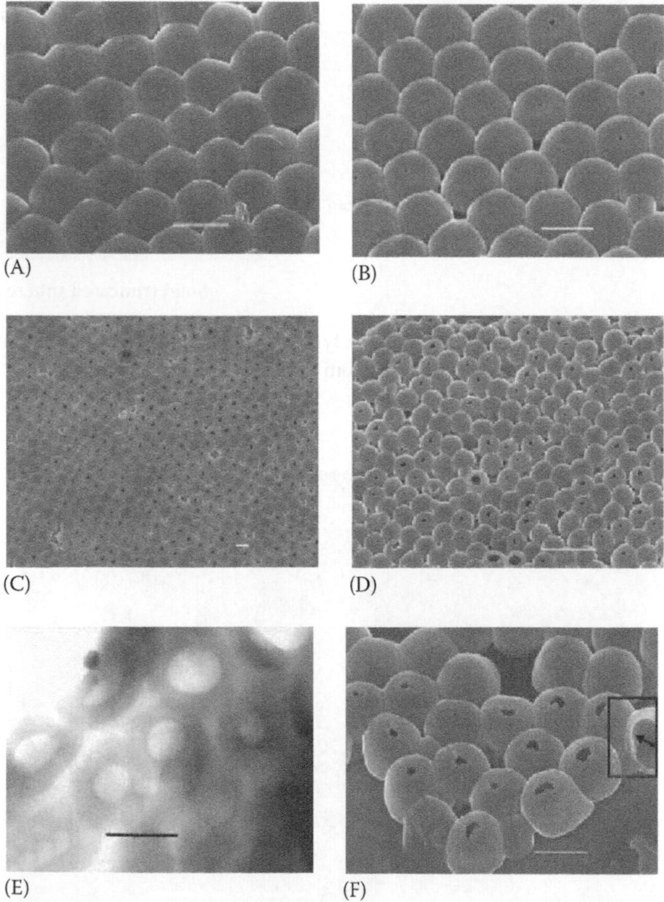

FIGURE 7.26 FESEM images of 2D PVA hollow sphere arrays based on the FeO(OH) ordered spherical pore array shown in Figure 7.25. Pore size is 1000 nm for (A– C) and 350 nm for (D). The concentrations of polymer precursor are (A) 50 g/L, (B) 30 g/L, (C, D) 20 g/L, (E) is the corresponding TEM image of sample (A) (by shaving the products from the glass substrate), and (F) is the magnified SEM image in the edge region of sample (C). The scale bar is 1 μm. (Reproduced from Li, Y. et al., *J. Mater. Res.*, 20, 338, 2005. With permission.)

the precursor concentration range used in our study, when the precursor solutions of the polymer evaporate and solidify, the polymer molecules will adhere to the pore wall of the second template and blockade channels in the sidewall. So the final polymer spheres have the closed sidewall (shell). When we change the periodicity of a second template to 350 nm, using a lower PVA concentration (20 g/L), the hollow sphere array (Figure 7.26D) can be also obtained, which is similar to that with 1000 nm in pore diameter (Figure 7.26C). The hollow spheres with small holes on the top could be useful in microreactor devices that can endure acidic and alkaline conditions, selective permeability, and nutrient and drug delivery. More interestingly, such a hollow sphere with a small hole on its top could be used for microacetabula because of the elastic polymer. In addition, it could also be an ideal model for studying black body radiation in micro- or even nanospace because the small holes on top of the hollow spheres correspond to the nano-sized black body. Further, when the PVA concentration is kept unchanged, the morphology of the hollow sphere array film is mainly determined by that of the second template. Figure 7.27A illustrates the second template prepared with a precursor concentration of 0.05 M, which shows the pore array with regular hexagon from top view. The corresponding hollow array film is shown in Figure 7.27B.

It has been indicated that the solid truncated sphere arrays can also be created by such a method. After the cycle of infiltration with 30 g/L PVA precursor and drying for five times, the solidified PVA is full of pores in the second template synthesized by 1000 nm PS and 0.08 M $Fe(NO_3)_3$ solution (Figure 7.28A). After removal of the template, a hemisphere or convex structure array film was fabricated, as shown in Figure 7.28B. TEM examination has demonstrated that the hemispheres are solid (inset in Figure 7.28B). Such arrays could be used for microlens arrays in miniaturized optical devices [55,56].

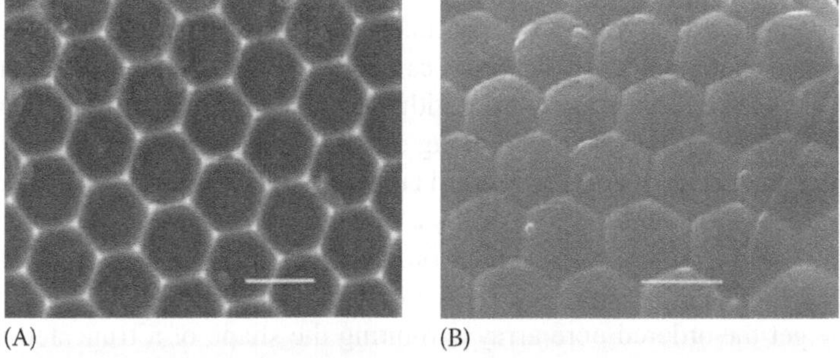

FIGURE 7.27 FESEM images of (A) the 2D FeO(OH) porous film prepared with 0.05 M precursor and (B) the corresponding 2D ordered PVA hollow sphere array fabricated by a 50 g/L polymer precursor. The scale bar is 1 μm. (Reproduced from Li, Y. et al., *J. Mater. Res.*, 20, 338, 2005. With permission.)

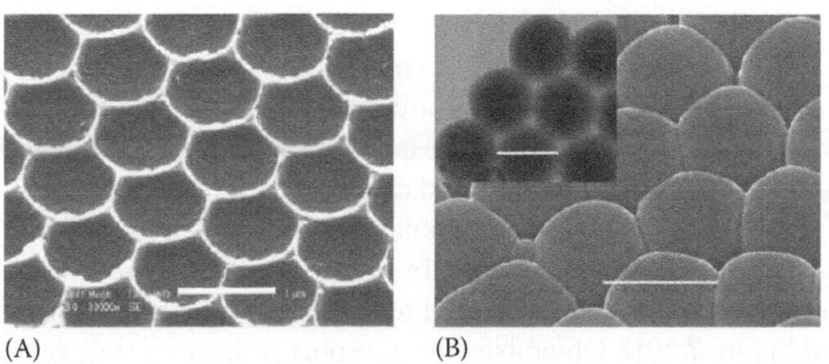

FIGURE 7.28 FESEM images of (A) the 2D FeO(OH) pore array prepared with 0.08 M precursor and (B) the corresponding PVA solid truncated sphere array fabricated by five times of cycle of infiltration with 30 g/L PVA precursor and drying. The inset in (B) is its corresponding TEM image. The scale bar is 1 μm. (Reproduced from Li, Y. et al., *J. Mater. Res.*, 20, 338, 2005. With permission.)

Similarly, we can also synthesize the PMPS and PVP 2D ordered hollow sphere array film and the convex structure array film with similar morphology to that of PVA. Further experiments indicate that many soluble polymer materials, which can solidify when being concentrated, can be used to fabricate the hollow sphere arrays by this strategy.

The mechanism for hollow sphere formation can easily be understood. The solvents can evaporate through the rim cracks during drying. As the solvent evaporates gradually, the polymer will concentrate further and adhere to the pore wall of the second template. After the second template is removed, the sphere shell will be left. The holes on the top of the hollow spheres are associated with precursor concentration. The formation of the holes can be attributed to evaporation of solvent. During drying, the level of the solution surface within pores of the second template decreases, and the concentration increases. Due to the affinity between polymer molecules and pore walls, some polymer molecules will attach on the pore wall. Obviously, the shell formed on top of the spherical or the truncated spherical hollow is the thinnest. If the concentration of polymer precursor solution is low enough, a hole will be formed on the top because of the very thin shell.

7.2.3.3.4 Ordered Nanopillar Arrays Based on Heating-Induced Template Deformation

It is well known that the patterned 1D nanostructures (nanorod or nanopillar) with a small aspect ratio for potential applications in sensor arrays, piezoelectric antenna arrays, optoelectronic devices, interconnects, and superhydrophobic and self-cleaning surfaces. Using the colloidal monolayer as a mask, metal periodic nanoparticle arrays can be fabricated, and then using them as catalysts or new masks, large-area nanorod/nanopillar arrays were fabricated by epitaxial approach on different substrates or active plasma etching [34,57]. Recently, based on the 2D heat-deformed colloidal monolayers, we have presented a simple approach to prepare large-scale hexagonal nanopillar arrays by the solution-dipping method [58]. In this method, PS colloidal monolayers are first synthesized on the substrates and heated at 120°C for a certain time, the colloidal monolayer is deformed so much due to heating that the triangular prism channels can be formed at the interstices among them. And then, using such deformed colloidal monolayer with channels as template, periodic nanopillar array film can be directly fabricated by solution dipping or sol dipping, as described in Figure 7.29. This method makes the applications of the PS colloidal crystal templates more plentiful, and opens a new way to produce patterned 1D nanostructure arrays.

Figure 7.30 shows FESEM images of Fe_2O_3 ordered structures by the heat-induced, deformed colloidal monolayer templates. Using the original colloidal monolayer without heating, as a template, we can only get the ordered pore array exhibiting the shape of a truncated hollow sphere if using 0.5 M precursor solution, as shown in Figure 7.30A. However, when such PS colloidal monolayer is sintered at 120°C for 15 min, its morphology exhibits a large change. Due to heating at the temperature above the glass transition of PSs ($T_g \approx 100°C$) [59,60], PSs are obviously deformed, leading to contact evolution between neighboring PSs from quasi point contact (see the inset in Figure 7.30A) to facet contact (see the inset in Figure 7.30B). Correspondingly, triangular-shaped channels are formed among PSs. Based on such deformed templates, one obtained the ordered nanopillar array film, as illustrated in Figure 7.30B and C. Its morphology is very different from those obtained by templates without heating. Such nanopillars are of triangular prism shape with an aspect ratio of about 1.5:1. The whole arrays also exhibit hexagonal alignment. The nanopillars do not grow directly on the substrate, but on the nodes of the skeleton network surrounding the previous bottoms of the sintered PSs. However, if the colloidal monolayer is overheated (say 25 min), the channels will almost disappear in the templates (see the inset in Figure 7.30D). Correspondingly, no pillars will be formed due to non-channels in the overheated template, and the morphology takes on a regular network, as shown in Figure 7.30D. Obviously, the appropriate heating time of the PS colloidal templates is a key factor in fabricating periodic nanopillar arrays by this strategy. In addition, if we use the monolayer polymer colloidal template with smaller PSs (350 nm), higher density of Fe_2O_3 nanopillar arrays can be also prepared, as displayed in Figure 7.31. Besides Fe_2O_3, other material (such as SiO_2) ordered nanopillar arrays could be also fabricated by the strategy presented here.

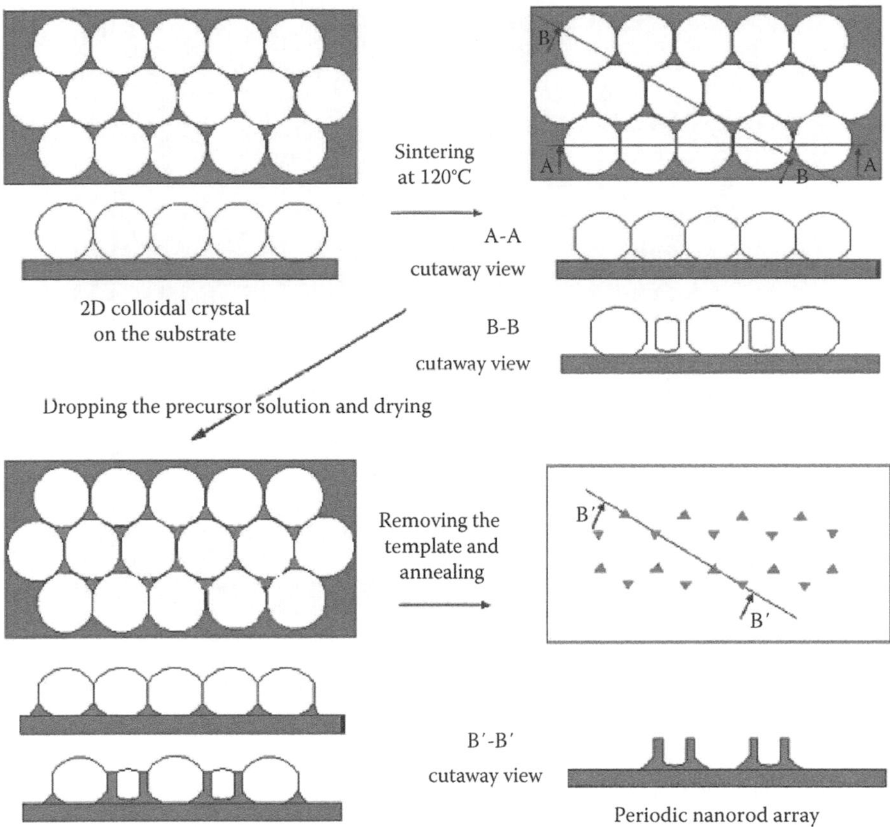

FIGURE 7.29 Schematic illustration of the periodic hexagonal nanopillar array. (Reproduced from Li, Y. et al., *Polymer*, 46, 12033, 2005. With permission.)

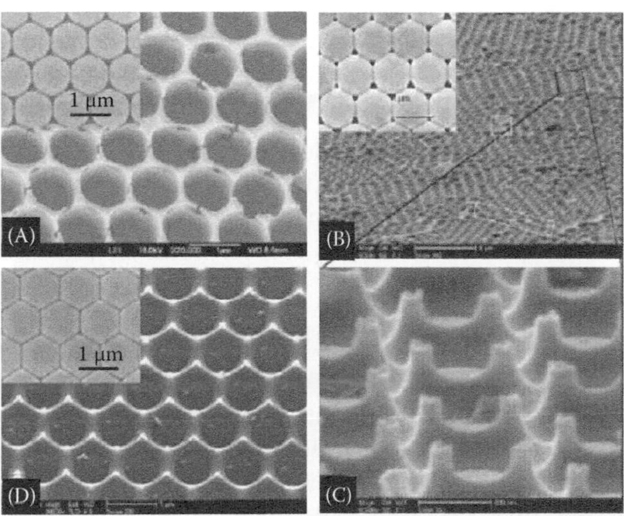

FIGURE 7.30 FESEM images of Fe_2O_3 ordered patterns fabricated with 0.5 M $Fe(NO_3)_3$ precursor solution, based on a colloidal monolayer with 1000 nm PSs in diameter sintered for different time at 120°C. The insets in A, B, and D are colloidal monolayer sintered for 0, 15, and 25 min, respectively, and A, B, and D are corresponding periodic nanostructures prepared using such templates. D is the local magnification of B. (Reproduced from Li, Y. et al., *Polymer*, 46, 12033, 2005. With permission.)

FIGURE 7.31 The morphology of a Fe_2O_3 nanopillar array fabricated by a colloid monolayer with 350 nm PSs in diameter sintered at 120°C for 6 min. The inset is a local magnified image. (Reproduced from Li, Y. et al., *Polymer*, 46, 12033, 2005. With permission.)

7.2.4 Electrodeposition Strategy

In past years, electrodeposition technique has drawn more and more interests in the field of film fabrication due to its low cost in equipment, facility in manipulation, and controllability in surface morphology of the film. Combining the colloidal monolayer with the electrodeposition strategy, one can obtain many kinds of micro/nanostructured films with controllable morphology. In this section, we will review the progress in micro/nanostructured films derived from such combinations. The main contents come from a recent work of our group.

7.2.4.1 Direct Electrodeposition

Direct electrochemical deposition (electrodeposition) on the PS colloidal monolayers has proved to be an effective method for the fabrication of regular micro/nanostructured array film with diverse morphologies such as ring, pore, hollow sphere, and hierarchical structure. In addition, this method allows large area in film scale, and universality in materials [17,61–66].

7.2.4.1.1 Synthesis Route

A typical electrodeposition method based on colloidal monolayer is illustrated in Figure 7.32. First, a centimeter-sized monolayer colloidal crystal is fabricated on a cleaned glass substrate by spin-coating or dip-coating, and transferred to a conductive substrate (Figure 7.32B), such as indium tin oxides coated glass (ITO glass). It is then heated for a given time at a certain temperature to sinter it firmly on the new substrate (Figure 7.32C). Finally, this conductive substrate is used as the working electrode in a three-electrode electrolytic cell, with a graphite plate as the auxiliary electrode, a saturated calomel electrode (SCE) as the reference electrode, and the aqueous electrodeposition solution as the electrolyte. After electrodeposition at a constant electric potential (or current density) for a certain time, the colloidal monolayer template is removed; different 2D ordered nanostructures can be obtained depending on deposition parameters. The morphology can be controlled by heating time before deposition and deposition time.

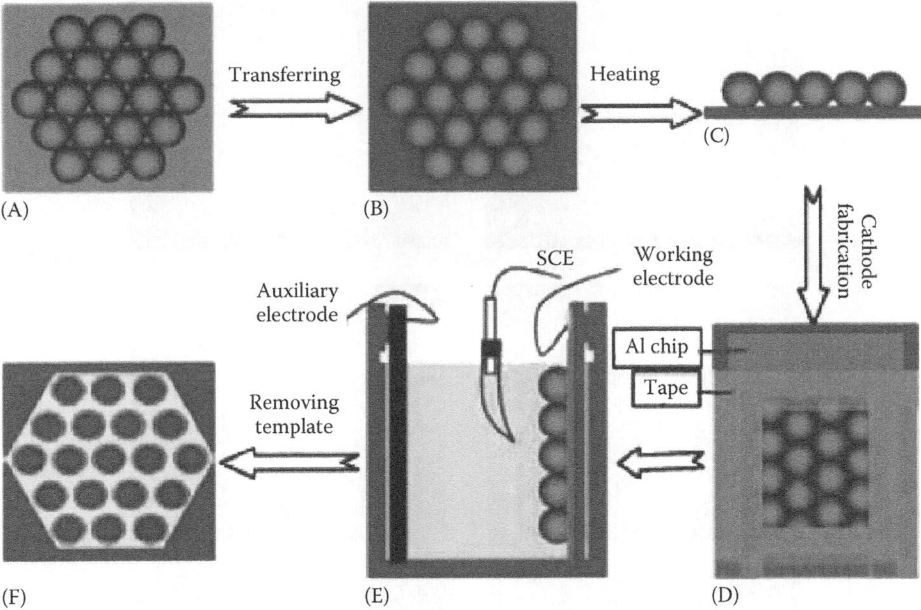

FIGURE 7.32 Schematic illustration of the electrodeposition approach based on the colloidal monolayer. (A) Colloidal monolayer on the glass substrate; (B) the monolayer on the conductive substrate (such as ITO glass or silicon); (C) the monolayer fixed on the conductive substrate by heating; (D) design of the cathode; (E) electrodeposition in a custom-built cell; (F) Ordered through-pore array after removal of the monolayer template. (Reproduced from Sun, F. et al., *Adv. Mater.*, 16, 1116, 2004. With permission.)

7.2.4.1.2 Ordered Pore Array Films of Metallic Materials

Taking gold as the model material of metallic materials, ordered pore array film is demonstrated. Figure 7.33 shows gold through-pore arrays fabricated after electrodeposition for different times and the removal of the PS colloidal templates by dissolution in methylene chloride (CH_2Cl_2). A solution composed of $HAuCl_4$ (12 g/L), EDTA (5 g/L), Na_2SO_3 (160 g/L), and K_2HPO_4 (30 g/L) was used as the electrolyte; the pH value was 5. The deposition was carried out at 45°C and 0.7 V vs. SCE. With increase of heating time for the PS colloidal monolayer, the opening pores at the bottom evolve from an irregular shape (Figure 7.33A and B) to circles (Figure 7.33C and D), and the pore size also increases due to a heating-induced rise in contact area between the PSs and the substrate. Additionally, the film thickness and openings at the film surface can be controlled by the electrodeposition time. With a shorter electrodeposition time (8 min, for instance), the film thickness is thinner and the pore diameters at the film surface are smaller than those of the PSs, as shown in Figure 7.33C. However, if the electrodeposition is long enough, the film thickness will be greater than the distance between the center of the PSs and the substrate, and the pore diameters at the film surface will also be smaller than those of the PSs, as displayed in Figure 7.33E. Similar results can be obtained with monolayer templates composed of much smaller PSs, as shown in Figure 7.33F: the ordered pore arrays fabricated by a 350 nm PS monolayer with heating and electrodeposition time of 3 and 5 min, respectively, indicating that the pore distribution density can be tuned by the colloidal monolayer templates with different colloidal sphere sizes [17].

The formation of gold porous film was also investigated. An edge region of the sample shown in Figure 7.33D was observed carefully, as shown in Figure 7.34. We can see that the pores are not orderly arranged, but it can tell us more details of the film growth. In the area without any PSs, deposited gold films were not found. It means that the film only grows around PSs. When a single PS is found on the substrate (zone A in Figure 7.34), the gold just grows around the sphere and forms a hemisphere-like shell with a fixed thickness. However, when two spheres are arranged closed to each other (zone B), the region between the two spheres and the substrate is obviously different from

FIGURE 7.33 Ordered gold through-pore arrays by an electrodeposition method based on a colloidal monolayer. The sphere sizes in the monolayer are 1000nm for (A–E) and 350nm for (F). The polystyrene templates were heated at 110°C at an electrodeposition potential of 0.7V vs. SCE. The heating time and the deposition time are, respectively, (A) 2 and 15min, (B) 5 and 15min, (C) 16 and 8min, (D) 40 and 8min, (E) 16 and 30min, (F) 3 and 5min. (Reproduced from Sun, F. et al., *Adv. Mater.*, 16, 1116, 2004. With permission.)

others in property due to the surface charge of the spheres: the amount of adsorbate is less, and as a result, the growth in such regions will be faster. For the same reason, when the three spheres are close to each other (zone C), the growth in the region between the three spheres and the substrate is also faster than in other regions. A more clear case is that where the spheres are arranged neither too close nor too far from each other (zone D). The pores (from top view) are interconnected by a film of a certain thickness, and the thickness of the film between the pores is smaller than that surrounding the pores. Based on this point, it is easy to understand the formation of the closely packed pore array shown in Figure 7.33 if all the PSs are closely packed in a monolayer.

By the electrodeposition method, other metal pore array films can also be fabricated, such as nickel, tin, zinc, and silver, as shown in Figure 7.35, using the following electrolyte and depositing conditions, respectively:

A. Ni, electrolyte: mixture of $NiCl_2$ (0.01M) and $(NH_4)_2SO_4$ (0.03M); pH value=8; heating time of PS colloidal monolayer: 2min; current density: 0.25mA/cm^2; auxiliary electrode: nickel sheet; deposition time: 30min.
B. Sn, electrolyte: mixture of $SnCl_2 \cdot 2H_2O$ (40g/L), NaF (20g/L), $N(CH_3COO)_3$ (15g/L), and polyglycol (M=4000~6000) (6g/L); pH=4.5; heating time of PS colloidal monolayer: 15min; voltage: 1.5V (vs. SCE); deposition time: 30min.

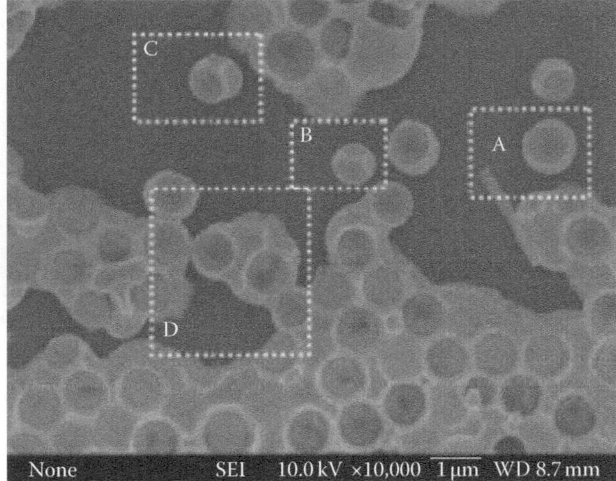

FIGURE 7.34 Morphology in the edge region of the sample shown in Figure 7.33D. (A) A pore from one PS. (B) Pores from two spheres in close contact. (C) Pores from three spheres in close contact. (D) Pores from spheres not in very close contact. (Reproduced from Sun, F. et al., *Adv. Mater.*, 16, 1116, 2004. With permission.)

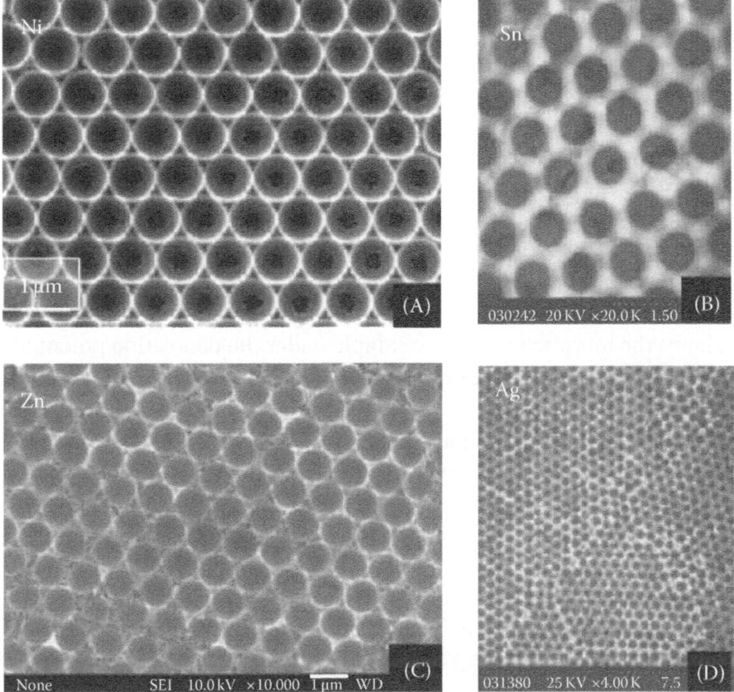

FIGURE 7.35 Other metal ordered pore array films by an electrodeposition method based on a colloidal monolayer with PS size of 1000 nm. Materials: (A) Ni; (B) Sn; (C) Zn; (D) Ag.

C. Zn, electrolyte: mixture of $ZnSO_4 \cdot 7H_2O$ (80 g/L), H_3BO_3 (20 g/L), pH = 5; heating time of PS colloidal monolayer: 15 min; voltage: 0.7 V (vs. SCE); deposition time: 30 min.

D. Ag, electrolyte: mixture of $AgNO_3$ (45 g/L), $(NH_4)S_2O_3$ (250 g/L), NH_4AC (20 g/L), and Na_2SO_3 (80 g/L); pH = 5; heating time of PS colloidal monolayer: 3 min; voltage: 1.5 V (vs. SCE); deposition time: 30 min.

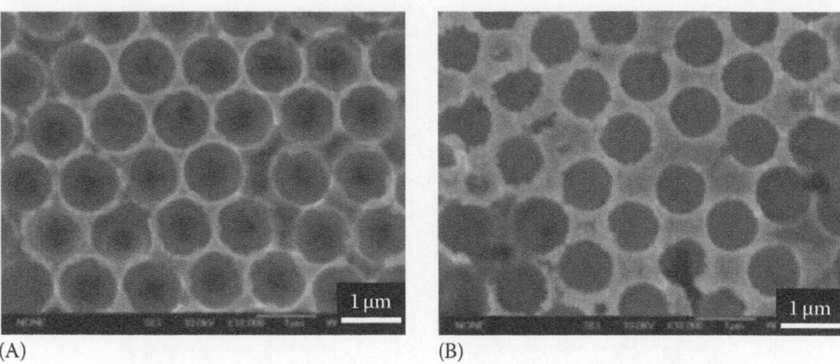

FIGURE 7.36 FESEM images of a Au–Cu bimetal pore array film. Deposition time of Au–Cu: (A) 10/2 min; (B) 10/5 min.

Also, using this strategy, the bimetal pore array film can be fabricated by two-step electrodeposition via changing different electrolytes. For example, 2D gold ordered pore array was first electrodeposited on the conducting substrate in gold precursor electrolyte for 10 min under voltage of 0.7 V (vs. SCE), and then transferred to the electrolyte composed of EDTA (44 g/L), $CuSO_4 \cdot 5H_2O$ (28 g/L), Na_2SO_3 (5 g/L), and CH_4N_2S (0.5 g/L) under voltage of 1.0 V (vs. SCE). Figure 7.36 shows the samples electrodeposited for different times for various compositions: Au (10 min)/Cu (2 min) and Au (10 min)/Cu (5 min), respectively. Corresponding XRD spectrum (not shown here) indicates that such pore array film is composed of two kinds of metals: Au and Cu. Fabrication of bimetal pore array films opens a new way to devise micro- or nanodevices.

7.2.4.1.3 Ordered Pore Array Films of Semiconductor Materials

Besides metallic pore array films, metal oxide, and some semiconductor array films can be prepared by electrodeposition based on the PS colloidal monolayer. Representatively, zinc oxide pore array films can be created on an ITO glass and (111)-oriented Au/Si substrate, using the electrolyte of zinc nitride and pure zinc sheet as a counter electrode, as demonstrated in Figure 7.37 [61,62].

Figure 7.37A shows the morphology for the sample under the deposition potential of 1.0 V, exhibiting truncated spherical hollow array. Corresponding XRD data (curve b in Figure 7.38) indicates that all diffraction peaks of the sample can be identified to those of wurtzite ZnO, and there exists a slightly preferred orientation of (002) in the polycrystalline ZnO film. For the samples on the Au/Si substrates, the deposition currents have little effect on the film morphology, showing similar morphology in a current range of 0.7–1.2 mA, as shown in Figure 7.37B through D corresponding to the applied current of 0.9 mA. XRD spectrum (curve (a) in Figure 7.38) shows only two peaks at 34.4° and 38.2°, indexed by ZnO (002) and face-centered-cubic gold (111) peaks, respectively, indicating that the film is highly oriented in crystal structure and most of ZnO (001) and Au (111) planes are parallel to the substrate surface (the peak of Si (400) is at 69°). FESEM observation demonstrates that the pores are highly orderly arranged in the film, which is the reverse replica of the PS colloidal monolayer template. From its magnified SEM image in Figure 7.37C, the skeleton seems to be packed by the block-units ZnO nanosheets parallel to substrate surface, or the morphology shows step-structure from top view. However, the cross-section observation demonstrates that the skeleton is composed of densely aligned ZnO hexagonal nanocolumns with a small difference in heights (see Figure 7.37D). Such highly oriented ZnO ordered pore arrays should have excellent optical properties, such as strong photoluminescence at room temperature, etc., which could have potential applications in future functional optical devices.

The oriented growth of the porous ZnO film can be mainly attributed to the oriented substrate, which leads to the formation of oriented nuclei due to the lattice match between the substrate and

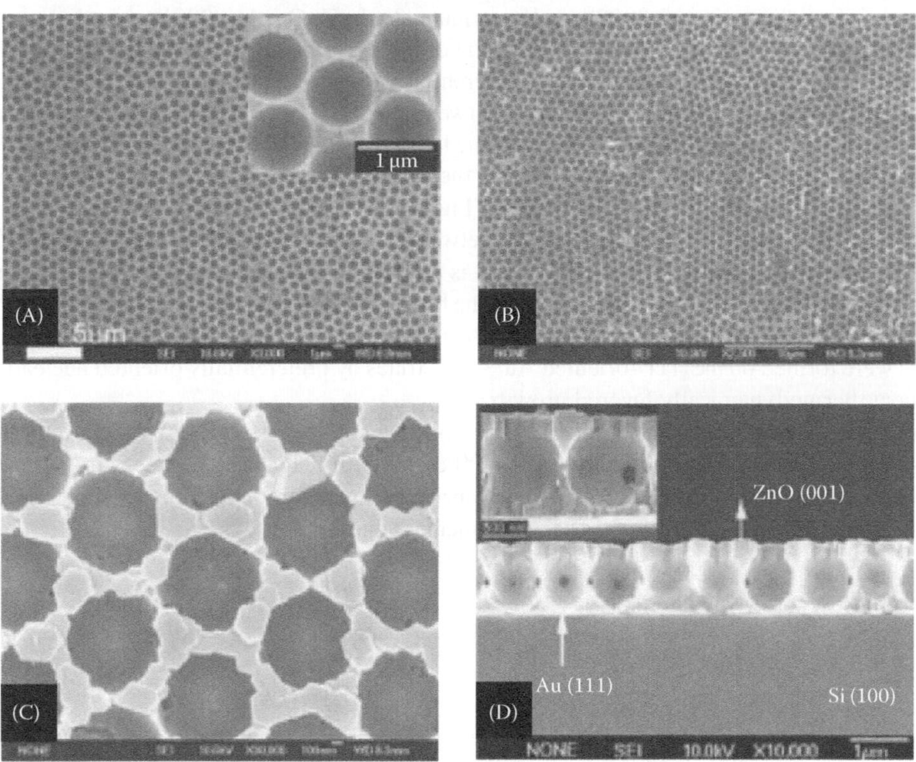

FIGURE 7.37 FESEM morphology of ZnO ordered pore arrays. (A) ZnO film on an ITO-glass substrate 1.0 V, 120 min, (B) ZnO film on a Au/Si substrate 0.9 mA, 120 min, (C) Magnified image of (B), (D) Cross-section image of (C). (Reproduced from Cao, B.Q. et al., *Electrochem. Solid-State Lett.*, 8, G237, 2005. With permission.)

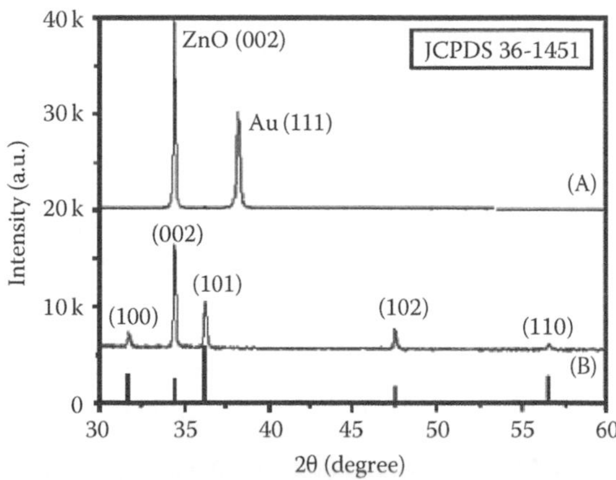

FIGURE 7.38 XRD spectra of as-synthesized films electrodeposited on different substrates (A) An/Si, 0.9 mA, 120 min, and (B) ITO glass, 1.0 V, 120 min. (Reproduced from Cao, B.Q. et al., *Electrochem. Solid-State Lett.*, 8, G237, 2005. With permission.)

ZnO. The electrodeposition conditions on the whole working electrode (substrate) in the electrolyte are homogeneous and, hence, the nucleation can occur at any site on the substrate that is not covered with the PS monolayer. Because the ITO-glass substrate is amorphous in structure, and there is no influence of epitaxy during the initial nucleation period, the crystal nuclei of ZnO will be randomly oriented on the substrate, leading to the ZnO skeleton without obvious preferred orientation. The slightly preferred orientation of (0002) can be attributed to the ZnO polar (0001) crystal plane, which has higher surface free energy compared with the other basal planes of (01$\bar{1}$0) and (2$\bar{1}\bar{1}$0) [67]. For the (111)-oriented Au/Si substrate, however, preferentially oriented ZnO nuclei will be formed on the substrate to reach the lowest interface energy between the ZnO and the gold film. It is well known that the interface energy is directly related with the lattice mismatch of interfaces. To lower the interface energy, the orientation relationship between the ZnO nuclei and the Au(111)/Si substrate can be deduced to be ZnO (0001) [11$\bar{2}$0]//Au(111) [$\bar{1}$10]. Figure 7.39 schematically shows such relationship, which results in the smallest lattice mismatch between the ZnO film and the substrate. The lattice mismatch along ZnO <11$\bar{2}$0> and the Au <$\bar{1}$10> is about 12.7%. In addition, fast growth along the *c*-axis direction is energetically favorable due to the higher surface free energy of polar (0001) crystal planes. Thus, the (001)-oriented ZnO ordered pore arrays with flat top surfaces and hexagonal nanocolumns were formed on the (111)-oriented Au/Si substrates by preferentially oriented nucleation and subsequent thermodynamically favored growth.

In addition to zinc oxide, other metal oxides or semiconductor pore array films can be also synthesized by electrodeposition based on the PS colloidal monolayer template. For instance, the Eu_2O_3, CdS, Fe_2O_3 pore array film can also be fabricated by electrodeposition, as shown in Figure 7.40, using the following electrolytes and experimental conditions:

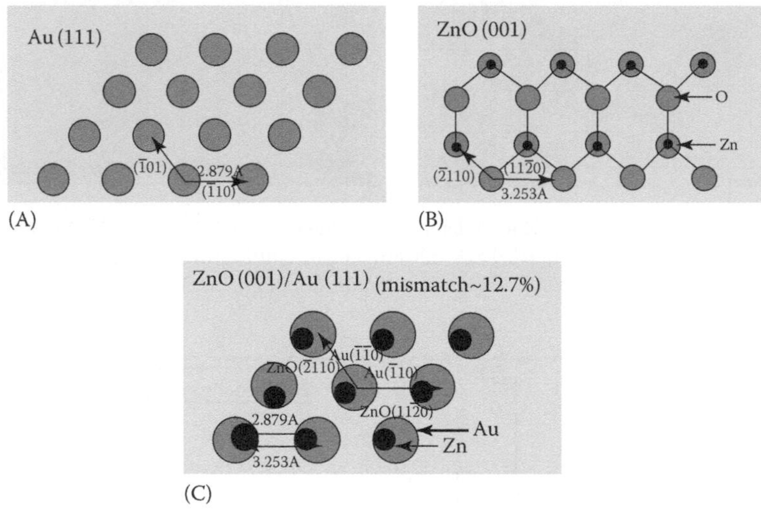

FIGURE 7.39 Schematic drawings of crystal planes: (A) Au (111), (B) ZnO (001), and (C) the epitaxial relationship of ZnO (001)/Au (111). (Reproduced from Cao, B.Q. et al., *Electrochem. Solid-State Lett.*, 8, G237, 2005. With permission.)

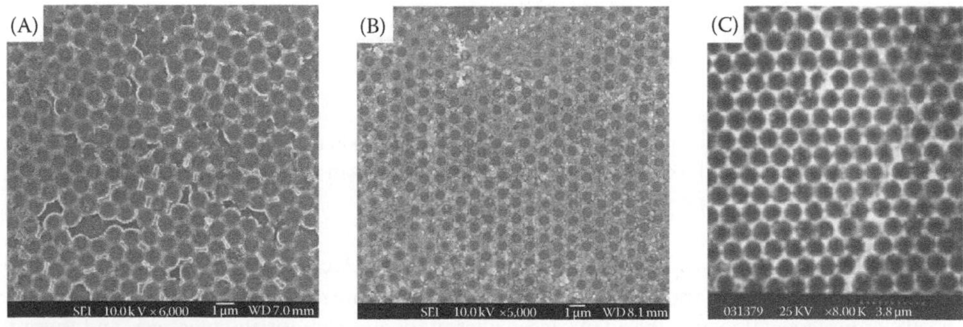

FIGURE 7.40 Ordered pore array films by an electrodeposition method based on a colloidal monolayer with PS size of 1000 nm. Materials: (A) Eu_2O_3, (B) CdS, (C) Fe_2O_3.

A. Eu_2O_3, electrolytes: mixture of $EuCl_3$ (0.1 M), sodium citrate (0.1 M), pH = 4; heating time of colloidal monolayer: 40 min, voltage: 1.0 V (vs. SCE); deposition time: 20 min; finally, keeping the sample in furnace with hydrogen ambience at 400°C for 2 h.
B. CdS, electrolytes: mixture of $CdSO_4$ (0.05 M), $Na_2S_2O_3$ (0.1 M), CH_4N_2S (0.05 M), pH = 5; heating time of colloidal monolayer: 10 min, voltage: 1.0 V (vs. SCE); deposition time: 10 min.
C. Fe_2O_3, electrolyte, $Fe(NO_3)_3$ (0.1 M); pH = 5; heating time of colloidal monolayer: 5 min, voltage: 1.0 V (vs. SCE); deposition time: 30 min; finally, keeping the sample in furnace at 500°C for 2 h.

7.2.4.1.4 Ordered Hollow Sphere Array Films

Besides common pore array films, ordered hollow sphere array films were also prepared based on colloidal monolayer. For example, Ni hollow sphere array films were successfully fabricated by the direct electrodeposition method at our group [63]. Figure 7.41 shows the whole synthesis procedures based on the PS colloidal monolayer. Figure 7.42 shows the morphology of the sample after electrodeposition for 90 min at a low current density (0.25 mA/cm^2) and removal of the PS template. The Ni hollow sphere arrays are formed. The thickness of the sphere shell is about 60 nm from the broken hollow spheres. Additionally, it has been revealed that the deposition time is important

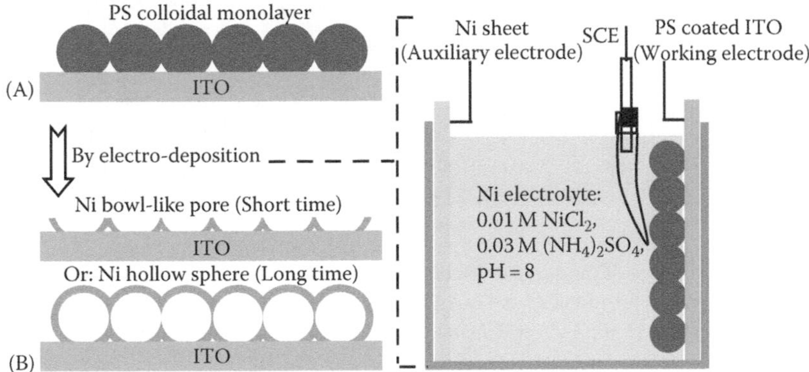

FIGURE 7.41 Schematic Ni electrodeposition procedures based on the colloidal monolayer. (A) A colloidal monolayer sintered on the ITO-glass substrate by heating. (B) Ordered bowl-like pore arrays or hollow sphere arrays after removal of the monolayer electrodeposited for short time or long time. The right: electrodeposition in a custom-built cell. (Reproduced from Duan, G.T. et al., *J. Phys. Chem. B*, 110, 7184, 2006. With permission.)

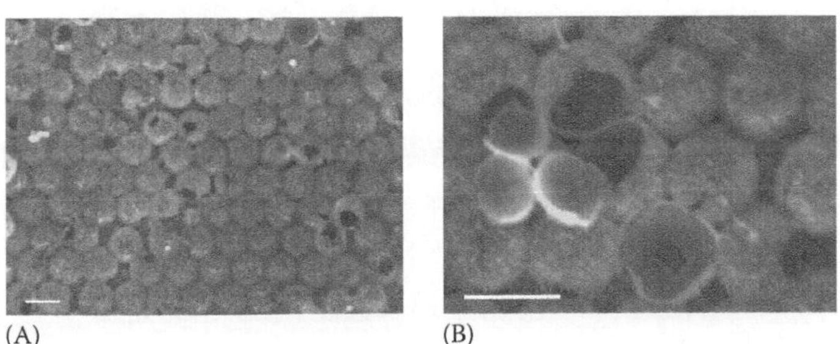

FIGURE 7.42 FESEM images of the as-deposited sample after removal of the colloidal monolayer. (B) is the magnification of (A). The sample was electrodeposited for 90 min at cathodic deposition current density 0.25 mA/cm^2. Scale bars are 1 μm. (Reproduced from Duan, G.T. et al., *J. Phys. Chem. B*, 110, 7184, 2006. With permission.)

to the formation of the hollow sphere arrays under the constant current. If the deposition time is decreased to 45 min, morphology of the sample evolves from a hollow sphere to a bowl-like structure. When the deposition time is further decreased to 15 min, only nearly spherical nanoparticles were obtained on the substrate.

Further experiments demonstrate that the cathodic deposition current density is crucial to the formation of a hollow sphere structure. When the current density is increased to 1.0 mA/cm^2, the bowl-like ordered porous arrays can still be formed after a short deposition time (such as, 12 min), while hollow sphere arrays cannot be obtained after increasing deposition time (see Figure 7.43A). A middle current density of 0.4 mA/cm^2 leads to a transition. A hollow spherical structure is still formed after a long deposition time, while the spherical shell are composed of coarse particles and thus forming a much rougher surface (see Figure 7.43B). These Ni hollow sphere array films show important morphology-related magnetic properties. This could be of importance both in fundamental magnetic researches and in device applications [63].

The formation of the hollow sphere arrays can be easily understood. When the cathodic current density, J, is low (to say 0.25 mA/cm^2), first, Ni nuclei were preferentially formed at the bottom interstitial sites between PSs and the substrate (see the solid arrow in Figure 7.44), and then bowl-like ordered porous arrays after shorter deposition will be formed and hollow sphere arrays will be

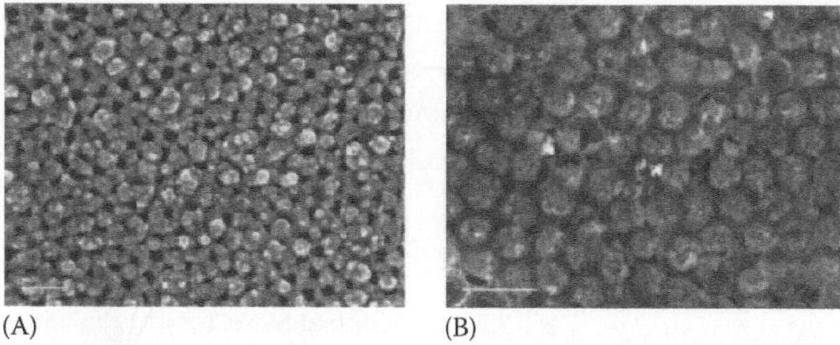

FIGURE 7.43 FESEM images of the samples electrodeposited for (A) 25 min at cathodic deposition current density $J = 1.0$ mA/cm^2 and (B) 60 min at $J = 0.4$ mA/cm^2 based on the PS monolayer. Both scale bars are 2 μm. (Reproduced from Duan, G.T. et al., *J. Phys. Chem. B*, 110, 7184, 2006. With permission.)

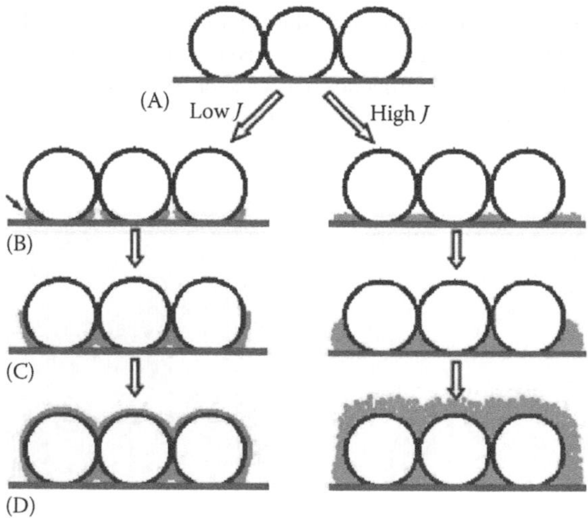

FIGURE 7.44 Schematic illustrations of the formation processes of ordered porous Ni arrays at different cathodic deposition current density (J). (A) colloidal monolayer sintered on the ITO-glass substrate by heating. (B) Ni nucleation. (C) After a short deposition time. (D) After a longer deposition time. (Reproduced from Duan, G.T. et al., *J. Phys. Chem. B*, 110, 7184, 2006. With permission.)

obtained along PSs' surface after longer deposition and removal of PSs. In our case, the PSs are surface negatively charged, which induces the Ni^{2+} ions (existing in the form of $[Ni(NH_3)_2]^{2+}$) [68] to adsorb on the PSs' surface easily. Thus, PSs' surface should be of lower barrier for Ni nucleation and growth compared with ITO substrate, leading to the preferential nucleation and growth on the PSs. Due to such preferential nucleation and growth along PSs' surface, there will exist small interstitials between the deposited film and the substrate, meaning partial contact between the film and the substrate, leading to a weak adherence force between substrate and the Ni film.

When the cathodic current density is high (1.0 mA/cm^2), high deposition rate should be large, which will result in homogeneous (or unselective) nucleation on the substrate. Obviously, when the thickness of a deposition film is smaller than the PSs' radius (after a short deposition time), bowl-like pore morphology would always be formed due to the PSs' geometry, irrespective of the current density. For the film with a thickness larger than the PSs' radius (after a longer deposition time), however, a high deposition current density (1.0 mA/cm^2) will lead to failure of the preferential growth surrounding PSs' surface; hence, the hollow spherical shell will not be formed, as illustrated in the right column of Figure 7.44 and Figure 7.43A. In addition, due to unselective nucleation on the substrate, the deposited film will contact fully with the substrate, leading to an increase of the adherence force between them.

However, a middle current density (0.4 mA/cm^2) induces a transition between the two modes above (the unselective nucleation on the substrate and the preferential growth along PSs) or coexistence of them. A hollow spherical shell is still formed due to preferential growth along PSs (Figure 7.43B).

7.2.4.1.5 Hierarchically Micro/Nanostructured Hollow Sphere Array Films

Recently, fabrication of the hierarchically structured micro/nanoarchitectures has been a research hot spot. By a mild hydrothermal process, some groups have fabricated various hierarchically structured hollow spheres with nanoparticles, nanorods, or nanosheets as building blocks [69–74]. Hierarchically structured hollow spheres with the shell composed of nano-sized building blocks (such as, nanoparticles, nanorods, or nanosheets) combine the features of hierarchical structure, hollow sphere, and patterned arrays. It should be of great importance to the next generation of nanodevices. Our group fabricated $Ni(OH)_2$ hierarchically micro/nanostructured hollow sphere array films successfully based on the direct electrodeposition in $Ni(NO_3)_2$ electrolyte on the PS colloidal monolayer. The morphology and size can be easily controlled.

Electrodeposition experiment is similar to the above case. Briefly, the electrolyte was composed of 1 M $Ni(NO_3)_2$ aqueous solution and its pH value was adjusted to 1.7 with nitric acid. A cleaned polycrystalline nickel sheet was used as the auxiliary electrode. The distance between the working electrode and the auxiliary electrode was about 6 cm. The electrodeposition was carried out at 60°C with cathodic current density of 1.2 mA/cm^2.

Figure 7.45A through C shows the morphologies of the as-prepared samples (two kinds of PSs' sizes) after removal of the PS monolayer by dissolution in CH_2Cl_2. It can be seen that the microspheres are uniformly packed into an array with hexagonal symmetry. The periodicity, i.e., the central distance between the adjacent spheres in the film, can be controlled by the PS's size (Figure 7.45A and C). The local magnification shows that the spheres in the array are of the hierarchical structure (Figure 7.45B). The spheres are composed of massive ultrathin sheets or nanoflakelets (or nanowall) nearly vertical to the spherical surface, showing high specific surface area. For reference, electrochemical deposition was also carried out on the ITO-glass substrate without the PS template, showing only a film consisting of net- (or wall-) like arranged nanoflakelets (nearly vertical standing) (Figure 7.45D). The corresponding x-ray diffraction (XRD) spectra show that the as-deposited samples are α-nickel hydroxide with the lattice parameters $a = 3.09$ Å and $c = 22.11$ Å (as shown in Figure 7.46).

For determination of the structural formula of such α-nickel hydroxide, infrared (IR) spectra, and thermogravimetric analysis (TGA) were conducted for the as-deposited monolayer hollow

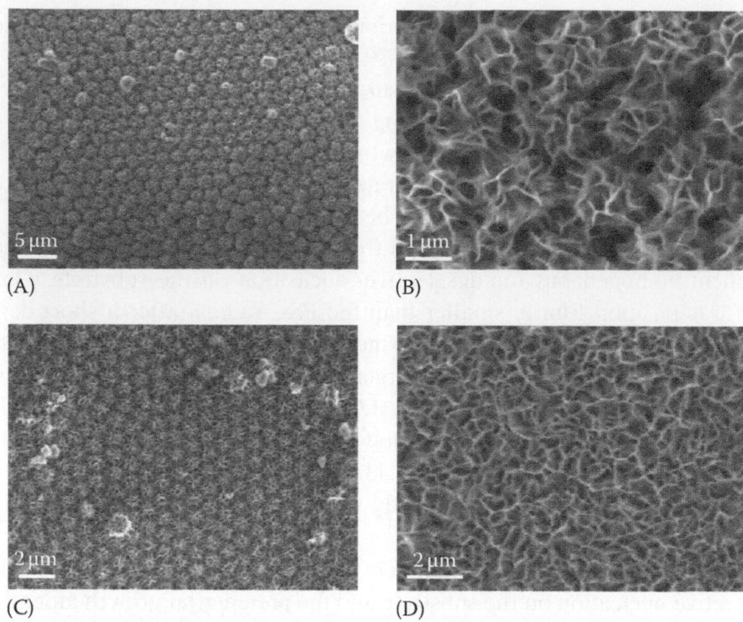

FIGURE 7.45 FESEM images of the as-deposited samples. The PSs' size and deposition time are, respectively, (A, B) 2 μm and 100 min, (C) 1 μm and 60 min, (D) no PSs used and 60 min. (Reproduced from Duan, G.T. et al., *Adv. Funct. Mater.*, 17, 644, 2007. With permission.)

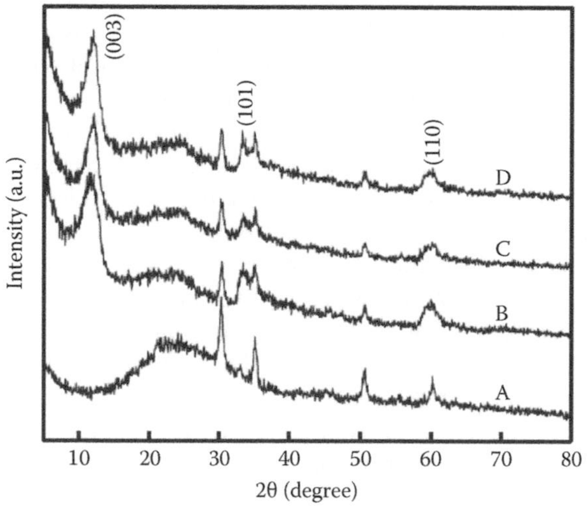

FIGURE 7.46 XRD spectra for ITO glass (A) and the as-deposited samples (B–D) PSs' size and the deposition time are, respectively, for (B) no PS used and 60 min, (C) 1 μm and 60 min, (D) 2 μm and 100 min. The indexes correspond to α-$Ni(OH)_2$. (Reproduced from Duan, G.T. et al., *Adv. Funct. Mater.*, 17, 644, 2007. With permission.)

sphere array based on PS (2 μm in diameter) colloidal monolayer (deposition time: 100 min). The corresponding measurements are shown in Figures 7.47 and 7.48. The IR spectrum includes (1) a broad peak centered around 3420 cm^{-1} corresponding to the interlamellar water and OH-bond vibration, (2) absorption at 2424 cm^{-1} and 2356 cm^{-1} due to CO_2 in air, (3) absorptions in the range of 1000–1500 cm^{-1} due to intercalated anions NO_3^-, and (4) absorptions at 648 and 485 cm^{-1} due to

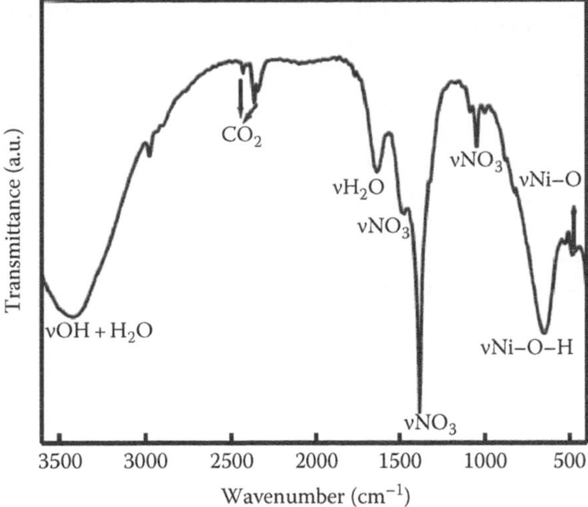

FIGURE 7.47 IR spectrum of the as-deposited $Ni(OH)_2$ hollow sphere array based on the PS (2 µm in diameter) colloidal monolayer (deposition time: 100 min). (Reproduced from Duan, G.T. et al., *Adv. Funct. Mater.*, 17, 644, 2007. With permission.)

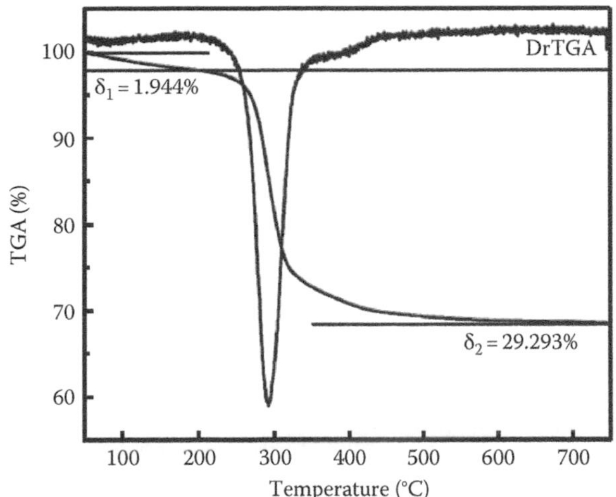

FIGURE 7.48 TGA curve of the sample corresponding to Figure 7.47. (Reproduced from Duan, G.T. et al., *Adv. Funct. Mater.*, 17, 644, 2007. With permission.)

the Ni–O–H bending and Ni–O stretching vibrations, respectively. Then the composition of hollow spheres can be marked with the approximate formulae $Ni(OH)_x(NO_3)_{2-x} \cdot yH_2O$ [75], in which the values of x and y can be determined by the TGA curve, with $x = 1.693$ and $y = 0.117$ [75], or formulae $Ni(OH)_{1.693}(NO_3)_{0.307} \cdot 0.117H_2O$.

TEM examination shows that the microspheres are hollow in the interior (see Figure 7.49A), which means that the shell layer of the sphere consists of incompact-arranged nanoflakelets. The selected-area electron diffraction (SAED) pattern reveals that the flakelet is not an intact single crystal because of weak diffraction rings penetrating the diffraction spots, according to the nature of α-$Ni(OH)_2$ [74]. However, both the diffraction spots and rings show that the flakelet has an orientation with the planar surface perpendicular to the c-axis (see inset in Figure 7.49B).

The hollow structure can be more clearly seen in the edge region of the sample shown in Figure 7.45A, as illustrated in Figure 7.50. There are many broken hollow spheres. Also, we can see that

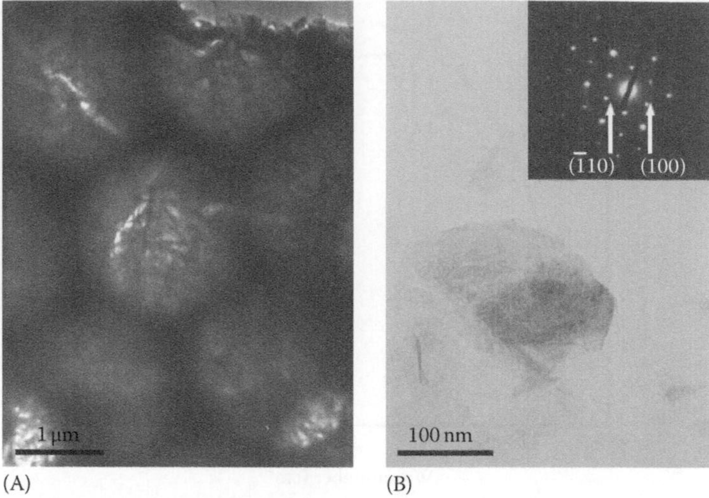

FIGURE 7.49 TEM images of the as-deposited sample shown in Figure 7.45A. (A) A detached layer of the hollow sphere array, (B) a single nanoflakelet. Inset in (B) SAED pattern corresponding to (B). (Reproduced from Duan, G.T. et al., *Adv. Funct. Mater.*, 17, 644, 2007. With permission.)

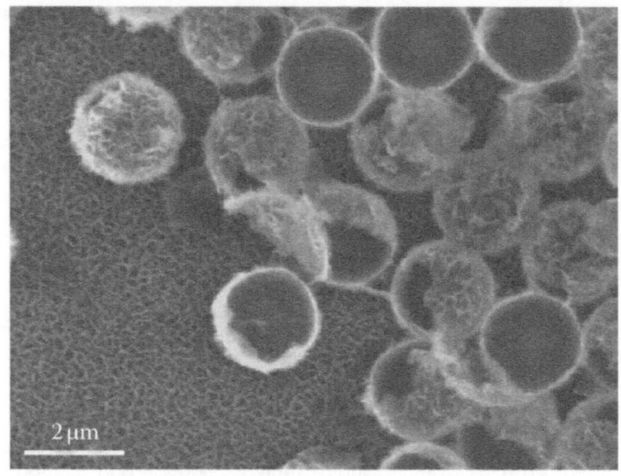

FIGURE 7.50 FESEM image in the edge region for the sample shown in Figure 7.45A. (Reproduced from Duan, G.T. et al., *Adv. Funct. Mater.*, 17, 644, 2007. With permission.)

$Ni(OH)_2$ grows both on the place of the substrate without PSs and along PSs in the existence of the PS colloidal monolayer. Such hollow sphere monolayer film can be detached from the substrate by a blade, as illustrated in Figure 7.51A, corresponding to the sample in Figure 7.45A and the partial disconnection of the hollow sphere monolayer. After detachment of the hollow sphere layer, a layer of $Ni(OH)_2$ with a bowl-like pore array on the ITO substrate is left due to the PSs' geometry (Figure 7.51B). Obviously, the thickness of the pore array is much smaller than the radius of PS, indicating that the growth on ITO will be depressed after the formation of a layer of net-like film in the existence of the PS colloidal monolayer.

Further experiment reveals that the deposition time is very important to form the hollow sphere array film with hierarchical structure, as shown in Figure 7.52. For the template with PSs of 2 μm, a short deposition time (30 min) only induces a ring-like array film of $Ni(OH)_2$ on the ITO substrate (see Figure 7.52A), meaning nucleation and growth start at the sites between PSs and the substrate. Increasing the deposition time to 50 min, ordered bowl-like pore array film was formed

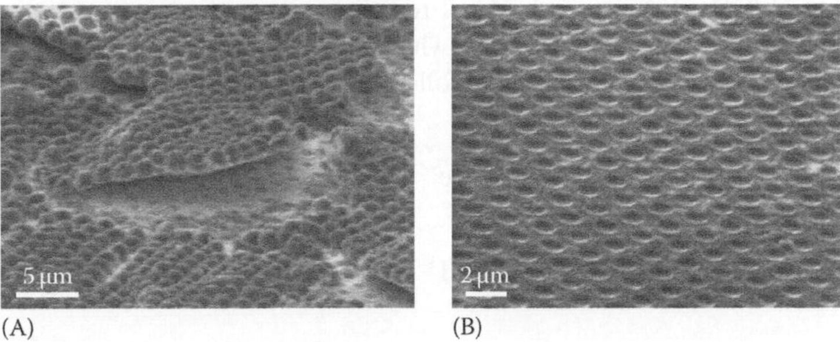

(A) (B)

FIGURE 7.51 FESEM images of the sample shown in Figure 7.45A with partial detachment (A) and after detachment of the hollow sphere array (B). (Reproduced from Duan, G.T. et al., *Adv. Funct. Mater.*, 17, 644, 2007. With permission.)

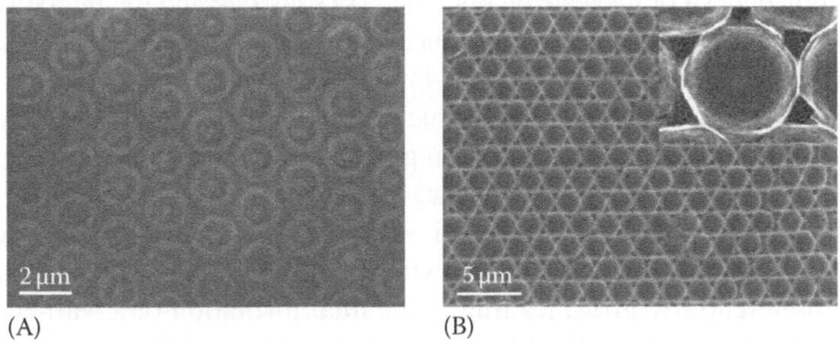

(A) (B)

FIGURE 7.52 FESEM images of ordered $Ni(OH)_2$ arrays based on the PS (2 μm in diameter) colloidal monolayer. The deposition time is, respectively, (A) 30 min, (B) 50 min. (Reproduced from Duan, G.T. et al., *Adv. Funct. Mater.*, 17, 644, 2007. With permission.)

(A) (B)

FIGURE 7.53 FESEM images of $Ni(OH)_2$ hierarchical structures electrodeposited with current density 2 mA/cm^2 and 100 min without template. (B) is the local image of (A). The existence of the PS colloidal monolayer also shows a similar morphology. Also, hollow sphere array cannot be obtained at such current density by decreasing deposition time. (Reproduced from Duan, G.T. et al., *Adv. Funct. Mater.*, 17, 644, 2007. With permission.)

(see Figure 7.52B). We can also see the net-like film at the interstitial sites among PSs (the inset of Figure 7.52B), which demonstrated that the film growth on the ITO substrate was finished before the complete formation of hollow spheres. After the formation of the hollow sphere array, on further increasing the deposition time, the shell of the hollow spheres would become denser with more nanoflakelets. In addition, it has been shown that the formation of hollow-structured spheres was sensitive to the cathodic deposition current density (J). With J increasing from 1.2 to 2.0 mA/cm^2, a uniform hollow sphere array cannot be obtained, but aggregation of nanoflakelets takes place (see Figure 7.53). Also, too small J is not appropriate due to the low deposition efficiency.

Formation of α-$Ni(OH)_2$ by electrochemical reaction was studied in the earlier work [75,76]. Metal ions Ni^{2+} in the solution react with OH^-, which was produced by the reduction of NO_3^- close to cathode, forming $Ni(OH)_2$, according to the following reactions:

$$NO_3^- + H_2O + 2e^- \rightarrow NO_2^- + 2OH^- \tag{7.2}$$

$$Ni^{2+} + 2OH^- \rightarrow Ni(OH)_2 \tag{7.3}$$

In the reaction process, Ni cannot form because the reaction potential of Ni^{2+} to Ni is much more negative than that of NO_3^- to NO_2^- (−0.23 V vs. 0.01 V) [76]. It has been reported that negatively charged PSs can be covered with a smooth inorganic layer by the hydrolysis of metal ions [77]. The PSs used in our experiment were also negatively charged. The $Ni(OH)_2$ formed by reactions (7.2 and 7.3) will subsequently deposit or nucleate on both the ITO substrate and the interstitial sites between the PSs and the substrate. For the latter, deposition along the PS's surface will take place, leading to a final spherical shell, as schematically illustrated in Figure 7.54.

Since the $Ni(OH)_2$ crystal has a layered structure of the CdI_2 type, which shows the weak interaction between layers and the strong binding in the layered planes, i.e., the surface energy of the layered plane (001) is the lowest. Nickel hydroxide will preferentially grow along the layered plane after the formation of its nuclei. Also, due to the directional deposition under the electric potential, only the oriented nuclei with the (001) plane vertical to substrate and the PS's surface will preferentially grow, leading to the final formation of a wall-like fine structure of $Ni(OH)_2$ nanoflakelets nearly vertical to the substrate and the PS's surface (see Figure 7.54). Here it should be mentioned that partial NO_3^- anions and H_2O molecules will be inserted into the $Ni(OH)_2$ layers during electrodeposition, forming α-type structure. As for the effect of J, obviously, a high J value is not beneficial to the preferential deposition due to the fast deposition rate. In the case of $J = 2.0\,mA/cm^2$, homogeneous nucleation takes place close to the electrode, and finally the $Ni(OH)_2$ grew into the aggregation of nanoflakelets without spherical hollows (Figure 7.53).

Besides $Ni(OH)_2$, Co-based hierarchically micro/nanostructured films can also be obtained by such direct electrodeposition strategy.

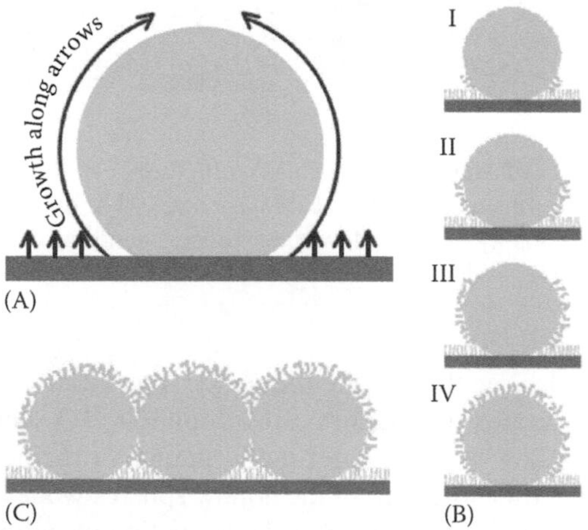

FIGURE 7.54 Schematic illustration for the formation of hierarchical $Ni(OH)_2$ hollow sphere array. (A,C) Schematic initial and final state; (B) schematic growth process. (Reproduced from Duan, G.T. et al., *Adv. Funct. Mater.*, 17, 644, 2007. With permission.)

7.2.4.1.6 Transferability of Micro/Nanostructured Films

Interestingly, it was found that if the operating parameters are well controlled for the direct electrodeposition based on the colloidal monolayer, some as-prepared micro/nanostructured films can be transferred from the original substrate onto any desired one, such as, insulting substrate (ordinary glass, mica, ceramic, etc.), even curved ones, which will overcome the restriction of electrodeposition to conducting substrates. The ordered array (film) prepared by electrodeposition can also be fabricated indirectly on some insulating substrates by its transferability. Therefore, this transferability extends applications of these micro/nanostructured films in the fabrication of micro- or nanodevices [17,64].

A typical example is transferability of the gold pore array film on the glass ITO substrate (see Figure 7.33) electrodeposited based on the colloidal monolayer template [17]. In experiments, gold ordered pore array films can be transferred integrally from ITO substrate onto another one by lifting it off on water surface and picking it up with another substrate, as shown in Figure 7.55. These substrates can be flat or curved according to their application requirements. In addition, the front and back surface of the film can also be chosen. Figure 7.56A is a photograph of a ceramic tube

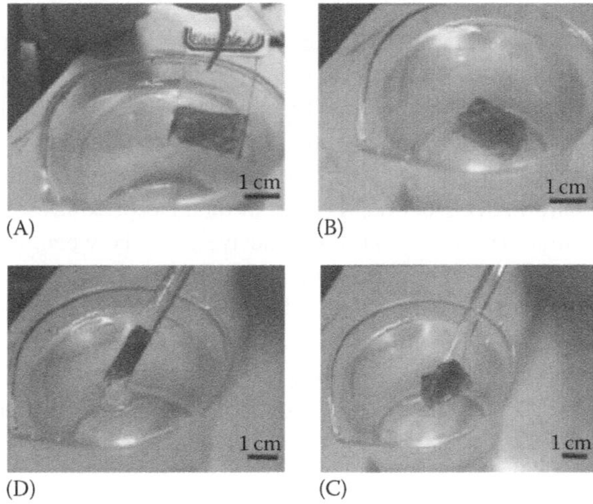

FIGURE 7.55 Photos showing the transfer of the ordered gold pore array film on ITO glass to a curved surface (glass rod). (A) The ITO glass coated with gold porous film is dipped into water. (B) The gold film is floating on the water surface. (C) The gold film is picked up with a glass rod. (D) The gold film is transferred onto the curved surface of the glass rod. (Reproduced from Sun, F. et al., *Adv. Mater.*, 16, 1116, 2004. With permission.)

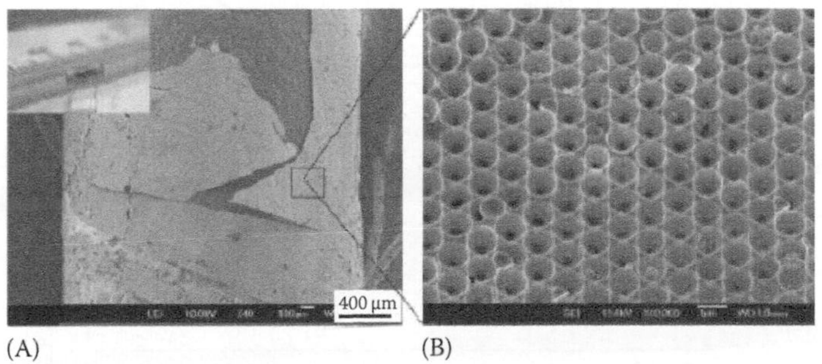

FIGURE 7.56 Ordered gold pore array film on the curved surface of a ceramic tube by lifting off the films from the ITO glass. (A) A FESEM image with a lower magnification of the gold pore array on the curved surface of a ceramic tube, the inset is a photo of the tube covered with a gold porous film. (B) A magnified image of the zone shown in (A). (Reproduced from Sun, F. et al., *Adv. Mater.*, 16, 1116, 2004. With permission.)

covered with gold porous film by such transferring, and Figure 7.56B is an enlargement of the zone marked in Figure 7.56A showing orderly arranged nanostructured pores. It seems that the gold porous array film cannot adhere tightly on a new substrate, which can be conquered by heating at a proper temperature.

The experiments showed that the transferability is related to the electrodeposition rate (or deposition potential) and additive agents in the electrolyte, although it is mainly determined by the film-growth mechanism, which itself depends on the nature of the electrolyte. By this method, the electrolyte contains $HAuCl_4$, Na_2SO_3, $C_{10}H_{14}N_2O_8Na_2 \cdot 2H_2O$ (EDTA), and K_2HPO_4. The $HAuCl_4$ is the source of the initial Au^{III}, some of which immediately forms Au [78,79]. Na_2SO_3 acts as a complexing and buffering agent. When the ITO glass covered with the PS monolayer is dipped into the electrolyte, HPO_4^{2-} is adsorbed onto the substrate before deposition. However, the substrate has already been structured with PSs. The PSs possess some negative charge: the surfaces of the PSs were previously treated with SO_4^{2-} and this will influence the properties of ITO around the sphere. As a result, the HPO_4^{2-} will be adsorbed onto the substrate away from the spheres, as illustrated in Figure 7.57A. When the electrodeposition begins, the Au crystal nuclei will naturally be formed first in the wedge-shaped regions (or corners) between the PSs and the substrate due to the low energy barrier of nucleation in this area (Figure 7.57B). Since the applied voltage during electrodeposition is not high enough to desorb the adsorbates from the substrate, as the gold grows along the substrate and the PSs from the corners, it will also cover the adsorbed area on the substrate. Therefore, the thickness of the film at the edge of the pores (at the film surface) will always be thicker than that in the interstitial region between three closely packed spheres, as is clearly illustrated by the tilt view shown in the inset of Figure 7.33B. Due to existence of an adsorption layer in some areas between the Au film and the substrate (Figure 7.57), the adhesive force between them will be weakened, enabling the gold film to be stripped by the surface tension of water. If the potential is increased, the gold film does not float off on the water surface. The reason is that such a potential is high enough to remove the adsorbates quickly during the initial deposition, therefore the gold film has a much stronger adhesive force with the substrate, resulting in difficulty to be stripped. For the same reason, if we do not add K_2HPO_4 to the electrolyte, the gold film cannot be stripped at any potential, due to the absence of adsorption on the substrate. This fact has also been found for the electrodeposition of copper.

Additionally, it was found that the Ni hollow sphere array film [64], fabricated by electrodeposition as described above, can also be transferred to the other desired substrates. Figure 7.58 shows

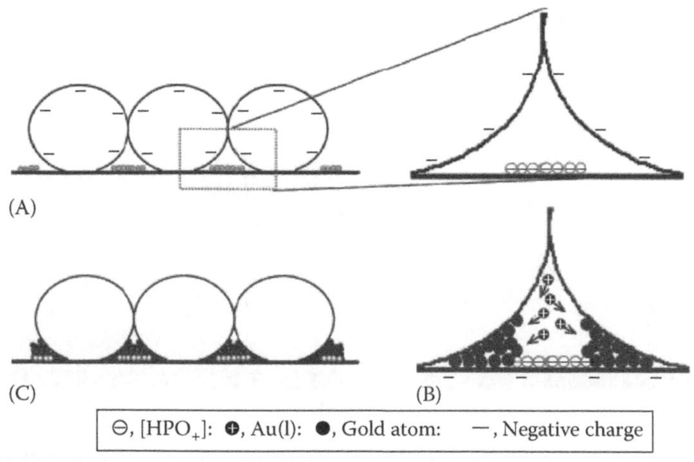

FIGURE 7.57 Schematic illustrations of the formation process of ordered gold porous films. (A) Monolayer coated with ITO glass in the electrolyte before electrodeposition. (B) During initial electrodeposition. (C) After electrodeposition. (Reproduced from Sun, F. et al., *Adv. Mater.*, 16, 1116, 2004. With permission.).

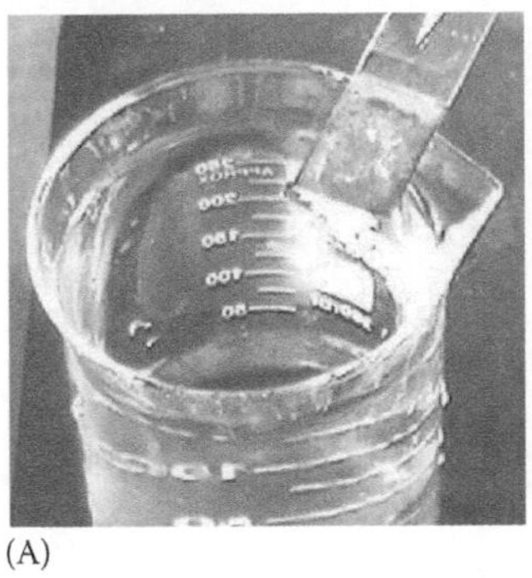
(A)

(B)
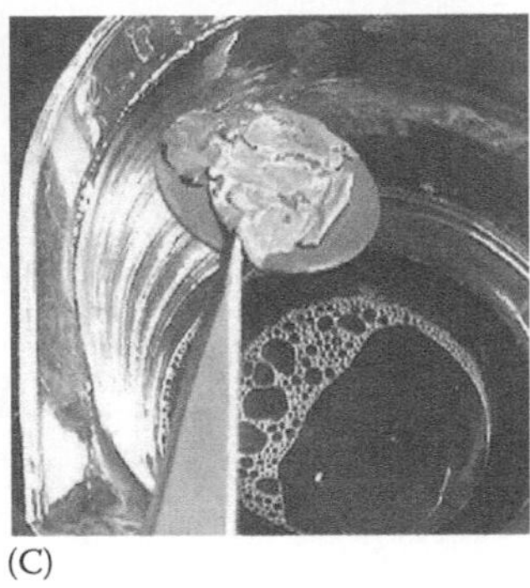
(C)

FIGURE 7.58 Photos taken by camera depicting transferring Ni hollow sphere arrays on an ITO glass to a silicon wafer. (A) Ni hollow sphere arrays on ITO substrate, (B) lifting off on the water surface, (C) picking up of the hollow sphere array with a silicon wafer. (Reproduced from Duan, G.T. et al., *J. Phys. Chem. B*, 110, 7184, 2006. With permission.)

the whole transferring process of Ni hollow sphere array film. However, the transferable mechanism of Ni hollow sphere arrays is different from that of gold porous films. As mentioned above, using a low cathodic deposition current density, Ni will preferentially nucleate at the bottom interstitial sites between PSs and the substrate and then grows along PSs' surface during deposition. Ni^{2+} ions are adsorbed on the PSs surface due to the negative charge on the sphere surface, leading to lower barrier for Ni nucleation and growth on PS surface. Thus, there should be small interstitials between the Ni hollow sphere array film and the substrate, leading to a very weak force between the film and the substrate, and finally resulting in the transferability.

From existing results, we believe that, if controlling the experimental conditions, some other 2D ordered micro/nanostructured films induced by electrodeposition could also be transferred to the desired substrates, including insulating ones and curved surfaces.

7.2.4.2 Second Templated Electrodeposition

As mentioned above, ordered through-pore array films can be prepared directly on the PS colloidal monolayer. If the through-pore arrays are insulating materials on a conductive substrate, and used as a working electrode for electrodeposition, the electrochemical reaction will be restricted within pores, which could induce many novel ordered micro/nanostructured array films. In this case, the template for electrodeposition is of pore structure and the secondary product from the PS colloidal monolayer. Bartlett et al. fabricated diverse ordered array films of metal dots based on poly(pyrrole) (PPy) pore array from colloidal monolayer [16]. In their case, PPy was electrodeposited first, and then changed from conductive to insulating by electrochemical treatment. Recently, we chose Al_2O_3 through-pore array as secondary template to fabricate various micro/nanostructured films. In this section, we will demonstrate the details of such second templated-synthesis method based on our recent work.

7.2.4.2.1 Hierarchical Surface Rough Au Particle Array Film [65]

The fabrication strategy is shown in Figure 7.59. Centimeter-sized ordered PS colloidal monolayer was first prepared (step I), and bonded onto ITO substrate with an area contact by heating (step II). A droplet of 0.2 M $Al(NO_3)_3$ aqueous solution was put on the monolayer based on the solution-dipping technique [63]. Amorphous Al_2O_3 ordered through-pore arrays can thus be obtained (step III) after drying and removal of PSs. Finally, Au particle arrays with the hierarchical structure can be obtained by electrodeposition on the substrate with the ordered through-pore Al_2O_3 template and its subsequent removal (steps IV–VI). Here Al_2O_3 was chosen due to four main reasons: (a) amorphous Al_2O_3 has a high crystallized temperature, and thus a high-quality uniform ordered pore template

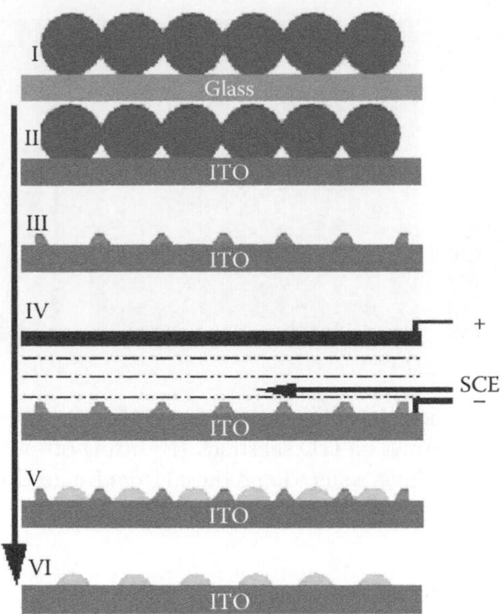

FIGURE 7.59 An illumination for the fabrication of Au particle array. (I) PS colloidal monolayer on a common glass substrate. (II) The monolayer is transferred to an ITO substrate and bonded to the substrate by heating. (III) Ordered alumina through-pore array is formed by solution dipping and heat treatment. (IV) Electrodeposition is carried in a three-electrode electrolytic cell. (V) Au is formed within the alumina pores. (VI) Au particle array is obtained after the removal of the alumina pore template in 5% NaOH solution.

can easily be obtained; (b) it is a insulating material which can restrict the deposition to the selected areas (or within the pores), instead of everywhere; (c) it can be removed easily in acid or alkaline solution; and (d) Al_2O_3 can induce the formation of particle array with hierarchical surface roughness, as detailed in the following.

Figure 7.60A is the typical FESEM image of periodically arranged amorphous Al_2O_3 through-pore array film induced by the solution-dipping technique. The pores are through to the ITO substrate directly, which is induced by heating the colloidal monolayer on the substrate. Such through-pores are crucial to the formation of Au particle array on the substrate during the following electrodeposition. Correspondingly, the ordered Au particle array film was obtained. Figure 7.60B and C show its morphology after electrodeposition and removal of the Al_2O_3 template. The nearly micro-sized Au particles are isolated from each other and arranged hexagonally with the periodicity of 1000 nm, corresponding to pores' arrangement in the Al_2O_3 template. Local magnification reveals that the individual building blocks in the array are composed of many smaller particles with the size close to 60 nm, showing a nanoscaled surface roughness. Thus, such array exhibits the hierarchical surface roughness: the non-close-packed microparticles in the array forming a microscaled roughness on the substrate, and the nanoscaled surface roughness on the microparticles. Corresponding XRD measurement has confirmed that the as-prepared particle array is gold crystal with face-centered cubic lattice structure, and the grain size is about 50 nm (on the basis of the broadening of diffraction peaks), close to FESEM observation (as shown in Figure 7.60D).

The formation of such hierarchical structure is associated with the insulating porous Al_2O_3 template, which restricts Au deposition within the through-pores and leads to the formation of the isolated orderly arranged Au particles. As mentioned above about the Au ordered pore array film based on electrodeposition on PS colloidal monolayer [17], gold grows along PSs' surface (negatively charged), forming a compact skeleton structure. In this work, however, Au deposition along the Al_2O_3 pore walls did not occur. Figure 7.61 demonstrates the morphology of the as-prepared sample before removal of the Al_2O_3 template. Au particles are obviously isolated from the pore walls and

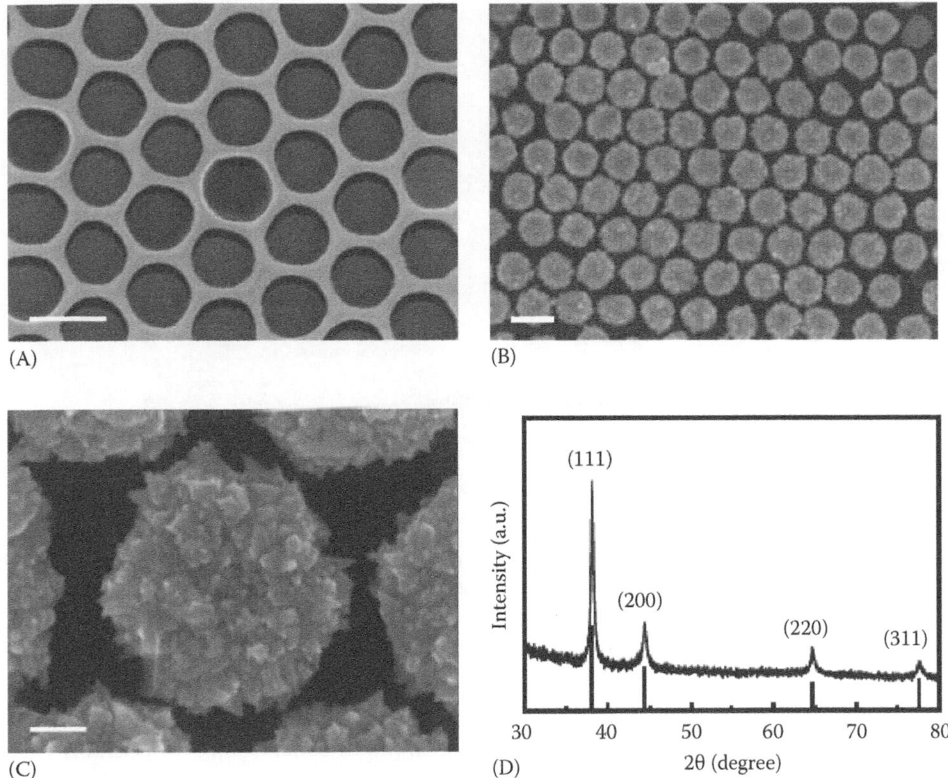

FIGURE 7.60 FESEM images of (A) ordered alumina through-pore array induced by 0.2 M $Al(NO_3)_3$ solution dipping on PS (1000 nm in diameter) colloidal monolayer, and (B) Au particle array deposited for 20 min at 1 mA/cm^2 based on the template shown in (A). (C) A local magnification of (B). Scale bars are 1000 nm for (A) and (B), and 200 nm for (C). (D) The corresponding XRD pattern for sample (B). (Reproduced from Duan, G.T. et al., *Appl. Phys. Lett.*, 89, 181918, 2006. With permission.)

located within Al_2O_3 pores, showing preferential deposition on the substrate around the middle area of each pore. No growth along the pore walls was found. Because alumina is insulating, Au nucleation can only occur on the conducting ITO substrate. Such nucleation should be unselective on the substrate within the Al_2O_3 pores because of amorphous ITO (or more than one nucleus on the substrate within a single pore). Also, since gold is unsoluble in Al_2O_3 and has bad wetting with it [80], it is unfavorable in energy for Au nuclei on the substrate to grow along Al_2O_3 pore walls. Finally the nuclei grow within pores and away from the walls, constituting a microparticle within each pore and leading to nanoscaled surface roughness.

Further experiments have revealed that the current density (J) is of importance in the formation of such hierarchically rough particle array films. With the decrease of J down to 0.3 mA/cm^2, only low percentage (<20%) of Al_2O_3 pores are deposited with Au microparticles. If further decreasing J to 0.1 mA/cm^2, no Au deposition within alumina pores was found, indicating the Au^0 atoms or clusters from the reduction of $[AuCl_4]^{-1}$ cannot nucleate on the substrate within the Al_2O_3 pores due to the nucleation barrier. Also, too high J is not appropriate because the rapid deposition rate leads to Au deposition everywhere including onto Al_2O_3 porc walls, and hence the failure to the formation of the isolated microparticles.

7.2.4.2.2 *Hierarchical Ni Nanoring and Hollow Sphere Array Films [64]*

The nanoring-structured film has received much attention due to its importance in the fundamental studies of persistent current [81], bi-domain state in magnetics [82], and novel optical properties

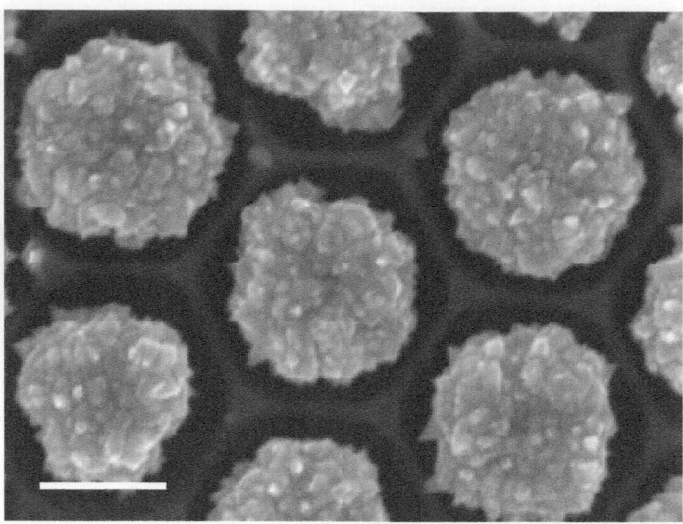

FIGURE 7.61 FESEM images of Au particle array before removal of the alumina template. (Reproduced from Duan, G.T. et al., *Appl. Phys. Lett.*, 89, 181918, 2006. With permission.)

[83], and also due to its technical application in, such as recording media [84]. The hollow sphere is another special structure with applications ranging from versatile microreactors to advanced functional materials, as depicted above [85,86]. Generally, the nanoring is synthesized by electron-beam lithography [82], molecular beam epitaxy [87], chemical modification [88], self-assembly [89], and suitable structured-template technique [90,91], while the hollow sphere is usually fabricated by templating colloidal sphere core [92,93] or self-assembly strategy [86]. Such structures and their patterns could be the base of nanodevice design [70,94]. However, morphology-controlled synthesis of ordered nanoring-like and hollow spherical-like structured arrays in a simple way is still in challenge.

Recently, our group have demonstrated the fabrication of hierarchical Ni micro/nanostructured films based on electrodeposition on an ordered Al_2O_3 through-pore template induced by solution-dipping colloidal monolayer. The Ni films show ring-like or hollow spherical surface structure, depending on the template geometry. Importantly, the skeleton of the arrays is of floc- or flakelet-like fine structures on nanoscale due to morphology inheritance in the electrochemical process. The fabrication strategy is illustrated in Figure 7.62.

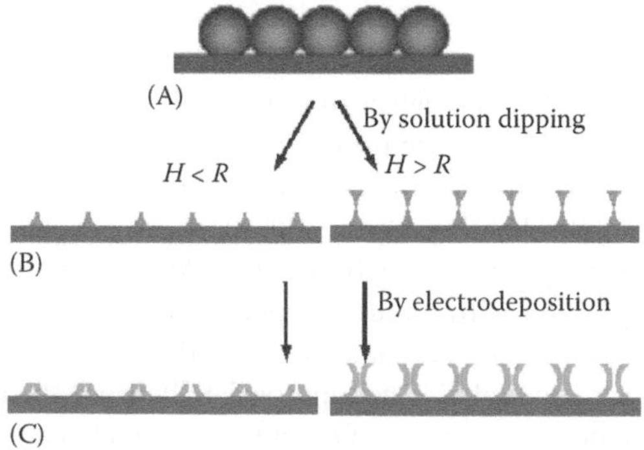

FIGURE 7.62 Schematic cross-section illustration of the fabrication procedures for ring-like and hollow spherical-like micro/nanostructured films. (A) PS colloidal monolayer on an ITO substrate by heating. (B) Ordered alumina through-pore array with pore depth (H) less (left) or larger (right) than the radius of the PS (R). (C) Ordered Ni ring-like (left) or hollow spherical-like (right) film. (Reproduced from Duan, G.T. et al., *J. Phys. Chem. B*, 110, 15729, 2006.)

Based on the PS colloidal monolayer and $Al(NO_3)_3$ solution dipping [15], we can obtain alumina ordered through-pore array films after removing the PS monolayer, as described above. The morphologies can be controlled just by the concentration (C) of $Al(NO_3)_3$. Figure 7.63 shows the typical FESEM images of alumina ordered pore films on ITO glass with different pore depths (H) corresponding to the colloidal monolayer of 1000 nm PSs. When $C = 0.2$ M, the pores in the films are of bowl-like morphology with pore depth less than the radius (R) of PSs due to the geometry of PSs and show circular openings from top view (Figure 7.63A). When C is up to 0.5 M, $H > R$. There exists a hole in the wall between two adjacent pores, resulting from area contact of PSs due to heating the colloidal monolayer before solution dipping, as seen in Figure 7.63B. Further characterization shows that the pore wall or skeleton is amorphous Al_2O_3.

Based on such Al_2O_3 ordered pore arrays and electrodeposition in the electrolyte, Ni ordered patterns were obtained. Figure 7.64 shows the morphologies of the samples on ITO glass, based on the template shown in Figure 7.63A, after electrodeposition with different potentials vs. SCE and

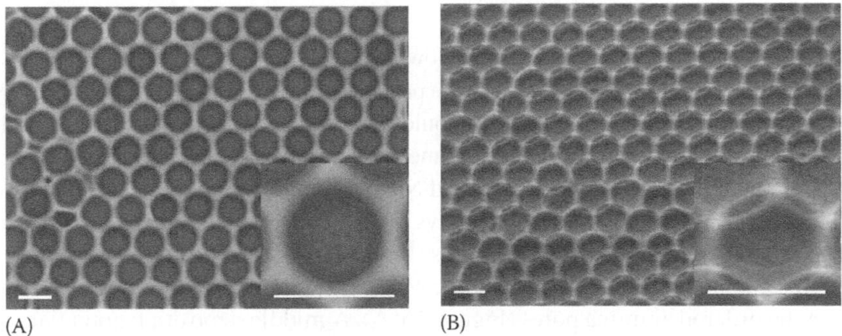

FIGURE 7.63 FESEM images of ordered alumina through-pore arrays. (A) The array formed with 0.2 M $Al(NO_3)_3$ ($H < R$); (B) formed with 0.5 M $Al(NO_3)_3$ ($H > R$). All scale bars are 1 μm. (Reproduced from Duan, G.T. et al., *J. Phys. Chem. B*, 110, 15729, 2006.)

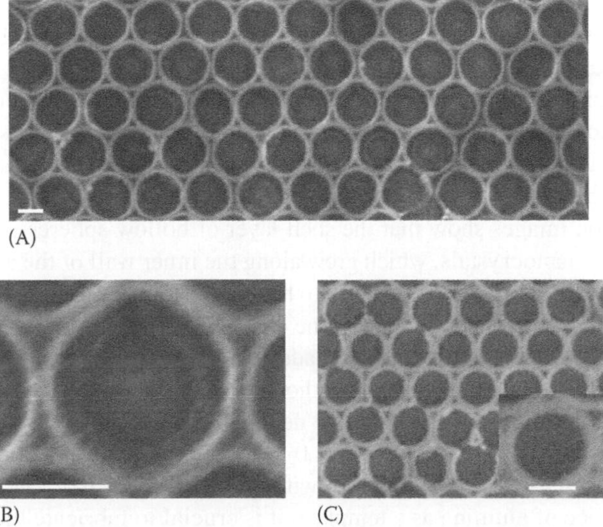

FIGURE 7.64 FESEM images of ordered Ni nanoring array film using the template shown in Figure 7.63A. The electrodeposition potential vs. SCE and the deposition time are, respectively, (A) −0.8 V and 20 min, (C) −1.0 V and 16 min. (B) The local enlarged image of sample (A). Inset in (C) local enlarged image. All scale bars are 500 nm. (Reproduced from Duan, G.T. et al., *J. Phys. Chem. B*, 110, 15729, 2006.)

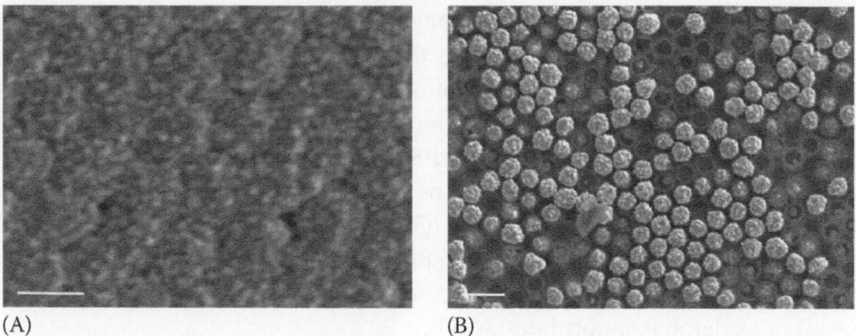

FIGURE 7.65 FESEM images of ordered Ni array film using the template shown in Figure 7.63A. The electrodeposition potential vs. SCE and the deposition time are, respectively, (A) −2.0 V and 8 min, (C) −1.2 V and 14 min. Scale bars are 500 nm for (A) and 2 μm for (B). (Reproduced from Duan, G.T. et al., *J. Phys. Chem. B*, 110, 15729, 2006.)

removal of alumina in 5% NaOH solution. At a low negative deposition potential (−0.8 V vs. SCE), Ni grows preferentially along the inner wall of the pores, forming a ring-like structure (Figure 7.64A and B). The thickness of the film or *H* value should be close to but less than *R*. The high-magnification FESEM image shows that there are innumerable floc-like (or flakelet-like) nanocrystals on the ring surface. The morphology of such ordered Ni ring-like array can be controlled by deposition potential. When deposition potential was −1.0 V vs. SCE, the floc-like Ni nanocrystal grew both on the ITO substrate and the inner walls of the pores (Figure 7.64C). If further increasing the potential to −2.0 V vs. SCE, the floc (or flakelet) vanished, while spherical particles with rough surface were formed within the ordered alumina pores (Figure 7.65A). A middle deposition potential of −1.2 V vs. SCE leads to a transitional morphology (Figure 7.65B). Both spherical particles and nanorings with flakelet-like structure are formed in or along the pores. The corresponding XRD spectra indicate that all the as-deposited samples are nickel crystal with a face-centered structure.

As mentioned above, the low negative deposition potential induces the preferred growth of Ni along the pore wall. So, if we use the template shown in Figure 7.63B, truncated hollow sphere-like Ni array film should be obtained by using a low negative deposition potential, which has been confirmed by our further experiments. Figure 7.66A and B show the morphology of the sample electrodeposited at −0.8 V vs. SCE based on the template shown in Figure 7.63B, after the removal of the template. The circular openings on the top of the hollow spheres can be adjusted by the alumina template morphology, which depends on the concentration of the precursor solution $Al(NO_3)_3$. Further, if using the template prepared by 1.0 M $Al(NO_3)_3$, ordered Ni hollow sphere array film with much smaller top openings were obtained after deposition at −0.8 V vs. SCE (Figure 7.66C). The high-magnification images show that the shell layer of hollow spheres is composed of a large quantity of flakelet-like nanocrystals, which grew along the inner wall of the alumina pores, and the thickness of the nanoflakets is estimated nearly to be 20 nm (Figure 7.66B).

Further experiments show that pH value in the electrolyte and alumina template are important to the morphology of the deposited products, in addition to the deposition potential. Low pH value (say, 4.5) leads to the spherical-particle array without flakelet-like fine structure (see Figure 7.67A). Without the ordered alumina pore array, at low deposition potential (−0.8 V vs. SCE), Ni cannot be deposited onto the bare ITO substrate. At −1.0 to −2.0 V vs. SCE, the Ni is deposited onto the ITO substrate, forming particle chain-like film without any flakelet-like fine structure (see Figure 7.67B). As for the choice of alumina as a template, it is crucial to fabricate the Ni arrays with hierarchical structure in this work because transitional product $Ni(OH)_2$ can grow along the pore wall of alumina (see the following). Alumina is insulating and can be removed in alkaline solution, which is important since ITO and Ni can be dissolved in acidic solution. Our experiments have demonstrated that ordered pore arrays of many other materials, such as, Fe_2O_3, ZnO, TiO_2, CeO_2,

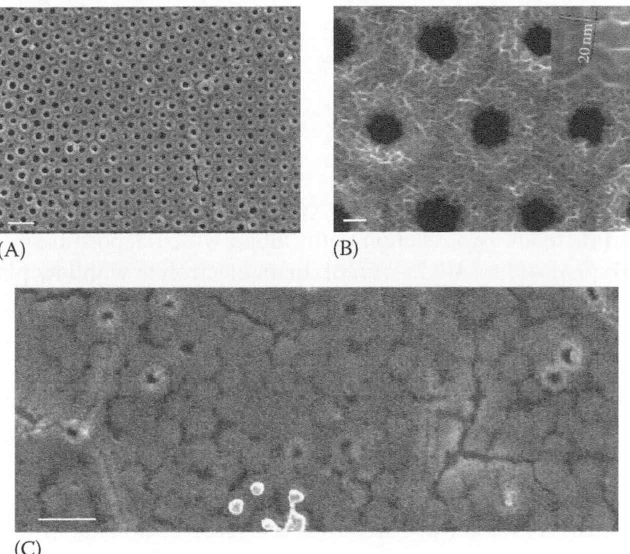

FIGURE 7.66 FESEM images of the samples electrodeposited at −0.8 V vs. SCE based on the ordered alumina templates with $H > R$, prepared by the precursor $Al(NO_3)_3$ with (A) 0.5 M and (C) 1.0 M. (B) The local magnification of (A). The deposition time is 40 min for (A) and 50 min for (C). Inset in (B) thickness measurement of the nanoflakelet. Scale bars are 2 µm for (A) and (C), and 200 nm for (B). (Reproduced from Duan, G.T. et al., *J. Phys. Chem. B*, 110, 15729, 2006.)

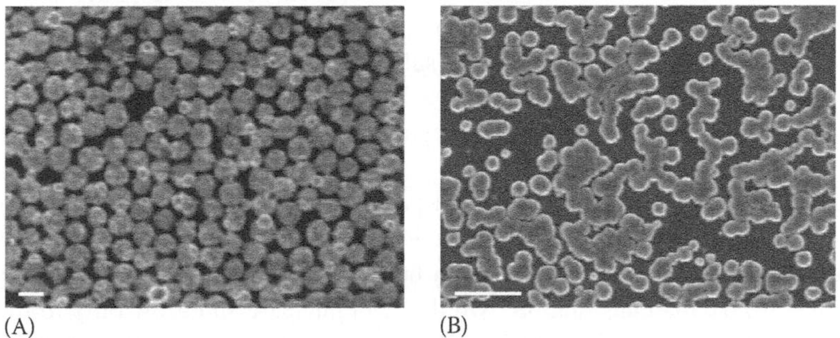

FIGURE 7.67 FESEM images of the sample (A) electrodeposited at −0.8 V vs. SCE with 20 min and pH = 4.5 using the template shown in Figure 7.63A and Ni particle film (B) electrodeposited at −1.0 V vs. SCE with 16 min and pH = 8.5 without the template. Both scale bars are 1 µm. (Reproduced from Duan, G.T. et al., *J. Phys. Chem. B*, 110, 15729, 2006.)

and In_2O_3, induced by the solution-dipping colloidal monolayer [15] cannot be used as template for the Ni films.

The Ni deposition can be described by the following reactions:

$$[Ni(NH_3)_2]^{2+} = Ni^{2+} + 2NH_3 \tag{7.4}$$

$$Ni^{2+} + 2e^- \rightarrow Ni \quad (E_1^o = -0.23\ V) \tag{7.5}$$

$$NH_3 + H_2O = NH_4^+ + OH^- \tag{7.6}$$

$$Ni^{2+} + 2OH^- \rightarrow Ni(OH)_2 \tag{7.7}$$

$$Ni(OH)_2 + 2e^- \rightarrow Ni + 2OH^- \quad (E_2^\circ = -0.66\,V) \tag{7.8}$$

where E° is the standard electrode potential. Ni^{2+} ions in the electrolyte exist in the form of $[Ni(NH_3)_2]^{2+}$ [68], and there are two reaction paths along which deposition of Ni will occur [path I: (7.4)→(7.5), path II: (7.4)→(7.6)→(7.7)→(7.8)]. In an electrolyte with low pH value, the reaction will go along path I. When the pH is high, the reactions along path II should occur, in addition to path I. During the initial electrodeposition, the concentration of Ni^{2+} close to the substrate will be increased due to the movement of Ni^{2+} ions to cathode, which leads to the formation of $Ni(OH)_2$ [see reaction (7.7)] because of high $[OH^-]$. At a low potential, reaction (7.5) will thus be restricted, and the reaction along path II will be dominant. However, when the deposition potential is very high, reaction (7.2) will be accelerated and the reaction along path I will be dominant again. In our cases (high pH value and low negative deposition potential), the reactions along path II were dominant. $Ni(OH)_2$ was first formed. As we know, $Ni(OH)_2$ is a layered compound of CdI_2 type. Its lamellar 2D structure easily leads to the formation of flakelet-like nanocrystals [74]. So deposited $Ni(OH)_2$ would be of flakelet-like fine structure, and then decomposed into Ni during the reaction, but the flakelet- or floc-like morphology should be kept. We call it a morphology inheritance process.

Then, formation of the ring- or hollow sphere-like structure can be understood. Al_2O_3 used here is an insulating material and can be removed easily, which is the main reason that we chose it as a secondary template. When the potential is low negative, corresponding to the reactions along path II, the transitional product $Ni(OH)_2$ preferentially grows along the alumina pore wall because of the low barrier in energy, leading to ring-like or hollow sphere-like structured patterns due to the geometry of the Al_2O_3 template. A small increase of deposition potential can promote growth, resulting in the growth along both the pore wall and ITO substrate after nucleation at the bottom of the pore wall (see Figure 7.64C).

As mentioned above, low pH value or high negative deposition potential will lead to the deposition reactions along path I, which results in disappearance of the preferential growth along pore wall and the flakelet-like fine structure, and spherical-particle arrays are formed (see Figures 7.67A and 7.65A). In addition, without Al_2O_3 template, reaction (7) or the reactions along path II will be suppressed and only island-shaped particles without flakelet-like fine structure are formed on the substrate (see Figure 7.67B). It means that the $Ni(OH)_2$ can nucleate and grow only in the existence of Al_2O_3 template. Al_2O_3 is crucial and indispensable to get such hierarchically structured patterns.

On these bases, the reactions along path II occur only at low negative deposition potential, high pH value in the electrolyte, and with Al_2O_3 template, resulting in the flakelet-like fine structure by morphology inheritance from $Ni(OH)_2$. Otherwise, the reactions along path I will appear, leading to the final morphology of spherical particles. So, if the case, the reactions along path I and II should coexist at high pH value in the electrolyte and with the Al_2O_3 template but at a moderate deposition potential. This has been confirmed by our experiment, as shown in Figure 7.65B, which corresponds to the potential −1.2 V vs. SCE and the template ($H < R$) shown in Figure 7.63A. Both spherical particles and nanorings with flakelet-like structure are formed, which should originate from the reactions along paths I and II, respectively. As for the occurrence of reaction (7.5), because of the deposition potential used in our case is more negative than E_2° (−0.66 V), the reaction (7.8) is feasible from a thermodynamical point of view since the $Ni(OH)_2$ can be formed on the alumina wall at high pH value.

7.2.4.2.3 Binary Ordered Micro/Nanostructured Porous Films [95]

Thus far, the existing fabrication techniques, based on colloidal lithography strategy, are commonly concerning the periodically structured arrays (or films) with the building blocks of unitary size and shape, or mono-block arrays. Binary array is another kind of member in the family of ordered patterns, which is the bi-periodic array consisting of two kinds of building blocks with hierarchical sizes or shapes. The most typical binary array is the binary colloidal crystal induced by self-assembly strategy [96–99], which is composed of two kinds of monodisperse colloidal spheres with different sizes. This structure is of significance for both fundamental researches in low-dimensional physics, and applications in optoelectric devices and functional solid surfaces including the SERS active substrate, the modified electrolyte, and the superhydrophobic surface. Although some advanced lithography techniques may be available for preparation of the binary arrays, the colloidal lithography strategy is the widely used method due to its low cost and advantage for large-area fabrication. Up to date, in addition to the binary colloidal crystals by self-assembly [96–99] and their inverse opals [100], there are also some reports of binary nanoparticle arrays prepared by colloidal lithography strategy [44]. However, the design and fabrication of the binary ordered micro/nanostructured arrays with controlled structures and morphologies are still in challenge. Especially, there is limited report on binary pore arrays, and their structure and morphology control. It is well known that the small-sized pore array is of high activity, due to high specific surface area, and thus advantageous for enhancing detection signal for its device application, such as catalysis, sensing, SERS, and so on, but of weak transportation of detected molecules, leading to the low response rate for device application; and vice versa for the large-sized pore array. Hierarchically sized binary pore array would combine both the advantages of small-sized and large-sized pore arrays, which will be of high activity and strong transportation, and thus lead to the potential superiority in device application.

Here, combining two sizes of PSs (2000 and 200 nm used here), we report a methodology, named step-by-step colloidal lithography strategy, to sculpture binary pore array films with controllable morphologies and structures. Based on our designed fabrication routes, four kinds of hierarchically micro/nanostructured 2D ordered binary porous array films were thus obtained by electrodeposition technique. As an exemplification, gold was used as a model material to demonstrate the designing routes and fabrication process, while this strategy can also be extended to other material arrays or multiplex systems by combination of PSs with different sizes.

Figure 7.68 presents the strategy and shows whole fabrication routes. Large-sized (>1 cm^2) PS colloidal monolayer can be synthesized on a cleaned glass substrate, and easily be transferred onto any desired substrate, such as conducting ITO substrate, as previously demonstrated [15,51]. On this basis, one can design four kinds of step-by-step colloidal lithography routes. Combining these routes with electrochemical deposition technique, four varieties of 2D binary pore arrays (structures A–D in Figure 7.68) can thus be obtained.

Route 1 [I→II→III→A in Figure 7.68A]: The colloidal monolayer with larger PSs size is transferred onto an ITO substrate and heated for several minutes to induce an area contact between PSs and the substrate. Based on the heated PS colloidal monolayer on the ITO substrate, the ordered alumina through-pore array film can be obtained by solution-dipping technique after the removal of the PS template (I→II) [65]. Smaller PSs can then self-assemble on the substrate within the alumina through-pores by spin-coating method (II→III). And then using such composite system as template and work electrode, the hierarchically structured binary porous array will thus be obtained after electrodeposition and the removal of alumina and PSs. Due to the insulated feature of alumina, the designed structure should be of binary microparticle/nanopore-structured films (Structure A) that consist of hexagonally non-close-packed microparticles containing nanopore arrays.

Route 2 [I→II→III→B in Figure 7.68B]: Based on the colloidal monolayer composed of larger PSs on the ITO substrate, ordered bowl-like gold pore array can be obtained after electrodeposition and the removal of PSs [17]. If a colloidal monolayer consisting of smaller PSs is transferred onto the preformed ordered bowl-like gold pore array film (II→III), using such composite system as the template, a binary hierarchically micro/nanostructured porous array film can be formed. Because the smaller PSs above the large pores do not touch with the conductive substrate, the smaller pores will only be formed on the bank and not on the inside of the large pores (Structure B).

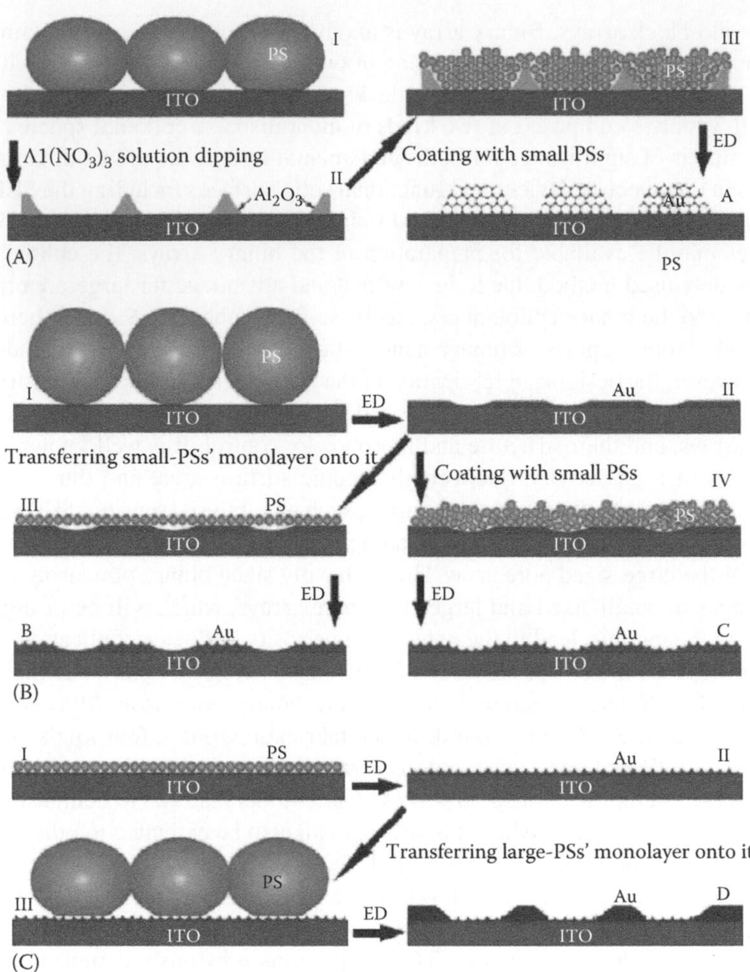

FIGURE 7.68 An illumination for the whole designed fabrication routes. (Reproduced from Duan, G.T. et al., *Langmuir*, 25, 2558, 2009. With permission.)

Route 3 [I→II→IV→C in Figure 7.68B]: If a multilayer of small PSs self-assembles directly, instead of being transferred as Route 2, onto the surface of the preformed large-bowl-like pore array by spin-coating, the small PSs will touch on both the bank and the wall of the large pores. Using such composite system as template and working electrode, another binary hierarchically micro/nanostructured porous array can thus be obtained via an appropriate electrodeposition with limited time. The small pore will be formed anywhere on the surface of a large-pore array (Structure C).

Route 4 [I→II→III→D in Figure 7.68C]: Furthermore, the small-pore array film only on the bottom of the large pores (Structure D) can also be obtained by using another kind of step-by-step route. First, small-pore ordered array film is formed on the ITO substrate by electrodeposition on the colloidal monolayer of smaller PSs and removal of it (see I→II in Figure 7.68C). Then, an as-prepared colloidal monolayer of larger PSs is transferred onto such preformed small-pore array, followed by heating to form an area contact between the PSs and the substrate covered with a small-pore array (III). Finally, using it as the working electrode and template, the intending binary pore array film will thus be obtained after electrodeposition and removal of the PSs (see Structure D in Figure 7.68A).

Based on the designed step-by-step colloidal lithography routes above, four kinds of binary periodic porous arrays with controlled structure and morphology were thus obtained by electrodeposition in $HAuCl_4$ electrolyte under certain conditions. Figure 7.69A is the typical FESEM image for the Al_2O_3 through-pore array induced by solution-dipping technique based on the 2000 nm PS colloidal monolayer. The pores are directly through to the conductive ITO substrate, which is vital for the following formation of Structure A. By spin-coating smaller PSs (200 nm in size) on this through-pore array, the colloidal spheres will self-assemble into multilayer colloidal crystals inside each Al_2O_3 pore. Thus, based on such composite template and Route 1, a gold porous microparticle array film (Structure A) was formed by electrochemical deposition. Figure 7.69B shows the corresponding morphology after the removal of the smaller PSs and alumina templates by dissolution. The nearly micro-sized particles are isolated and packed hexagonally in array. Each microparticle in the array is characteristic of nano-sized porous structure. Both the microparticles and nanopores are determined in size by the corresponding PSs' diameters, showing a size-controllability according to the designed requirement. As a new member in the family of binary arrays, we call it "binary microparticle/nanopore-structured array." Owing to the feature of hierarchical micro/nanostructure, this binary array could be of importance in the applications of surface science (such as, superhydrophobicity and SERS active substrate). As an example, such structured array, with a lotus leaf-like surface, should easily lead to superhydrophobicity by certain surface modification. Our further experiment has demonstrated that its surface CA of water can reach to 165° after modification with a monolayer 1-hexadecanethiol, showing a nice superhydrophobicity [95]. This will bring such structure importance in the applications of waterproof coating, microfluidic device, and so on.

Further, we can adjust fabrication route in step-by-step colloidal lithography. Figure 7.70A shows the gold pore array induced by electrodeposition on the colloidal monolayer of larger PSs and removal of it, which is the transitional sample for preparation of Structures B and C. The array

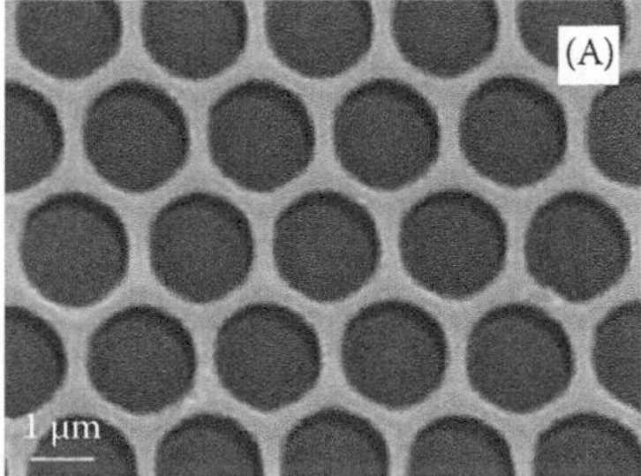

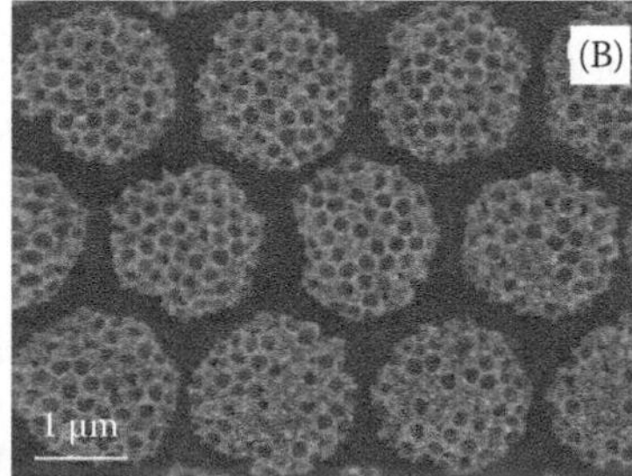

FIGURE 7.69 FESEM images for the samples (A) Al_2O_3 through-pore template and (B) hierarchical microparticle/nanopore binary array [Structure A] based on Route 1 shown in Figure 7.68A. (Reproduced from Duan, G.T. et al., *Langmuir*, 25, 2558, 2009. With permission.)

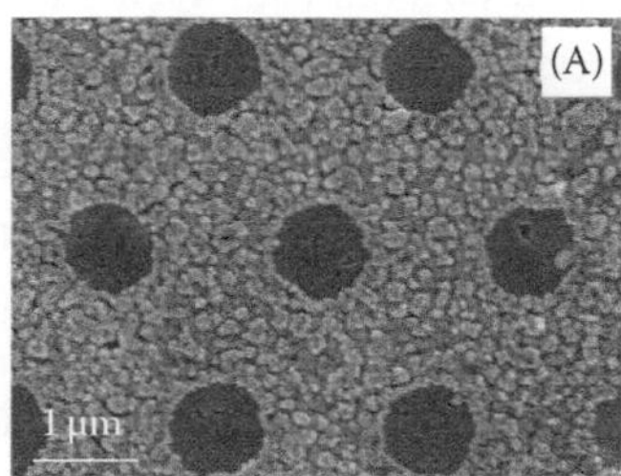

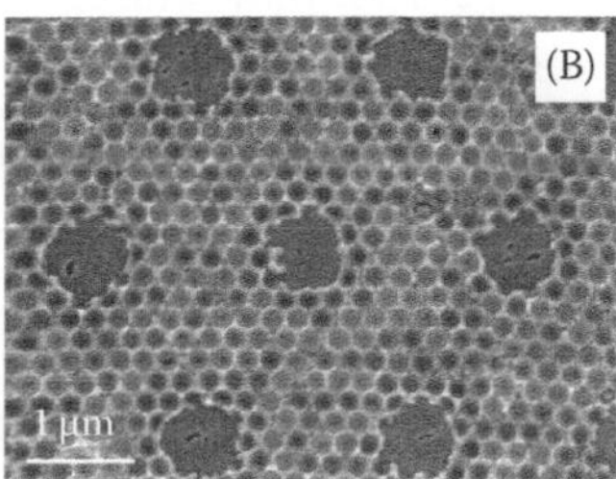

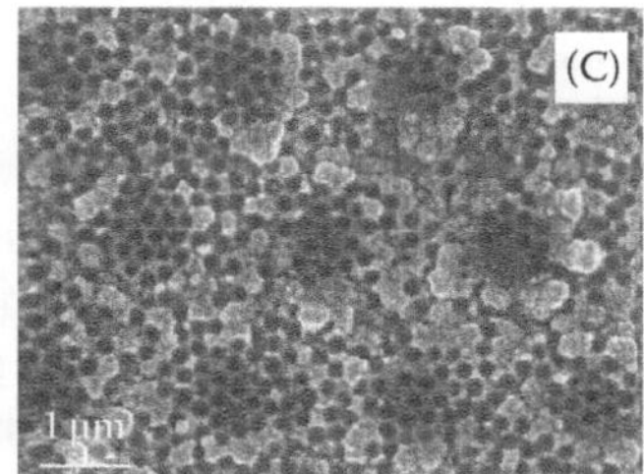

FIGURE 7.70 FESEM images of the samples corresponding to the strategy shown in Figure 7.68B. (A) The ordered pore array of gold after ED on 2000 nm PS colloidal monolayer for 20 min and removal of it. (B) The result corresponds to Route 2. (C) The sample based on Route 3. The time of second deposition is 6 min for both (B) and (C). (Reproduced from Duan, G.T. et al., *Langmuir*, 25, 2558, 2009. With permission.)

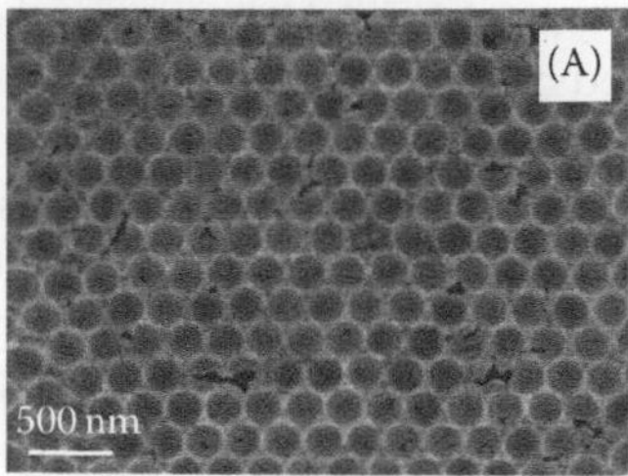

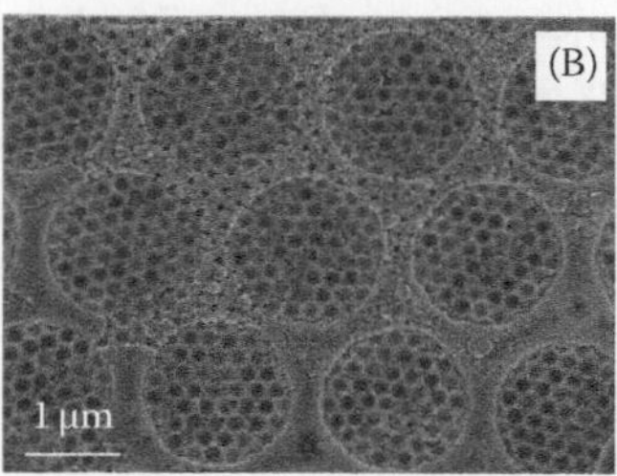

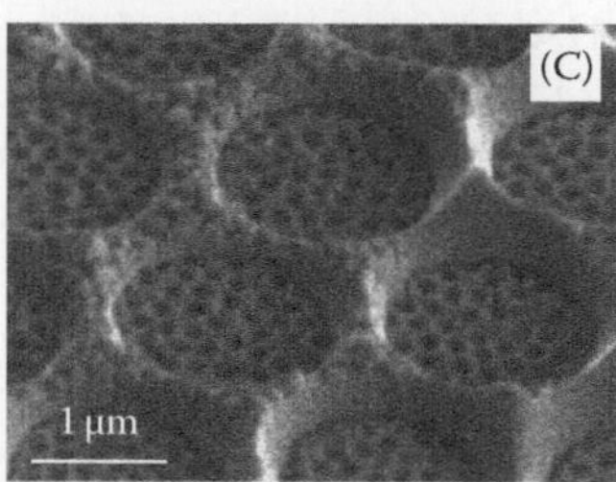

FIGURE 7.71 FESEM images of the samples corresponding to the strategy shown in Figure 7.68C. (A) The gold ordered pore array film after ED on 200 nm PS colloidal monolayer and removal of it. (B) The hexagonal binary porous array film by Route 4. The time of the first and second depositions is 6 and 20 min, respectively. (C) The corresponding tilted FESEM images for (B). (Reproduced from Duan, G.T. et al., *Langmuir*, 25, 2558, 2009. With permission.)

is packed hexagonally with bowl-like pores. Based on this gold porous array and Route 2, the hierarchically binary porous array (Structure B) was thus obtained by electrodeposition and the subsequent removal of the smaller PSs, as shown in Figure 7.70B. The whole pattern consists of two ordered porous arrays (large- and small-pore arrays), and both of them are hexagonally packed, with their lattice constants controlled by the corresponding PS colloidal monolayers. As a new kind of hierarchically structured binary array, we call it "binary micropore/nanopore-structured array." It should be mentioned here for Structure B that only the bank of the large pores be covered with small pores, which is due to the part contact of small-sized PS colloidal monolayer with conductive substrate and the preferential growth of gold along PSs' surface.

Similarly, according to Route 3, if a colloidal monolayer of smaller PSs self-assemble directly on the surface of the preformed gold large-pore array by spin-coating, instead of being transferred onto it like Route 2, another kind of binary micropore/nanopore-structured array (Structure C) can thus be obtained, as illustrated in Figure 7.70C. Whole preformed gold large-pore array is covered with small-pore array induced by 200 nm PSs. Unlike Structure B, the walls and insides of the large pores are also covered with smaller-pore array.

According to Route 4, the morphology and structure of the binary porous arrays can be further manipulated. A hexagonally packed small-pore array was first fabricated on the ITO substrate based on the colloidal monolayer of 200 nm PSs (see Figure 7.71A). The final binary porous array corresponding to Structure D in Figure 7.68C was thus formed based on Route 4, as shown in Figure 7.71B. The small-pore array is only located at the bottom of the large-pore array, which is different from Structures B and C. By transferring the colloidal monolayer of 2000 nm PSs onto the preformed small-pore array, an area contact will exist between larger PSs and the small-pore array after heating. After further deposition of gold, the small pores in area-contacting regions will be unchanged due to the shelter from the larger PSs, while other areas will be covered with the net-like skeleton due to the deposition.

As is demonstrated above, we can easily manipulate and control the morphology and structure of the resulting binary pore arrays. Furthermore, the size of pores in the resulting arrays can also be controlled by the used PSs. This structure- and morphology-controllable fabrication should also be the unique advantage of the stepwise fabrication strategy presented in this study, in addition to the low cost, flexibility, and suitability for large-area fabrication.

It should be mentioned here that the presenting step-by-step colloidal lithography strategy should be independent of deposition materials, thus providing a versatile methodology for preparation of diverse ordered binary arrays. In addition, owing to the stepwise feature of the fabrication routes, we can also design the arrays in binary or complex composite materials. For example, in Routes 2–4, if different material species are deposited in different deposition stages, the resulting binary arrays will be of heterogeneity in composition, in addition to structure and morphology. The hierarchical micro/nanostructure and their patterns have shown unique properties and applications in such as SERS, modified electrodes, superhydrophobicity, and micro/nanodevices. The hierarchically structured binary porous films will be advantageous for both fundamental research related to micro/nanosystems and their device applications.

7.3 WETTABILITY OF MICRO/NANOSTRUCTURED FILMS

The wettability of solid surfaces is a very important property depending on both chemical composition and surface structure. Generally, the wettability of a surface is associated with its roughness for a certain material. There are two well-established models to describe the water dewetting behavior on rough films. When a water droplet can dip into the pores, Wenzel gave a wettability description of a porous film [101],

$$\cos\theta_r = r\cos\theta \tag{7.9}$$

where

r is the roughness factor, which is the ratio of the total surface area to the projected area on the horizontal plane

θ_r and θ are the CAs on the porous film and a native film with smooth surface, respectively

For Wenzel's type surface, obviously, high roughness can enhance both hydrophobicity of hydrophobic surface and hydrophilicity of hydrophilic surface

When a water droplet cannot dip into the pores, Cassie presented the following equation [102]:

$$\cos\theta_r = f_1\cos\theta - f_2 \tag{7.10}$$

where

f_1 is the contact area fraction of a water droplet in contact with the skeleton of the film surface

$f_2 = 1 - f_1$ is the area fraction of a water droplet in contact with air in pores of the film surface

Obviously, increasing f_2 can lead to larger θ_r, that is, the area fraction on pores in the surface is important to the superhydrophobicity for Cassie's type surface.

Especially, superhydrophobic surfaces (or CA larger than 150°) and superhydrophilic surfaces (or CA smaller than 5°) have attracted much attention for fundamental interest and practical applications, such as the prevention of the adhesion of snow, fog, and raindrops to antennas and windows, the reduction of the friction drag, and the creation of self-cleaning, antioxidation, and microfluidic devices. Generally, for fabrication of superhydrophobic surfaces or films, it is necessary to create a rough structure on a hydrophobic surface (CA > 90°) and to lower the surface energy of a rough surface by chemical modification, such as coating with fluoroalkylsilanes or thiol.

As we know, many orderly structured films synthesized using the colloidal monolayers as templates, such as ordered pore array film, ordered pillar array film, etc., are rough on the nano- and microscales and could also be superhydrophobic or superhydrophilic. If in the case, the nanodevices based on such micro/nanostructured films would be waterproof and even self-cleaning in addition to their special device functions.

7.3.1 WETTABILITY OF ORDERED PORE ARRAY FILMS

As we introduced above, ordered pore array films can be prepared based on PS colloidal monolayers and their morphologies are closely dependent on the experimental conditions, for example, precursor concentrations. Recently, our group have found that the morphologies of ZnO ordered pore

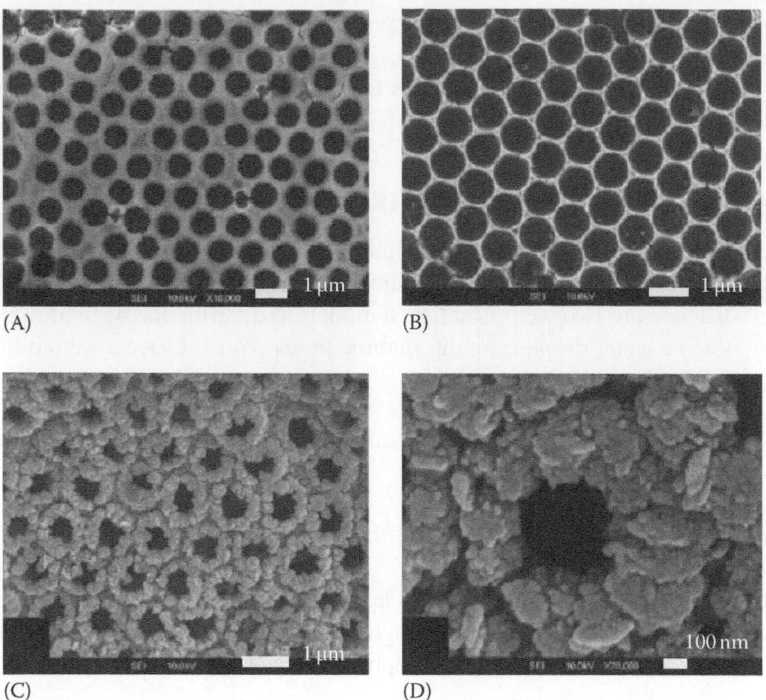

FIGURE 7.72 FESEM images of ZnO ordered pore array films prepared by the solution-dipping method using different precursor concentration: (A) 0.3 M, (B) 0.5 M, (C, D) 1.0 M. (Reproduced from Li, Y. et al., *J. Colloid Interface Sci.*, 287, 634, 2005. With permission.)

arrays can be well controlled by increasing the precursor concentrations [103]. The surface roughness increases with the increase of precursor concentration. Therefore, it is expected to use this phenomenon to control the surface wettability. Figure 7.72 shows ZnO ordered pore arrays with different morphologies prepared with different precursor concentrations. These three kinds of surface microstructures correspond to different precursor (zinc acetate) concentrations. At a low precursor concentration (0.3 M, Figure 7.72A), the pores in the film demonstrate truncated hollow spheres, and the pore sizes are smaller than the diameter of the colloidal sphere of the template. The depth is also smaller than the radius of the template. With increase of precursor concentration to 0.5 M, each pore looks like a hollow hemisphere and the pore size increases to about the PS diameter (Figure 7.72B). If the concentration is further increased (1.0 M), the pores show a noncircular shape from the top view. The surface morphology exhibits a hierarchical structure, which is composed of close-packed rough wreaths and some small particles 30 nm in size, as shown in Figure 7.72C and D.

The wettability of the as-prepared ZnO porous films with different surface morphologies was measured by carefully dropping water droplets upon their surface in a dark chamber. Figure 7.73A through C gives the photographs of the water droplets on different films and the corresponding water CAs are 125°, 131°, and 143°, respectively. These indicate that such ordered porous structures can effectively increase the hydrophobicity in comparison with the relatively flat ZnO films (CA, 109°). Moreover, the wettability shows clearly dependence on surface microstructures, which is determined by the precursor concentration. Moreover, after chemical modification with low-surface-energy materials, fluoroalkylsilanes, the above-mentioned three samples demonstrate superhydrophobicity, and the corresponding water CAs increase to 152°, 156°, and 165°, respectively, as shown in Figure 7.72D through F, where the water droplets display rather spherical shapes.

Interestingly, the water droplets on the surfaces of the modified samples shown in Figure 7.72A and B did not slide even when the surfaces were almost tilted vertically. But, slightly tilted (<5°), the

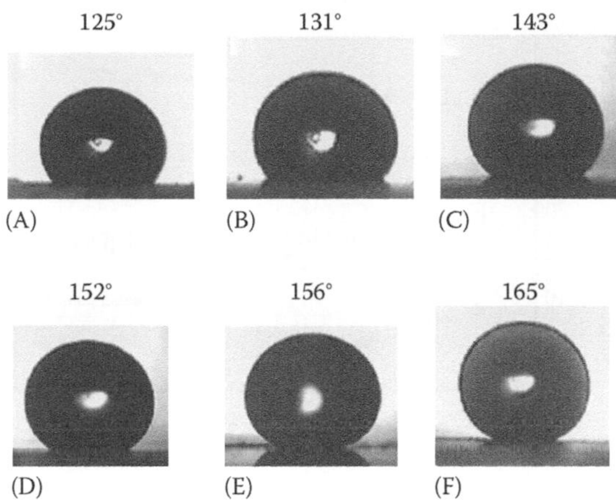

FIGURE 7.73 Water droplet shapes on ZnO pore arrays and the corresponding static CAs on the as-prepared samples (A–C), corresponding to Figure 7.72A through C; (D–F) show the water droplets and CAs on the same samples after chemical modifications, corresponding to (A–C), respectively. (Reproduced from Li, Y. et al., *J. Colloid Interface Sci.*, 287, 634, 2005. With permission.)

water droplets on the modified surface of Figure 7.72C rolled off quickly. This phenomena can be described more accurately by the dynamic CAs, the advancing CA (θ_A), and the receding CA (θ_R). The dynamic CAs of the modified ordered pore array films fabricated by 1.0 M precursor are too small to be measured [sliding angle (SA), <5°]. Nevertheless, its CA hysteresis θ_H ($=\theta_A - \theta_R$) is the smallest due to its largest water CA and smallest SA. The results indicate that with rise of precursor concentration, the CA hysteresis of such as-prepared and modified nanostructured films both decrease. Moreover, the as-prepared samples have a relatively larger CA hysteresis (more than 21°), while the modified films have a smaller one (<9°).

As we know, dewetting behavior of the rough surface that can be described with the Wenzel model (Wenzel surface) shows larger θ_H than surfaces that can be described with the Cassie model (Cassie surface) due to the bigger adhesive force between the water and the film. In this case, the as-prepared sample has a large θ_H (more than 21°). Consequently, this dewetting behavior principally prefers the Wenzel model. From Figure 7.72, it can be found that the roughness of the porous array film increases with increase of precursor concentration. Correspondingly, according to the Wenzel model, θ_r should increase, which agrees well with these results. After modification with fluoroalkylsilane, these ordered pore array films exhibit superhydrophobicity (CA > 150°) with a small θ_H (<9°), which corresponds to Cassie surface. After modification with lower surface free energy materials, air can be trapped into such ordered porous film. For the modified samples, since the parameter f_s decreases with the precursor concentration, corresponding water CA also increases.

With increase of precursor concentration, CA hysteresis decreases for both the as-prepared and modified samples, which is possible due to the gradual reduction of the adhesive forces between water droplets and films induced by the concentration-dependent roughness. When the precursor concentration is not high (say, 0.3 M or 0.5 M), although the modified samples are of superhydrophobicity, water droplets on their surface do not roll off when the films are tilted. This is mainly attributed to the continuous, stable three-phase (air–liquid–solid) contact line for such net-like ordered pore array films. However, when the precursor concentration is high enough (say, 1.0 M), the surface of the ordered porous film is much rougher and shows hierarchical structure (Figure 7.72C and D), which is similar to that of the well-known lotus leaves, leading to superhydrophobicity with large water CA (165°) and small SA (<5°). This film could be expected to show self-cleaning effect.

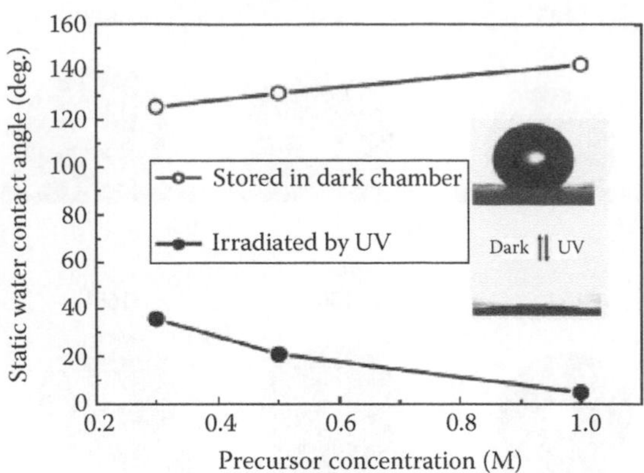

FIGURE 7.74 Transition between hydrophobicity and hydrophilicity induced by UV irradiation for the as-prepared samples. (Reproduced from Li, Y. et al., *J. Colloid Interface Sci.*, 287, 634, 2005. With permission.)

Our group also found a reversible wettability for ZnO pore array films. Very interestingly, ultraviolet (UV) light irradiation for ZnO pore array films can lead to transition of the wettability from hydrophobicity to hydrophilicity for the as-prepared samples. Figure 7.74 demonstrates the results for the as-prepared sample kept in the dark chamber for 7 days before and after irradiation for 2 h by UV light from a 500 W Hg lamp with a 400 nm filter. The differences between hydrophobic and hydrophilic CA are increased from 89° to 138° with rise of precursor concentration. If we alternate keeping the samples in dark for 7 days and then exposing them to UV light, this wettability transition is reversible.

This also can be explained. As we know, the electron-hole pairs generated by UV irradiation will move to the surface, and these holes will react with the lattice oxygen to form the surface oxygen vacancies. Meanwhile, water and oxygen may compete to adsorb on them. The defective sites are kinetically more favorable for hydroxyl adsorption than oxygen adsorption. Such reason and the rough surface can lead to hydrophilicity. Due to the instability of the hydroxyl absorption and the thermodynamical favorite of the oxygen adsorption, oxygen atoms will replace the hydroxyl groups adsorbed on the defective sites gradually when the UV-irradiated ordered porous films were placed in the dark chamber. So when the original state of the surface is recovered, the wettability is reconverted from hydrophilicity to hydrophobicity. According to the Wenzel model, surface roughness enhances both hydrophobicity of hydrophobic surfaces and hydrophilicity of hydrophilic ones. The increase of the porous film roughness factor induced by increasing precursor concentration can enhance both the hydrophobicity and hydrophilicity of the surface with two contrary states, leading to ever-increasing CA differences of reversible wettability transition with rise in precursor concentration, shown in Figure 7.74. Such a surface is very important in the application of microfluidic devices.

7.3.2 Wettability of Hierarchically Structured Array Films

Hierarchically micro/nanostructured films have a bionic locus surface, which can easily lead to superhydrophobicity after certain surface modification with low energy material. As mentioned above, colloidal lithography strategy can produce various hierarchically micro/nanostructured array films. Thus, one can realize surface superhydrophobicity via such strategy. Recently, Li, et al. have performed this strategy successfully [104].

In one of their cases, they prepared silver nanoparticles on the PS colloidal monolayer and thus obtained hierarchically micro/nanostructured array films successfully. The fabrication process of the bionic surfaces is illustrated in Figure 7.75. The monolayer PS colloidal crystal with the area of about 2 cm^2 was prepared on a glass substrate by spin-coating. Subsequently, AgAc aqueous solution

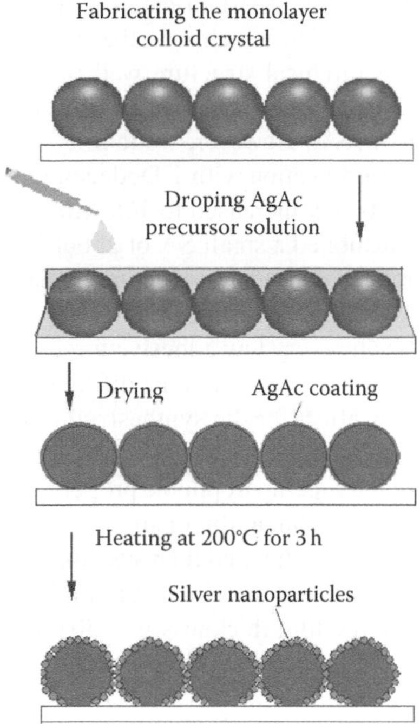

FIGURE 7.75 Scheme for the fabricating process for the bionic surfaces with the hierarchical microsphere/nanoparticle composite arrays.

was dripped onto the colloidal crystals, forming a thin AgAc coating on the PS colloidal crystal. The colloidal crystal with the AgAc coating was heated in an oven, leading to the formation of comparatively uniform decoration of silver nanoparticles on the surfaces of the PSs. Consequently, hierarchical structures consisting of ordered PS microspheres and silver nanoparticles were formed.

Figure 7.76 shows the as-synthesized hierarchical micro/nanostructures. The monolayer colloidal crystal has the periodicity of 5 μm, and the nanoparticles on the colloidal crystal have an average

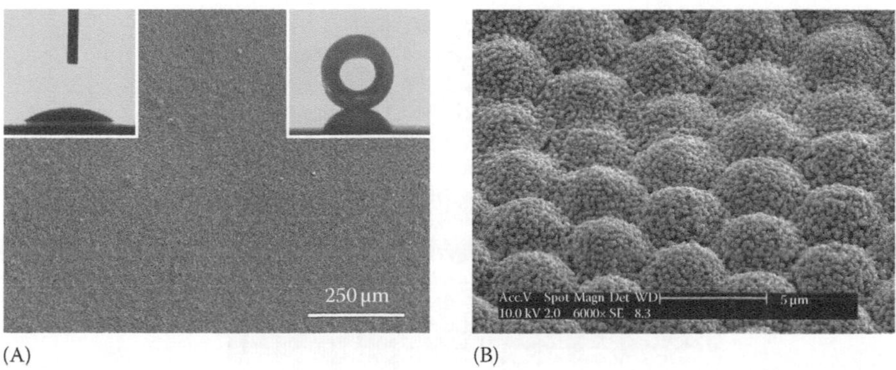

FIGURE 7.76 FESEM images of the bionic surface fabricated using the colloidal monolayer (the diameter of PS: 5 μm) with AgAc coating after heating at 200°C for 3 h and water droplet shapes on its surface. (A) FESEM image of the large-area surface. The insets on the left and the right top are water droplet shapes before and after modification, respectively. The water CAs were 29° (left) and 168° (right), respectively. (B) Magnified FESEM image of the bionic surface. (Reproduced from Li, Y. et al., *J. Phys. Chem. C*, 111, 14813, 2007. With permission.)

size of 180 nm. The hierarchical structure was fabricated with a precursor solution of 0.5 M and at a heating temperature of 200°C for 3 h, which was the optimized experimental condition for the fabrication. The synthesized hierarchical structure well mimicked the surface of a lotus leaf. The water CA of the as-prepared bionic structure was measured to be 29°, showing hydrophilicity (the inset at left top in Figure 7.76A). However, wettability of the structure was changed into superhydrophobicity after surface modification with 1-Dodecanethiol, a kind of low-surface-energy material. The water CA was dramatically increased to 168° (the inset at right top in Figure 7.76A). In addition, the modified surface exhibited a small SA of about 2°. These results indicated that the fabricated surface with the hierarchical micro/nanostructure had a typical self-cleaning property. In the process of CA measurement, it was difficult to add a water droplet on the fabricated bionic surface, demonstrating that the surface has very low adhesive force and very small CA hysteresis. The facts of superhydrophobicity with a very low SA and the difficulty in dropping water on the surface provide strong evidence of the lotus effect for the synthesized bionic surface with the hierarchical micro/nanostructure.

In order to identify the origin of the superhydrophobic properties of the synthesized structures, Li, et al. fabricated a flat and uniform silver film, a film of silver nanoparticles, and a monolayer PS colloidal crystal coated with a uniform silver film, and then systematically investigated CAs of the films before and after the modification of surfaces with thiol (Figure 7.77). The uniform silver films were prepared by plasma sputtering, and the film thickness was 30 nm. The film of silver nanoparticles

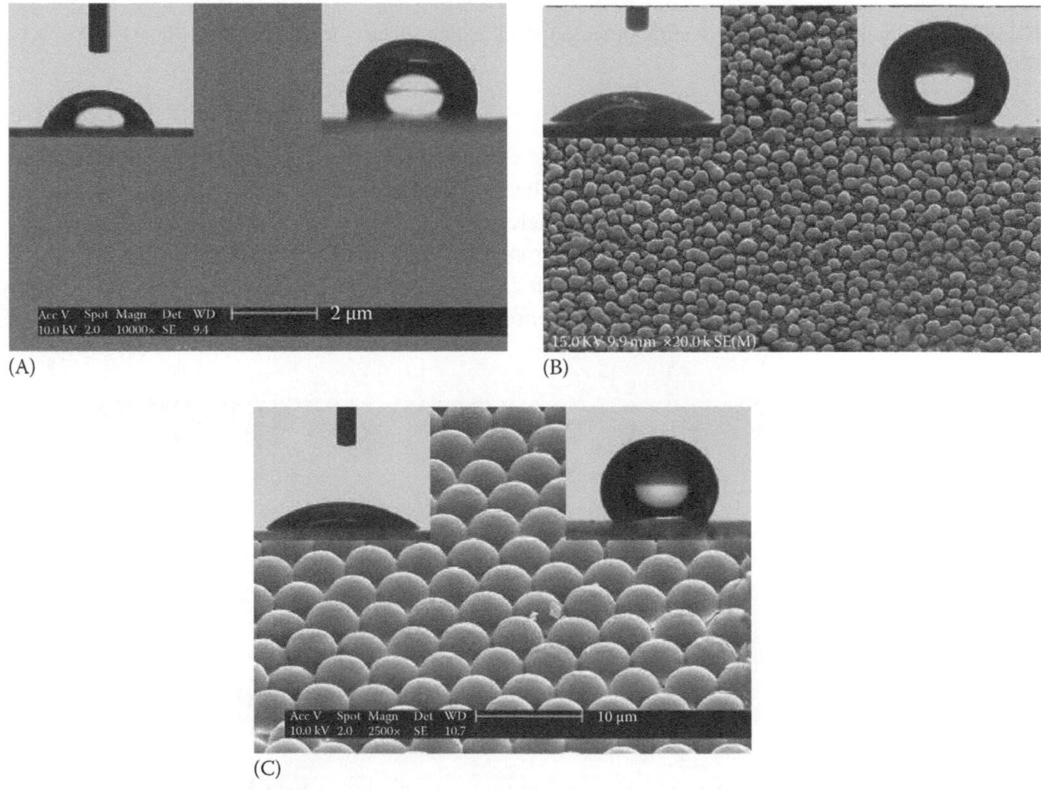

FIGURE 7.77 FESEM images and water droplets on (A) a flat silver film surface, (B) a silver nanoparticle film, and (C) a colloid monolayer with a thin silver coating. Insets at the left (right) top of the figures are the droplets before (after) the chemical modification. The water CAs before and after the modification were (A) 68° and 110°, (B) 41° and 135°, (C) 43° and 129°, respectively. (Reproduced from Li, Y. et al., *J. Phys. Chem. C*, 111, 14813, 2007. With permission.)

was provided on a flat substrate by the thermal decomposition of AgAc. The water CA of the flat silver film was 68°, and the film was transformed into hydrophobicity with a water CA of 110° after the surface modification (Figure 7.77A). The silver nanoparticle film had a water CA of 41°, which was lower than the flat silver film. However, its CA was enhanced to 135° after the surface modification (Figure 7.77B). The silver-coated monolayer PS colloidal crystal showed the water dewetting behavior similar to the silver nanoparticle film: the CA was 43° and increased to 129° after the surface modification (Figure 7.77). These results show that both the silver nanoparticle film and the PS colloidal crystal do not induce superhydrophobicity, and suggest that the strong superhydrophobicity of the synthesized bionic surface originates from the hierarchically combined micro/nanostructure of the surface.

Besides the hierarchically micro/nanostructured array films, by decorating 0D (or nanoparticle) nanostructures on the microsized PSs, Li et al. also presented a facile and alternative method to create superhydrophobic bionic surfaces with hierarchically micro/nano combined structures by loading 1D nanostructures (carbon nanotubes, CNTs) on the microsized PSs [105]. The microstructure was produced by the PS colloidal monolayer on a glass or silicon substrate, and the nanostructure was induced by single-walled carbon nanotubes (SWCNTs) decorated on the microstructures by wet chemical self-assembly [105–107]. The morphology and the distribution density of the nanostructure can be easily controlled by concentration of the SWCNT solution. The presented route well exhibits the concept of bionic fabrication. The morphology of the resultant product bears more resemblance to the natural lotus leaf, and consequently shows strong superhydrophobicity.

The fabrication process is illustrated in Figure 7.78. The PS monolayer colloidal crystals with centimeter square size were prepared on glass substrates and then heated at a temperature of 130°C for 40 min, which strongly increases the adherence of the PS crystals to the substrate [59,108]. Subsequently, a gold layer of 30 nm thickness was coated on the microsphere surface by plasma sputtering. The sample was then dipped into 0.1 mol/L aqueous solution of mercaptoethylamine for the functionalization of amino group on the gold surface. SWCNTs with carboxylic acid functionality (–COOH) solution was prepared by ultrasonating the raw CNTs (Iljin Nanotech Co., Ltd., Korea) in the mixture of concentrated sulfuric and nitric acids with a volume ratio of 3:1 for 6 h, followed by redispersing the SWCNTs into acetone after filtration by sonication. When the prepared SWCNTs solution was dropped on the microsphere surface, the SWCNTs can be self-assembled on the PSs due to the condensation reaction between the –COOH and the $-NH_2$ as well as the electrostatic attraction and Van der Waals interactions between the CNTs [109]. The SWCNTs were then decorated randomly on the surfaces of PS microspheres with almost uniform distribution. As a result, the bionic surface consisting of PSs and SWCNTs was created.

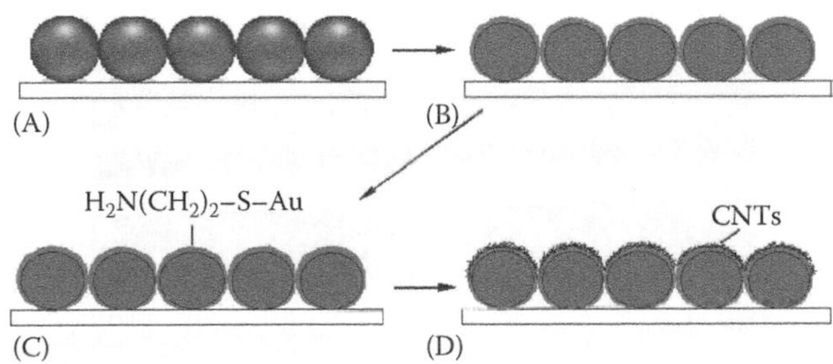

FIGURE 7.78 Schematic fabrication process of bionic surface with micro/nanostructures. (A) Fabricating colloidal monolayer on a glass substrate; (B) coating gold layer on the microsphere surface; (C) modifying the surface with $NH_2(CH_2)_2$; (D) self-assembling CNTs with –COOH on the sphere surface. (Reproduced from Li, Y. et al., *Langmuir*, 23, 2169, 2007. With permission.)

Figure 7.79A through C shows the bionic structure fabricated by 2.0 mg/L SWCNTs solution and the colloidal monolayer with periodicity of 5.0 μm. The images clearly show that dense SWCNTs were assembled on the microspheres' surface by the wet chemical self-assembly method. SWCNTs on the microspheres take on interlaced "net" structures and they tightly adhere to the microsphere surfaces. The whole hierarchical microsphere/SWCNTs composite arrays take on hexagonally close-packed arrangement. The FESEM image of a natural lotus leaf surface was also shown for comparison (Figure 7.79D) [110]. It can be seen that the synthesized hierarchical structure well mimic the surface of a lotus leaf.

The wettability of such hierarchical structures was investigated in both static water CAs and SAs. The water CA of the as-prepared samples was measured to be 33° (Figure 7.80A), exhibiting hydrophilicity. After the chemical modification with fluoroalkylsilane (immersing the sample in a hexane solution of 20 mM 1H, 1H, 2H, 2H-perfluorodecyltrichlorosilane for 30 min and subsequently

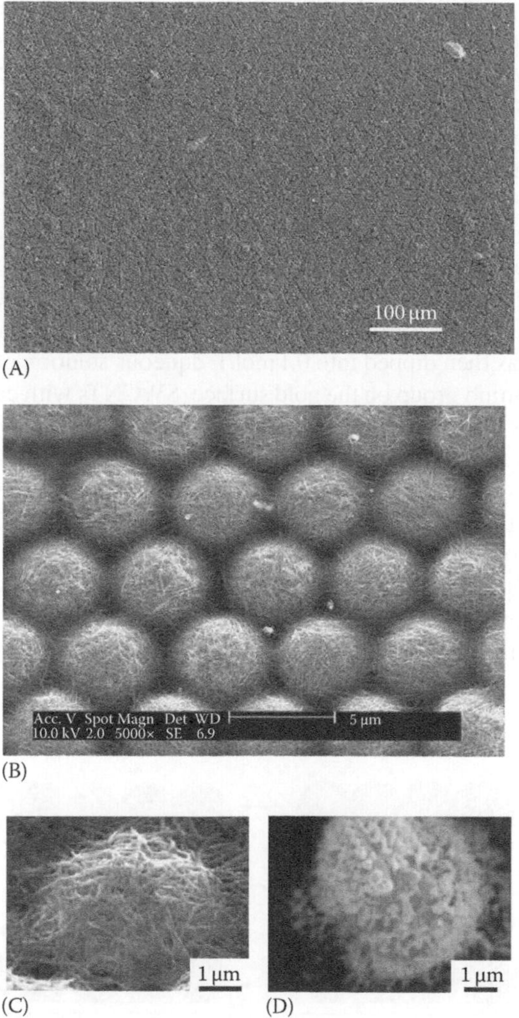

FIGURE 7.79 (A) FESEM image with large area of the bionic surface with microsphere/SWCNTs composition array. (B) FESEM image with higher magnification of bionic surface. The concentration of SWCNTs was 2.0 mg/L, (C) is the feature picture of the bionic surface obtained with a tilting angle of 40° and (D) is the surface microstructure of a natural lotus leaf. (Reproduced from Li, Y. et al., *Langmuir*, 23, 2169, 2007. With permission.)

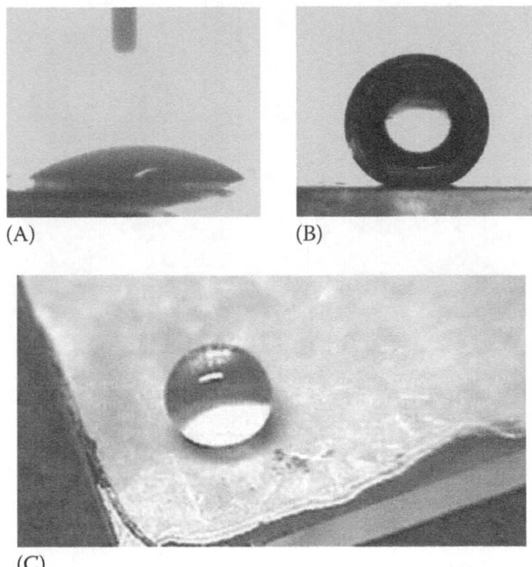

FIGURE 7.80 Shapes of water droplets on the fabricated bionic surface (A) before and (B) after the surface modification. The corresponding CAs are 33° and 165°, respectively. (C) A photograph of the water droplet on the as-prepared bionic surface. (Reproduced from Li, Y. et al., *Langmuir*, 23, 2169, 2007. With permission.)

drying in an oven at 50°C for 30 min), the CA of the sample was dramatically increased to 165° and the water droplet was nearly spherical (Figure 7.80B and C). In addition, the surface exhibited a small SA of about 5°. The presence of superhydrophobicity with a very low SA provides strong evidence of the lotus effect for the synthesized hierarchical micro/nanostructures.

Further, Li et al. systematically investigated the CAs of a monolayer PS colloidal crystal coated with gold film and a SWCNTs film on flat substrate (Figure 7.81). A SWCNT film was prepared on gold-coated silicon substrate by a wet chemical self-assembly as described above. The CA of the gold-coated PS colloidal monolayer was 94° and increased to 138° after the surface modification

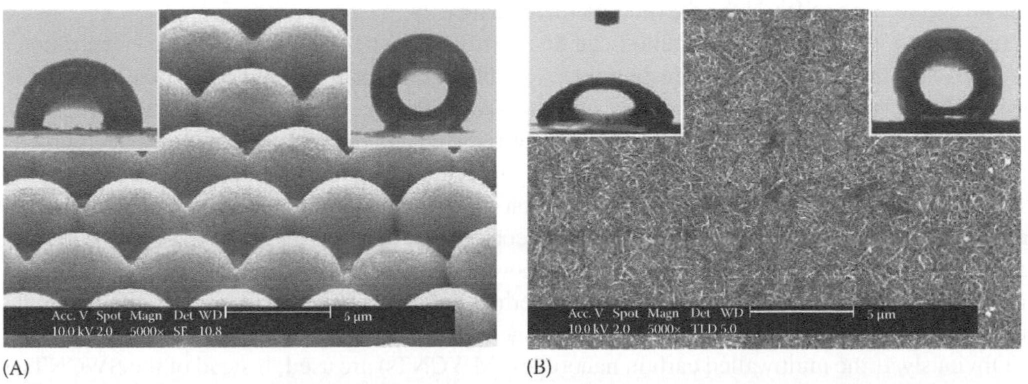

FIGURE 7.81 Images (A) and (B) are a PS array with gold coating and SWCNT film on the substrate, respectively. Both insets on the left top in a and b are the water drop shapes on the surface and the CAs are 94° and 63°, respectively. The insets on right in a and b are water drop shapes after the surface modification and the CAs are 138° and 132°, respectively. (Reproduced from Li, Y. et al., *Langmuir*, 23, 2169, 2007. With permission.)

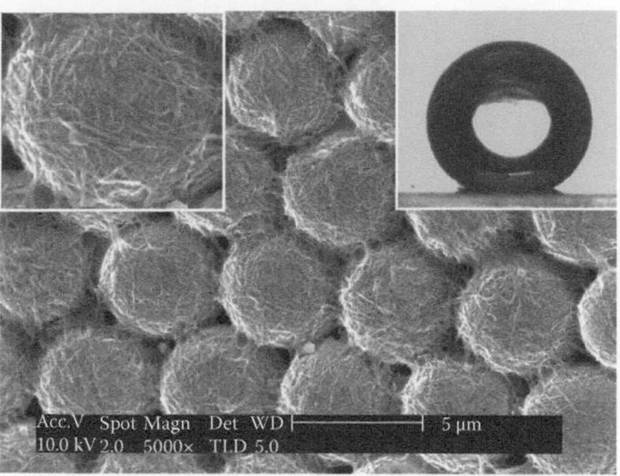

FIGURE 7.82 A FESEM image of the microsphere/SWCNTs composition array at a SWCNTs concentrations of 1.0 mg/L. The left inset is the magnified FESEM image of the structure and the right inset shows the water droplet on the surface. (Reproduced from Li, Y. et al., *Langmuir*, 23, 2169, 2007. With permission.)

(Figure 7.81A). Moreover, the SWCNT film on a flat substrate exhibited hydrophilicity with a water CA of 63° and showed hydrophobicity with the CA of 132° after the surface modification (Figure 7.81B). These results indicate that the strong superhydrophobicity of the fabricated bionic surface originates from its unique hierarchical structure combined by the nanoscaled net structure of SWCNTs and the microscaled PSs.

Before surface modification, the wettability of the SWCNT film on the flat surface was hydrophilic (CA: 63°), which was attributed to –COOH groups on the CNT surfaces. Evidently, SWCNTs on the PS colloidal monolayers make the surface roughness increase. According to the Wenzel model [111], a rough surface is more hydrophilic than a flat surface, and accordingly the hydrophilicity increases with the increasing roughness, which can explain why the water CA (33°) of the SWCNTs on the colloid monolayer is smaller than the CA (63°) of the SWCNT film on the flat substrate. When the bionic surfaces are modified with the low surface-energy material, the air can be trapped in grooves or interstices on the bionic surfaces. In this case, the Cassie model can be used to explain this phenomenon. Since the given water CAs of the flat graphite surface [112,113] and the bionic surface modified with fluoroalkylsilane are 108° and 165°, respectively, f_2 is calculated to be 0.95. This indicates that such strong superhydrophobicity of the bionic surface is mainly due to the air trapped in the rough surface by combining the microstructure of the ordered PS arrays and the nanostructure of the SWCNTs adsorbed on their surface.

In the case, the SWCNT distribution can be controlled on the PSs' surfaces by concentration of the SWCNT solution, and the corresponding wettability can also be tuned due to the morphology change. For a typical example, by decreasing the SWCNT concentration from 2.0 to 1.0 mg/L, the corresponding distribution density of SWCNT was reduced and the PS surfaces were not completely coated by SWCNTs, as displayed in Figure 7.82. This reflects that at a lower SWCNT concentration, a smaller quantity of SWCNTs was assembled on the PSs during the wet chemical self-assembly process. Consequently, the surface roughness becomes less, resulting in lower CA of 156° after the surface modification. The experimental results showed that the CAs decreased further with decreasing the SWCNT concentration and that the superhydrophobic bionic surfaces with the CA higher than 150° were obtained at the concentration range of 2.5–0.7 mg/L.

Obviously, if the multiwalled carbon nanotubes (MWCNTs) are used, instead of the SWCNTs, to fabricate the bionic surfaces, the costs will be largely reduced. Figure 7.83 shows the morphology of the hierarchical microsphere/MWCNT composite arrays. The hierarchical structure was obtained from the MWCNT solution with concentration of 2.0 mg/L and PSs of 5.0 μm in diameter. The MWCNTs combined with the monolayer colloidal crystals also exhibited superhydrophobicity with a CA of 166° and an SA of 5° after surface treatment (see the inset in Figure 7.83).

FIGURE 7.83 FESEM image of the microsphere/MWCNT composite arrays and the water droplet on the surface. (Reproduced from Li, Y. et al., *Langmuir*, 23, 2169, 2007. With permission.)

It should be noted that, compared to the lotus leaves with randomly distributed microstructures and nonuniform unit sizes, the prepared bionic surfaces consist of regularly ordered microsphere/CNT composite arrays. Therefore, the synthesized bionic surfaces had a very uniform wettability on the whole surface, which has been confirmed by measuring water CA at different areas on the whole fabricated surface with only deviation around 1°.

7.4 GAS SENSORS BASED ON MICRO/NANOSTRUCTURED FILMS

As demonstrated in Section 7.2, the micro/nanostructured films based on colloidal monolayer are of macroscopic size (centimeter square), periodic arrangement, high specific surface area, and hierarchical fine structure. These features provide practical realization for some nanostructured devices, such as gas sensors. Here we give two examples for such realization.

7.4.1 Multilayer Homopore Sized Porous Film Gas Sensors

As described above, the ordered porous films can be directly synthesized on curved surfaces based on the strategy presented at our group. Therefore, the gas sensor with the porous film on commercially available ceramic tubes was also constructed directly [51]. Based on the strategy shown in Figure 7.22, the PS colloidal monolayer floating on the surface of $SnCl_4$ solution is directly picked up by a commercially ceramic tube (2 mm in outer diameter and 5 mm in length) with a relative rough surface. After subsequent heat treatments, the SnO_2 ordered pore array film grows directly on the surface of the ceramic tube, which is usually used for a gas-sensor substrate. Figure 7.84 shows a schematic depiction, a photograph, and a microstructure of the gas sensor with a monolayer porous film. There are two gold electrodes at the ends of the ceramic tube, and a Pt resistance heater is placed inside the tube to control the temperature. The ordered porous film can cover the whole surface of the tube and connect the two electrodes. However, as shown in Figure 7.84D, there are many gaps (or breaks) in the pore walls (or skeletons) because of mismatched thermal-expansion coefficients of SnO_2 and the ceramic tube. These breaks could adversely influence the conductivity of the gas sensor. The resistance of a monolayer ordered porous film on the ceramic tube is higher than 30 MΩ, which leads to no signal while measuring the sensing properties. To increase the conductivity, we have adopted a layer-by-layer procedure. By using a tube that has already been covered with a monolayer porous film as the substrate, and repeating the procedures shown in Figure 7.22, we can obtain a second ordered porous layer on the tube. The first layer can act as a buffer layer. Using this method, there are fewer gaps in the pore walls of the second layer. Importantly, the breaks in the

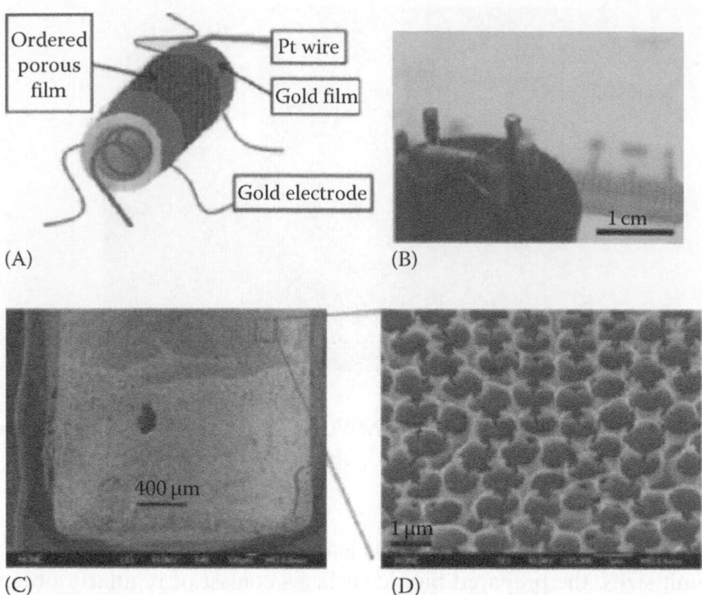

FIGURE 7.84 A gas sensor fabricated with a SnO_2 ordered pore array film on a ceramic tube. (A) The structure of the gas sensor; (B) photograph of a final gas sensor; (C) a low-magnification FESEM image of the gas sensor surface; and (D) the magnified image of the area marked in (C). (Reproduced from Sun, F.Q. et al., *Adv. Mater.*, 17, 2872, 2005. With permission.)

lower (or first) layer can be remedied during formation of the second layer due to the presence of the solution. In this way, the microstructures and hence the conductivities of the gas sensors are much improved. Also, we can fabricate the multilayer porous film on the tube with any number of layers according to the practical requirements. It has been shown that the resistance of the nanostructured porous film with four layers on the tube is below 3 MΩ, which is an acceptable scale for the conventional gas-sensing measurement. A nanostructured gas sensor is thus successfully manufactured.

Figure 7.85A shows the surface morphology of a gas sensor consisting of four ordered porous layers of SnO_2 on the ceramic tube, prepared by a 1000 nm PS colloidal monolayer. To increase the specific surface area of the film, we can use PSs with smaller diameters. Figure 7.85B and C are the corresponding sensors with four layers of SnO_2 porous films on the tube, fabricated with the 350 and 200 nm PSs, respectively. The surface morphologies look quite homogeneous, compared with the monolayer porous film (Figure 7.84D). The gas-sensing performance of the four-layer porous films (sensors) is demonstrated in Figure 7.85D through F, which are the response curves to the 100 ppm ethanol gas at 300°C for these three types of multilayer porous films. The response time, t_R, is defined as the time required for the resistance of the sensor to change to 90% of the saturation value after exposure to the test gas. The gas sensitivity is defined as $S = R_{air}/R_{gas}$, where R_{air} and R_{gas} are the resistances of the sensor in air and ethanol gas, respectively. It can be found that the response time decreases and the sensitivity increases with a reduction in the pore size. This implies that we can easily control the response time and sensitivity of such micro/nanostructured sensors by changing the size of PSs.

As for gas-sensing properties, the increased sensitivity and decreased response time with reduction of PS size can be attributed to the increased surface areas. For SnO_2 gas sensors, the changes in resistance are mainly induced by the adsorption and desorption of gas molecules from the surfaces of the sensing structures. In an air environment, oxygen molecules are generally chemisorbed onto the surface of SnO_2 as O_2^-, O^-, and O^{2-}, which exist in an equilibrium state [114]: $O_2 \Leftrightarrow O_2^- \Leftrightarrow O^- \Leftrightarrow O^{2-}$. Because of the adsorbents, some electrons in SnO_2 are captured [115], and

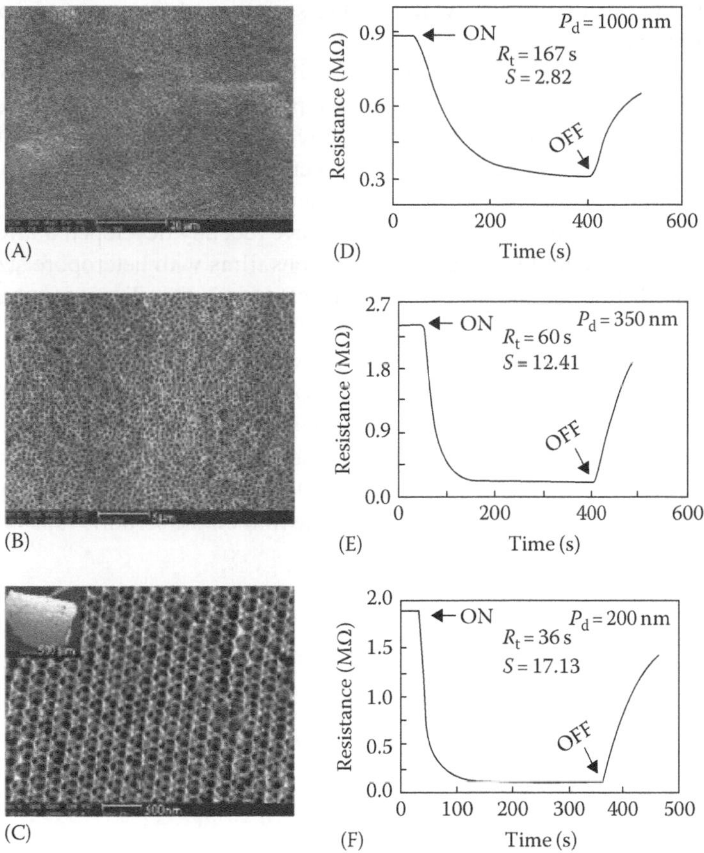

FIGURE 7.85 FESEM images of SnO_2 ordered porous films with four layers fabricated on the curved surface of the gas sensors and their corresponding sensing properties in ethanol gas. The sensors have been fabricated with (A) 1000 nm, (B) 350 nm, and (C) 200 nm in the diameter of a PS colloidal monolayer. The inset in Figure 7.85C is a low-magnification image of a part of the gas sensor. Graphs D–F show the corresponding measured resistance changes vs. time for 100 ppm ethanol gas at 300°C. "ON" means injection of gas into the test chamber and "OFF" means venting of the gas from the chamber. P_d is the pore diameter, R_t is the response, and S is the sensitivity. (Reproduced from Sun, F.Q. et al., *Adv. Mater.*, 17, 2872, 2005. With permission.)

hence the porous films show a higher resistance. When the film is exposed to reductive ethanol gas, the surface concentration of oxygen ions will be decreased, and hence, the conductivity of the SnO_2 ordered porous films will be increased [116,117]. Obviously, pore array films with a higher surface area (or smaller pore size) can adsorb more molecules, and thus possess a higher concentration of oxygen ions. When such films are exposed to the gas, more adsorbed oxygen is desorbed, leading to a higher variation in resistance, or higher sensitivity.

7.4.2 Multilayer Heteropore Sized Porous Film Gas Sensors

As mentioned in Section 7.4.1, the sensitivity and response time of the homopore sized porous film-based sensors could easily be controlled by the pore size based on colloidal monolayer templates (Figure 7.85). However, the two main parameters, sensitivity and response time, are usually correlated with each other. It means that a large pore size corresponds to the fast response but the low sensitivity due to a low surface–volume ratio, and vice versa for a small pore size. The preparation of sensing elements with both high sensitivity and fast response, as well as separately controllable sensing parameters in a large range, are still in challenge based on such micro/nanostructured porous films. Solution of this problem will be a big step toward the practical application of the micro/nanostructured porous film sensors.

It can be imagined that if two monolayer porous films, with big and small pore sizes respectively, are assembled into a multilayer heteropore sized porous film, it could combine the advantages of these two monolayer porous films, or high sensitivity (for the film with small pore size) and fast response (for the film with large pore size). Further, reasonable combination of the monolayer porous films with different pore sizes would obtain the sensing elements with controllable sensing parameters in a large range. On this basis, our group have recently developed a layer-by-layer strategy for fabrication of 2D ordered multilayer oxide porous films with heteropore sized micro/nanostructures on a desired substrate (flat or curved surfaces) by solution dipping and alternately using the monolayer colloidal crystal with different sizes of PSs as templates, as schematically shown in Figure 7.86 [118]. Based on this method, we can directly fabricate the double-layered hierarchically micro/nanostructured porous film on the surface of a commercial ceramic tube with two preformed gold electrodes at the parts close to its two ends and one Pt wire inside the tube for heating, as illustrated in Figure 7.87A (the length, the outer diameter, and the inner diameter of the tube are about 5, 2, and 1 mm, respectively), corresponding to the In_2O_3 film. In this way, we can get a series of gas sensors with different pore sizes. The morphology of the film on the ceramic tube is representatively illustrated in Figure 7.87B.

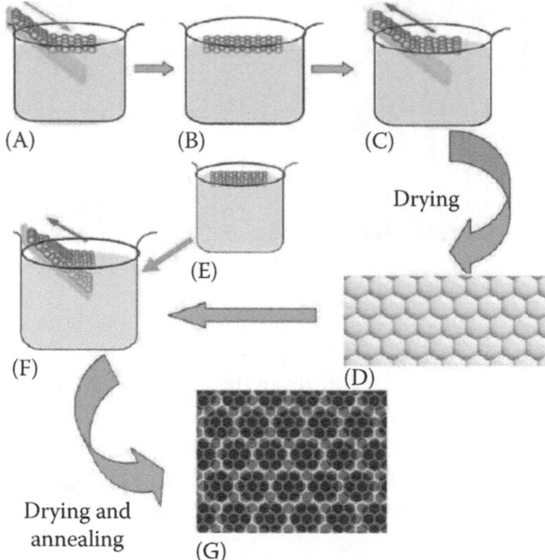

FIGURE 7.86 Outline of the layer-by-layer synthesis strategy for preparing ordered hierarchically micro/nanostructured porous films.

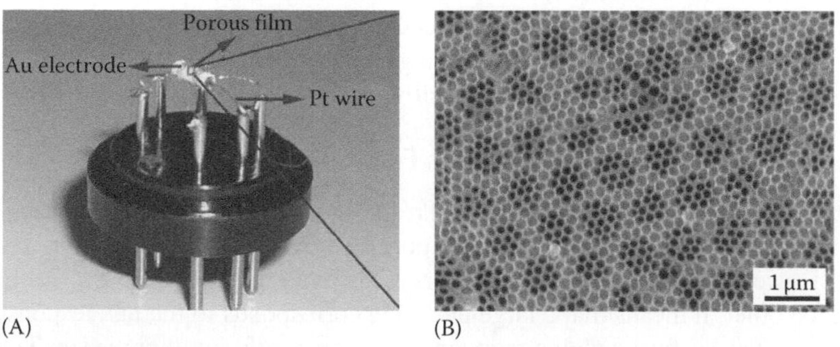

FIGURE 7.87 The photo of the hierarchically structured porous In_2O_3 film (1000/200 nm)-based sensor on a ceramic tube (A) and FESEM image of the film (B) marked in (A).

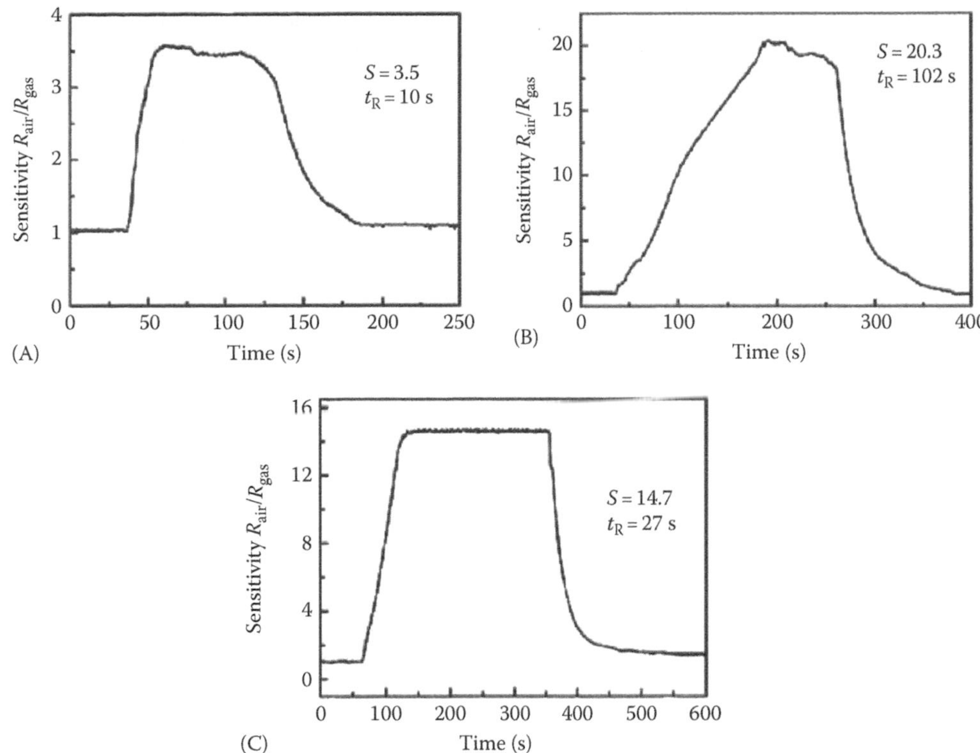

FIGURE 7.88 Response curves to 500 ppm NH_3 test gas at 60°C for different hierarchically micro/nanostructured porous In_2O_3 films. (A) 1000/1000 nm, (B) 200/200 nm, (C) 1000/200 nm.

The electric response measurements in the NH_3 atmosphere have shown that the gas sensitivity and response time of the sensors are significantly dependent on the structure and morphology of the films. Figure 7.88 presents the response curves to 500 ppm NH_3 gas at the working temperature of about 60°C for the samples (or sensors) based on the double-layered micro/nanostructured porous films of 1000/200 nm (the pore size of the first layer is 1000 nm, while the top layer 200 nm), 200/200 nm, and 1000/1000 nm on the ceramic tubes. For the sample with the homopore sized porous film of 200/200 nm, the gas-sensing sensitivity (S) and response time, t_R, are about 20.3 and 102 s, respectively, whereas 3.5 and 10 s for the 1000/1000 nm, as shown in Figure 7.88A and B. Obviously, for the samples with the homopore sized porous structure, the big pore size (1000 nm) corresponds to the fast response (10 s) but a very low sensitivity (3.5), while the small pore size (200 nm) leads to the opposite situation or high sensitivity (20.3) but a slow response (102 s). For the hierarchically porous structured sample with 1000/200 nm film, however, the values of S and t_R are 14.7 and 27 s respectively (Figure 7.88C). Undoubtedly, the double-layered hierarchically structured porous film-based sensor combines the advantages of both big- and small-sized pores, and has a better sensing performance with both high sensitivity and fast response. Furthermore, it is much more sensitive (nine times higher) than the sensing film of the In_2O_3 nanoparticles with 30 nm in average size, which was fabricated on the ceramic tube with about 1 mm in film thickness by a simple printing method in addition to the comparable t_R.

The gas-sensing performance of such films can be understood. For In_2O_3-based sensing elements, the change in resistance is mainly caused by the redox reaction on the surface of the In_2O_3 film in gas atmosphere. When the film is exposed to air, oxygen in air is chemisorbed as O^{2-} ions on its surface, some electrons in the film are localized in the film surface, leading to an increase of the resistance [119–121]. When exposed to a reducing gas, NH_3 for instance, the redox reaction occurs, at a certain temperature, between the reducing gas and the adsorbed oxygen molecules, and the amount of the adsorbed oxygen will thus be decreased, inducing release of the surface-trapped electrons back to the In_2O_3 film, and hence significant reduction of resistance. Obviously, for a porous film, the high specific surface area exposed to environmental atmosphere should correspond to the high S value. Also, the big pore size will lead to a fast transportation of gas molecules and hence fast response or small t_R value. Then, the results in this study are easily understood. When decreasing the pore size in the top layer, the S value will increase due to an increase of the specific surface area, but the response rate is slowed down because of the small-pore-induced difficult transportation of the test gas. Additionally, the peculiar hierarchical micro/nanostructure will lead to significantly more dangling bonds or higher surface activity, for the reaction between the reducing gas and adsorbed oxygen molecules, than the bulk structure [122,123], and thus remarkably decrease the reaction temperature. In this study, our working temperature is only 60°C, much lower than that (about 300°C) of the other In_2O_3 sensors [124,125], which is very important for the practical application, especially for detection of some flammable gas.

Further, if combining the monolayer porous structure with different pore sizes into multilayered hierarchically micro/nanostructured porous films, we can modulate both S and t_R values separately in a large range, and obtain the sensors with the desired sensing-performance, or both high sensitivity and fast response, to meet practical requirements. Such sensors are superior to those of the nanoparticle-based ones and can overcome the shortages of the homopore sized porous films, and can improve the S and t_R values to the desirable values.

7.5 OPTICAL PROPERTIES AND DEVICES OF MICRO/NANOSTRUCTURED FILMS

It was found in many cases that micro/nanostructured films exhibit important morphology-correlated optical properties. Based on such properties, many novel devices have been designed according to practical applications. In this section, we will mainly introduce the recent related work at our group.

7.5.1 Morphology-Dependent Localized Surface Plasmon Resonance of Nanoparticle Arrays

The unique optical property of a metallic nanoparticle is the localized surface plasmon resonance (LSPR). This resonance occurs when the correct wavelength of light strikes a metallic nanoparticle, causing the plasmon of conduction electrons to oscillate collectively. The term LSPR is used because this collective oscillation is localized within the near-surface region of the nanoparticle. The two consequences of exciting the LSPR are (1) selective photon absorption and (2) generation of locally enhanced or amplified electromagnetic (EM) fields at the nanoparticle surface. The LSPR for noble metal nanoparticles in the 20 to a few hundred nanometer-size regime occurs in the visible and IR regions of the spectrum and can be measured by UV–visible–IR extinction spectroscopy. Works by Van Duyne et al. show that Ag nanoparticle arrays from colloidal lithography exhibited important LSPR properties [26]. The LSPR of the array is correlated with the shape and size of the building blocks in the array (see Figure 7.89).

Our work has revealed that the morphologies of the periodic nanoparticle arrays (films) can be modified by annealing or laser irradiation. The former will spheroidize the particles of the whole sample and the particles are easily aggregated during heating. The latter will modify the morphology of the array in a selected area as required. As an example, Nd:YAG laser is used, operating at 1 Hz at the third harmonic wavelength of 355 nm with a nominal pulse width of 7 ns. The laser pulses were unfocused with an energy density of 15 mJ/cm^2. The incident light beam is perpendicular to the substrate of samples. It can be found that the morphology of nanoparticles in the array

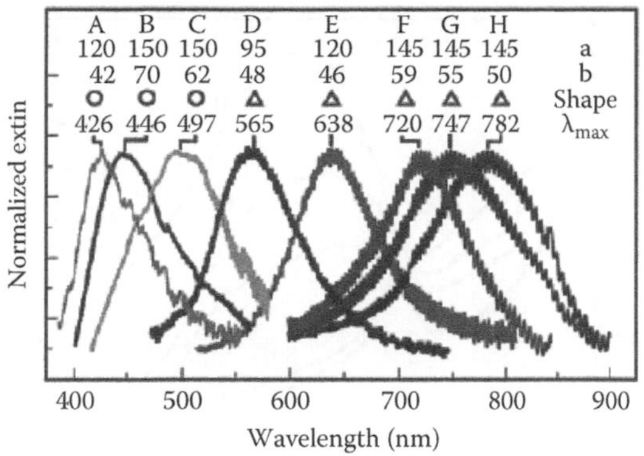

FIGURE 7.89 UV-visible extinction spectra of Ag periodic particle array film on mica substrates. (Reproduced from Haynes, C.L. and Van Duyne, R.P., *J. Phys. Chem. B*, 105, 5599, 2001. With permission.)

evolved from truncated triangle (from top view) to completely spherical shape with increase of the number of laser pulses. Figure 7.90 is the corresponding results for Au nanoparticle array film (the morphology of original sample is similar to that shown in Figure 7.11). After about 40 laser pulses, three sharp corners of each particle become separated from the main body of the particle and three nanogaps of about 30 nm are formed in each particle, as demonstrated in Figure 7.90A. This morphology is particularly intriguing because it might be possible for such an array to be used as a substrate for molecular switching devices. As the number of laser pulses is increased to 100, the nanoparticles at the corners become smaller and almost disappear, while the main body of the particle evolves from a polyhedron to a rounded and finally to a nearly spherical shape, as shown in Figure 7.90B through D. This demonstrates that the morphology of the nanostructured arrays can be manipulated by laser radiation through the appropriate number of pulses. Applying more than 100 laser pulses did not induce any further changes but complete disappearance of the nanoparticles at the corners and the edge sides of the original particles. Figure 7.90E shows that a sample has been irradiated by more than 500 pulses, and its morphology is similar to that of the sample irradiated for about 100 pulses indicating that the particle has reached its equilibrium shape after 100 pulses.

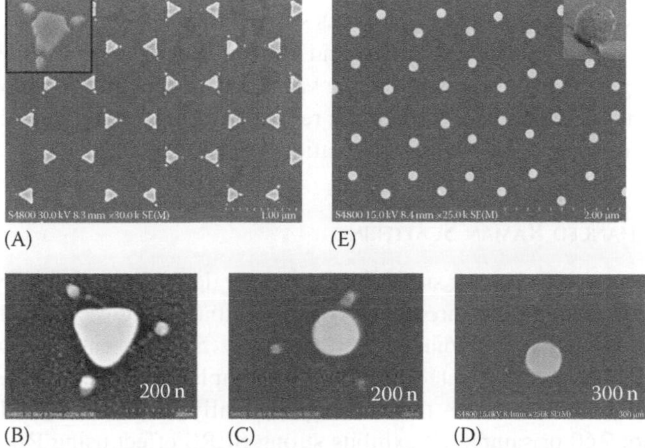

FIGURE 7.90 Morphology of gold particle arrays on ITO substrates after 355 nm laser irradiation (15 mJ/cm^2 per pulse) for different number of pulses. (A) 40 times; (B) 60 times; (C) 80 times; (D) 100 times, and (E) 500 times. (Reproduced from Sun, F. et al., *Appl. Phys. B: Lasers Opt.*, 81, 765, 2005. With permission.)

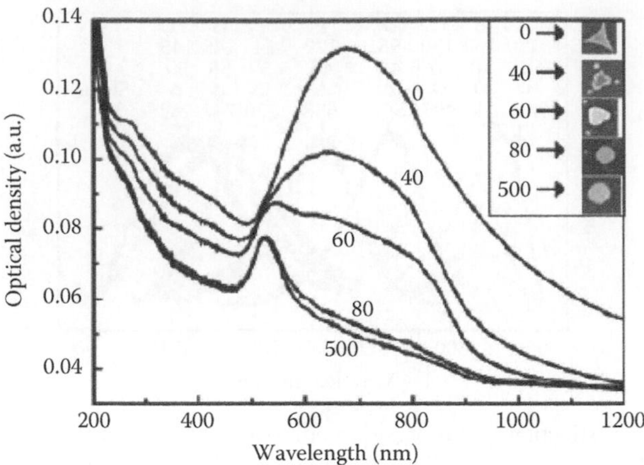

FIGURE 7.91 Evolution of the optical spectra of gold particle arrays on quartz vs. laser irradiation with the indicated number of laser pulses. Inset: The corresponding morphology of individual dots at each pulse number. (Reproduced from Sun, F. et al., *Appl. Phys. B: Lasers Opt.*, 81, 765, 2005. With permission.)

Further, tilted observation has shown that the final particles are nearly spherically shaped, as demonstrated in the inset of Figure 7.90E.

Importantly, the LSPR of such nanoparticle array films shows strong morphology dependence. The corresponding evolution of optical absorption spectra is shown in Figure 7.91. The as-prepared samples show a very broad absorption peak centered around 680 nm together with a shoulder extending well into the near-IR region, which indicates that the peak is composed of at least two peaks. The peak at 680 nm decreases and disappears as the laser irradiation is increased up to 100 pulses. In addition, after irradiation by about 60 pulses, another peak emerges around 550 nm. After about 100 pulses, the peak shifts to 530 nm. Jin et al.'s work [127] and Mie theory [128] indicate that triangular gold particles should exhibit SPR containing one out-plane resonance and two in-plane resonances. The out-plane resonance should be at a shorter wavelength that is too weak to be discerned. One of the in-plane resonances should be around 530 nm and another at a longer wavelength (depending on a morphology factor). Generally, spherical particles have only a single SPR band at about 530 nm. In the spectra shown in Figure 7.91, the broad absorption band around 680 nm for the sample before irradiation can be attributed to the superposition of the two bands of the triangle-shaped particles. Laser irradiation makes particles spheroidize, leading to a decrease and eventual disappearance of the shape-dependent band around 680 nm, leaving to only the single SPR band at 530 nm corresponding to a spherical particle. The variations of spectra can reflect the morphological changes of the particle arrays and, further, the related information of the laser, which shows the potential application of such method in the fabrications of data storage devices.

7.5.2 Surface-Enhanced Raman Scattering

SERS effect has proven to be of strong surface-morphology dependence. In many cases, it has been found that ordered mirco/nanostructured array films exhibit strong SERS activity. Recently, we have found that hierarchically micro/nanostructured array film presents a structurally enhanced SERS effect. The sample was fabricated by a second template strategy, as described above (see Figures 7.59 and 7.60). It can be seen that the hierarchically rough structured Au particle array film (shown in Figure 7.60 or sample C) exhibits strong SERS effect using Rhodamine 6G (R6G) as a probe molecule, as shown in Figure 7.92. For comparison, the results for the smooth Au film (sample A), prepared by vacuum physical vapor deposition on an ITO substrate, and a rough Au film (sample B), prepared by electrodeposition on the ITO substrate without a template using the

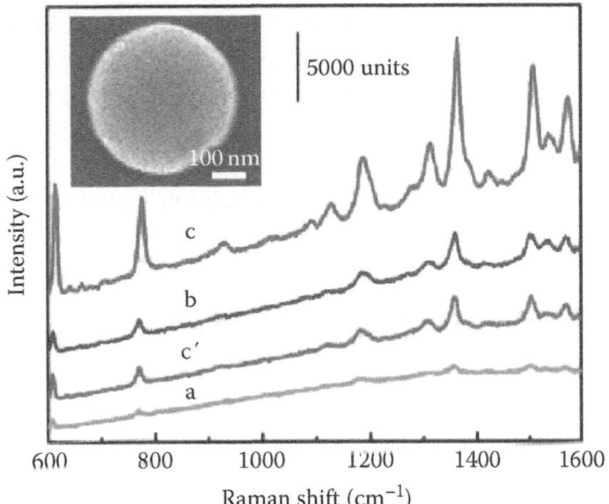

FIGURE 7.92 SERS spectra of R6G on different substrates. (a) Smooth Au film (sample A); (b) rough Au film (sample B); (c and c′) Au particle array before and after laser irradiation (sample C). Raman measurement is conducted with excitation wavelength of 514.5 nm from an air-cooled argon ion laser. All of the spectra were the results of a single 15 s accumulation. Inset: the FESEM image of a single Au particle in the array after 532 nm laser irradiation (15 mJ/cm^2 per pulse) for 800 shots. (Reproduced from Duan, G.T. et al., *Appl. Phys. Lett.*, 89, 181918, 2006. With permission.)

same deposition parameters as sample C (it exhibits similar surface roughness to sample C), are also given in Figure 7.92. Before Raman examination, the samples were dipped into a solution with 10^{-6} M R6G with stirring for 10 min, rinsed with deionized water, and dried with high-purity flowing nitrogen. Sample A only gives a very weak signal (curve a). As for sample B, the signal remains relatively weak but stronger than that of sample A due to its nanoscaled surface roughness (curve b). Sample C, however, exhibits very strong SERS signal (see curve c), which is much stronger than that of sample B. For a confirmation of the effect of surface roughness on the SERS, laser irradiation, which can modify the morphology of Au particles and decrease the surface roughness [126], was conducted for sample C. The morphology of Au particle shows spherical shape with smooth surface after proper irradiation, as shown in the inset of Figure 7.92. The corresponding SERS intensity was decreased dramatically (curve c′ in Figure 7.92), suggesting that nano-sized surface roughness of the microparticles is also a favorable factor for the SERS. In contrast to many conventional SERS substrate of colloidal film, there is no observable activity loss within one month for the SERS from the gold particle array with hierarchical roughness, showing high stability and good reproduction, which could be due to the stability of its special morphology. Here, it should be mentioned that although same dipping conditions do not necessarily mean the same number of molecules attached and detected, we believe that the difference of R6G adsorption amounts on different substrates in this study is not the main reason for the difference of SERS signals. Samples B and C have similar nanoscaled surface roughness, and the surface area should be larger for the former than sample C due to the non-close arrangement of the microparticles. But the opposite is true for the SERS signal (i.e., signals for sample C is stronger than sample B). Also, our further experiments have demonstrated that when the R6G concentration is down to 10^{-8} M, sample C still shows obvious Raman signals while no signal can be detected for sample B or sample C after laser irradiation.

The strong SERS for the array (sample C) can be attributed to both the periodic structure and the hierarchical surface roughness. First, according to Gaponenko [129], the redistribution of photon density of states may easily occur in a periodic structure, resulting in an increase of the density of optical modes and thus the enhancement of the Raman scattering of the detected molecules. Second, the microparticles have a roughly sphere-like shape with diameter on the order of the laser wavelength (514.5 nm). Therefore, the localized plasmon mode can contribute to the Raman scattering enhancement [130,131]. The incident light excites plasmons trapped at the crevices in the long-ranges ordered array, which can produce significant SERS. Third, the nanoscaled surface roughness and the size of Au particles in this study could be optimal for SERS based on Nie's founding that the most efficient SERS should occur on the nanoparticles in several tens of nanometers in size [132].

Strong SERS effects of diverse micro/nanostructured films have also been reported by many other groups. This allows the important practical application in the field of SERS-based sensors, biotechnology, and micro/nanodevices.

7.5.3 Equivalent Photonic Crystal of Hollow Sphere Arrays and Optical Gas Sensors

In Section 7.2.4, we demonstrated a novel micro/nanostructured film, i.e., hierarchically micro/nanostructured hollow sphere array films. It has been found that such films shows novel morphology and size-correlated optical properties as follows.

7.5.3.1 Tunable Optical Transmission Stop Band

The monolayer hollow sphere array can demonstrate tunable optical properties in a large region. Figure 7.93 shows the optical transmission spectra with the incident light aligned perpendicularly to the as-prepared $Ni(OH)_2$ monolayer hollow sphere arrays on ITO. A size-dependent optical transmission stop band exists in the hierarchically structured hollow sphere arrays, which red-shifts in a large range from 455 to 1595 nm with the size increase of the hollow spheres in the array from 1000 to 4500 nm. For the sample of hollow sphere array from 4500 nm PSs, in addition to the main stop band around 1595 nm, there is another weaker band located around 700 nm together with a shoulder at 450 nm. As known, the periodicity of the array is controlled by the PSs' size, one can thus easily adjust the optical transmission stop band in a large region from the visible to near-IR region simply by PSs' size. On the contrary, no such stop band is detected for the net-like film [see curve (a) in Figure 7.93A corresponding to the sample shown in Figure 7.45D], or for the hierarchically structured pore array film [see curves (a) and (f) in Figure 7.93B]. This demonstrates that the monolayer hollow sphere structured array is crucial to produce such tunable optical bands.

Interestingly, the position of stop band can also be fine-adjusted by deposition time without the change of periodicity. Increase of the deposition time leads to red-shift of the band for the hollow sphere arrays, as shown in curves (b)–(e) of Figure 7.93B, which could be due to the denser shell of the hollow sphere. It means that we can very flexibly control the optical transmission stop band

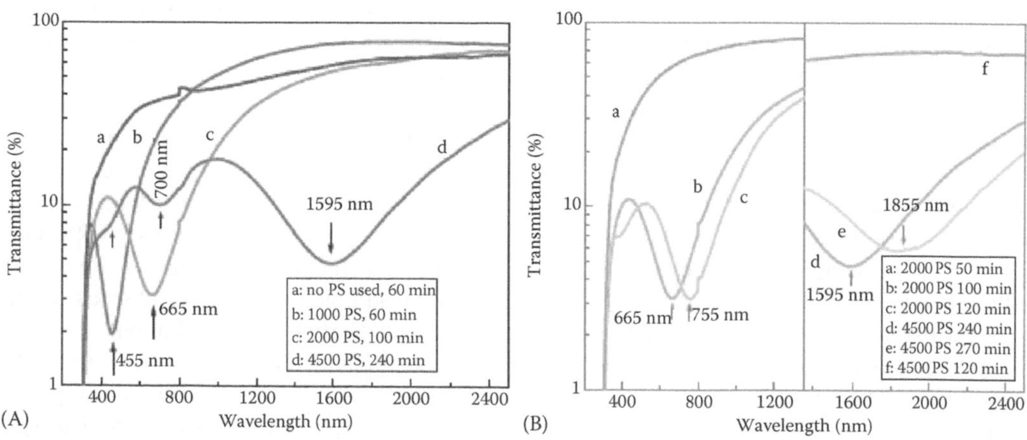

FIGURE 7.93 Optical transmission spectra of $Ni(OH)_2$ samples on ITO glass with the incident light perpendicular to the substrate. (Reproduced from Duan, G.T. et al., *Adv. Funct. Mater.*, 17, 644, 2007. With permission.)

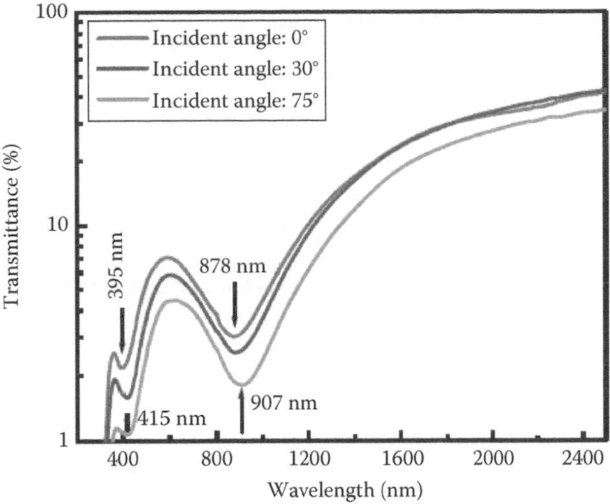

FIGURE 7.94 Optical transmission spectra of $Ni(OH)_2$ sample on ITO glass with different incident angles (the angle between incident light and the normal to the sample plane, θ). The sample was prepared by electrodeposition with $J = 1.2\,mA/cm^2$ for 140 min based on a 2000 nm PS monolayer. It has two clear transmission stop bands due to the first and second diffractions, which is beneficial to the understanding of the shift of stop band. (Reproduced from Duan, G.T. et al., *Adv. Funct. Mater.*, 17, 644, 2007. With permission.)

in a large region, by PS's size for coarse-tuning and deposition time for fine adjustment, which is undoubtedly important both in applications and in fundamental researches.

Further experiments show that the position of the transmission stop band is almost independent of the incident angle θ (the angle between the incident light beam and the normal to the sample plane). For the monolayer hollow sphere array induced by 2000 nm PS monolayer, the center of transmission stop band red-shifts less than 30 nm with change of the incident angle from 0° to 75°, as shown in Figure 7.94. This is quite small and almost negligible compared with the three-dimensional (3D) photonic crystals, which is very sensitive to θ according to [133–135]

$$m\lambda_{min} = 2d(n^2 - \sin^2\theta)^{1/2} \tag{7.11}$$

where

- m is the order of the diffraction
- λ_{min} is the wavelength of the stop band
- n is the mean refraction index of the 3D crystal
- d is the periodical constant along the normal of the sample plane

It is well known that the θ-dependent position of the stop band is a disadvantage of the 3D photonic crystal in application. If one wants to prevent transmission of light from any incident angle, the fully photonic crystal with a θ-independent stop band is needed. This is a challenge for 3D photonic crystals. The monolayer hollow sphere array here could be a good candidate with a θ-independent stop band.

7.5.3.2 Equivalent Double Layer Photonic Crystal

Generally, an optical transmission stop band will be found in 3D colloidal crystal (or inverse opal structure) but cannot be produced in a 2D monolayer colloidal crystal. The monolayer hollow sphere

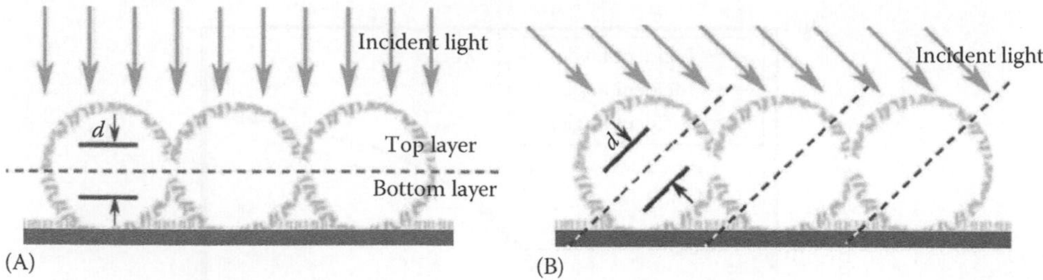

FIGURE 7.95 Schematic illustration of the double-layer photonic crystal approximation for the hollow sphere array. (A) Incidence of light perpendicular to the sample plane ($\theta=0$); (B) aslant incidence of light ($\theta > 0$). (Reproduced from Duan, G.T. et al., *Adv. Funct. Mater.*, 17, 644, 2007. With permission.)

array in our case, however, can equivalently be considered as symmetrical double layers (top layer and bottom layer) with interspacing d, as shown in Figure 7.95, indicating a photonic crystal with double layers. Under perpendicular incidence of a light beam (or $\theta=0$, see Figure 7.95A) Equation (7.11) can be written as $m\lambda_{min}=2nd$, where n corresponds to the mean refractive index of the layers (consisting of sphere shell and interstice). Although the exact values of n and d are unknown, the d value should increase with rise in PSs' size, and n depends on the material species and the structure of the shell. Obviously, the denser the sphere shell (or the longer the deposition time), the larger the n value. Thus, for the first-order diffraction ($m=1$), the transmission stop band should red-shift with an increase of the sphere size or deposition time (see Figure 7.93). Also, there should be a multiple relation between different diffraction orders ($m=1, 2, 3$), which is in rough agreement with the result (see curve d in Figure 7.93). The slight deviation should origin from the wavelength-dependent optical absorption of the $Ni(OH)_2$, which superimposes on the transmission spectra and hence change the measured band position.

As for the θ-independent stop band, it can be attributed to the special structure of monolayer hollow sphere. Due to the symmetry of the hollow spheres, for the incident light with different θ, d value should be similar (see Figure 7.95B), and thus the position of the stop band is almost independent of θ.

7.5.3.3 Optical Gas Sensors

Since the position of the stop band on $Ni(OH)_2$ hollow sphere array films is associated with n, the factors influencing n value will lead to the position change of the stop band. In the case, the hollow spheres with the shells composed of massive nanoflakelets possess high specific surface area. Adsorption of environmental gases will occur on the surface of nanoflakelets, which will lead to a change of n value and hence shift of the stop band. It means that such monolayer hollow sphere array with hierarchical structure could be a good optical gas sensor. Preliminary experiments have confirmed a significant shift of the stop band induced by exposure to alcohol atmosphere, and such shift can be recovered completely by subsequent heating at 100°C in air (see Figure 7.96). Alternate exposure and heating shows a reversible shift of the band. It is due to $Ni(OH)_2$ that we can produce the hollow spheres with nanoflakelets. The monolayer hollow sphere array with such hierarchical structure provides the practical possibility of a new optical sensor for gas detection.

7.5.4 Optical Wavemeter in Transmission Mode

Ordered pore array films can be fabricated based on colloidal lithography, as described above. Due to the periodical structure of the ordered pore arrays, when a beam of parallel incident light with a wavelength λ perpendicularly passes through the arrays, diffraction will occur and the spots (or ring) will be seen if a screen is placed behind the sample (see the following text) due to the interference among cells of the arrays, which has been observed on ordered colloidal monolayer [136].

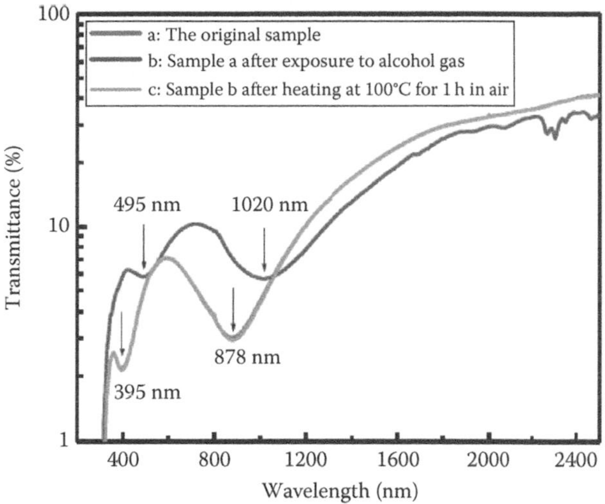

FIGURE 7.96 Optical transmission spectra for the sample shown in Figure 7.94, with the incident light perpendicular to the substrate. After exposure of the original sample to alcohol gas (in a sealed cup with alcohol) for 30 min, the first and second stop bands red-shift more than 140 and 100 nm, respectively (curve b). After subsequent heating at 100°C for 1 h in air (curve c), the bands are recovered completely. (Reproduced from Duan, G.T. et al., *Adv. Funct. Mater.*, 17, 644, 2007. With permission.)

The diffraction angle is associated with the wavelength of the incident light. Based on this, one can determine the wavelength of incident light by measuring the diffraction angle. So the 2D ordered pore array film on a transparent substrate could be used for the miniature optical frequency splitter based on such ordered arrays. The conventional light wavelength measurement is based on the Michelson interference principle, which is of high resolution and has been widely used in optical communication. However, the setup is very expensive and big sized.

When the periodic parameter of the ordered arrays is comparative to the wavelength of incident light, interference will occur between the neighboring uni-cells and diffraction will be very significant [137]. If the beam size, Φ, of incident light is smaller than the domain size, d, of the 2D ordered pore array, diffraction spots will appear on the screen behind the array after the incident light passes through. If $d \ll \Phi$, however, diffraction rings will be observed on the screen (as schematically shown in Figure 7.97) [138].

Obviously, when the pores in an array are arranged in a hexagonal structure, diffraction spots should also be of hexagonal pattern with the spacing m between adjacent spots or diffraction ring

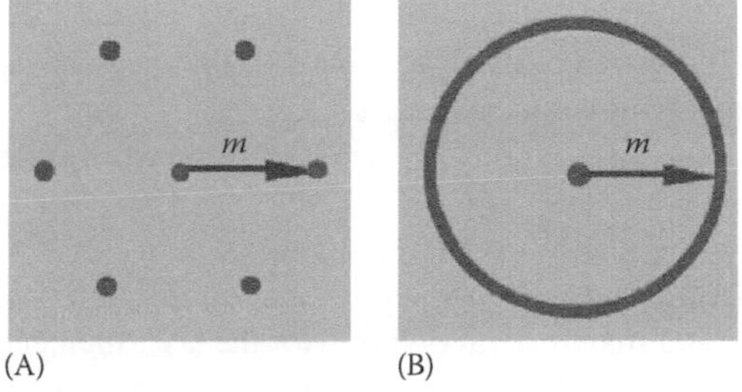

FIGURE 7.97 Schematic illustrations of diffraction patterns when a beam of incident light perpendicularly passes through the 2D ordered pore array with hexagonal closed packed pores. (A) $\Phi < d$, (B) $\Phi \gg d$. (Reproduced from Duan, G.T. et al., *J. Nanosci. Nanotechnol.*, 6, 2474, 2006. With permission.)

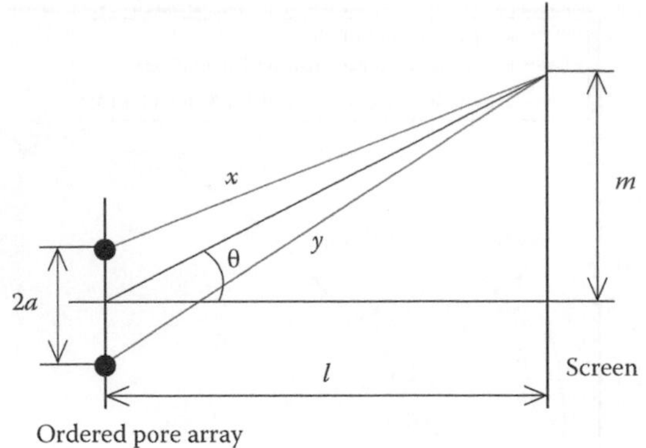

FIGURE 7.98 A sketch of the interference between neighboring uni-cells. (Reproduced from Duan, G.T. et al., *J. Nanosci. Nanotechnol.*, 6, 2474, 2006. With permission.)

with radius m, depending on the relative size of d and Φ, where m is the distance between optical axis and diffraction spots, as illustrated in Figure 7.98.

Generally, $l \gg a$, we have

$$y - x = 2a \sin \theta \tag{7.12}$$

where

l is the distance between the screen and the array
$2a$ is the spacing between neighboring uni-cells (pores) in the array (or lattice parameter)
x and y are the distance between pores and the diffraction spot on the screen
θ is the diffraction angle

According to the interference condition $y - x = k\lambda$, one obtains

$$\sin \theta = k \frac{\lambda}{2a} \tag{7.13}$$

where k is interference order with integer 0,1,2,..., also, $\theta \infty (0, \pi/2)$ and $\lambda < 2a$. From Figure 7.98 and letting $k = 1$, we have the relation

$$\lambda = \frac{2am}{\sqrt{m^2 + l^2}} \tag{7.14}$$

Obviously, if we measure the m value, the wavelength of the incident light can be determined by Equation 7.14. Such ordered arrays can be used for optical frequency splitters. Therefore, a new type of wavelength meter can be designed completely different from the conventional wavemeter in mode.

Figure 7.99 shows the typical photos taken by a digital camera corresponding to the 532 nm laser light (Figure 7.99A) and natural light (Figure 7.99B) passing through the ferric oxide ordered pore arrays. There is a very clear and sharp diffraction ring for the laser incidence with single wavelength, and colored spectra from blue to red for the natural incident light, showing good optical frequency splitting. For other ordered pore arrays, we get the same results. Such diffraction is only determined by the wavelength of incident light and pore spacing, but independent of the material species. Table 7.1 presents the ring sizes and the corresponding wavelength calculated from Equation 7.14 for all

FIGURE 7.99 Diffraction patterns corresponding to the incident light (A) 532 nm laser beam and (B) natural light passing through ferric oxide ordered pore arrays. The periodic parameter is 1000 nm. (Reproduced from Duan, G.T. et al., *J. Nanosci. Nanotechnol.*, 6, 2474, 2006. With permission.)

TABLE 7.1
The Diffraction Ring Sizes and Corresponding Wavelength Calculated from Equation 7.14 (l = 12 cm and 2a = 1000 nm) [138]

Incident Light (nm)	Samples	Ring Radius m (cm)	λ from Equation 7.14 (nm)
532	PVA pore array	7.6±0.2	535±10
532	α-Fe_2O_3 pore array	7.5±0.2	530±10
355	PVA pore array	4.6±0.2	357±14
355	α-Fe_2O_3 pore array	4.5±0.2	351±14

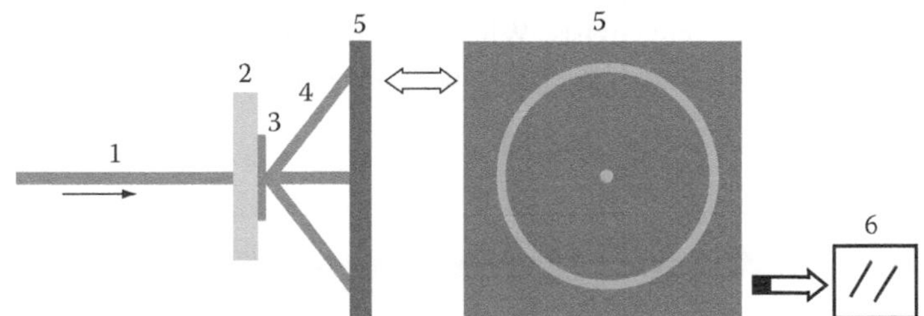

FIGURE 7.100 The sketch of a wavemeter in transmission mode. The numbers respectively represent the following: (1) a parallel incident beam obtained by an optical component; (2) glass substrate; (3) ordered pore arrays on a substrate; (4) diffraction light beam; (5) screen; and (6) controller. (Reproduced from Duan, G.T. et al., *J. Nanosci. Nanotechnol.*, 6, 2474, 2006. With permission.)

samples when l = 12 cm and 2a = 1000 nm using laser incidence. In addition, it was found the ring thickness and the size of zero-order diffraction spot are always nearly equal to the laser beam size Φ, independent of the l value.

Based on the mention above, a new type of setup for wavelength measurement can be designed in transmission mode, as schematically illustrated in Figure 7.100. The ordered pore array film is a key component. A beam of light passes through a special optical component (for example, waveguide and lens) to form a beam of parallel light with a certain beam size, which makes incidence on the arrays perpendicularly, leading to diffraction ring pattern on the screen. Then the ring size can be measured and the wavelength of the incident light can be determined according to Equation 7.14.

Herein, we only demonstrate the possibility that the 2D ordered pore film with multi-domains can be used for wavemeter with a certain precision. It has been shown that the measured wavelength of incident light is in good agreement with the real value, confirming the reliability of our setup. Further, from Figure 7.98A, the diffraction ring has a thickness. This can mainly be attributed to the monochromatic quality and beam size of incident light, and both of them affect the resolution and precision for determination of the wavelength.

7.5.4.1 Resolution of Wavelength Measurement

It is well known that the monochromatic quality of a beam of incident light will influence the thickness of the diffraction ring. Obviously, higher resolution corresponds to the thicker ring at a given monochromatic quality.

From Equation 7.14, we have

$$\frac{\partial m}{\partial \lambda} = K \tag{7.15}$$

where

$$K = \frac{l}{2a\left[1-\left(\frac{\lambda}{2a}\right)^2\right]^{3/2}} \tag{7.16}$$

or the differential form $\delta m = K\delta\lambda$, K determines the effect of monochromatic quality on the ring's thickness, depending on the parameters, such as λ, l, and a. Higher K value corresponds to larger effect. Figure 7.101 shows the K vs. λ at different l values. In the long wavelength region (especially close to the value $2a$), a high K value exists. When $\lambda/2a < \sim 0.6$, the K value can be kept lower. In addition, larger l value corresponds to higher K value. It means that the wavelength $\lambda/2a < \sim 0.6$ and

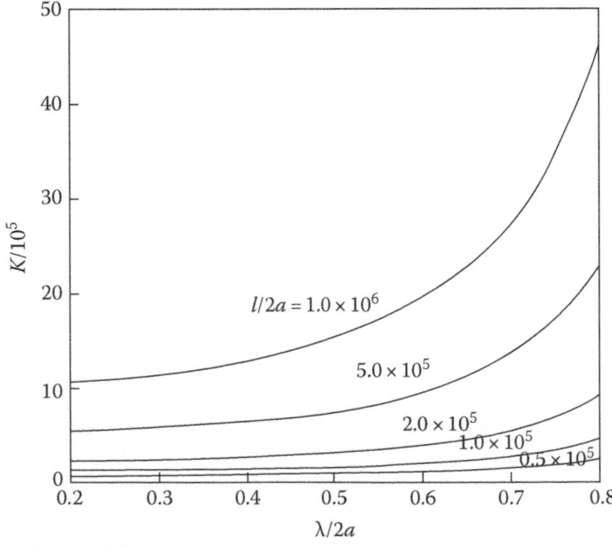

FIGURE 7.101 Curves of K vs. λ for different $l/2a$ values. (Reproduced from Duan, G.T. et al., *J. Nanosci. Nanotechnol.*, 6, 2474, 2006. With permission.)

smaller l value will correspond to a sharper ring at a given monochromatic quality. On the other hand, the larger K value should correspond to a higher resolution to determine the wavelength of incident light. For understanding, letting minimum discernable spacing on the screen $\delta m = 0.5$ mm, $2a = 1000$ nm, $\lambda/2a = 0.8 \sim 0.6$, and $l = 20$ cm, $K = 0.93 \sim 0.39$ mm/nm and the resolution for incident light wavelength $\delta\lambda = 0.6 \sim 1.3$ nm. It means that the wavelength resolution could reach <1 nm in the wavelength region close to $2a$. It should be mentioned that this is a theoretic resolution without consideration of the effect from the beam size.

7.5.4.2 Precision of Wavelength Determination

Since a beam size with Φ will lead to the ring thickness $\delta m \approx \Phi$, from Equation 7.14, the corresponding uncertainty of the wavelength for a monochromatic incident light $\delta\lambda = \Phi/K$. It is evident that the wavelength $\lambda/2a > \sim 0.6$ and large l value should lead to the low uncertainty or high precision for determining λ. If taking the values of $\lambda/2a$, l, and $2a$, the same as above, respectively, $\delta\lambda = (1.1 \sim 2.6)\Phi$, in which Φ and $\delta\lambda$ are in millimeter and nanometer units respectively. If using $\Phi = 0.5$ mm, $\delta\lambda = 0.6 \sim 1.3$ nm. This indicates that the precision of the wavelength determined by our wavelength meter could reach <1 nm in the wavelength region close to $2a$ and beam size smaller than 0.5 mm.

It should be mentioned that there inevitably exist some defects or disorderly arranged areas in the 2D ordered pore arrays with multi-domains, such as, boundaries between domains, vacancies, and dislocations. Fortunately, the percentage of such defects is low compared with the regular uni-cells in the arrays. The diffraction ring is produced only by the domains with lattice periodic arrangement. The disorderly arranged areas or defects cannot form such diffraction rings or spots but background on the screen, which leads to a decrease of contrast of the diffraction ring produced by the domains but not to a significant change of the ring thickness or spot size. Obviously, when the percentage of defects in the 2D ordered pore array is high or the crystallinity is low, the contrast of the diffraction ring produced by the domains would be weak.

In general, the designed wavemeter is simple in structure, low in cost, convenient in measurement, and especially, small in size. It could be used for online demarcation and real-time measurement of light wavelength in such as, mine operation, seabed exploration, and even space exploration due to its miniature.

7.6 CONCLUSIONS AND OUTLOOK

In conclusion, we have demonstrated that a colloidal monolayer can be used as a flexible template to create diverse micro/nanostructured films. Combining it with other assistant techniques, such as evaporation deposition, solution dipping (or sol dipping) and electrodeposition, a serious of ordered micro/nanostructured films with centimeter size, including nanoparticle arrays, pore arrays, hollow sphere arrays, nanoring arrays, nanopillar arrays, hierarchically structured array, etc., can be easily synthesized on any desired substrate for different materials (metal, semiconductor, even polymers). The morphology and size of these array films can be well controlled by experiment conditions, such as the diameter of colloidal spheres, the concentration of precursor solution, the heating time of colloidal monolayer on substrate, recipe of electrolyte, and the electrodeposition time. These micro/nanostructured films have demonstrated some important morphology-correlated novel properties in optics, wettability, magnetics, SERS, gas sensitivity, etc. Based on these attractive properties and macro-dimension, such films provide practical possibility for realization of micro/nanostructured devices, such as gas sensors, SERS-based biosensors, photonic crystals, microfluidic devices, and wavemeter.

As a growing member in films, the micro/nanostructured film remains in challenges in some aspects: (1) fabrication and structure optimization as desired; (2) the relation between structure and properties; (3) further design, realization, and optimization of devices; (4) application exploration in some new fields; (5) surface-related fundamental physical and chemical researches. The opportunities and challenges coexist, and inspire more and more researchers into these interesting and important fields.

ACKNOWLEDGMENTS

The authors acknowledge the financial supports from the National Natural Science Foundation of China (Grant Nos. 50831005, 10704075, 10874184, and 50601026), and the Major State research program of China "Fundamental Investigation on Micro-Nano Sensors and Systems based on BNI Fusion" (Grant No. 2006CB300402) and the Knowledge Innovation Program of the Chinese Academy of Sciences (Grant No. KJCX2-SW-W31).

REFERENCES

1. Cai, W.; Li, Y.; Duan, G. 2007. Nanostructured arrays based on colloidal monolayer. In *Advanced Materials Research Trends*, Ed. Basbanes, L. V. Nova Science Publishers, Inc., New York.
2. Howard, R. E.; Liao, P. F.; Skocpol, W. J.; Jackel, L. D.; Craighead, H. G. 1983. Microfabrication as a scientific tool. *Science* 221: 117–121.
3. Ito, T.; Okazaki, S. 2000. Pushing the limits of lithography. *Nature* 406: 1027–1031.
4. Pease, R. F. 1992. *Nanostructures and Mesoscopic Systems*, Eds. Kirk, W. P. and Reed, M. A. Academic Press, Boston, MA, pp. 37–50.
5. Pease, R. F. W. 1992. Nanolithography and its prospects as a manufacturing technology. *J. Vac. Sci. Technol. B* 10: 278–285.
6. Nakayama, Y. 1990. Electron-beam cell projection lithography: A new high-throughput electro-beam direct-writing technology using a specially tailored Si aperture. *J. Vac. Sci. Technol. B* 8: 1836–1840.
7. Berger, S. D. 1991. Projection electron-beam lithography: A new approach. *J. Vac. Sci. Technol. B* 9: 2996–2999.
8. Pfeiffer, H. C.; Stickel, W. 1995. PREVAIL—An E-beam stepper with variable axis immersion lenses. *Microelectron. Eng*. 27: 143–146.
9. Xia, Y.; Rogers, J.; Paul, K. E.; Whitesides, G. M. 1999. Unconventional methods for fabricating and patterning nanostructures. *Chem. Rev*. 99: 1823–1848.
10. Jeon, N. L.; Choi, I. S.; Whitesides, G. M. 1999. Patterned polymer growth on silicon surfaces using microcontact printing and surface-initiated polymerization. *Appl. Phys. Lett*. 75: 4201.
11. Nagayama, K.; Takeda, S.; Endo, S.; Yoshimura, H. 1995. Fabrication and control of 2-dimensional crystalline arrays of protein molecules. *Jpn. J. Appl. Phys*. 34: 3947–3954.
12. Matsushita, S.; Miwa, T.; Fujishima, A. 1997. Preparation of a new nanostructured TiO_2 surface using a two-dimensional array-based template. *Chem. Lett*. 309: 925–926.
13. Matsushita, S. I.; Miwa, T.; Tryk, D. A.; Fujishima, A. 1998. New mesostructured porous TiO_2 surface prepared using a two-dimensional array-based template of silica particles. *Langmuir* 14: 6441–6447.
14. Tessier, P. M.; Velev, O. D.; Kalambur, A. T. 2000. Assembly of gold nanostructured films templated by colloidal crystals and use in surface-enhanced Raman spectroscopy. *J. Am. Chem. Soc*. 122: 9554–9555.
15. Sun, F.; Cai, W.; Li, Y.; Cao, B.; Lei, Y.; Zhang, L. 2004. Morphology-controlled growth of large-area two-dimensional ordered pore arrays. *Adv. Funct. Mater*. 14: 283–288.
16. Bartlett, P. N.; Baumberg, J. J.; Coyle, S.; Abdelsalam, M. E. 2004. Optical properties of nanostructured metal films. *Faraday Discuss*. 125: 117–132.
17. Sun, F.; Cai, W.; Li, Y. Cao, B.; Lu, F.; Duan, G.; Zhang, L. 2004. Morphology control and transferability of ordered through-pore arrays based on electrodeposition and colloidal monolayer. *Adv. Mater*. 16: 1116–1121.
18. Stöber, W.; Fink, A.; Bohn, E. 1968. Controlled growth of monodisperse silica spheres in the micron size range. *J. Colloid. Interface Sci*. 26: 62–69.
19. Pieranski, P. 1983. Colloidal crystals. *Contemp. Phys*. 24: 25–73.
20. Piirma, I. 1982. *Emulsion Polymerization*. Academic Press, New York.
21. Murray, C. A.; Winkle, D. H. V. 1987. Experimental observation of two-stage melting in a classical two-dimensional screened Coulomb system. *Phys. Rev. Lett*. 58: 1200–1203.

22. Skjeltorp, A. T.; Meakin, P. 1988. Fracture in microsphere monolayers studied by experiment. *Nature*, 335: 424–426.
23. Yang, S. M.; Jang, S. G.; Choi, D. G.; Kim, S.; Yu, H. K. 2006. Nanomachining by colloidal lithography. *Small* 2: 458–475.
24. Li, Y.; Cai, W. P.; Duan, G. T.; Sun, F. Q.; Cao, B. Q.; Lu, F. 2005. 2D Nanoparticle arrays by partial dissolution of ordered pore films. *Mater. Lett.* 59: 276–279.
25. Li, Y.; Cai, W. P.; Cao, B. Q.; Duan, G. T.; Li, C. C.; Sun, F. Q.; Zeng, H. B. 2006. Morphology-controlled 2D ordered arrays by heating-induced deformation of 2D colloidal monolayer. *J. Mater. Chem.* 16: 609–612.
26. Haynes, C. L.; Van Duyne, R. P. 2001. Nanosphere lithography: A versatile nanofabrication tool for studies of size-dependent nanoparticle optics. *J. Phys. Chem. B* 105: 5599–5611.
27. Micheletto, R.; Fukuda, H.; Ohtsu, M. 1995. A simple method for the production of a two-dimensional, ordered array of small latex particles. *Langmuir* 11: 3333–3336.
28. Dimitrov, A. S.; Nagayama, K. 1996. Continuous convective assembling of fine particles into two-dimensional arrays on solid surfaces. *Langmuir* 12: 1303–1311.
29. Burmeister, F.; Schäfle, C.; Matthes, T.; Böhmisch, M.; Boneberg, J.; Leiderer, P. 1997. Colloid monolayers as versatile lithographic masks. *Langmuir* 13: 2983–2987.
30. Jiang, P.; McFarland, M. J. 2005. Wafer-scale periodic nanohole arrays templated from two-dimensional nonclose-packed colloidal crystals. *J. Am. Chem. Soc.* 127: 3710–3711.
31. Fischer, U. C.; Zingsheim, H. P. 1981. Submicroscopic pattern replication with visible light. *J. Vac. Sci. Technol.* 19: 881–885.
32. Burmeister, F.; Schafle, C.; Keilhofer, B.; Bechinger, C.; Boneberg, J.; Leiderer, P. 1998. From mesoscopic to nanoscopic surface structures: Lithography with colloid monolayers. *Adv. Mater.* 10: 495–497.
33. Burmeister, F.; Badowsky, W.; Braun, T.; Wieprich, S.; Boneberg, J.; Leiderer, P. 1999. Colloid monolayer lithography: A flexible approach for nanostructuring of surfaces. *Appl. Surf. Sci.* 461: 144–145.
34. Wang, X.; Summers, C. J.; Wang, Z. L. 2004. Large-scale hexagonal-patterned growth of aligned ZnO nanorods for nano-optoelectronics and nanosensor arrays. *Nano Lett.* 4: 423–426.
35. Hulteen, J. C.; Treichel, D. A.; Smith, M. T.; Duval, M. L.; Jensen, T. R.; Van Duyne, R. P. 1999. Nanosphere lithography: Size-tunable silver nanoparticle and surface cluster arrays. *J. Phys. Chem. B* 103: 3854–3863.
36. Jensen, T. R.; Duval, M. M.; Haynes, C. L.; Van Duyne, R. P. 2000. Nanosphere lithography: Tunable localized surface plasmon resonance spectra of silver nanoparticles. *J. Phys. Chem. B* 104: 10549–10556.
37. Winzer, M.; Kleiber, M.; Dix, N.; Wiesendanger, R. 1996. Fabrication of nano-dot and nano-ring-arrays by nanosphere lithography. *Appl. Phys. A* 63: 617–619.
38. Rybczynski, J.; Ebels, U.; Giersig, M. 2003. Large-scale 2D arrays of magnetic nanoparticles. *Colloids Surf. A Physicochem. Eng. Aspects* 219: 1–6.
39. Bullen, H. A.; Garrett, S. J. 2002. TiO_2 Nanoparticle arrays prepared using a nanosphere lithography technique. *Nano Lett.* 2: 739–745.
40. Lenzmann, F.; Li, K.; Kitai, A. H.; Stover, H. D. H. 1994. Thin-film micropatterning using polymer microspheres. *Chem. Mater.* 6: 156–159.
41. Hulteen, J. C.; Van Duyne, R. P. 1995. Nanosphere lithography: A materials general fabrication process for periodic particle array surfaces. *J. Vac. Sci. Technol. A* 13: 1553–1558.
42. Haynes, C. L.; McFarland, A. D.; Smith, M. T.; Hulteen, J. C.; Van Duyne, R. P. 2002. Angle-resolved nanosphere lithography: Manipulation of nanoparticle size, shape, and interparticle spacing. *J. Phys. Chem. B*, 106: 1898–1902.
43. Kosiorek, A.; Kandulski, W.; Glaczynska, H.; Giersig, M. 2005. Fabrication of nanoscale rings, dots, and rods by combining shadow nanosphere lithography and annealed polystyrene nanosphere masks. *Small* 1, 439–444.
44. Zhang, G.; Wang, D.; Möhwald, H. 2007. Ordered binary arrays of Au nanoparticles derived from colloidal lithography. *Nano Lett.* 7: 127–132.
45. Li, Y.; Cai, W. Duan, G.; Sun, F.; Cao, B.; Lu, F.; Fang, Q.; Boyd, I. W. 2005. Large-area In_2O_3 ordered pore arrays and their photoluminescence properties. *Appl. Phys. A* 81: 269–273.
46. Li, Y.; Cai, W. P.; Cao, B. Q.; Duan, G. T.; Sun, F. Q.; Li, C.; Jia, L. C. 2006. Two-dimensional hierarchical porous silica film and its tunable superhydrophobicity. *Nanotechnology* 17: 238–243.
47. Sun, F.; Yu, J. C.; Wang, X. 2006. Construction of size-controllable hierarchical nanoporous TiO_2 ring arrays and their modifications. *Chem. Mater.* 18: 3774–3779.
48. Cai, W. P.; Zhang, L. D.; Zhong, H. C.; He, G. 1998. Annealing of mesoporous silica loaded with silver nanoparticles within its pores from isothermal sorption. *J. Mater. Res.* 13: 2888–2895.

49. Holland, B. T.; Abrams, L.; Stein, A. 1999. Dual templating of macroporous silicates with zeolitic microporous frameworks. *J. Am. Chem. Soc.* 121: 4308–4309.
50. Sen, T.; Tiddy, G. J. T.; Casci, J. L.; Anderson, M. W. 2004. Synthesis and characterization of hierarchically ordered porous silica materials. *Chem. Mater.* 16: 2044–2054.
51. Sun, F. Q.; Cai, W. P.; Li, Y.; Jia, L. C.; Lu, F. 2005. Direct growth of mono- and multilayer nanostructured porous films on curved surfaces and their application as gas sensors. *Adv. Mater.* 17: 2872–2877.
52. Li, Y.; Cai, W. P.; Duan, G. T.; Cao, B. Q.; Sun, F. Q. 2005. Two-dimensional ordered polymer hollow sphere and convex structure arrays based on monolayer pore films. *J. Mater. Res.* 20: 338–343.
53. Djalali, R.; Samson, J.; Matsui, H. 2004. Doughnut-shaped peptide nano- assemblies and their applications as nanoreactors. *J. Am. Chem. Soc.* 126: 7935–7939.
54. Dinsmore, A. D.; Hsu, M. F.; Nikolaides, M. G.; Marquez, M.; Bausch, A. R.; Weitz, D. A. 2002. Colloidosomes: Selectively permeable capsules composed of colloidal particles. *Science* 298: 1006–1009.
55. Lu, Y.; Yin, Y.; Xia, Y. 2001. A self-assembly approach to the fabrication of patterned, two-dimensional arrays of microlenses of organic polymers. *Adv. Mater.* 13: 34–37.
56. Gu, E.; Choi, H. W.; Liu, C.; Griffin, C.; Girkin, J. M.; Watson, I. M.; Dawson, M. D.; McConnell, G.; Gurney, A. M. 2004. Reflection/transmission confocal microscopy characterisation of single-crystal diamond microlens arrays. *Appl. Phys. Lett.* 84: 2754.
57. Kuo, C. W.; Shiu, J. Y.; Chen, P. 2003. Size and shape controlled fabrication of large-area periodic nanopillar arrays. *Chem. Mater.* 15: 2917–2920.
58. Li, Y.; Cai, W. P.; Cao, B. Q.; Duan, G. T.; Sun, F. Q. 2005. Fabrication of the periodic nanopillar arrays by heat-induced deformation of 2D polymer colloidal monolayer. *Polymer* 46: 12033–12036.
59. Mazur, S.; Beckerbauer, R.; Buckholz, J. 1997. Particle size limits for sintering polymer colloids without viscous flow. *Langmuir* 13: 4287–4294.
60. Dinsmore, A.; Hsu, M.; Nikolaides, M.; Marquez, M.; Bausch, A.; Weitz, D. 2002. Colloidsomes: Selectively permeable capsules composed of colloidal particles. *Science* 298: 1006–1011.
61. Cao, B. Q.; Cai, W. P.; Sun, F. Q.; Li, Y.; Lei, Y.; Zhang, L. D. 2004. Fabrication of large-scale zinc oxide ordered pore arrays with controllable morphology. *Chem. Comm.* 14:1604–1605.
62. Cao, B. Q.; Cai, W. P.; Sun, F. Q. 2005. Electrodeposition-induced highly oriented zinc oxide ordered pore arrays and their ultraviolet emissions. *Electrochem. Solid-State Lett.* 8: G237–G240.
63. Duan, G. T.; Cai, W. P.; Li, Y.; Li, Z. G.; Cao, B. Q.; Luo, Y. Y. 2006. Transferable ordered Ni hollow sphere arrays induced by electrodeposition on colloidal monolayer. *J. Phys. Chem. B* 110: 7184–7188.
64. Duan, G. T.; Cai, W. P.; Luo, Y. Y.; Li, Z. G.; Lei, Y. 2006. Hierarchical structured Ni nanoring and hollow sphere arrays by morphology inheritance based on ordered through-pore template and electrodeposition. *J. Phys. Chem. B* 110: 15729–15733.
65. Duan, G. T.; Cai, W. P.; Luo, Y. Y.; Li, Y.; Lei, Y. 2006. Hierarchical surface rough ordered Au particle arrays and their surface enhanced Raman scattering. *Appl. Phys. Lett.* 89: 181918.
66. Duan, G. T.; Cai, W. P.; Luo, Y. Y.; Sun, F. Q. 2007. A hierarchically structured $Ni(OH)_2$ monolayer hollow-sphere array and its tunable optical properties over a large region. *Adv. Funct. Mater.* 17: 644–650.
67. Yang, S.; Chen, G.; Megens, M.; Ullal, C. K.; Han, Y. J.; Rapaport, R.; Thomas, E. L.; Aizenberg, J. 2005. Functional biomimetic microlens arrays with integrated pores. *Adv. Mater.* 17: 435–438.
68. Natarajan, C.; Matsumoto, H.; Nogami, G. 1997. Improvement in electrochromic stability of electrodeposited nickel hydroxide thin film. *J. Electrochem. Soc.* 144: 121–126.
69. Liu, B.; Zeng, H. C. 2004. Mesoscale organization of CuO nanoribbons: Formation of “Dandelions.” *J. Am. Chem. Soc.* 126: 8124–8125.
70. Liu, B.; Zeng, H. C. 2004. Fabrication of ZnO “Dandelions” via a modified Kirkendall process. *J. Am. Chem. Soc.* 126: 16744–16746.
71. Li, J.; Zeng, H. C. 2005. Nanoreactors-size tuning, functionalization, and reactivation of Au in TiO_2 nanoreactors. *Angew. Chem. Int. Ed.* 44: 4342–4345.
72. Zhang, Z.; Shao, X.; Yu, H.; Wang, Y.; Han, M. 2005. Morphosynthesis and ornamentation of 3D dendritic nanoarchitectures. *Chem. Mater.* 17: 332–336.
73. Mo, M.; Yu, J. C.; Zhang, L.; Li, S. K. A. 2005. Self-assembly of ZnO nanorods and nanosheets into hollow microhemispheres and microspheres. *Adv. Mater.* 17: 756–760.
74. Wang, D.; Song, C.; Hu, Z.; Fu, X. 2005. Fabrication of hollow spheres and thin films of nickel hydroxide and nickel oxide with hierarchical structures. *J. Phys. Chem. B* 109: 1125–1129.
75. Jayashree, R. S.; Kamath, P. V. 1999. Factors governing the electrochemical synthesis of α-nickel(II) hydroxide. *J. Appl. Electrochem.* 29: 449–454.

76. Therese, G. H. A.; Kamath, P. V. 2000. Electrochemical synthesis of metal oxides and hydroxides. *Chem. Mater.* 12: 1195–1204.
77. Shiho, H.; Kawahashi, N. 2000. Titanium compounds as coatings on polystyrene lattices and as hollow spheres. *Colloid Polym. Sci.* 278: 270–274.
78. He, A.; Djurfors, B.; Akhlaghi, S.; Ivey, D. G. 2002. Pulse plating of gold-tin alloys for microelectronic & optoelectronic applications. *Plat. Surf. Finish.* 89: 48–53.
79. Ghanem, M. A.; Bartlett, P. N.; de Groot, P.; Zhukov, A. 2004. A double templated electrodeposition method for the fabrication of arrays of metal nanodots. *Electrochem. Commun.* 6: 447–453.
80. Weast, R. C. 1989. *CRC Handbook of Chemistry and Physics*, 70th ed. CRC Press, Boca Raton, FL.
81. Matveev, K. A.; Larkin, A. I.; Glazman, L. I. 2002. Persistent current in superconducting nanorings. *Phys. Rev. Lett.* 89: 096802.
82. Rothman, J.; Klaui, M.; Lopez-Diaz, L.; Vaz, C. A. F.; Bleloch, A.; Bland, J. A. C.; Cui, Z.; Speaks, R. 2001. Observation of a bi-domain state and nucleation free switching in mesoscopic ring magnets. *Phys. Rev. Lett.* 86: 1098–1101.
83. Aizpurua, J.; Hanarp, P.; Sutherland, D. S.; Kall, M.; Bryant, G. W.; de Abajo, F. 2003. Optical properties of gold nanorings. *Phys. Rev. Lett.* 90: 057401.
84. Li, S. P.; Peyrade, D.; Natali, M.; Lebib, A.; Chen, Y.; Ebels, U.; Buda, L. D.; Ounadjela, K. 2001. Flux closure structures in cobalt rings. *Phys. Rev. Lett.* 86: 1102–1105.
85. Caruso, F. 2000. Hollow capsule processing through colloidal templating and self-assembly. *Chem. Eur. J.* 6: 413–419.
86. Hu, Y.; Jiang, X.; Ding, Y.; Chen, Q.; Yang, C. 2004. Core-template-free strategy for preparing hollow nanosphere. *Adv. Mater.* 16: 933–937.
87. Warburton, R. J.; Schaflein, C.; Haft, D.; Bickel, F.; Lorke, A.; Karrai, K.; Garcia, J. M.; Schoenfeld, W.; Petroff, P. M. 2000. Optical emission from a charge-tunable quantum ring. *Nature* 405: 926–929.
88. Sano, M.; Kamino, A.; Okamura, J.; Shinkai, S. 2001. Ring closure of carbon nano-tubes. *Science* 293: 1299–1301.
89. Li, F.; Dong, Y.; Gao, P.; Xin, X.; Wang, Z. 2004. Single-crystal hexagonal disks and rings of ZnO: Low-temperature, large-scale synthesis and growth mechanism. *Angew. Chem. Int. Ed.* 43: 5238–5242.
90. Hobbs, K. L.; Larson, P. R.; Lian, G. D.; Keay, J. C.; Johnson, M. B. 2004. Fabrication of nanoring arrays by sputter redeposition using porous alumina templates. *Nano Lett.* 4: 167–171.
91. Pearson, D. H.; Tonucci, R. J.; Bussmann, K. M.; Bolden, E. A. 1999. Parallel patterning of mesoscopic ring arrays using nanochannel glass replica masks. *Adv. Mater.* 11: 769–773.
92. Imhof, A. 2001. Preparation and characterization of titania-coated polystyrene spheres and hollow titania shells. *Langmuir* 17: 3579–3585.
93. Tartaj, P.; Gonzalez-Carreno, T.; Serna, C. J. 2001. Single-step nanoengineering of silica coated maghemite hollow spheres with tunable magnetic properties. *Adv. Mater.* 13: 1620–1624.
94. Huang, Y.; Duan, X. F.; Wei, Q. Q.; Lieber, C. M. 2001. Directed assembly of one-dimensional nanostructures into functional networks. *Science* 291: 630–633.
95. Duan, G. T.; Cai, W. P.; Luo, Y. Y.; Lv, F. J.; Yang, J. L. 2009. Design and electrochemical fabrication of gold binary ordered micro/nanostructured porous arrays via step-by-step colloidal lithography. *Langmuir* 25: 2558–2562.
96. Velikov, K. P.; Christova, C. G.; Dullens, R. P. A.; Van Blaaderen, A. 2002. A. layer-by-layer growth of binary colloidal crystals. *Science* 296: 106–109.
97. Leunissen, M. E.; Christova, C. G.; Hynninen, A. P.; Royall, C. P.; Campbell, A.; Imhof, A.; Dijkstra, M.; van Roij, R.; Van Blaaderen, A. 2005. Ionic colloidal crystals of oppositely charged particles. *Nature* 437: 235–240.
98. Kitaev, V.; Ozin, G. A. 2003. Self-assembled surface patterns of binary colloidal crystals. *Adv. Mater.* 15: 75–78.
99. Bartlett, P.; Campbell, A. I. 2005. Three-dimensional binary superlattices of oppositely charged colloids. *Phys. Rev. Lett.* 95: 128302.
100. Wang, J.; Li, Q.; Knoll, W.; Jonas, U. 2006. Preparation of multilayered trimodal colloid crystals and binary inverse opals. *J. Am. Chem. Soc.* 128: 15606–15607.
101. Wenzel, R. N. 1936. Resistance of solid surfaces to wetting by water. *Ind. Eng. Chem.* 28: 988–994.
102. Cassie, A. B. D. 1948. Contact angles. *Discuss. Faraday Soc.* 3: 11–16.
103. Li, Y.; Cai, W. P.; Duan, G. T.; Cao, B. Q.; Sun, F. Q.; Lu, F. 2005. Superhydrophobicity of 2D ZnO ordered pore arrays formed by solution-dipping template method. *J. Colloid Interface Sci.* 287: 634–639.
104. Li, Y.; Lee, E. J.; Cho, S. O. 2007. Superhydrophobic coatings on curved surfaces featuring remarkable supporting force. *J. Phys. Chem. C* 111: 14813–14817.

105. Li, Y.; Huang, X. J.; Heo, S. H.; Li, C. C.; Choi, Y. K.; Cai, W. P.; Cho, S. O. 2007. Superhydrophobic bionic surfaces with hierarchical microsphere/SWCNT composite arrays. *Langmuir* 23: 2169–2174.
106. Huang, X. J.; Li, Y.; Im, H. S.; Yarimaga, O.; Kim, J. H.; Jang, D. Y.; Cho, S. O; Cai, W. P.; Choi, Y. K. 2006. Morphology-controlled SWCNT/polymeric microsphere arrays by a wet chemical self-assembly technique and their application for sensors. *Nanotechnology* 17: 2988–2993.
107. Huang, X. J.; Li, Y.; Choi, Y. K. 2008. A chestnut-like hierarchical architecture of a SWCNT/microsphere composite on an electrode for electroanalysis. *J. Electroanal. Chem.* 617: 218–223.
108. Gates, B.; Park, S. H.; Xia, Y. N. 2000. Tuning the photonic bandgap properties of crystalline arrays of polystyrene beads by annealing at elevated temperatures. *Adv. Mater.* 12: 653–656.
109. Wu, B.; Zhang, J.; Wei, Z.; Cai, S. M.; Liu, Z. F. 2001. Chemical alignment of oxidatively shortened single-walled carbon nanotubes on silver surface. *J. Phys. Chem. B* 105: 5075–5078.
110. Feng, L.; Li, S.; Li, Y.; Li, H.; Zhang, L.; Zhai, J.; Song, Y.; Liu, B.; Jiang, L.; Zhu, D. 2002. Superhydrophobic surfaces: From natural to artificial. *Adv. Mater.* 14: 1857–1860.
111. Wenzel, R. N. 1949. Surface roughness and contact angle. *J. Phys. Colloid Chem.* 53: 1446.
112. Feng, L.; Yang, Z. L.; Zhai, J.; Song, Y. L.; Liu, B. Q.; Ma, Y. M.; Yang, Z. Z.; Jiang, L.; Zhu, D. B. 2003. Superhydrophobicity of nanostructured carbon films in a wide range of pH values. *Angew. Chem. Int. Ed.* 42: 4217–4220.
113. Lau, K. K. S.; Bico, J.; Teo, K. B. K.; Chhowalla, M.; Amaratunga, G. A. J.; Milne, W. I.; McKinley, G. H.; Gleason, K. K. 2003. Superhydrophobic carbon nanotube forests. *Nano Lett.* 3: 1701–1705.
114. Huang, X. J.; Meng, F. L.; Pi, Z. X.; Xu, W. H.; Liu, J. H. 2004. Gas sensing behavior of a single tin dioxide sensor under dynamic temperature modulation. *Sens. Actuators B* 99: 444–450.
115. Chang, S. C. 1983. Sensing mechanism of thin film tin oxide. In *Proceedings of the International Meeting on Chemical Sensors*, Elsevier, Amsterdam, the Netherlands, pp. 78–83.
116. Zhao, Y.; Feng, Z. C.; Liang, Y. 1999. SnO_2 gas sensor films deposited by pulsed laser ablation. *Sens. Actuators B* 56: 224–227.
117. Ying, Z.; Wan, Q.; Song, Z. T.; Feng, S. L. 2004. SnO_2 nanowhiskers and their ethanol sensing characteristics. *Nanotechnology* 15: 1682–1684.
118. Jia, L. C.; Cai, W. P.; Wang, H. Q.; Sun, F. Q.; Li, Y. Multi-layer hetero-pore sized porous films with hierarchical micro/nanoarchitectures and their large structurally induced controllability of gas sensing performance. Unpublished.
119. Alexander, G.; Ralf, R. 2007. In situ and operando spectroscopy for assessing mechanisms of gas sensing. *Angew. Chem. Int. Ed.* 46: 3826–3848.
120. Korotcenkov, G. 2004. In_2O_3 films deposited by spray pyrolysis as a material for ozone gas sensors. *Sens. Actuators B* 99: 297–303.
121. Franke, M. E.; Koplin, T. J.; Simon, U. 2006. Metal and metaloxide nanoparticles in chemiresistors: Does the nanoscale matter? *Small* 2: 36–50.
122. Du, N.; Zhang, H.; Chen, B.; Ma, X.; Liu, Z.; Wu, J.; Yang, D. 2007. Porous Co_3O_4 nanotubes derived from Co-4(CO)(12) clusters on carbon nanotube templates: A highly efficient material for Li-battery applications. *Adv. Mater.* 19: 4505–4509.
123. Liu, Z. F.; Yamazaki, T.; Shen, Y.; Kikuta, T.; Nakatani, N.; Kawabata, T. 2007. Room temperature gas sensing of p-type TeO_2 nanowires. *Appl. Phys. Lett.* 90: 173119.
124. Pinna, N.; Neri, G.; Antonietti, M.; Niederberger, M. 2004. Nonaqueous synthesis of nanocrystalline semi-conducting metal oxides for gas sensing. *Angew. Chem. Int. Ed.* 43: 4345–4349.
125. Soulantica, K.; Erades, L.; Sauvan, M.; Senocq, F.; Maisonnat, A.; Chaudret, B. 2003. Synthesis of indium and indium oxide nanoparticles from indium cyclopentadienyl precursor and their application for gas sensing. *Adv. Funct. Mater.* 13: 553–557.
126. Sun, F.; Cai, W.; Li, Y.; Duan, G.; Nichols, W. T.; Liang, C.; Koshizaki, N.; Fang, Q.; Boyd, I. W. 2005. Laser morphological manipulation of gold nanoparticles periodically arranged on solid supports. *Appl. Phys. B Lasers Opt.* 81: 765.
127. Jin, R. C.; Cao, Y. W.; Mirkin, C. A.; Kelly, K. L.; Schatz, G. C.; Zheng, J. G. 2001. Photoinduced conversion of silver nanoparticles to nanoprisms. *Science* 294: 1901–1903.
128. Kreibig, U.; Vollmer, M. 1995. *Optical Properties of Metal Clusters*. Springer, Berlin, Germany.
129. Gaponenko, S. V. 2002. Effects of photon density of states on Raman scattering in mesoscopic structures. *Phys. Rev. B* 65: 140303.
130. Garcia-Vidal, F. J.; Pendry, J. B. 1996. Collective theory for surface enhanced Raman scattering. *Phys. Rev. Lett.* 77: 1163–1166.
131. Shalaev, V. M. 2000. *Nonlinear Optics of Random Media*. Springer, New York.

132. Krug, J. T.; Wang, G. D.; Emory, S. R.; Nie, S. M. 1999. Efficient Raman enhancement and intermittent light emission observed in single gold nanocrystals. *J. Am. Chem. Soc.* 121: 9208–9214.
133. Fudouzi, H.; Xia, Y. 2003. Colloidal crystals with tunable colors and their use as photonic papers. *Langmuir* 19: 9653–9658.
134. Richel, A.; Johnson, N. P.; McComb, D. W. 2000. Observation of Bragg reflection in photonic crystals synthesized from air spheres in a titania matrix. *Appl. Phys. Lett.* 77: 1062–1063.
135. Mayoral, R.; Requena, J.; Moya, J. S.; López, C.; Cintas, A.; Miguez, H.; Meseguer, F.; Vázquez, L.; Holgado, M.; Blanco, Á. 1997. 3D long range ordering of submicrometric SiO_2 sintered superstructures. *Adv. Mater.* 9: 257–260.
136. Prevo, B. G.; Velev, O. D. 2004. Controlled rapid deposition of structured coatings from micro- and nanoparticle suspensions. *Langmuir* 20: 2099–2107.
137. Sommerfeld, A. 1964. *Optics*. Academic Press, San Diego, CA.
138. Duan, G. T.; Cai, W. P.; Li, Y.; Cao, B. Q. 2006. Measurement of light wavelength based on nanostructures ordered pore arrays. *J. Nanosci. Nanotechnol.* 6: 2474–2478.

8 Thin Film Shape Memory Alloy for Microsystem Applications

K.P. Mohanchandra and G.P. Carman

CONTENTS

8.1 INTRODUCTION

Shape memory alloys (SMAs) have fascinated researchers for the last few decades in a range of industrial and medical fields due to their intrinsic properties such as large stress output, recoverable strain, excellent damping, and biocompatibility. SMAs exhibit a thermally induced crystalline transformation between a martensitic phase (a low-temperature phase) and an austenitic phase (a high-temperature phase). Upon cooling below the martensitic finish (M_f) temperature, undeformed

SMAs have a twinned martensitic structure. Under a relatively low applied stress, the twinned martensitic structure is rearranged along the stress direction and produces deformation. When heated above the austenitic finish (A_f) temperature, the SMAs recover their highly ordered austenite phase and regain their original shape in which they were crystallized. During shape recovery, they generate high pressures and large strokes. This effect is typically referred to as the shape memory effect (SME). This was first reported in gold–cadmium alloy in 1932 [1] and subsequently in equiatomic nickel–titanium (Ni–Ti) by Buehler and his coworkers at the Naval Ordnance Laboratory, United States, in 1962 [2,3]. The discovery of Ni–Ti provided an SMA material with many desirable attributes including large recoverable strain and force output, corrosion resistance, excellent damping, and biocompatibility. The Ni–Ti discovery has led to extensive research and development efforts during the last four decades.

Shape memory and superelastic properties of Ni–Ti have been studied extensively since its initial discovery [4–6]. In addition, researchers have also studied the corrosion or wear resistance [7,8], damping and fatigue behaviors [9,10], and biocompatibility [11,12] of Ni–Ti alloys. All these properties are sensitive to the ratio of the two elements, nickel (Ni) and titanium (Ti), as well as thermal processing [13,14]. The large force output provided by Ni–Ti made this as an excellent candidate for actuator design. Several actuators based on bulk Ni–Ti are now commercially available. However, research on microelectromechanical systems (MEMS) looking for a suitable actuation material is still the source of research activities. In 1990, Walker et al. [15] successfully micromachined Ni–Ti thin film, sputter deposited on a spun and cured polyimide layer, to make a coil on a silicon (Si) wafer using a conventional micromachining release process. In the same year, Busch et al. [16] demonstrated shape memory behavior (SMB) in Ni–Ti thin film deposited using the magnetron sputtering technique. Since this initial research on thin film Ni–Ti, studies into deposition, characterization, and MEMS fabrication has been substantial. In this chapter, we review deposition techniques to fabricate Ni–Ti and Ni–Ti-based thin film SMAs, characterization and properties of the deposited films, the influence of different process parameters on their properties, and a few applications in microsystem.

8.2 FABRICATION OF Ni–Ti THIN FILMS

Various types of physical vapor deposition (PVD) techniques have been used extensively to fabricate thin film Ni–Ti [17–22], i.e., sputter deposition, electron beam deposition [23], flash evaporation pulse laser deposition (PLD) [20,21], molecular beam epitaxial (MBE) [24], etc. The major challenges in all these techniques are accurate composition control and uniformity throughout the fabricated film. These issues are extremely important because a small shift in the composition, particularly in Ni-rich films, leads to a large shift in the transformation temperatures. Similarly, nonuniformity in film composition results in the widening of transformation hysteresis. Researchers have attempted to resolve these problems by modifying various deposition parameters. In the following paragraphs, we focus on two of the more common techniques to produce thin film Ni–Ti, namely, sputter deposition and PLD.

The first report on sputter-deposited Ni–Ti film was released in 1985. Moine and his coworkers [25,26] co-sputtered Ni and Ti on a rotating substrate to fabricate Ni–Ti thin films. While Moine was the first to deposit, successful SMB in sputter-deposited Ni–Ti thin film was first reported in 1990 by Busch et al. [16]. Following their success, many researchers began studying the sputter deposition of Ni–Ti. Currently sputter deposition has become the most common PVD technique to produce Ni–Ti thin films because of certain advantages such as high deposition rate, large area deposition, and compatibility with the MEMS process when compared to other techniques. Many researchers use a direct-current (DC) magnetron sputtering method to produce Ni–Ti thin films; however, the major disadvantage of this method is that the precise control of the film composition is challenging. This is mainly due to the different sputtering rates of Ni and Ti that lead to nonstoichiometric film. In addition, in a conventional DC magnetron sputtering, maximum sputtering takes place along a

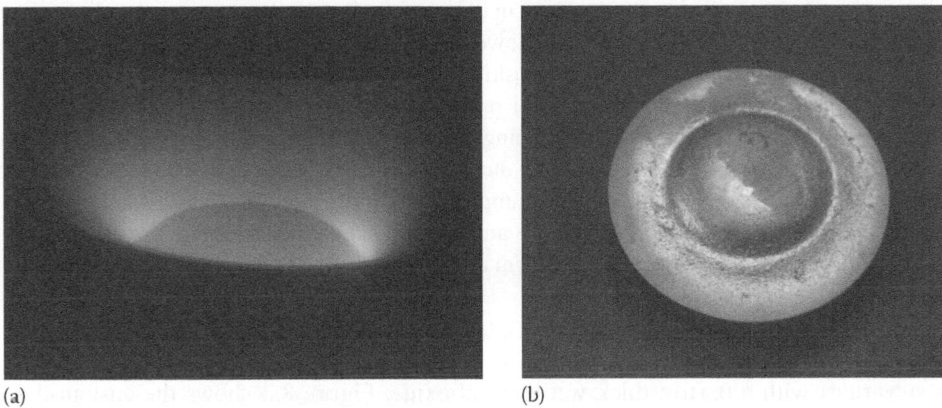

(a) (b)

FIGURE 8.1 (a) Ar^+ plasma over Ni–Ti target. (b) Groove present in used Ni–Ti target.

circular ring where there is a large concentration of confined electrons causing nonuniform sputtering. Similarly, the nonuniform plasma density in front of the target develops a circular groove on the target due to erosion, as shown in Figure 8.1, and the depth of the groove further increases as the target wears after prolonged use. This leads to nonuniformity in the film composition. To overcome this problem, Martynov et al. [20] modified the existing conventional magnetron sputtering technique by changing the magnetic field configuration. They utilized a rotating adjustable magnet assembly to avoid nonuniform sputtering from a target surface, in which, as the magnetic assembly rotates, the plasma moves in front of the target surface along with the magnets that causes sputtering of the target material from the area swept by the plasma configuration. This sputtering provides more homogeneous erosion in the target. However, they have also reported difficulty in producing uniform composition across the thickness and the film surface.

Ni–Ti thin film composition control is essential to produce the desired martensitic transformation temperature. This is due to the strong influence that Ni-content has on martensitic transformation temperatures [27–31]. In conventional sputter deposition, a single alloyed Ni–Ti target is used. However, it is difficult to produce a specified composition due to different sputtering rates of Ni and Ti. To address this issue, several researchers used multiple targets to control compositional variation. The researchers employed individual elemental targets or alloy and elemental targets [21,32–35]. Variable target power is used to either individually or simultaneously adjust film composition. By using this method, Ni–Ti thin film compositional control would appear to be easier in comparison with the alloyed target method, since the composition of the Ni–Ti thin film can be adjusted by varying the sputtering powers for both metal targets.

Another approach is to fabricate the Ni/Ti multilayer thin film composed of Ti and Ni layers by using an alternating sputter method [36]. Yaakoubi et al. [33] produced Ni–Ti thin films by sequential sputtering Ni and Ti elemental layers of similar thickness followed by metal interdiffusion via thermal annealing, whereas Lehnert et al. [32] obtained Ni–Ti thin film from Ni/Ti multilayer structures formed by 50 bilayers, each consisting of 10 nm Ni and 19 nm Ti. The average composition of the multilayer thin film is estimated by the thicknesses of Ti and Ni layers. The thicknesses of Ti and Ni layers can be adjusted by setting both sputtering power and sputtering time. Cho et al. [37] studied the alloying process of the Ni/Ti multilayer film using x-ray diffractometry (XRD) and transmission electron microscopy (TEM).

Ishida et al. [35] used a carousel-type magnetron sputtering apparatus with independently controlled multiple targets and a rotating substrate holder placed at the center with the surrounding planar magnetron cathodes to control better composition and uniformity in sputtered Ni–Ti thin film. Takeda et al. [38] applied a magnetic field using both permanent magnets and a solenoid coil to sputter from a composite Ni–Ti target (i.e., Ni ring mounted on a Ti disc) to control the composition to less than 0.5 at%. They observed a decrease in the erosion ring on the target as the electric current flowing through the solenoid coil increases. However, all these methods appear to be inadequate in producing a precise control of uniform composition.

In 2000, Ho and Carman adopted a novel process parameter "hot" target to control the film composition [39]. This was achieved by reducing thermal past between the target material and the copper chill block allowing the target to heat during deposition. The film obtained by heating the target during the sputter deposition showed composition almost equal to that of the target and was attributed to changes in Ti and Ni distribution angles. The experimental method was used recently by Mohanchandra et al. [40] to produce uniform composition Ni–Ti films, and is described in more detail below.

A custom-designed Denton DC magnetron sputtering system equipped with a Stanford Research Systems® residual gas analyzer (RGA) is used to fabricate Ni–Ti thin films on 10 cm diameter silicon substrates with a 0.5 μm thick wet thermal oxide. Figure 8.2 shows the custom-designed sputtering system used. The films are deposited on the substrate under "hot" target conditions. The target temperature profile as a function of sputtering time is shown in Figure 8.3. During sputtering, the target temperature ramps sharply to 546°C in approximately 240 s before reaching a steady temperature of 625°C. Prior to each deposition run, residual gas pressures are measured with RGA to make sure no significant contamination in the chamber. Water, carbon dioxide, and carbon monoxide gas pressures are kept below 10^{-8} Torr because they are highly reactive with Ti. An argon (Ar) scrubber is used to further clean the 99.999% purity argon gas used in this process. During deposition, the base pressure is kept below 4×10^{8} Torr and the Ar pressure of 1.5×10^{3} Torr is maintained. Substrate-to-target distance of 4 cm and 300 W of DC power are used. A shutter placed between the substrate and the target is used to shield the substrate from the target. The substrate is shielded from the target during the first 4 min of sputtering. The film is deposited on the SiO_2/Si substrate for 10 min. Following this deposition, the target is cooled back to room temperature, which takes approximately 90 min and the deposition is repeated to produce thicker films. The films produced show composition similar to the target composition in addition to composition uniformity. In general, Mohanchandra et al. [40] suggest that compositional control less than 0.5 at% can be achieved.

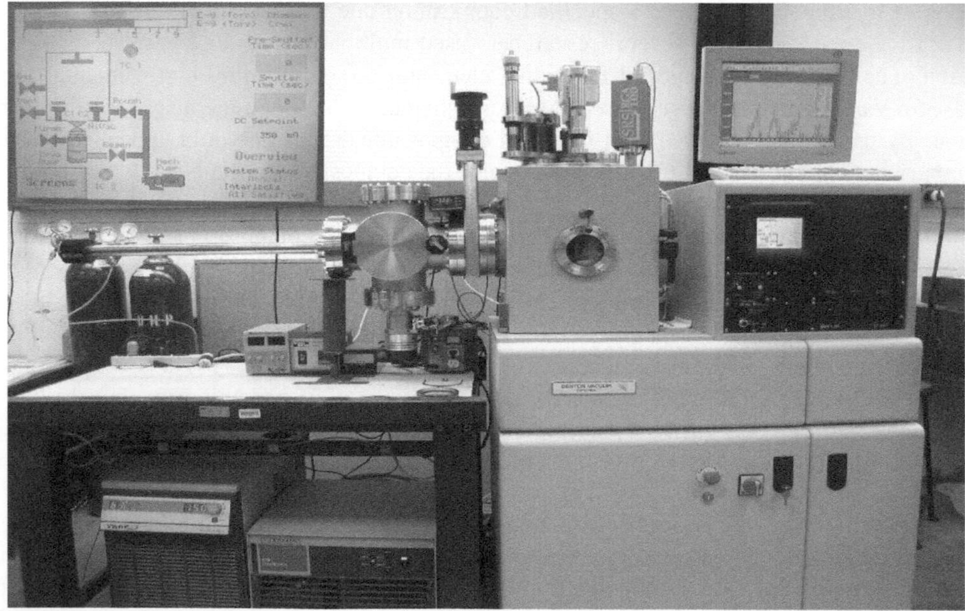

FIGURE 8.2 Custom-designed Denton sputtering system used at UCLA.

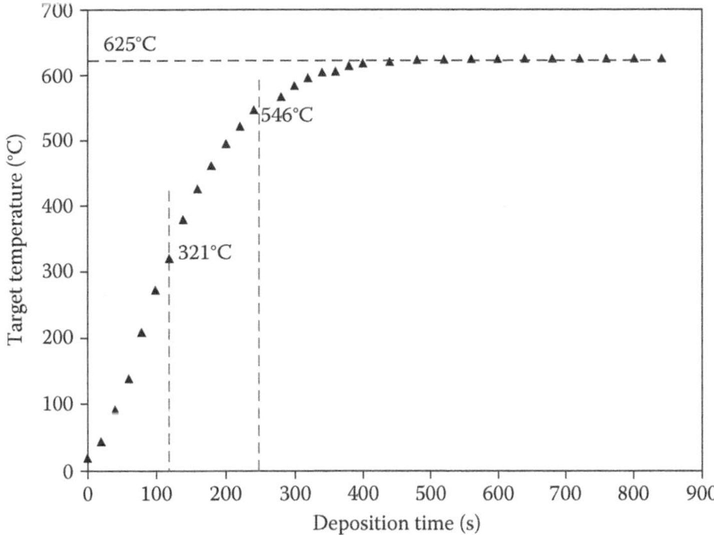

FIGURE 8.3 Plot of target temperature versus deposition time.

Pulsed laser deposition is another frequently used deposition method to produce Ni–Ti thin films. The PLD method provides better compositional control than sputtering because it is based on an opto-chemical reaction. The deposition is typically carried out at a pressure below 10^{-7} Torr and at relatively large laser fluence of the order 4–8 J/cm^2. The method was first suggested by Ikuta et al. [41,42] to reduce the contamination, and match the composition between the target and the Ni–Ti film. Later, Ciabattari et al. [43] deposited the Ni–Ti film on different substrates and showed the SMB in their PLD films. Gu et al. [44] investigated the optimum deposition parameters, the target-to-substrate distance, and the rotation speed of the target to fabricate sufficiently thick Ni–Ti thin films in PLD under vacuum. Because the film composition is very sensitive to the deposition parameters, mainly the target-to-substrate distance, it is difficult to optimize the deposition parameters. Lu et al. [45] altered the composition of the PLD Ni–Ti film through depositing the film at higher substrate temperatures by considering the fact that increase in surface mobility of the laser-ablated species at higher substrate temperature changes the film composition [46]. They also showed that the Ni–Ti films deposited at room temperature are Ni-rich and as the substrate temperature increases, the films become Ti-rich until the substrate temperature reaches 800°C where it reverts to Ni-rich due to oxidation. They attributed this composition change to the film–substrate interaction at the interface. The substrate temperature has been suggested as playing an important role in both the composition control and the crystallization of the films produced using PLD.

Ahn et al. [47] studied the effect of Ar gas on the composition control and crystalline properties of Ni–Ti films fabricated by the PLD method. They used KrF excimer laser with a wavelength of 248 nm as source and the laser pulse repetition rate of 16 Hz. To investigate the ambient gas effect on the composition and the growth rate of Ni–Ti thin films, they performed the PLD both in 5×10^{-6} Torr vacuum and in 200 mTorr Ar gas atmosphere. They varied the distance between the target and the substrate in the range of 25–50 mm. They also varied the substrate temperatures ranging from room temperature to 600°C while depositing the film. In the case of PLD under Ar gas atmosphere they observed conical plume with a spherical tip. However, under vacuum, the plume boundary was not observed due to large plume expansion. Their results indicated that the composition of Ni–Ti film deposited under vacuum significantly different from the target composition, whereas the film and target compositions are nearly equal when the film deposited under Ar gas atmosphere.

Even though both sputtering and PLD are the more common methods to produce Ni–Ti thin films, most researchers use the sputtering method. The main reason is that controlling thickness and composition uniformity over a large area is extremely challenging in PLD processes. In addition, sputtering is a preferred deposition method in the MEMS community. In most Ni–Ti microsystem applications, compatibility with the batch-processing technology of silicon micromachining is important and as of today, the sputtering method is the primary method to produce Ni–Ti thin films.

8.2.1 Crystallization

Most Ni–Ti films produced are amorphous and must be crystallized in order to show the SMB since the crystal structure represents a critical physical parameter in the SME of Ni–Ti, i.e., thermoelastic martensitic transformation between austenitic phase and martensitic phase. The austenitic phase in Ni–Ti has two interpenetrating simple cubic lattices displaced by half a cubic unit cell along three directions, B2, with lattice parameter $a=3.015$ Å and the martensitic phase has distorted B19 type with a monoclinic unit cell, B19′, with lattice parameters $a=2.889$ Å, $b=4.120$ Å, $c=4.622$ Å, and $\beta=96.8°$. During the transformation, the B2 matrix transforms into an unstable B19 first and then into a stable B19′ by shearing the B19 along the (100) direction [48]. However, Huang et al. [49] theoretically indicated that B19′ is an unstable structure and it is not responsible for the SME in Ni–Ti film. They suggested that the SME in Ni–Ti results from the storage of the shape information in the internal stresses arising from the microstructure, apparently at twin and grain boundaries. In addition, they also believed that the internal stresses are also responsible for the observed stable B19′ structure. In the following paragraphs, a review of crystallization process and their impact on the crystal structure is presented.

As deposited Ni–Ti thin films are usually amorphous in nature irrespective of their mode of fabrication, amorphous Ni–Ti thin films will not exhibit any shape memory or pseudoelastic behavior. Hence, the crystallization of the film is necessary to show these behaviors. Crystallization can be achieved at high temperature. However, if the temperature is too high or the heating time is too long, the Ni–Ti film will interact with the substrate to form an interfacial layer, which may change the properties of thin films. As a result, factors such as activation energy and crystallization temperature are very critical for the fabrication of Ni–Ti thin films. In addition, crystal nucleation and growth velocity are also critical quantities required to define the final microstructure of the thin film. Wang and Vlassak [50] used a combined approach adopted by Kalb et al. [51] for crystallizing in a high precision furnace and microscopically tracking individual crystallites to determine the temperature dependence of the crystal nucleation and growth velocities in amorphous Ni–Ti thin films.

The first report on the crystallization of Ni–Ti thin films was by Kim et al. in 1986 [26]. They determined the crystallization temperatures for $Ni_{68}Ti_{32}$, $Ni_{56}Ti_{44}$, and $Ni_{30}Ti_{70}$ thin films are 443°C, 518°C, and 436°C, respectively, using differential scanning calorimeter (DSC) by observing the onset temperature of the exothermic peaks. Several researchers extensively studied the dependence of film thickness [52], thermal history [53,54], residual stresses [55,56], Ar pressure, and rate of heating [57] on crystallization behavior in sputter-deposited Ni–Ti thin films. Chen and Ting [57] showed that the crystallization temperature increases with rate of heating as well as with Ar pressure. They also determined the crystallization activation energy, E_a, using a Kissinger relationship [58]

$$\ln\left[\frac{\alpha}{T_p^2}\right] = C - \frac{E_a}{RT_p} \tag{8.1}$$

where

α is the heating rate
C is a constant
R is the gas constant
T_p is the crystallization temperature

Chen and Ting [57] observed a linear increase in the crystallization activation energy with Ar pressure. Lower pressure favors lower activation energy due to the initial presence of higher compressive stress [59], smaller nuclei, and higher defect concentration. They also observed that at the same heating rate, crystallization temperature increases with pressure, indicating a higher barrier for the crystallization obtained at higher pressure. From their studies, it is clear that a minimum of 430°C is needed to crystallize Ni–Ti thin films. This makes it difficult if not impossible to deposit Ni–Ti thin film on substrates, which are unstable at 430°C unless other measures are taken. Liu and Duh [60] studied the crystallization kinetics in $Ni_{50.54}Ti_{49.46}$ using DSC through continuous heating and isothermal annealing under nitrogen atmosphere. They observed a shift in both crystallization peak temperature, T_p, and onset crystallization temperature, T_0, toward higher temperature side with increasing heating rate. They determined the crystallization activation energy as 411 kJ/mol using Kissinger's equation (Equation 8.1).

Several attempts have been made to reduce the crystallization temperature by adding energy from other sources. In 1992, Gisser et al. [53] showed Ni–Ti films deposited on heated Si substrate (~460°C) crystallized during the deposition. The resulting films were highly oriented with the (110) of B2 face parallel to the (100) face of the Si substrate. These findings led researchers to use a heated substrate to crystallize Ni–Ti films. At higher substrate temperatures, atoms acquire high energies and act as efficient crystal nucleation sites. Substrate temperatures also enhance the diffusion process promoting crystalline film. Isalgue et al. [61] found that as received Ti-rich Ni–Ti film deposited on a substrate at 200°C was crystallized without further annealing. Mohanchandra and Carman [62] observed the crystallization in as-grown Ti-rich Ni–Ti film (Ti – 47 at% Ni) deposited under a hot target condition. Where, the target temperature increases the substrate temperature to just above 200°C at the end of the deposition process. However, these approaches, substrate temperature and heated target, are difficult to adopt in MEMS fabrication techniques where the annealing of local area in micrometer scale is desired. To overcome this problem, He et al. [63] used continuous wave CO_2 laser, 10.6 μm wavelength and 25 W power, to crystallize Ni–Ti thin film locally by penetrating the laser beam through silicon substrate, which produced a localized high temperature region at the Si/Ni–Ti interface thus promoting localized crystallization.

Cho et al. [64] studied crystallization in Ni/Ti multilayer films using XRD. Figure 8.4 shows the XRD profiles of the as-sputtered multilayer thin film and multilayer thin films heated up to various

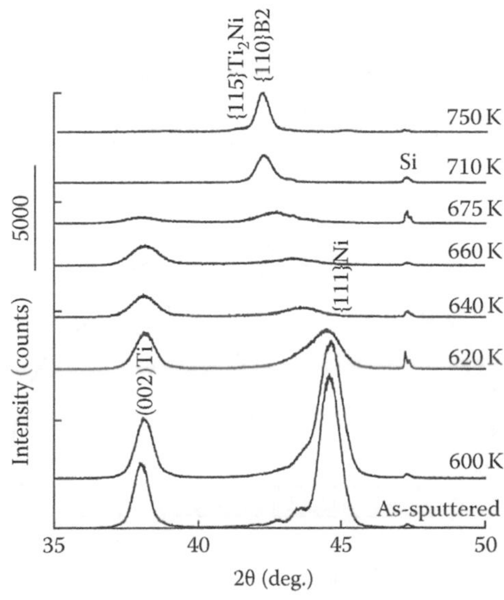

FIGURE 8.4 XRD profiles of the as-sputtered multilayer thin film and multilayer thin films heated up to various temperatures. (From Cho, H. et al., *Mater. Sci. Eng. A*, 699, 2006. With permission.)

temperatures. As-deposited thin films shows two peaks, (002) and (111), correspond to Ti and Ni. Above 347°C, both Ti and Ni peaks become broader and their intensity decreases as temperature increases. As the temperature increased to 367°C the (111) peak shifts toward lower diffraction angle. On the other hand, the (0 0 2) peak intensity decreased gradually by heating up to 402°C with no change in its position. At 402°C, a broad diffraction peak corresponding to the Ti–Ni B2 phase appeared and the peak from Ni disappeared in the multilayer thin film. The intensity of the peak from the Ni–Ti B2 increased and the peak from Ti disappeared in the multilayer thin film when heated to 437°C. When the specimen is heated to 477°C, the diffraction peak corresponding to the Ti_2Ni appeared.

To observe SMB in Ni–Ti thin films, irrespective of deposition process, it is necessary to crystallize the as-deposited thin film, which is in amorphous state, by heating. From the above-mentioned literatures, it is concluded that a minimum temperature of 430°C is necessary to crystallize the Ni–Ti film grown under ambient condition. However, the crystallization temperature can be reduced by growing the film on heated substrates and using energy from the other sources (e.g., mechanical) to induce the kinetic transformation.

8.2.2 Film Composition

The Ni–Ti thin film properties depend highly on film composition. In sputtering, the film can be produced either from a single multicomponent target or from individual elemental targets. Various Researchers have shown that sputtering from a multicomponent target will produce a different sputtering flux for each constituent [65–67] and the sputtering yield of each species varies at different polar angles, but the net atomic flux integrated over the hemisphere should be equal to the target stoichiometry. Neshev et al. [68,69] investigated the sputtering profile of Ni–Ti using 4 keV Ar ion beam to sputter a Ni–Ti target. They collected the sputtered flux on an aluminum foil and measured the composition using a Rutherford backscattering spectrometer (RBS). They observed Ti polar angular distribution wider than that of the Ni such that Ni preferentially emitted closer to the surface normal. Chernysh et al. [70] also measured the angular distribution of Ni and Ti sputtered from an equiatomic Ni–Ti target using an RBS and showed that Ni atoms are preferentially ejected perpendicular to the target surface while the sputtered flux is enriched with Ti for larger polar angle from the target normal. Ho et al. [71] deposited Ni–Ti on several $1 \times 1\,cm^2$ Si (100) wafers placed on a semicircular arc, as shown in Figure 8.5, around the Ni–Ti target to study the angular distribution of sputtered species. RBS was used to determine the compositional profile of the sputtering trajectory at 18° intervals. Composition was found to vary from a lower Ti concentration at 0° to higher Ti concentration at larger polar angles. This indicated that, in relative to Ni flux, the titanium flux is greater at larger polar angles, thus producing relatively more Ni at the center of a wafer ($\theta = 0°$)

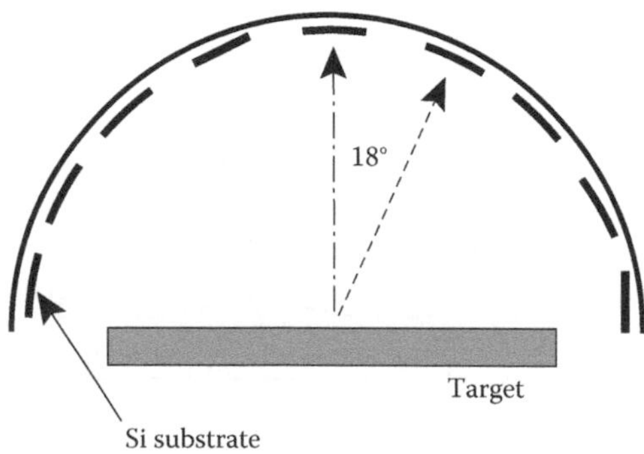

FIGURE 8.5 Schematic diagram of the substrate arrangement during sputtering to capture the sputtering profile. (From Ho, K.K. et al., *Thin Solid Films*, 413, 1, 2002. With permission.)

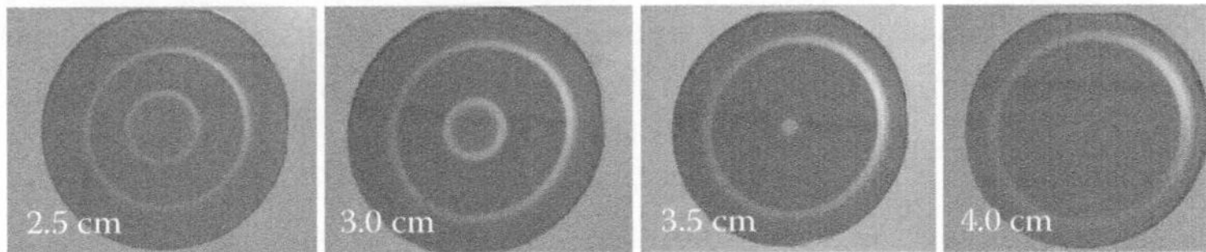

FIGURE 8.6 Fogy rings due to compositional variations, i.e., martensitic phase, when Ni–Ti film sputtered on Si wafers.

and less at the edges of a wafer ($\theta > 20°$). In general, all three of these studies were in quantitative agreement regarding the sputtering angular profiles of Ni and Ti from a Ni–Ti target.

In a DC magnetron sputtering, the maximum sputtering of Ni and Ti occurs along a circular ring due to the confined electrons on the target producing variable Ni and Ti compositions in the plane of deposited film. To verify this, the authors used a conventional sputter deposition to produce Ni–Ti films on 4 in. Si wafers at different substrates to target distances. They observed two foggy rings, as shown in Figure 8.6. The fogginess is due to the texture of martensitic phase at room temperature. As the distance between the substrate to target increases from 2.5 to 4 cm, the radius of the outer foggy ring increases due to increases in the Ti composition as the substrate to target distance increases to 4.0 cm whereas, the inner foggy ring reduces to a point and disappears between a target distance of 3.5 and 4.0 cm. The foggy ring formations are attributed to difference in the deposited polar angles of Ni and Ti constituents observed by previous researchers and illustrated in Figure 8.6.

Besides RBS, electron probe for microanalyses (EPMA) and energy dispersive spectroscopy (EDS) are two additional techniques used to measure film composition. Several researchers used these techniques to evaluate the process parameters effect on film composition. Bendahan et al. [72,73] investigated the influence of Ar gas pressure on Ni–Ti film composition. They showed that the Ni concentration in the film formed under high Ar pressure was lower than that of the target. They attributed this observation to the easy thermalization of Ni atoms due to their larger scattering cross section to Ar atoms when compared to that of the Ti. This attribution suggests that it is difficult for the thermalized atoms to arrive at the substrate. They also showed that the Ni concentration increases with a decrease in the pressure and that the film composition depends on target-to-substrate distance. Ikeda et al. [74] also reported similar results. Yamazaki et al. [75] developed a simulation where the influence of angular distribution of the sputtered atoms as well as the influences of the scattering by Ar gas are considered. Chen and Ting [57] used EPMA to evaluate the effect of target-to-substrate distance, and Ar pressure on the composition of the sputter-deposited Ni–Ti film. They found that Ti concentration increases with increasing Ar pressure and the rate of increase is higher at lower Ar pressure than at higher pressure. They also attributed this observation to thermalization phenomena, a mass transport phenomenon, where the phenomena is characterized by a thermalization distance $D = n\lambda$, where n is the number of collisions between sputtered atoms and the Ar ions, and λ is the mean free path of the sputtered atoms given by the following equation [76,77]:

$$\lambda^{-1} = \frac{8.34 \times 10^{14} \rho (\sigma_m + \sigma_g)^2}{4} \left[1 + \frac{M_m}{M_g} \right] \tag{8.2}$$

where

ρ is the pressure (Pa)
σ_m is the atomic size (nm) of Ti or Ni
σ_g is the atomic size (nm) of Ar
M_m is the molar mass (kg/mol) of Ti or Ni
M_g is the molar mass (kg/mol) of Ar

Thermalization distance was found to increase with the energy and mass of the sputtered atoms, and decrease with pressure. This helps explain the difference in the composition of the constituent elements when compared to the target. All these studies indicate that Ti content in the sputtered film is much lower than that in the target material. From these studies, it is concluded that the Ti content in the film can be varied by varying Ar pressure and substrate to target distance.

As previously described, the Ni–Ti film composition can be adjusted by sputtering from multiple sources, either simultaneously or by alternating sputtering between targets [78]. A number of researchers used alternate methods to solve the apparent loss of Ti while sputtering during the last 18 years with moderate success [27,39,79–81]. Miyazaki and Ishida [79] placed Ti mesh covering portion of the equiatomic target to compensate for Ti loss during the deposition. Quandt et al. [81] employed a Ti-rich target to obtain equiatomic Ni–Ti thin film. Ho and Carman adopted a heated target approach to increase the Ti content in the sputtered deposited film [39]. Figure 8.7 shows the Ni–Ti film deposited under both conventional, cold and heated target at a substrate to the target distance of 4 cm. The absence of a foggy ring on the film deposited from a heated target shows qualitatively that the film has a more uniform composition of Ni and Ti and also the film composition is similar to the composition of the target. This was confirmed with EDS measurements.

The compositional uniformity can also be improved by translating the substrate parallel with the target surface during the deposition. For example, Mohanchandra et al. [40] translated the substrate at 45° arc with lengths of 80 mm during the deposition process. EDS was used to measure the composition of the film along both the direction of translation motion (L) and the direction perpendicular to it (T). Figure 8.8 shows the plot of Ti composition, in at%, versus distance for

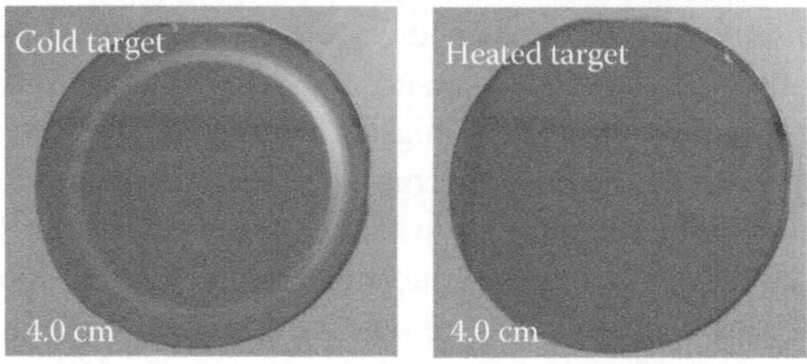

FIGURE 8.7 Comparison of Ni–Ti film formed from cold and heated target.

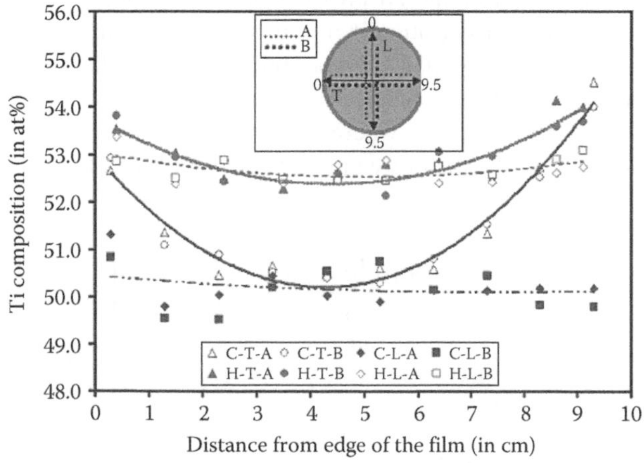

FIGURE 8.8 Ti composition along "L" and "T" directions. (From Mohanchandra, K.P. et al., *Mater. Lett.*, 62, 3481, 2008. With permission.)

the films deposited under both "hot" and "cold" target conditions. The composition along the "L" direction is more uniform due to substrate translation. However, there is a variation in the composition along "T" direction especially in the film deposited under "cold" target condition. The Ti composition at the center of the wafer, 45 mm from the edge, of the film is 49.9 at% for "cold" while it is 52.5 at% for film deposited under "hot" conditions. The film deposited at "hot" target condition shows on average 2.5 at% higher Ti composition compared to the one deposited under "cold" target condition [40].

All the above-mentioned studies show, irrespective of fabrication method, Ni–Ti film composition is different than that of the target. While several attempts have been made to minimize compositional differences, all of these approaches have had varying degree of success. The fundamental problem of compositional control remains a critical issue in fabricating Ni–Ti thin film devices. The approaches currently worth note are supplementing the target, heating the target, and translating the substrate.

8.2.3 Annealing

In many instances Ni–Ti thin films are also annealed at the temperature above the crystallization temperature for long periods of time to promote precipitates for enhancing material properties, may be most importantly pseudoelasticity [79,82]. Commonly produced precipitates in Ni-rich and Ti-rich Ni–Ti thin films are Ni_4Ti_3 and Ti_2Ni, respectively. Instead of employing two-step heat treatments, crystallization and then annealing, Ogawa et al. [83] heated Ti-rich Ni–Ti thin film near the crystallization temperature for 1–4 h to obtain extremely thin plates of coherent Ti-rich precipitates in the B2 matrix [84] and showed excellent SMB. Tadaki et al. [85] were the first to report the very thin plate precipitates formed on the {100} B2 planes in sputter-deposited Ti 48.2 at% Ni thin film annealed at 500°C for 10 min. Ishida et al. [86] reported that under further annealing, plate-like precipitates disappear and facilitate the formation of a new Ti_2Ni equilibrium phase. Due to the existence of these precipitates in the B2 phase, the critical stress for slip in the parent phase is significantly increased and shape memory properties of 6% recoverable transformation strain with a 670 MPa recoverable stress were achievable [87].

Lehnert et al. [88] studied the influence of annealing parameters, temperature, and time on the martensitic phase transformation in sputter-deposited Ti-rich film by DSC. They showed annealing at temperatures in the range from 600°C to 700°C for 30 min produced multiple martensite/austenite transformation via rhombohedral (R) phase. In the temperature range between 750°C and 900°C a direct transformation from martensitic to austenitic phase is observed. Similarly, annealing at 700°C for 10 min up to 3 h shows direct martensitic to austenitic transformation during heating and austenitic to martensitic phase via R-phase during cooling. However, the sharpness of the peak as measured by DSC increased as the annealing time increases.

Characteristics of Ni–Ti thin films are strongly influenced by the presence of Ti_2Ni precipitates in Ti-rich and Ni_4Ti_3 precipitates in Ni-rich matrix. A uniform distribution of the small-sized precipitates in the matrix helps prevent the dislocation of slip-producing films with large recovery force and deformation [89]. The formation of Ni_4Ti_3 precipitates in Ni-rich film introduces a strain field in the surrounding matrix and also influences the matrix composition. Ni_4Ti_3 precipitates near grain boundaries lead to martensitic to austenitic transformation via R-phase [90]. Similarly, nonuniform distribution of Ni and Ti in the sputtered film produces an uneven distribution of the precipitates that leads to embrittlement in the film.

From the above research works, one concludes that annealing is an important process parameter to control the precipitates and fundamental properties of Ni–Ti films. In general, Ti_2Ni precipitates form in Ti-rich films and Ni_4Ti_3 precipitates form in Ni-rich films when annealed at temperature above crystallization temperature for long periods of time. While this is a simple statement, the process and the precipitates that form in generally regarded as a complex system.

8.2.4 Film Thickness

One of the few studies on the influence of film thickness on the sputter-deposited Ni–Ti thin film properties was reported in 2003 by Ishida and Sato [91]. They deposited multilayered Ni–Ti films with thickness varying from 0.5 to 7 μm separated by sputter-deposited Cu films to avoid any effect of deposition conditions on the film properties other than thickness. Each separated film was crystallized by annealing at 500°C for 5 min. They observed a strong influence of thickness on transformation strain and residual strain with a gradual increase of strain as the thickness decreases from 5 to 2 μm and rapid decrease as the thickness decreases from 2 to 0.5 μm. However, the influence of thickness on transformation temperature is negligible. They attributed the influence of thickness on transformation strain and plastic strain to changes in the resistance against deformation, constraints from both neighboring grains [92] and the surface oxide layer. The effect of constraints from neighboring grains increases with increasing thickness whereas, the constraint from surface oxide effect increases with decreasing thickness. The constraints from the neighboring grains saturate when the film thickness is more than the average grain size leading to thickness independence. Ni–Ti thin film deposited using PLD shows the thickness dependence on transformation temperatures [93] at thickness lower than 0.5 μm. When the film thickness is less than 0.1 μm, the temperature hysteresis loop disappears and as the thickness increases phase stabilization phenomenon occurs at 0.5 μm.

8.2.5 Film–Substrate Interface

Single crystal Si wafers are commonly used as substrates to deposit Ni–Ti thin film because of its compatibility with MEMS process. However, several other materials such as glass, mica, and polymer have also been used as substrates. Ni–Ti adheres well to the Si substrate, if the substrate is clean, while the film deposited on glass, copper, or gold substrates can easily be peeled off suggesting poor adhesion or lack of reaction. Hence, to obtain free-standing film glass, gold- or Cu-coated Si wafers are generally used.

For Ni–Ti film deposited directly on Si, the high crystallization temperature leads to inter diffusion and chemical reactions at the interface, which favors the formation of Ni and Ti silicides. Fu et al. [94] showed a total intermixing diffusion layer, Ni–Ti into Si and Si into Ni–Ti, of approximately about 120 nm. Frantz et al. [95] observed six ternary silicides, Ni–$TiSi_2$, Ni_6Ti_5Si, $Ni_{16}Ti_6Si_7$, $Ni_{49}Ti_{14}Si_{37}$, Ni_3Ti_2Si, and Ni–Ti–Si in sputter-deposited Ni–Ti on Si substrates [96]. Hung and Mayer [97] reported that reaction between a Ni–Ti film and a Si substrate takes place at 600°C via the migration of Si into the alloy matrix. When the film is annealed at 700°C for 30 min, 400 nm thin layer of Ni_6Ti_5Si compound and 400 nm thin layer of $Ni_4Ti_4Si_7$ compound along with $TiSi_2$ precipitates are formed on the Si substrate. Inside the Si substrate, only Ni and Si are detected which confirms the presence of a Ni_xSi_x compound in the Si substrate. Several researchers [96,98,99] showed a triangle microstructure in the Si substrate of an annealed Ni–Ti film and heat-treated Ni/Ti bilayer. They identified the microstructure as nickel disilicide along with little Ti solid solution resulting from the diffusion of Ni to silicon during heat treatment. Due to similar lattice parameters of Si (diamond structure $a = 5.43$ Å) and $NiSi_2$ (face-centered cubic, $a = 5.41$ Å) the nickel disilicide grows epitaxially and stress-free on silicon.

To circumvent the reaction between Si and Ni–Ti, many researchers have grown SiO_2 layer on the Si substrate prior to the Ni–Ti deposition. This sacrificial layer provides electrical and thermal insulation to the film and also prevents silicide formation at the film–substrate interface. This leads to relatively poor adhesion (or lack of reaction), compared to bare Si, due to the formation of a fragile and brittle TiO_2 layer [94]. To improve the adhesion of Ni–Ti on SiO_2, Wolf and Heuer [27] deposited 0.1 μm Ni–Ti on SiO_2 at 700°C to promote inter-diffusion of the elements prior to the standard deposition at room temperature. The macroscopic phase transformation behaviors of relatively thick Ni–Ti film are almost independent of substrate used. However, the film deposited on (100) Si single crystal wafer shows a preferred orientation of B19′ along (002) where as the film on amorphous SiO_2 shows the preferred orientation along (111) direction. Fu et al. [94] quantitatively measured the adhesion of Ni–Ti on Si and SiO_2 using scratch test. They observed adhesion failures at 150–180 mN and 60–90 mN for Ni–Ti/Si and Ni–Ti/SiO_2 interface, respectively. Martins et al. [100] studied the effect of a SiO_2 buffer layer on sputter-deposited Ni–Ti thin films. They used both naturally oxidized Si (100) wafer and thermally oxidized Si (100) wafer as substrates. *In situ* XRD measurements reveal that on naturally oxidized Si (100) wafer the B2 phase starts by stacking onto (*h*00) planes (where h = positive integers) and as the thickness increases it changes to (110) preferentially stacking parallel to the substrate surface. Whereas, the film deposited on a thermally oxidized Si (100) wafer shows a strong preferential stacking on (*h*00) planes and as the thickness increases the B2 (200) intensity saturates at a thickness 540 nm.

Li et al. [101] studied the effect of substrate temperature on the surface and interface oxidation of Ni–Ti thin films deposited on Cu substrate at ambient temperature, 300°C and 450°C using x-ray photoelectron spectroscopy (XPS). They showed that the film deposited at room temperature has film composition Ti-51.06 at% Ni whereas, the film deposited at 450°C shows Ti-52.0 at% Ni. More stable rutile TiO_2 layer is formed on the surface of all Ni–Ti films whereas, at the film/substrate interface Ti_2O_3 and NiO layers are formed when the film deposited at higher substrate temperatures and TiO and metallic Ni are observed at the film/substrate interface when the film is deposited at ambient temperature.

From the above works, it is very clear that Ni–Ti film adhere well to the Si substrates due to the formation of different types of thick Ni and Ti silicide layers by reaction and interdiffusion. Whereas, if SiO_2 and Cu are used as substrates, ultra thin natural oxide layer on the substrates react with Ni and/or Ti and form extremely brittle and thin Ni–O and/or Ti–O layers. These layers help to remove the film from the substrate without any difficulty.

8.2.6 Microstructure

The martensitic transformation of Ni–Ti thin films is directly associated to the microstructure that emerges during the crystallization process [88,102]. Better understanding of microstructural development enables better control of the thin film properties. It is very well known that the crystallization behavior in Ni–Ti exhibits a compositional sensitivity. The compositional sensitivity lies in the narrow intermetallic region of the Ni–Ti phase diagram. From the Ni–Ti phase diagram [103], it is clear that the solubility limit of Ni increases with increasing temperature on the Ni-rich side, whereas no change in solubility limit of Ti with the temperature. Hence, it is possible to improve the parent phase of Ni–Ti using metastable Ti_3Ni_4 precipitates in Ni-rich side but it is impossible to improve the parent phase in Ti-rich side using precipitate hardening. However, Ti can be absorbed freely in the amorphous state leading to possible precipitation hardening on Ti-rich side. This is an advantage in Ni–Ti thin films where as-deposited film is typically in amorphous state. In the following paragraphs a brief review on Ni–Ti microstructure is given.

In *MRS Bulletin*, 2002 issue, Ishida and Martynov [104] summarized the work of several researchers on microstructure of Ni-rich Ni–Ti thin films, where conventional heat treatment (i.e., annealing) is used to stabilize the SMB, in the same manner as for bulk materials. In Ni-rich thin films, heat treatment produces the same microstructure as that in bulk materials. In contrast, Ti-rich Ni–Ti thin film crystallized from amorphous state shows a peculiar fine microstructure. When an amorphous Ti—48.2 at% Ni thin film is heat treated at various temperatures, the microstructure in the B2 matrix changes drastically in the sequence of plate precipitates along (100) plane of the matrix, plate precipitates and spherical Ti_2Ni precipitates, and spherical Ti_2Ni precipitates with increasing heat treatment temperatures. The distribution of plate precipitates changes depending on heat treatment temperatures.

It has been demonstrated that Ti-rich Ni–Ti thin films show different microstructures and SMB from those of bulk materials [86]. Two different types of precipitates can be seen in a Ti-48.2 at% Ni thin film after annealing at 500°C for 1 h, plate-like precipitates formed along <100> directions of the Ni–Ti phase and spherical precipitates, known to be Guinier–Preston (GP) zones and Ti_2Ni precipitates. It is important to know the microstructure that forms during the phase transformation in order to successfully predict the SMB. It is well known that phase transformation and mechanical properties can be significantly effected by various defects or precipitates, such as dislocations in thermal cycling [105] and Ti_3Ni_4 precipitates in aging-treated Ni-rich alloys [106]. When amorphous thin film is crystallized, the microstructure will change in the sequence of GP zones, GP zones and Ti_2Ni precipitates within the grain, Ti_2Ni precipitates within the grain, and Ti_2Ni precipitates along the grain boundaries. In the case of bulk Ni–Ti only the last structure, Ti_2Ni precipitates along the grain boundaries, has been reported [107]. This difference is considered to originate from the thermal history. In bulk Ni–Ti, Ti_2Ni precipitates already form along the grain boundaries when the specimens are cooled from the liquid state.

It is also very well known that Ni-50 at% Ti thin film has a typical <001> type II twining substructure in the martensite phase. However, if the volume fraction of GP zones or Ti_2Ni precipitates exceeds a certain limit, the substructure of martensite turns into (001) compound twinning from the <011> type II twinning. Zhang et al. [90] studied the microstructure of stress-induced martensite in Ti-rich Ti–Ni thin film. They deformed the film at temperatures higher than M_s and lower than A_s. When stress-induced martensite appears, it does not retransform into austenite. Therefore, pseudoelasticity does not occur in this temperature range $M_s < T < A_s$. They observed two types of martensite with different substructures in the grain. Type one has the characteristic of a lamellar structure, <011> type II twin, and the other is relatively tiny fine plates, (001) compound twin. The martensite with (001) twin shows morphology similar to that of thermally formed martensite whereas, martensite with <011> type II twin has a preferentially oriented structure. However, microstructure does not show the typical triangular self-accommodation morphology commonly found in the thermally formed martensite. The martensite variants nucleate under stress, a feature typical of stress-induced martensite. The deformation of the parent phase initiates <011> type II twinning in preference to (001) twinning. From the study of preferential orientation, oriented domains of <011> type II twinning suggest that it is the main twinning mode of the stress-induced martensite. Applied stress changes the twinning mode from (001) compound twinning into <011> type II twinning. Zhang et al. [108] also studied the martensitic deformation mechanism by observing microstructural martensitic morphology in different stages of tensile deformation. The main challenge in developing shape memory thin films is the control of the microstructure. In Ni–Ti bulk alloys, the (001), (100), and (201) deformation twins are introduced after the detwinning of initial <011> type II twins. In Ti-rich Ni–Ti thin films, thermally produced martensite has (001) compound twins [109,110]. In a thin film without GP zones or with less GP zones the martensite substructure is predominantly <011> type II twinning. When there are sufficient GP zones (001) compound twinning of martensite is observed in thin films. The microstructural evolution involves the reorientation of (001) transformation twin and the formation of deformation twins (100), (201), and (113). At the beginning of tensile deformation, the strain is attributed to elastic deformation of the martensite. As the tensile stress increases, (001) twinning domain boundaries along the stress direction begin to move and the domains of the other two orientation shrink. In the well-oriented (001) twinning region, (201) and (100) twins form and grow one by one. As the tensile stress increases further, (001) twin disappears by detwinning and a new twin (113) forms in addition to (201) and (100) twins. Whether the thin films with initial (001) compound twin or <011> Type II twins, the deformation mechanism are the same. The (001) and (100) twins in martensite can reverse transform into the parent phase with complete shape recovery. However, the deformation-induced (201) and (113) twins are lost during heating.

The referenced articles on Ni–Ti thin film microstructure show the microstructure is strongly dependent on composition and fabrication process. In Ni-rich Ni–Ti thin film, they form a microstructure similar to the one observed in bulk Ni–Ti alloys, whereas Ti-rich Ni–Ti thin film shows substantially different types of precipitates, i.e., spherical Ti_2Ni precipitates and plate-like zones. Specifically, the Ti_2Ni precipitates in Ti-rich film produced a relatively smaller transformation strain, whereas Ti-rich films with GP zones show a larger transformation strain. In addition, for Ti-rich films the grain size decreases with increasing Ti content. For example, 1 µm for Ti-45.2 at% Ni, 3 µm for 47.0 at% Ni, and 5 µm for 47.9 at% Ni [111].

8.2.7 Residual Stress

Sputter-deposited SMA thin films onto the substrates have inherent mechanical stress known as residual stress. The residual stress in the film has a strong influence on the transformation temperatures and other properties of the film. It is an important factor to be considered when the films are used in MEMS actuators to prevent failure while also allowing large deformation. The residual stress in the Ni–Ti film is a combination of intrinsic stress, thermal stress, and transformation stress.

The intrinsic stress is formed during the deposition process and it depends on the lattice mismatch between the substrate and the film, working gas pressure, substrate to target distance, and impurity in the film. Intrinsic stress is also influenced by both gas pressure and the substrate to target distances. Windischmann [112] explained the dependence of intrinsic stress in polycrystalline or amorphous films on the sputtering gas pressure as follows. At low pressures, due to very few collisions in the vapor phase, high energy sputtered atoms reach the substrate leading to "surface peening" where they are pushed into the interior of the film producing density increase and creating compressive stresses in the film. As the gas pressure increases, the momentum of the sputtered atoms reaching the substrate decreases due to an increase in the number of collisions in the vapor phase leading to a decrease in compressive stress. As the gas pressure continues to increase, the film density decreases and the films eventually develop intrinsic tensile stresses. Further decreases in the bombarding energy by further increase in gas pressure, the arriving atoms have such a low surface mobility that the film grows in a columnar fashion with areas of very low density between the columns that leads to decrease in tensile stress. Hence, sputter-deposited films formed at low gas pressure or low substrate to target distance, or the combination of both have intrinsic compressive stresses. The film becomes tensile as the working gas pressure increases. At some critical pressure, the tensile stress will be maximum and decrease with further increase in pressure. It is important to note that these comments are directed only to the intrinsic contribution. In general, the intrinsic stress is more significant compared to thermal and transformation stress because it determines the state of stress, tensile or compression, in the film. In addition, the intrinsic stress also depends on film thickness. Grummon and Zhang [113] showed that intrinsic stress of a Ni–Ti thin film decreases with increasing film thickness. Later, in 2004, Liu and Huang [114] determined that a stress gradient also exists thorough the thickness of Ni–Ti film deposited on Si wafer.

The thermal stress contribution originates from the difference between the coefficients of thermal expansion of the substrate and the film. The built-up thermal stress in the film will be relaxed during the martensitic transformation. Grummon et al. [115] attributed the observed increase of 1.5 MPa per degree, on cooling from the deposition temperature of the Ni–Ti film on Si to room temperature, to the difference in thermal expansion coefficients. Several other researchers investigated the effect of residual stress on the properties of the sputter-deposited Ni–Ti thin films. Shih et al. [34] have shown that the film deposited on Si wafer at Ar pressure around 3.5 mTorr have the highest tensile stress and the film exhibits extensive delamination at the Ni–Ti/Si interface. The observed wavy fracture morphology on the Si wafer indicates the propagation of interface cracks through the Si substrate near the Ni–Ti/Si interface. This is an indication of a large Ni–Ti/Si interface toughness compared to the bulk toughness of the single crystal silicon substrate [116]. Shih et al. [34] also showed that the substrate temperature has a strong influence on the residual stresses in the film. The film deposited at a substrate temperature of 230°C has a lower intrinsic stress without cracking when compared to the film deposited on a substrate at ambient temperature. The increase surface mobility at elevated substrate temperatures allowed the incoming atoms to form an amorphous low energy structure, which is somewhat crack resistant. Thermal expansion mismatch between the film and the Si substrate causes the intrinsic stress to decrease on heating until around

300°C, where the stress begins to increase. The increase in stress indicates the onset of the crystallization of the amorphous Ni–Ti film and the maximum stress, the yield stress in the B2 phase of Ni–Ti 240 MPa [117] is achieved at 420°C.

Residual stress relaxation and recovery also takes place during phase transformation, i.e., transformation stress. Fu and Du [118] showed that in Ti-rich Ni–Ti film, the transformation stress increases significantly due to phase transformation from martensite to austenite while heating, and while cooling the transformation stress drops significantly due to the phase transformation from austenite to martensite. These large changes in the stress during cooling are due to formation and alignment of twins, shear-variant boundary motion, and stress induction of low-symmetry phases [119,120]. In both cases, the recovery stress is quite high and is about 500 MPa. They also showed similar behavior for transformation stress in Ni-rich Ni–Ti film. However, the intrinsic stress in the film obtained by extending the stress values to the deposition temperature of 450°C for the Ti-rich film is 55 MPa and that of Ni-rich film is 150 MPa. They also attributed the high residual stress in the Ni-rich film due to R-phase transformation.

The residual stresses/strains in Ni–Ti thin film are typically measured with wafer curvature method. The measurement represents contributions from all three mechanisms, i.e., intrinsic, thermal, and transformational stress. The wafer curvature methods measures substrate curvature as a function of temperature, which is used to determine the phase transformation temperature, recoverable stress, residual stress in shape memory alloy thin films [22,113–115,119,121]. The residual stress in the film is calculated using Stoney's equation [113]

$$\sigma = \frac{1}{6}\frac{Eh^2}{(1-\vartheta)\Delta R t} \tag{8.3}$$

where

E and ϑ are Young's modulus and Poisson's ratio of the substrate
h is the substrate thickness
t is the film thickness
ΔR is the change in the radius curvature before and after deposition

8.3 PROPERTIES

Ni–Ti thin films have fascinated many researchers in industrial and medical fields due to its intrinsic properties such as SME, pseudoelasticity, large stress output, large recoverable strain, excellent damping, and biocompatibility. We review both SME and pseudoelastic properties in the following section.

8.3.1 Mechanical Properties

The mechanical properties of Ni–Ti thin film have been widely investigated in the last two decades. It is well known that the mechanical response of Ni–Ti is highly temperature dependent and it is related to the distinct micro-mechanisms involved in the deformation processes of either the high temperature B2 phase or the martensitic B19′ phase. In the full martensitic phase ($T<M_f$), uniaxial loading gives rise to an elastic deformation with pseudo yielding produced by twin boundary motion. The strain produced by pseudo yielding, up to 8%, can be fully recovered by heating the specimen above A_f (i.e., SME). In the parent phase and in the temperature range between A_f and M_d (M_d is the temperature above which pseudo yielding vanishes) an increasing stress typically induces an elastic deformation until induced martensitic phase transformation occurs at nearly constant stress (i.e., pseudoelasticity). This produces relatively large deformations (~8%) that are fully recoverable and hysteretic. Inside the hysteresis cycle, both processes can be present at the same

time and it may be hard to distinguish the deformation mechanisms from macroscopic stress–strain behavior. The temperatures at which these phenomena occur depend on the Ni and Ti contents and film fabrication methods. There are several novel methods to characterize the mechanical properties of thin films. Tensile test, nanoindentation, and bulging test are the most common mechanical characteristic methods used for Ni–Ti thin films with varying degrees of correlation between the methodologies.

8.3.1.1 Tensile Test

Several researchers studied the mechanical properties of both Ni-rich and Ti-rich Ni–Ti thin films using conventional tensile testing methods [16,53,122–124]. Among them, Busch et al. [16] were the first to report mechanical properties of sputter-deposited Ni–Ti thin films in 1990. A detailed study of stress–strain behavior of both Ni-rich and Ti-rich film using tensile testing was reported by Ishida et al. [125] in 2000. They showed trends similar to bulk Ni–Ti [126], i.e., like bulk Ni–Ti thin film also has four stages in the stress–strain curves. When the tensile test is carried out below M_f temperature, the first linear stage corresponds to the elastic deformation of martensitic phase followed by a large increase of strain greater than 5%, which corresponds to the rearrangement of martensite variants. The third stage is again a linear region, which corresponds to the elastic deformation of reoriented martensite and the final stage corresponds to the dislocation-induced plastic deformation of the martensitic phase. Whereas when the test is carried out above A_f temperature, the first stage is a linear region and it corresponds to elastic deformation of the parent phase. This is followed by a large strain change due to stress-induced martensitic transformation. The third stage represented by the elastic deformation of martensitic phase and in the final stage dislocation induced plastic deformation of the martensite occurs until its fracture. Ni-50 at% Ti thin film elongates more than 40% before fracture, whereas, a Ti-48.3 at% Ni thin film annealed at low temperature (500°C) shows an elongation of almost 20% before fracture. However, it should be pointed out that while large deformations are possible only about 10% is recoverable and the remaining 10%–40% is permanent. Similarly, Ti-51.5 at% Ni thin film containing a large amount of Ti_3Ni_4 precipitates shows a tensile fracture stress as high as 1.6 GPa and is comparable to those in bulk materials. It is known that a bulk material with the same composition does not show any plastic deformation. This improvement is due to the presence of coarse grain boundary precipitates, which causes premature failure in the bulk material. In general, the yield stress for plastic deformation in Ni–Ti thin films is higher than that of bulk materials due to small grain size.

Ishida et al. [124] studied the temperature dependence of stress–strain curves of both Ni-rich and Ti-rich Ni–Ti thin film. Since, at higher temperature the parent phase becomes stable the critical stress for the martensitic transformation increases is explained using the Clausius–Clapeyron equation.

$$\frac{d\sigma}{dM_s} = \frac{-\Delta H}{T\varepsilon_0} \tag{8.4}$$

where

σ is the applied stress
M_s is the martensite start temperature
ΔH is the transformation latent heat
T is the deformation temperature
ε_0 is the transformational strain resolved along the direction of the applied stress

In both cases, Ishida et al. [124] observed an increase in critical stress to induce martensite phase and decrease in maximum elongation as the deformation temperature moved away from M_s temperature. In addition, they also showed decrease in yield stress as deformation temperature increases.

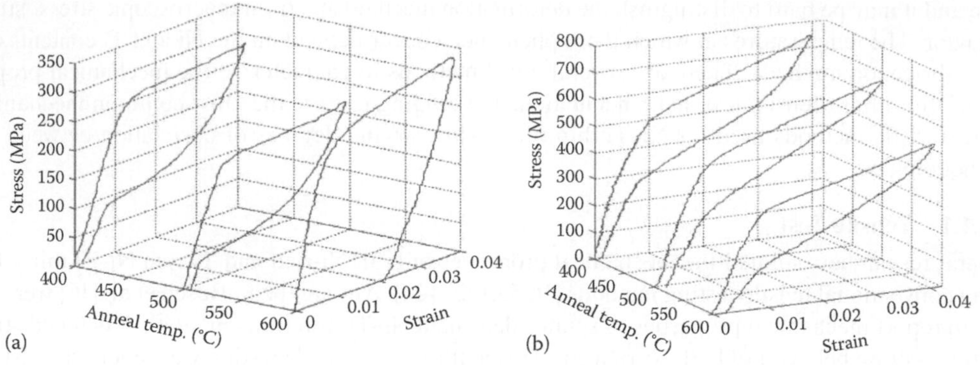

FIGURE 8.9 Stress strain plot for (a) 47.5 at% Ni (b) 50.5 at% Ni. (From Woolman, J. et al., *SPIE Proc.*, 5053, 9, 2003.)

Hou and Grummon [127] also observed similar behavior in Ti-50.8% Ni thin film deposited on a substrate at 430°C. In 2003, Woolman et al. [128] reported the influence of annealing temperature on the critical stress necessary to induce the martensite phase transformation in sputter-deposited Ni–Ti films with two different compositions—47.5 at% Ni and 50.5 at% Ni. Both the films were crystallized and annealed for 2 h at three different temperatures of 400°C, 500°C, and 600°C prior to the isothermal tensile tests just above A_f temperature for evaluating the mechanical response. Results obtained are shown in Figure 8.9a and b. For the 47.5 at% Ni film, increasing the annealing temperature from 400°C to 500°C the critical stress for the stress-induced martensite decreased by 55 MPa, while the film annealed at 600°C failed to demonstrate complete pseudoelastic behavior. For the 50.5 at% Ni film, increasing the annealing temperature from 400°C to 600°C, the critical stress for the induced martensite is decreased by 138 MPa. These results indicate the annealing temperature and time are critical for pseudoelasticity in Ni–Ti thin films.

8.3.1.2 Nano-Indentation

Nano-indentation offers another measurement approach to study thin film properties where, the continuous variation of indentation load as a function of penetration depth of indenter is measured. The test commonly uses a three-sided pyramidal diamond Berkovich-type indenter tip forced into the material being tested. As the indenter is driven into the material, both elastic and plastic deformation occur. The indentation depth (h_c) allows the calculation of the contact area between the indenter and the tested material. This method is used to evaluate the mechanical properties, particularly to calculate the hardness and Young's modulus of the tested material [129,130] from the following equations:

$$H = {}^{F}\!/_{A} \quad \text{with} \quad A = 24.5\,h_c^2 + 170\,h_c + 94.89 h_c^{1/2} + 0.89 h_c^{1/4} \tag{8.5}$$

$$E_r = \frac{\sqrt{\pi}}{2\sqrt{A}}\frac{dF}{dh} \quad \text{and} \quad \frac{1}{E_r} = \frac{1-\vartheta^2}{E} + \frac{1-\vartheta_i^2}{E_i} \tag{8.6}$$

where

A is the projected contact area
F is the applied load, indenter
E_i is the modulus of the indenter
E is the modulus of the material being studied
E_r is the modulus of the composite indenter and the film
ϑ_i and ϑ are the Poisson ratio for the indenter and the material under test, respectively

Since deformation under unloading is assumed to be purely elastic, the unloading path of the indenter follows Kick's law [129]:

$$F = a(h - h_f)^m \tag{8.7}$$

where

α and m are constants

h and h_f are the instantaneous and final displacement of the indenter

Moyne et al. [131] showed that for a penetration depth of 500–1000 nm, the E value is similar to that obtained in bulging tests. The nano-indentation generates high stresses near the indenter tip and hence enhances the film phase transformation. Kumar et al. [132] determined Young's modulus of both amorphous and crystallized sputter-deposited 0.5 µm thick Ni–Ti thin films as 138 and 150 GPa, respectively. These values are much higher than bulk Ni–Ti alloy, i.e., 85 GPa for austenitic phase. Similarly, the hardness value for as-received and crystallized films are $H = 7.11$ and 8.42 GPa, respectively. Both elastic modulus and hardness values are much higher compared to the bulk values. They attributed this difference to the thickness effect along with the composition change and the phase segregation of Ni_3Ti precipitates in the thin film. This may also be attributed to a stiff interface layer or the underlying substrate because similar reports for uniaxial tensile tests suggest much lower stiffness values (e.g., 30–50 GPa). They also showed that the elastic modulus of the film decreases with indentation depth. The mean value of film hardness obtained is 3.1 GPa, which is comparable to those reported by Moyne et al. [131] and Liu et al. [133]. Tall et al. [134] also determined the hardness and Young's modulus of Ni–Ti thin films sputter deposited on silicon substrates using a nanoindenter equipped with a Berkovich tip. They measured Young's modulus of 91 GPa for the film, which is predominantly austenitic while Young's modulus of 63.2 GPa obtained for the film with martensitic Ni–Ti after stress induced deformation. Young's modulus of 73.4 GPa obtained for the film, which is intermediate between austenitic and martensitic Ni–Ti films. The experimental value for nanoindentation tests are always higher than values reported for thin film Ni–Ti using uniaxial test.

8.3.1.3 Bulging Test

Bulging test is usually used for measuring residual stress and Young's modulus of thin film material by applying pressure on a film membrane [129,135–138]. The general load-deflection relationship $p = f(h)$ in the center of the membrane is given by

$$p = \frac{8E}{3(1-\vartheta)}\frac{e^3}{a^4}\left(\frac{2}{(1+\vartheta)} + C\frac{h^2}{e^2}\right) + \frac{4\sigma_0 eh}{a^2} \tag{8.8}$$

where

$C = Ca(\beta_0 - \beta_1\vartheta)$

C is the coefficient obtained from the analytical solution for the elasticity of plate

β_0, β_1 are two correction factors adjusted by the numerical solutions when the ratio $h/e \rightarrow \infty$

P is the applied pressure

E and ϑ are Young's modulus and the Poisson ratio

e is the thickness of the membrane

h is the deflection at the center

a is the radius of the circular membrane

σ_0 is the residual stress

In Equation 8.8, the first term corresponds to a small deflection, the second term corresponds to large deflection, and the third term corresponds to the influence of residual stresses. Yuan et al. [139] adopted this test to characterize superelastic Ni–Ti film by measuring film deflection due to pressure using an optical microscope. They measured the stress to induce martensitic transformation, which is much higher than the one obtained from uniaxial tensile test. They attributed the difference to the presence of biaxial loading in the bulge test. One advantage of the bulge test is that the sample experiences uniform force and does not neck as occurs in tensile tests. Small and Nix [135] report that Young's modulus of the film from the pressure-deflection plot is $E = 17.8$ GPa (on loading) and $E = 10.8$ GPa (on unloading) which is in general agreement with some uniaxial loading tests on thin film Ni–Ti.

8.3.2 Oxidation Behavior of Ni–Ti Thin Films

Ni–Ti thin films are often annealed and aged at elevated temperature [140,141] to enhance specific mechanical characteristics. These heat treatments at high temperatures in atmosphere containing oxygen will result in the oxidation of Ni–Ti thin films. The high-temperature oxidation behavior of equiatomic bulk Ni–Ti alloys has been investigated by several researchers [142–144]. However, work on the oxidation behavior of Ni–Ti thin films are rare. Since the structure of Ni–Ti thin films differs from that of bulk Ni–Ti alloys, they exhibit different high temperature oxidation behavior. Sputter-deposited films generally consist of many micro-defects, which provides pathways for the diffusion of oxygen. In 2007, Zhang et al. [145] reported the oxidation behavior of sputter-deposited Ni–Ti thin film at 550°C, 600°C, and 650°C. They measured the composition-depth profiles of the as-deposited and oxidized Ni–Ti thin films specimens using Auger electron spectroscopy (AES) at a base pressure below 1×10^{-7} Torr. They used a static primary electron beam of 3 keV, 1 μA with ion incidence angle approximately 50° with respect to the normal of the specimen surface. They observed that the weight gain per unit surface area of Ni–Ti thin films increases with the exposure time according to a near-parabolic law. They determined the oxidation rate constant, K_p, using the following equation:

$$(\Delta W/A)^2 = K_p t \tag{8.9}$$

where

$(\Delta W/A)$ is the weight gain per unit area of a specimen
t is the oxidation time

Equation 8.9 was originally used to analyze the isothermal oxidation kinetics of bulk Ni–Ti materials [143,144]. Based on this equation, K_p value can be obtained from the slope of a linear regression line on $(\Delta W/A)^2$ versus t plot. Zhang et al. [145] observed a much higher oxidation rate for Ni–Ti thin film than that of bulk Ni–Ti materials. They attributed this to the structural difference between Ni–Ti thin film and bulk Ni–Ti materials. That is, bulk Ni–Ti materials are relatively compact and dense, showing higher oxidation resistance. Therefore, at the same oxidation temperatures, Ni–Ti thin films will oxidize more easily and their weight gain per unit surface area $(\Delta W/A)$ during the exposure time is larger than that of bulk Ni–Ti materials. Consequently, the oxidation rate of Ni–Ti thin film is much higher than that of bulk Ni–Ti materials. The oxidized Ni–Ti thin films are composed of TiO_2, $TiNi_3$, and parent B2 phase. A Ni oxide does not form in spite of the fact that Ni–Ti thin film contains a large amount of Ni. This may reflect the difference in the oxygen affinity between Ti and Ni, i.e., Ti is easily oxidized while Ni is more difficult to oxidize. Zhang et al. [145] observed only TiO_2 in the oxidized Ni–Ti thin films. This observation indicates that a high-temperature oxygen-rich environment converts all the transient Ti oxides, TiO, and Ti_2O_3, to the more stable rutile oxide, TiO_2. As the result of the TiO_2 formation, the Ti content of the matrix is expected to decrease and $TiNi_3$ phase is formed due to the depletion of Ti. They also noticed the growth of TiO_2 and $TiNi_3$ phases in a few preferred orientations. As isothermal oxidation temperature increases, the intensity of TiO_2 peaks gradually increases, while that of B2 (110) peak intensity decreases. Hence, in oxidized Ni–Ti thin films, B2 content decreases and TiO_2 content increases with the increasing oxidation temperature.

The depth profile determined by Zhang et al. [145] using AES reveal the diffusion of O and the distribution of Ti and Ni in different layers of Ni–Ti thin films during oxidation. They used relative elemental sensitivity factors $S_O=0.500$, $S_{Ti}=0.321$, and $S_{Ni}=0.270$ for O, Ti, and Ni, respectively, to quantify the Auger peak-to-peak heights of O (503 eV), Ti (387 eV), and Ni (848 eV). The outermost layer of oxidized Ni–Ti thin film consists of about 55–57 at% O, 38–40 at% Ti, and 5 at% Ni. At inner layers, O content gradually decreases to zero and Ni content increases to 50 at%. Ti content decreases initially and then increases to 50 at%. Due to the selective oxidation of Ti, the outermost layer of oxidized Ni–Ti thin films shows a high content of TiO_2 and very small percentage of $TiNi_3$ phase or B2 phase or mixture of these two. The distribution of TiO_2 in the outermost layer leads to a local enrichment of Ni at the subsurface. Furthermore, the inward diffusion of oxygen atoms weakens with the increasing depth. Therefore, below the outermost layer, TiO_2 gradually decreases and $TiNi_3$ gradually increases forming a Ni-rich sub-layer. When the contents of Ti and Ni are stable and kept equiatomic, the Ni-rich layer terminates, and TiO_2 and $TiNi_3$ phases disappear. As oxidation temperature increases from 550°C to 600°C, the thickness of the outmost layer and the Ni-rich layer increases. For Ni–Ti thin films oxidized at 650°C, thickness of the outmost layer significantly increases. Therefore, as the isothermal oxidation temperature increase, the diffuse depth of O in Ni–Ti thin films increases with increasing TiO_2 content. While this provides a general overview of the oxidation process, more work is required to fully understand all the mechanisms present for different compositions and different oxidation temperatures.

8.3.3 Biocompatibility

In addition to shape memory and superelasticity, the biocompatibility of Ni–Ti alloy is another important property especially for biomedical applications [11,12,117,146–148]. Certain compositions of Ni–Ti alloy exhibit a temperature range where superelasticity includes body temperature. Hence, it has become "material of choice" for designers of medical devices like stents. The stent uses the property of thermally induced elastic recovery to change a small compact Ni–Ti device to a larger one using body heat. The stent is introduced into the body in its deformed shape through an artery using a catheter. When deployed at the desired position, the stent expands to the appropriate diameter. The technology of self-expansion stents is used in millions of procedures each year and these numbers are steadily increasing. This philosophical approach is also used in a number of other body's conduits, including the esophagus, the trachea, the biliary system, and the urinary system. In all these applications only bulk Ni–Ti alloys are currently used and there does not exist any commercial devices that employ thin film Ni–Ti. While commercial thin film devices do not exist several studies on biocompatibility properties of thin film Ni–Ti have been performed by researchers [146,148,149]. Furthermore, research on medical devices based on thin film Ni–Ti alloy is increasing with researchers looking into the development of a prosthetic heart valve [150] and covered stents [151] using sputter-deposited Ni–Ti thin films followed by a lithographic process.

8.4 Ni–Ti BASE SMA THIN FILMS

Even though Ni–Ti alloy is a suitable candidate for many actuators and sensor applications, the range of transformation temperatures and the transformation hysteresis limits their use. In order to increase transformation temperatures and reduce hysteresis, research has focused on Ni–Ti-based ternary alloys. The addition of third elements, such as Pd, Pt, Au, Ag, and Hf are effective for increasing transformation temperatures, whereas, the addition of Cu is effective in decreasing transformation hysteresis. For sputter-deposited thin films, Pd and Hf are the most common third elements to increase the transformation temperatures [152–158]. The partial substitution of Ni by Pd [140,154], or Ti by Hf [157,158], or Ti by Pt [156,159] results in the increase of transformation temperatures up to or value greater than 500°C. However, Hf increases the transformation hysteresis [158], while Pd decreases the transformation hysteresis [153]. The addition of Cu in Ni–Ti leads to a significant reduction in the hysteresis width without significant changes in transformation temperatures compared to Ni–Ti [157].

8.4.1 Ni–Ti-Cu Thin Films

Among Ni–Ti-based SMAs, Ni–Ti–Cu system is the most widely studied ternary alloy because of its narrow temperature hysteresis. Many investigations have evaluated Ni–Ti–Cu thin film system ranging from fabrication to applications [22,152,160]. By substituting Cu for 5–10 at% Ni, the transformation hysteresis is narrowed from around 30°C to around 10°C without changing significantly the transformation temperatures. Hashinaga et al. [161] studied the effect of Cu substitution on the transformation temperatures of Ni–Ti–Cu thin films. They showed the transformation temperatures are the function of Cu content. The transformation peak temperatures M* and O*, B2-martensitic or orthogonal-martensitic, and B2-orthogonal, decreases slightly with increasing Cu content until 9.5 at% Cu. The 9.5 at% Cu alloy shows a two-stage transformation, the first transformation temperature O* is at 41°C and the second one M* is at −3°C. By adding additional Cu, M* decreases drastically, while O* increases slightly. A stronger Cu-dependence of the hysteresis is observed in a single-stage transformation region than in a two-stage transformation region. Hashinaga et al. [161] also showed a decrease in hysteresis from 27°C to 11°C with increasing Cu content from 0 to 9.5 at% and this property is comparable to bulk specimen.

Chen and Wu [162] studied the crystallization behavior of rf sputtered Ni–Ti–Cu thin films using a DSC technique. Matsunaga et al. [163] studied the effect of low temperature heat treatment on sputter-deposited Ni–Ti–Cu thin films. They determined the crystallization temperature, 475°C, for their sputtered Ni–Ti–Cu film by DSC and heat treated the films at both 475°C and 425°C for 1 h. They determined the average grain size of the heat treated Ni–Ti–Cu film is about 3 μm and each grain contains thin bct nano crystals with tetragonality, c and a ratio between 1.01 and 1.11, of size 1–2.5 nm. They also showed shape recovery stress of the film, 870 MPa, is about 10 times more than that of the bulk Ni–Ti-Cu. Lo et al. [117] reported that thin film Ni–Ti–Cu have low hardness and high damping capacity.

Ishida et al. [164] investigated the SMB of (Ni, Cu)-rich Ni–Ti–Cu thin films ($Ti_{48.9}Ni_{44.9}Cu_{6.2}$, $Ti_{48.5}Ni_{40}Cu_{11.5}$, $Ti_{48.6}Ni_{35.9}Cu_{15.52}$, $Ti_{48.3}Ni_{28.4}Cu_{23.2}$, $Ti_{48.3}Ni_{23.9}Cu_{27.8}$, and $Ti_{48.3}Ni_{23.9}Cu_{27.8}$). The films with 6.2, 11.5–15.5, and 23.3–33.5 at% Cu showed a single-stage deformation due to a B2 <-> B19′ transformation, a two-stage deformation due to the B2 <-> B19 <-> B19′ transformation and a single-stage deformation due to the B2 <-> B19 transformation, respectively. The M_s increases with increasing Cu content and then leveled off for more than 15 at% Cu, indicating a high M_s temperature of 72°C. Temperature hysteresis is approximately 15 K for all films with more than 10 at% Cu. The critical stress for slip increased with increasing Cu content and increased significantly for the $Ti_{48.5}Ni_{18}Cu_{33.5}$ film, whereas the maximum recoverable strain significantly decreased for the $Ti_{48.5}Ni_{18}Cu_{33.5}$ film. With decreasing annealing temperature, the critical stress for slip increases, but the M_s temperature decreases. It was found that films with 11.5 at% Cu or greater annealed at 600°C, and also showed a high martensitic transformation temperature and a high critical stress for slip.

8.4.2 Ni–Ti–Pd Thin Films

The substitution of Ni with Pd increases the transformation temperature of the film up to 550°C. Several researchers characterized the Ni–Ti–Pd thin film system [81,140,165,166]. Quandt et al.

[81] produced Ni–Ti–Pd thin films and showed that partial substitution of Pd for Ni leads to an increase of the lattice constant of the austenite phase B2, whereas, martensitic phase B19 is only slightly influenced by the Ni/Pd ratio. Schlossmacher [166] investigated the effect of crystallization temperature on the microstructure of the Ti–Ni–Pd thin film sputtered from a $Ti_{54}Ni_{36.8}Pd_{9.2}$ target by TEM and reported the existence of precipitates with Ti_2Ni-type structure and Ti_2Pd-type structure. He showed that low temperature crystallization leads to a high-density plate-like precipitates that causes severe embrittlement of the film, whereas, high temperature crystallization leads to films exhibiting the expected SMB. Miyazaki and Ishida [140] showed that the sputter-deposited Ni–T–Pd thin films have transformation temperature hysteresis down to 10°C as Pd content increases beyond 10 at%. Sawaguchi et al. [165] studied the SMEs in $Ti_{51.2}(Pd_{27.0}Ni_{21.8})$ and $Ti_{49.5}(Pd_{28.5}Ni_{22.0})$ films and showed that Ti-rich film has plate-like Ti_2Pd-type precipitates with tetragonal structure.

Baldwin et al. [167] studied the processing of Ni–Ti–Pd thin films using ion beam-assisted deposition. They showed that the substitution of Pd for Ni can increase the transformation temperature of Ni–Ti up to 527°C [168]. To compensate for the loss of Ti during sputtering they placed pure Ti sheet on the surface of the target. The following equations are used to determine the increase in Ti percentage:

$$\text{Ti}_{\text{at\%}} = \left(\frac{A_{\text{Ti}}}{A_{\text{Total}}}\right)\left(\frac{S_{\text{Ti}}}{S_{\text{Total}}}\right) \tag{8.10}$$

$$S_{\text{Total}} = S_{\text{Ti}}(\text{at\%Ti}) + S_{\text{Ni}}(\text{at\%Ni}) + S_{\text{Pd}}(\text{at\%Pd}) \tag{8.11}$$

where

Ti at% is the desired increase in atomic percentage of Ti in the film

A_{Ti} and A_{Total} are Ti surface area needed on the target and total target surface area, respectively

S_{Ti}, S_{Ni}, S_{Pd}, and S_{Total} are the sputter yield of Ti, Ni, Pd, and the alloy target, respectively

The annealed films exhibit delamination and microcracks due to decreased adhesion strength between the film and the substrate because of titanium oxide formation [169]. Several experimental methods are used to determine the martensitic transformation temperature of the Ni–Ti–Pd films. Mathews et al. [170] measured the damping, the modulus, and the bending in sputter-deposited Ni–Ti–Pd thin films using the clamped-free reed vibration method [55] to determine the martensitic transformation temperature in addition to XRD.

8.4.3 Ni–Ti–Pt Thin Films

High temperature SMA is also possible by adding platinum (Pt) to the Ni–Ti-based alloys. The gradual substitution of Pt for nickel (i.e., 10 at% Pt) increases the A_s temperature up to 1000°C for the bulk material. This is substantially higher than the 532°C for Ti–Pd. Fu and Du [159] were the first to report on Ni–Ti–Pt thin films, used co-sputtering from a Ni–Ti target and a Pt target. The composition of their film was $Ti_{50}Ni_{40}Pt_{10}$ and the transformation temperature was around room temperature. Later, Mohanchandra et al. [154] successfully sputter-deposited $Ti_{54}Ni_{26}Pt_{20}$ thin film from a single alloy target on a silicon substrate having a wet thermal oxide layer. The films are crystallized at 550°C and 600°C for an hour. The M_f temperatures are 329°C and 354°C, while the A_f temperatures are 419°C and 452°C for films annealed at 550°C and 600°C, respectively. Film crystallized at 600°C showed large precipitates and also shows an increase in grain size (Figure 8.10). The increase in A_f temperature at higher annealing temperature was attributed to the presence of large precipitates [156].

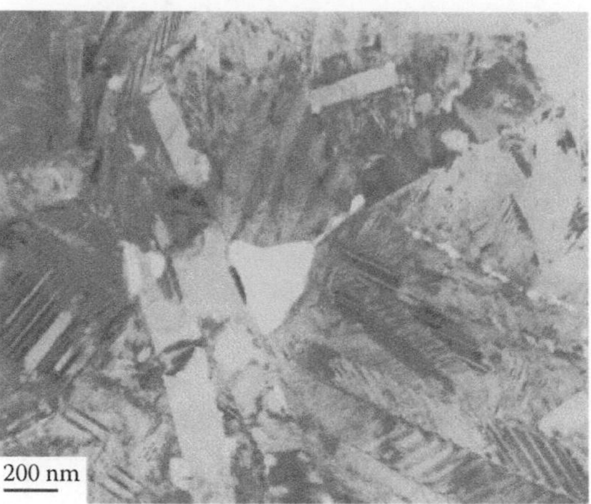

FIGURE 8.10 TEM micrograph of the $Ti_{54}Ni_{26}Pt_{20}$ film annealed at 600°C for 1 h showing precipitates at the grain boundaries. (From Mohanchandra, K.P. et al., *Thin Solid Films*, 515, 1938, 2006. With permission.)

8.4.4 Ni–Ti-Hf Thin Films

The substitution of Ti by Hf [171–173] is believed to be more realistic for engineering applications because of the lower costs associated with Hf compared to Pt. Several researchers fabricated Ni–Ti–Hf thin films using sputter deposition [174] and laser ablation from composite targets [175]. For Ni–Ti–Hf thin films, Gu et al. [175] reported that the crystallization temperature of $Ni_{50.04}Ti_{36.63}Hf_{13.33}$ thin film is slightly above 500°C. Zhang et al. [176] showed that transformation temperatures are strongly dependent on the deposition temperature and the annealing temperature. Tong et al. [177] studied the phase transformations in three different compositions, $Ni_{48}Ti_{37.7}Hf_{14.3}$, $Ni_{47.9}Ti_{35.7}Hf_{16.4}$, and $Ni_{45}Ti_{32}Hf_{23}$ of Ni–Ti–Hf films. They observed an increase of both crystallization temperatures and activation energy with increasing Hf content, which they attributed to the effect of atomic size difference between Hf and Ni and the stronger chemical bond among the constituent elements. The crystallized Ni–Ti–Hf thin films show one-stage martensitic transformation upon cooling. The addition of Hf is effective in increasing the martensitic transformation temperature. With increasing the annealing temperature or Hf content, martensitic transformation temperature increases. Wood et al. [178] used small diameter spherical indenter to study the mechanical properties of sputter-deposited Ni–Ti–Hf thin films. They observed superelasticity in the film with up to 20 at% of Hf. Above this level, a little or no deformation occurs.

8.5 APPLICATIONS

SMAs have been recognized as very promising materials as they can provide work output on the order of 10^7 J/m^3, which is substantially larger than can be achieved with other actuators (e.g., PZT is 5×10^4 J/m^3) [179]. Typically, the maximum available stress and elongation limits for Ni–Ti SMA actuators are 600 MPa and 5%, as compared with 40 MPa and 0.1% for piezoelectric actuators. In addition, the slow response, which is regarded as a drawback for SMA actuators of normal size, can be enhanced by increasing surface to volume ratio.

In recent years, with the development of MEMS, Ni–Ti-based thin films have received considerable attention due to their unique properties. Walker et al. [15] used Ni–Ti film in what is considered the first report of a Ni–Ti MEMS component in 1990. Later in the same year, Kuribayashi et al. [180] developed microactuators using a two-way SME of Ni–Ti thin film. As the demands for microactuators increased research on using Ni–Ti films in MEMS component increased. Several microactuators using thin film Ni–Ti have been developed during the last decades. Among them micro-pumps

[181], micro-gripper [182,183], micro-valves, and micro-optics switches [184–186] represent some micro-devices studied. Most of these devices use a Ni–Ti membrane (e.g., diaphragm/microbubble) for actuation [31,55,187,188].

One of the fundamental actuator concepts employing Ni–Ti thin film is a bilayer cantilever beam as described by Fu et al. [28]. They fabricated a SME microbeam made of rf-sputtered Ni–Ti thin film onto a 15 μm thick silicon cantilever. A cosine model proposed by Liang [189] was used to simulate the SMB of Ni–Ti thin film deposited on silicon beam under thermal cycling. During the simulations, the Ni–Ti/Si cantilever beams were heated up to 120°C and then cooled to room temperature to demonstrate the two-way shape memory effect (TWSME). Fundamentally, this concept could be used in a wide variety of actuator concepts including optical switches. Below we describe in more detail along with fabrication processes for a bubble actuator [190], a micro-pump [191], a microwrapper [192], and micro-valves [193] using sputter-deposited Ni–Ti thin film. These specific devices were chosen to illustrate specific fabrication processes as well as span the gap between simple devices to more complicated systems.

8.5.1 Bubble Actuator

In the late 1990s, an intrinsic two-way SME was observed in a functionally graded Ni–Ti thin film. This discovery spurred the concept of engineering a pseudo-monolithic membrane forming a bubble actuator [191] having two stable states, one geometry at high temperature and another geometry at low temperature. When this membrane is heated to 100°C either by Joule heating or with a hot plate, the membrane forms an approximately 3 mm diameter bubble shape, as shown in Figure 8.11. The researchers reported that the bubble shape vanishes when the material returns to room temperature [191]. When heated back to above 100°C, the membrane again regains its bubble shape, i.e., TWSME. The TWSME is attributed to compositional gradation of Ni/Ti through the thickness of the film. It is believed that this intrinsic approach may be a superior when compared to extrinsic bimorph designs relying on dissimilar materials (e.g., layered Si and Ni–Ti). The researchers report that the Ni-rich film adjacent to the Si wafer responds superelastically while the Ti-rich film along the top surface provides the SME in the structure. By combining these two features, a two-way shape memory actuator was developed. The deflection of the membrane and the actuation frequency can be optimized by providing proper cooling and Joule heating to the membrane.

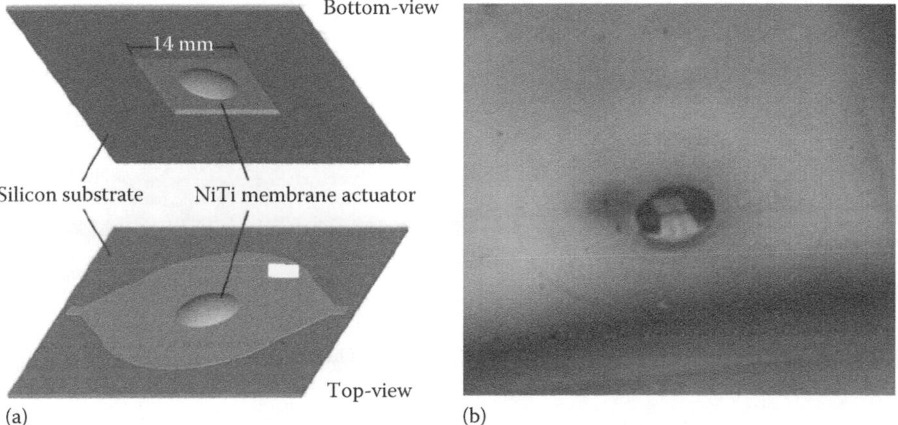

FIGURE 8.11 (a) Schematic diagram of bubble actuator, (b) picture of a bubble. (From Shin, D.D. et al., *Sens. Actuators A Phys.*, 111, 166, 2004. With permission.)

8.5.2 Micro-Pump

A concept similar to the bubble actuator was used to develop a compact fluidic pump. The pump demonstrated that Ni–Ti thin film combined with standard mechanical designs could produce a rather remarkable hydraulic-based actuator. The small hydraulic actuator produces large flow rates at high pressures [190], i.e., large power density. A picture of one pump and its schematic cross sections are given in Figure 8.12a and b, respectively. The pump consists of a reservoir, two conventional passive check valves attached to the inlet and outlet ports, a cubic pump chamber (note several versions were studied), and four Ni–Ti membranes on each cube face. The liquid entering the pump chamber from the pressurized reservoir through the inlet valve is divided among four cavities each containing a Ni–Ti membrane. The pressurized fluid causes spherical deformations in the Ni–Ti membrane. When a current is passed through the membrane, it is heated and the SME returns the deformed shape to the flat configuration and creates a net volume displacement in the pumping chamber. The fluid is subsequently pumped out, the outlet valve and the cycle is repeated. One key issue is the frequency (i.e., thermal cycling time) at which the Ni–Ti membrane can be actuated which is both related to the amount of energy supplied to the Ni–Ti membrane as well as the cooling time for the membrane.

8.5.3 Microwrapper

The previous two examples did not focus on small scale features i.e., features on the order of tens of microns. In this section, we describe a device that contains relatively small-scale features. One example is a MEMS microwrapper constructed from thin film Ni–Ti. The fabrication process for this device is considerably more complex than described in the previous two examples. The Ni–Ti microwrapper was suggested to be used for grabbing micro-size objects for *in situ* analysis in a living organism or be used in the removal of microscopic shards of glass during eye surgery. Micro-grabbing devices using electrostatic and pneumatic actuation mechanisms have been previously reported [194] but typically either produce small forces (electrostatic) or require substantial infrastructure (pneumatic lines). Ni–Ti actuation systems do not suffer from these problems and also represents a material compatible with embedding into the human body (e.g., stents). The particular microwrapper described in this section uses micromachined arms to form a cage structure (i.e., grabbing) in the room temperature state and the micromachined arms return flat (i.e., release) when current (i.e., Joule heating) is applied. Researchers have developed both bimorph MEMS-machined microwrappers consisting of Ni–Ti film with polyimide film as well as pseudo-homogeneous

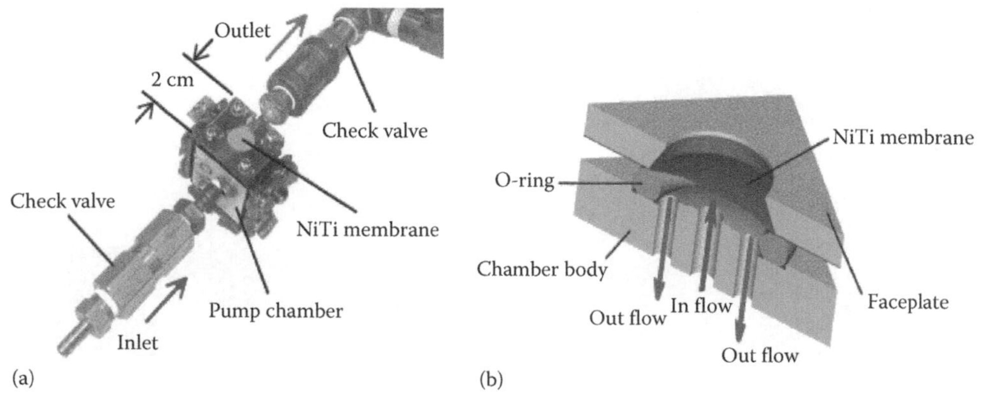

FIGURE 8.12 (a) A picture of a Ni–Ti pump, (b) schematic diagram of the pump's cross section. (From Shin, D.D. et al., *Sens. Actuators A Phys.*, 119, 151, 2005. With permission.)

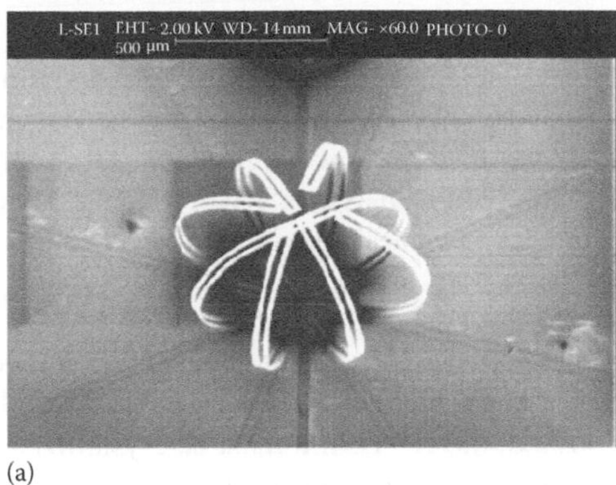

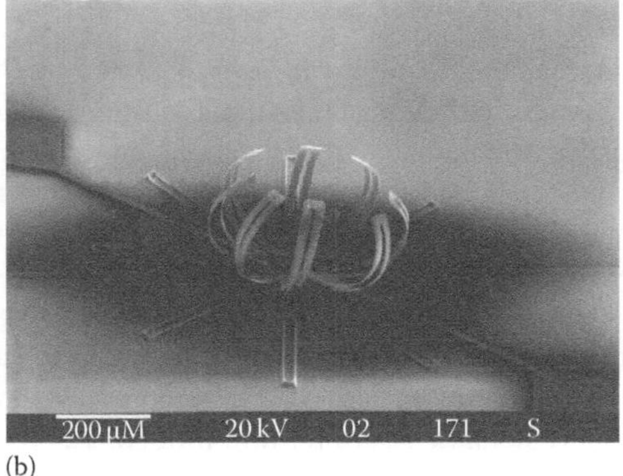

(a) (b)

FIGURE 8.13 Microwrapper using (a) thin film Ni–Ti and polyimide film (b) graded Ni–Ti thin film. (From Gill, J.J. et al., *Sens. Actuators A Phys.*, 93, 148, 2001. With permission.)

microwrappers using functionally graded Ni–Ti film [192]. Figure 8.13 shows two SEM micrographs of the two microwrappers developed.

8.5.4 Micro-Valves

One of the more common structures studied in the MEMS field are micro-valves with a few reports describing Ni–Ti thin film-based micro-valves [195,196]. Micro-valves are potentially useful in micro-fluidics, pumps, thermal switches, and a wide range of other applications. The main purpose of micro-valves is to open and allow fluid to flow or close and prevent fluid to flow. Both shape memory micro-valves (i.e., active check valves) and superelastic micro-valves (i.e., passive check valves) have been fabricated. Below, we provide a description of two such valves [193,197] containing either a tetragonal or a pentagonal flap. The tetragonal flap structure was fabricated using wet etching techniques while the pentagonal flap was constructed using a shadow masking technique, thus eliminating the need to wet etch Ni–Ti. Both valve sets use an array of micro-valves in parallel. Each micro-valve consists of a SMA TiNi flap connected with either four (tetragonal) or five (pentagonal) tethers anchored to the Si substrate, as shown in Figure 8.14. A through channel is made below each flap to allow fluid flow. The flap status determines the closing and opening of the channel. When the flap is flat, the closed channel restricts the fluid flow.

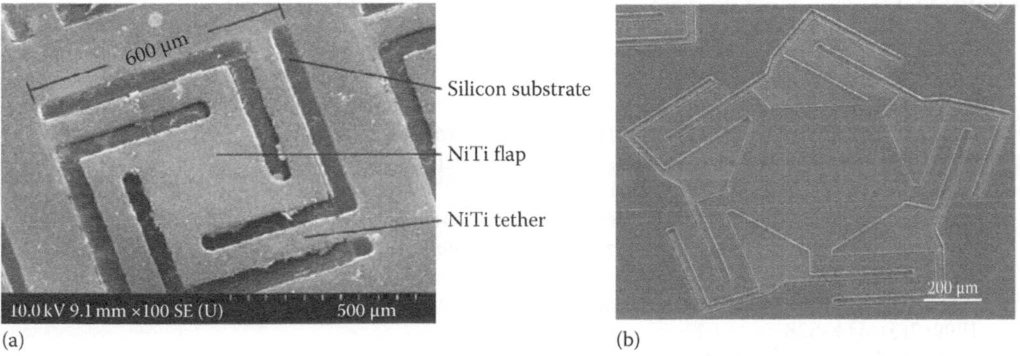

(a) (b)

FIGURE 8.14 Micrographs of (a) tetragonal flap (From Shin, D.D. et al., *Sens. Actuators. A Phys.*, 130–131, 37, 2006. With permission.); (b) pentagonal flap. (From Seong, M., et al., *J. Micromech. Microeng.*, 18, 10, 2008. With permission.)

8.6 SUMMARY

This chapter has provided a brief review of Ni–Ti shape memory alloy thin film fabrication, properties, and their applications in microactuators designs. As one can see, during the last two decades, the number of publications on thin film Ni–Ti has increased at a significant rate. While the research effort on thin film Ni–Ti has increased, the principal issues for producing device grade Ni–Ti thin film are still challenging. A brief summary of issues described in the chapter are highlighted below.

The advantages and issues related to sputtering and pulse laser deposition methods were described. While the majority of work has been conducted using the sputtering process, it is found that composition uniformity in deposited film represents to this day a major challenge. This remains one of the most problematic issues for thin film Ni–Ti. Once deposited, the Ni–Ti film must be crystallized at high temperature. The shape memory and pseudoelastic properties depend strongly on precipitates produced when the Ni–Ti films are annealed at high temperature. The precipitates found in thin films are sometimes different from bulk Ni–Ti alloy. For thin films, the crystallization and/or annealing process also produce undesirable reactions at substrate–film interface, which in turn influence the film properties. While several studies on substrate–film interface are available, more studies must be conducted to fully understand this interface. The oxidation behavior of Ni–Ti thin film is another important factor to be considered, especially in the context of biomedical applications. Although, detailed studies are available for bulk Ni–Ti alloys, few oxidation studies exist on Ni–Ti thin film.

Mechanical properties of the film have also been evaluated. Reported results vary widely depending upon the fabrication and measurement process used. Since most microsystems employ the mechanical properties of the film in a device, additional studies need to be conducted on thin film. In addition, important factors like fatigue behavior and training have not been adequately addressed by the community.

While basic materials properties are important, device fabrication is also a critical issue. During the last decades, a variety of reports on microsystems have been published. The fabrication of microsystems is fairly complex and difficult due to the etching properties of Ni–Ti. Typically, it is observed that the amorphous film etches qualitatively better than the crystalline forms of Ni–Ti. However, the isotropic nature of wet etchants on either polycrystalline or amorphous thin film Ni–Ti produces problems in device manufacturing. While many potential applications exist for thin film Ni–Ti, the biomedical area may represent the most productive area for future studies. This statement is based on applications available for bulk Ni–Ti. While a wide range of Ni–Ti applications exists, in general we find that substantial additional work needs to be conducted on depositing, testing, and fabricating thin film Ni–Ti for the community to use the material in microsystems.

REFERENCES

1. Olander, A. An electrochemical investigation of solid cadmium-gold alloys. *Journal of the American Chemical Society*, 1932, **54**, 3819–3833.
2. Buehler, W.J., Wiley, R.C., and Gilfrich, J.V. Effect of low-temperature phase changes on mechanical properties of alloys near composition TiNi. *Journal of Applied Physics*, 1963, **34**(5), 1475–1477.
3. Wang, F.E., Buehler, W.J., and Pickart, S.J. Crystal structure and a unique martensitic transition of TiNi. *Journal of Applied Physics*, 1965, **36**(10), 3232–3239
4. Tadaki, T., Nakata, Y., Shimizu, K., and Otsuka, K. Crystal-structure, composition and morphology of a precipitate in an aged Ti-51 at percent-Ni shape memory alloy. *Transactions of the Japan Institute of Metals*, 1986, **27**(10), 731–740.
5. Otsuka, K. and Ren, X.B. Recent developments in the research of shape memory alloys. *Intermetallics*, 1999, **7**(5), 511–528.
6. Andrzej, Z.A., Bogdan, R.A., and Miyazaki, S. Stress induced martensitic transformation kinetics of polycrystalline NiTi shape memory alloy. *Materials Science and Engineering A: Structural Materials: Properties, Microstructure and Processing*, 2004, **378**(1–2), 86–91.

7. Luo, Y.C. and Li, D.Y. New wear-resistant material: Nano-TiN/TiC/TiNi composite. *Journal of Materials Science*, 2001, **36**(19), 4695–4702.
8. Qian, L.M., Xiao, X.D., Sun, Q.P., and Yu, T.X. Anomalous relationship between hardness and wear properties of a superelastic nickel-titanium alloy. *Applied Physics Letters*, 2004, **84**(7), 1076–1078.
9. Hodgson, D.E. Damping applications of shape-memory alloys. *Materials Science Forum*, 2002, **394–395**(5), 69.
10. Hornbogen, E. Thermo-mechanical fatigue of shape memory alloys. *Journal of Materials Science*, 2004, **39**(2), 385–399.
11. Ryhanen, J. Biocompatibility of Nitinol. *Minimally Invasive Therapy & Allied Technologies*, 2000, **9**(2), 99–105.
12. Shabalovskaya, S.A. Surface, corrosion and biocompatibility aspects of Nitinol as an implant material. *Bio-Medical Materials and Engineering*, 2002, **12**(1), 69–109.
13. Gall, K. and Maier, H. Cyclic deformation mechanisms in precipitated NiTi shape memory alloys. *Acta Materialia*, 2002, **50**(18), 4643–4657.
14. Filip, P. and Mazanec, K. On precipitation kinetics in TiNi shape memory alloys. *Scripta Materialia*, 2001, **45**(6), 701–707.
15. Walker, J.A., Gabriel, K.J., and Mehregany, M. Thin-film processing of TiNi shape memory alloy. *Sensors and Actuators A: Physical*, 1990, **21**(1–3), 243–246.
16. Busch, J.D., Johnson, A.D., Lee, C.H., and Stevenson, D.A. Shape-memory properties in NI–TI sputter-deposited film. *Journal of Applied Physics*, 1990, **68**(12), 6224–6228.
17. Makino, E., Uenoyama, M., and Shibata, T. Flash evaporation of TiNi shape memory thin film for micro-actuators. *Sensors and Actuators A: Physical*, 1998, **71**(3), 187–192.
18. Sam T. and Davies, K.T. Ion beam sputter deposition of TiNi shape memory alloy thin films. *Proceedings of SPIE*, 1999, **3874**, 8.
19. Noh, H.Y., Lee, K.H., Cui, X.X., and Choi, C.-S. The composition and structure of TiNi thin film formed by electron beam evaporation. *Scripta Materialia*, 2000, **43**(9), 847–852.
20. Martynov, V., Johnson, A.D., and Gupta, V. Effect of magnetic field configuration on the properties of thin films sputter deposited from a dc magnetron TiNi target. *Journal De Physique Iv*, 2003, **112**, 845–848.
21. Akihiro O., Shekhar, B., Isao, K., and Akira U. Development of TiNi shape memory alloy film deposited by sputtering from separate Ti and Ni targets *Proceedings of SPIE* 1998, **3512**, 8.
22. Krulevitch, P., Ramsey, P.B., Makowiecki, D.M., Lee, A.P., Northrup, M.A., and Johnson, G.C. Mixed-sputter deposition of Ni–Ti-Cu shape memory films. *Thin Solid Films*, 1996, **274**(1–2), 101–105.
23. Adams, T.M., Kirkpatrick, S.R., Wang, Z., and Siahmakoun, A. NiTi shape memory alloy thin films deposited by co-evaporation. *Materials Letters*, 2005, **59**(10), 1161–1164.
24. Hassdorf, R., Feydt, J., Thienhaus, S., Boese, M., Buforn, L., Conte, N., and Moske, M. Microstructure, phase sequence, and superelasticity in highly oriented MBE-grown NiTiCu shape memory thin films. *Pricm 5: the Fifth Pacific Rim International Conference on Advanced Materials and Processing*, Pts 1–5, pp. 3827–3830, 2005.
25. Moine P., N.A., Kim, J.J., Marshall, A.F., and Stevenson D.A. *Proceedings of the International Conference on NCM3 Grenoble,* 1985, 533.
26. Kim, J.J., Moine, P., and Stevenson, D.A. Crystallization behavior of amorphous Ni–Ti alloys prepared by sputter deposition. *Scripta Metallurgica*, 1986, **20**(2), 243–248.
27. Wolf, R.H. and Heuer, A.H. TiNi (shape memory) films on silicon for MEMS applications. *Journal of Microelectromechanical Systems*, 1995, **4**(4), 206–212.
28. Fu, Y.Q., Huang, W.M., Du, H.J., Huang, X., Tan, J.P., and Gao, X.Y. Characterization of TiNi shape-memory alloy thin films for MEMS applications. *Surface & Coatings Technology*, 2001, **145**(1–3), 107–112.
29. Rout, S. and Kumar, M.S. Structure, morphology and magnetization of Fe-Pd thin films. *Journal of Nanoscience and Nanotechnology*, 2008, **8**(8), 4238–4242.
30. Zhu, J.J. and Liew, K.M. Description of deformation in shape memory alloys from DO3 austenite to 18R martensite by group theory. *Acta Materialia*, 2003, **51**(9), 2443–2456.
31. Kahn, H., Huff, M.A., and Heuer, A.H. The TiNi shape-memory alloy and its applications for MEMS. *Journal of Micromechanics and Microengineering*, 1998, **8**(3), 213–221.
32. Lehnert, T., Grimmer, H., Boni, P., Horisberger, M., and Gotthardt, R. Characterization of shape-memory alloy thin films made up from sputter-deposited Ni/Ti multilayers. *Acta Materialia*, 2000, **48**(16), 4065–4071.

33. Yaakoubi, N., Serre, C., Martínez, S., Pérez-Rodríguez, A., Morante, J.R., Esteve, J., Montserrat, J., Dufour-Gergam, E., Granchamp, J.P., Bosseboeuf, A., and Frantz-Rodriguez, N. Growth and characterization of shape memory alloy thin films for Si microactuator technologies. *Journal of Materials Science: Materials in Electronics*, 2001, **12**, 323–326.
34. Shih, C.L., Lai, B.K., Kahn, H., Phillips, S.M., and Heuer, A.H. A robust co-sputtering fabrication procedure for TiNi shape memory alloys for MEMS. *Journal of Microelectromechanical Systems*, 2001, **10**(1), 69–79.
35. Ishida, A., Sato, M., Tabata, O., and Yoshikawa, W. Shape memory thin films formed with carrousel-type magnetron sputtering apparatus. *Smart Materials & Structures*, 2005, **14**(5), S216–S222.
36. Li, Y., Cui, L.S., Zheng, Y.J., and Yang, D.Z. DSC study of the reverse martensitic transformation in prestrained TiNi shape memory alloy in different composites. *Materials Letters*, 2001, **51**(1), 73–77.
37. Cho, H., Kim, H.Y., and Miyazaki, S. Fabrication and characterization of Ti–Ni shape memory thin film using Ti/Ni multilayer technique. *Science and Technology of Advanced Materials*, 2005, **6**, 678–683.
38. Takeda, F., Yamazaki, T., and Nakajima, T. Composition control of NiTi shape memory alloy films formed by sputter deposition with a composite target. *Japanese Journal of Applied Physics Part 1-Regular Papers Short Notes & Review Papers*, 2000, **39**(10), 5992–5994.
39. Ho, K.K. and Carman, G.P. Sputter deposition of NiTi thin film shape memory alloy using a heated target. *Thin Solid Films*, 2000, **370**(1–2), 18–29.
40. Mohanchandra, K.P., Ho, K.K., and Carman, G.P. Compositional uniformity in sputter-deposited NiTi shape memory alloy thin films. *Materials Letters*, 2008, **62**(20), 3481–3483.
41. Ikuta, K., Hayashi, M., Matsuura, T., and Fujishiro, H. Shape memory alloy thin film fabricated by laser ablation. *Micro Electro Mechanical Systems, 1994, MEMS '94, Proceedings, IEEE Workshop* 1994, 355–360, Oiso, Japan.
42. Ikuta, K., Hayashi, M., Matsuura, T., and Fujishiro, H. *Proceedings of the IEEE Conference on Micro Electro Mechanical System*, New York, 1994.
43. Ciabattari, F., Fuso, F., and Arimondo, E. Pulsed laser deposition of NiTi shape memory effect thin films. *Applied Physics A: Materials Science & Processing*, 1997, **64**(6), 623–626.
44. Gu, H.D., Leung, K.M., Chung, C.Y., You, L., Han, X.D., Chan, K.S., and Lai, J.K.L. Pulsed laser deposition of NiTi shape memory alloy thin films with optimum parameters. *Thin Solid Films*, 1998, **330**(2), 196–201.
45. Lu, Y.F., Chen, X.Y., Ren, Z.M., Zhu, S., Wang, J.P., and Liew, T.Y.F. Pulsed laser deposition of TiNi thin films at various temperatures. *Japanese Journal of Applied Physics Part 1-Regular Papers Short Notes & Review Papers*, 2001, **40**(9A), 5329–5333.
46. Douglas, B.C. and Graham, K.H. *Pulsed Laser Deposition of Thin Films,* Wiley, New York, 1994.
47. Ahn, J.S., Cha, J.O., Shin, C.H., Yeo, S.J., Im, H.J., Sakai, J., Lee, K.B., Kim, H.M., and Nam, T.H. Effect of ambient Ar gas on the composition control and crystalline properties of TiNi thin films fabricated by using pulsed laser deposition. *Journal of the Korean Physical Society*, 2007, **50**, 1750–1754.
48. Otsuka, K., Sawamura, T., and Shimizu, K. Crystal structure and internal defects of equiatomic TiNi martensite. *Physica Status Solidi A*, 1971, **5**(2), 457–470.
49. Xiangyang, H., Graeme, J.A., and Karin M.R. Crystal structures and shape-memory behaviour of NiTi. *Nature Materials*, 2003, **2**, 307–311.
50. Wang, X. and Vlassak, J.J. Crystallization kinetics of amorphous NiTi shape memory alloy thin films. *Scripta Materialia*, 2006, **54**(5), 925–930.
51. Kalb, J., Spaepen, F., and Wuttig, M. Atomic force microscopy measurements of crystal nucleation and growth rates in thin films of amorphous Te alloys. *Applied Physics Letters*, 2004, **84**, 5240–5242.
52. Moberly, W.J., Busch, J.D., Johnson, A.D., and Berkson, M.H. In M. Chen, M.O. Thomson, R.B. Schwarz, and M. Libera (Eds.), *Phase Transformation Kinetics in Thin Films*, Materials Research Society: Pittsburgh, PA, 1992, 230, pp. 85–90.
53. Gisser, K.R.C., Busch, J.D., Johnson, A.D., and Ellis, A.B. Oriented nickel-titanium shape memory alloy-films prepared by annealing during deposition. *Applied Physics Letters*, 1992, **61**(14), 1632–1634.
54. Fujita, H., Komatsu, M., Sakata, T., and Fujita, N. Nucleation of crystals in amorphous materials. *Materials Transactions, JIM*, 1996, **37**(7), 1350–1355.
55. Benard, W.L., Kahn, H., Heuer, A.H., and Huff, M.A. Thin-film shape-memory alloy actuated micropumps. *Journal of Microelectromechanical Systems*, 1998, **7**(2), 245–251.
56. Madsen, J.S. and Jardine, A.P. Lowering the crystallization temperature of thin-film shape memory effect TiNi by cold-working for smart materials fabrication. *Scripta Metallurgica et Materialia*, 1994, **30**(9), 1189–1194.
57. Chen, P. and Ting, J.M. Characteristics of TiNi alloy thin films. *Thin Solid Films*, 2001, **398**, 597–601.

58. Kissinger, H.E. Reaction kinetics in differential thermal analysis. *Analytical Chemistry*, 1957, **29**(11), 1702–1706.
59. Hoffman, D.W. and Thornton, J.A. Internal stresses in sputtered chromium. *Thin Solid Films*, 1977, **40**, 355–363.
60. Liu, K.T. and Duh, J.G. Kinetics of the crystallization in amorphous NiTi thin films. *Journal of Non-Crystalline Solids*, 2007, **353**(11–12), 1060–1064.
61. Isalgue, A., Torra, V., Seguin, J.L., Bendahan, M., Amigo, J.M., and Esteve-Cano, V. Shape memory NiTi thin films deposited at low temperature. *Materials Science and Engineering A: Structural Materials: Properties, Microstructure and Processing*, 1999, **275**, 717–721.
62. Mohanchandra, K. and Carman, G.P. Grown crystalline Ni–Ti shape memory alloy thin films. *Proceedings of SPIE*, 2003, **5053**, 212–218.
63. He, Q., Hong, M.H., Huang, W.M., Chong, T.C., Fu, Y.Q., and Du, H.J. CO_2 laser annealing of sputtering deposited NiTi shape memory thin films. *Journal of Micromechanics and Microengineering*, 2004, **14**(7), 950–956.
64. Cho, H., Kim, H.Y., and Miyazaki, S. Alloying process of sputter-deposited Ti/Ni multilayer thin films. *Materials Science and Engineering A*, 2006, **438–440**, 699–702.
65. Olson, R.R. and Wehner, G.K. Composition variations as a function of ejection angle in sputtering of alloys. *Journal of Vacuum Science & Technology*, 1977, **14**(1), 319–321.
66. Nagatomi, T., Min, K.-Y., and Shimizu, R. Angular distributions of sputtered particles ejected from pure Cu, Pt and Cu–Pt alloy under 3 keV Ar+ ion bombardment *Japanese Journal of Applied Physics*, 1994, **33**, 6675–6679.
67. Liau, Z.L., Mayer, J.W., Brown, W.L., and Poate, J.M. Sputtering of Ptsi. *Journal of Applied Physics*, 1978, **49**(10), 5295–5305.
68. Neshev, I., Hamishkeev, V., Chernysh, V.S., Postnikov, S., and Mamaev, B. Angular-distributions of sputtered particles from NiTi alloy. *Vacuum*, 1993, **44**(11–12), 1131–1133.
69. Neshev, I., Vichev, R.G., Tzanev, S., and Todorov, S.S. Sputtering of NiTi alloys: A comparison of experiment and simulation. *Vacuum*, 1993, **44**(3–4), 209–212.
70. Chernysh, V.S., Tuboltsev, V.S., and Kulikauskas, V.S. Angular distributions of Ni and Ti atoms sputtered from a NiTi alloy under He+ and Ar+ ion bombardment. *Nuclear Instruments and Methods in Physics Research Section B: Beam Interactions with Materials and Atoms*, 1998, **140**(3–4), 303–310.
71. Ho, K.K., Mohanchandra, K.P., and Carman, G.P. Examination of the sputtering profile of NiTi under target heating conditions. *Thin Solid Films*, 2002, **413**(1–2), 1–7.
72. Bendahan, M., Canet, P., Seguin, J.L., and Carchano, H. Control composition study of sputtered Ni–Ti shape-memory alloy film. *Materials Science and Engineering B: Solid State Materials for Advanced Technology*, 1995, **34**(2–3), 112–115.
73. Bendahan, M., Seguin, J.-L., Canet, P., and Carchano, H. NiTi shape memory alloy thin films: Composition control using optical emission spectroscopy. *Thin Solid Films*, 1996, **283**(1–2), 61–66.
74. Ikeda, T., Ozeki, Y., Sakurai, D., Fujimoto, K., and Nakata, S. *Proceedings of the 3rd Symposium Microjoining and Assembly Technology in Electronics*, Yokohama, Japan, 1997.
75. Yamazaki, T., Yoshizawa, T., Takada, H., and Takeda, F. Dependence of composition distribution of NiTi sputtered films on Ar gas pressure. *Japanese Journal of Applied Physics Part 1-Regular Papers Short Notes & Review Papers*, 2001, **40**(12), 6936–6940.
76. Westwood, W.D. Calculation of deposition rates in diode sputtering systems. *Journal of Vacuum Science & Technology*, 1978, **15**(1), 1–9.
77. Tominaga, K., Iwamura, S., Shintani, Y., and Tada, O. Energy analysis of high-energy neutral atoms in the sputtering of Zno and $BaTiO_3$. *Japanese Journal of Applied Physics Part 1-Regular Papers Short Notes & Review Papers*, 1982, **21**(5), 688–695.
78. Chang, L. and Grummon, D. *Met. Res. Soc. Symp. Proc.*, 1992, **246**.
79. Miyazaki, S. and Ishida, A. Shape-memory characteristics of sputter-deposited Ti-Ni thin-films. *Materials Transactions, Japan Institute of Metals (JIM)*, 1994, **35**(1), 14–19.
80. Kawamura, Y., Gyobu, A., Horikawa, H., and Saburi, T. Martensitic transformations and shape memory effect in Ti-Ni sputter-deposited thin films. *Journal De Physique IV*, 1995, **5**(C8), 683–688.
81. Quandt, E., Halene, C., Holleck, H., Feit, K., Kohl, M., Schlomacher, P., Skokan, A., and Skrobanck, K.D. Sputter deposition of TiNi, TiNiPd and TiPd films displaying the two-way shape-memory effect. *Sensors and Actuators A: Physical*, 1996, **53**(1–3), 434–439.
82. Ishida, A., Sato, M., Takei, A., Nomura, K. and Miyazaki, S. Effect of aging on shape memory behavior of Ti-51.3 at pct Ni thin films. *Metallurgical and Materials Transactions A-Physical Metallurgy and Materials Science*, 1996, **27**(12), 3753–3759.

83. Ogawa, K., Kikuchi, T., Kajiwara, S., Matsunaga, T., and Miyazaki, S. Coherent subnanometric plate precipitates formed during crystallization of As-sputtered Ti-Ni films. *Journal De Physique IV*, 1997, **7**(C5), 221–226.
84. Kikuchi, T., Ogawa, K., Kajiwara, S., Matsunaga, T., Miyazaki, S., and Tomota, Y. High-resolution electron microscopy studies on coherent plate precipitates and nanocrystals formed by low-temperature heat treatments of amorphous Ti-rich Ti-Ni thin films. *Philosophical Magazine*, 1998, **78**(2), 467–489.
85. Tadaki, T., Nakata, Y., and Shimizu, K. Occupancy sites of constituent atoms and their effects on the martensitic transformations in some Cu-based and Ti-Ni-based ternary alloys. *Journal De Physique IV*, 1995, **5**(C8), 81–90.
86. Ishida, A., Sate, M., Takei, A., and Miyazaki, S. Effect of heat treatment on shape memory behavior of Ti-rich Ti-Ni thin films. *Materials Transactions JIM*, 1995, **36**(11), 1349–1355.
87. Kajiwara, S., Kikuchi, T., Ogawa, K., Matsunaga, T., and Miyazaki, S. Strengthening of Ti-Ni shape-memory films by coherent subnanometric plate precipitates. *Philosophical Magazine Letters*, 1996, **74**(3), 137–144.
88. Lehnert, T., Crevoiserat, S., and Gotthardt, R. Transformation properties and microstructure of sputter-deposited Ni–Ti shape memory alloy thin films. *Journal of Materials Science*, 2002, **37**(8), 1523–1533.
89. Vestel, M.J. and Grummon, D.S. Precipitates and lamellar microstructures in NiTi films. *Materials Science and Engineering A: Structural Materials: Properties, Microstructure and Processing*, 2004, **378**(1–2), 437–442.
90. Zhang, J.X., Sato, M., and Ishida, A. The effect of two types of precipitates upon the structure of martensite in sputter-deposited Ti-Ni thin films. *Smart Materials and Structures*, 2004, **13**(4), N37–N42.
91. Ishida, A. and Sato, M. Thickness effect on shape memory behavior of Ti-50.0 at.%Ni thin film. *Acta Materialia*, 2003, **51**(18), 5571–5578.
92. Miyazaki, S., No, V.H., Kitamura, K., Khantachawana, A., and Hosoda, H. Texture of Ti-Ni rolled thin plates and sputter-deposited thin films. *International Journal of Plasticity*, 2000, **16**(10–11), 1135–1154.
93. Camposeo, A., Puccini, N., Fuso, F., Allegrini, M., Arimondo, E., and Tuissi, A. Laser deposition of shape-memory alloy for MEMS applications. *Applied Surface Science*, 2003, **208**, 518–521.
94. Fu, Y.Q., Du, H.J., and Zhang, S. Adhesion and interfacial structure of magnetron sputtered TiNi films on Si/SiO2 substrate. *Thin Solid Films*, 2003, **444**(1–2), 85–90.
95. Frantz, N., Dufour-Gergam, E., Grandchamp, J.P., Bosseboeuf, A., Seiler, W., Nouet, G., and Catillon, G. Shape memory thin films with transition above room temperature from Ni-rich NiTi films. *Sensors and Actuators A: Physical*, 2002, **99**(1–2), 59–63.
96. Stemmer, S., Duscher, G., Scheu, C., Heuer, A.H., and Ruhle, M. The reaction between a TiNi shape memory thin film and silicon. *Journal of Materials Research*, 1997, **12**(7), 1734–1740.
97. Hung, L.S. and Mayer, J.W. Interactions of four metallic compounds with Si substrates. *Journal of Applied Physics*, 1986, **60**(3), 1002–1008.
98. Wu, S.K., Chen, J.Z., Wu, Y.J., Wang, J.Y., Yu, M.N., Chen, F.R., and Kai, J.J. Interfacial microstructures of rf-sputtered TiNi shape memory alloy thin films on (100) silicon. *Philosophical Magazine A: Physics of Condensed Matter Structure Defects and Mechanical Properties*, 2001, **81**(8), 1939–1949.
99. Falke, U., Fenske, F., Schulze, S., and Hietschold M. XTEM studies of nickel silicide growth on Si(100) using a Ni/Ti bilayer system. *Physica Status Solidi A*, 1997, **162**, 615–621.
100. Martins, R.M.S., Schell, N., Beckers, M., Mahesh, K.K., Silva, R.J.C., and Fernandes, F.M.B. Growth of sputter-deposited Ni–Ti thin films: Effect of a SiO_2 buffer layer. *Applied Physics A-Materials Science & Processing*, 2006, **84**(3), 285–289.
101. Li, Y.H., Li, L.M., Meng, F.L., Zheng, W.T., Zhao, J., and Wang, Y.M. Effect of substrate temperature on the surface and interface oxidation of NiTi thin films. *Journal of Electron Spectroscopy and Related Phenomena*, 2006, **151**(2), 144–148.
102. Takabayashi, S., Tanino, K., and Kitagawa, K. Heat treatment effect on transformation properties of TiNi shape memory alloy film. *Materials Science Research International*, 1997, **3**(4), 220–224.
103. Massalski, T.B., Okamoto, H., Subramanian, P.R., and Kacprzak, L. *Binary Alloy Phase Diagrams*, 2nd edn., Vol. 3, Ohio ASM International, Materials Park, OH, 1990.
104. Ishida, A. and Martynov, V. Sputter-deposited shape-memory alloy thin films: Properties and applications. *MRS Bulletin*, 2002, **27**(2), 111–114.
105. Krishnan, M. and Singh, J.B. A novel B19' martensite in nickel titanium shape memory alloys. *Acta Materialia*, 2000, **48**(6), 1325–1344.
106. Nishida, M., Wayman, C.M., and Chiba, A. Electron-microscopy studies of the martensitic-transformation in an aged Ti-51 at-percent-Ni shape memory alloy. *Metallography*, 1988, **21**(3), 275–291.

107. Lin, H.C., Wu, S.-K., and Lin, J.C. *Proceedings of the ICOMAT'92*. Perkins, J., ed., Monterey Institute of Advanced Studies, Monterey, CA, 1992, pp. 875–880.
108. Zhang, J.X., Sato, M., and Ishida, A. Deformation mechanism of martensite in Ti-rich Ti-Ni shape memory alloy thin films. *Acta Materialia*, 2006, **54**(4), 1185–1198.
109. Zhang, J.X., Sato, M., and Ishida, A. Structure of martensite in sputter-deposited Ti-Ni thin films containing Guinier-Preston zones. *Acta Materialia*, 2001, **49**(15), 3001–3010.
110. Zhang, J.X., Sato, M., and Ishida, A. Structure of martensite in sputter-deposited Ti-Ni thin films containing homogeneously distributed Ti2Ni precipitates. *Philosophical Magazine A: Physics of Condensed Matter Structure Defects and Mechanical Properties*, 2002, **82**(7), 1433–1449.
111. Ishida, A., Sawaguchi, T., and Sato, M. Shape memory behavior of Ti-rich Ti-Ni thin films formed by sputtering. *Journal de Physique IV France*, 2001, **11**, 6.
112. Windischmann, H. Intrinsic stress in sputter-deposited thin-films. *Critical Reviews in Solid State and Materials Sciences*, 1992, **17**(6), 547–596.
113. Grummon, D.S. and Zhang, J. Stress in sputtered films of near-equiatomic TiNiX on (100) Si: Intrinsic and extrinsic stresses and their modification by thermally activated mechanisms. *Physica Status Solidi A*, **186**, 17–39, 2001.
114. Liu, Y. and Huang, X. Substrate-induced stress and transformation characteristics of a deposited Ti-Ni-Cu thin film. *Philosophical Magazine*, 2004, **84**(19), 1919–1936.
115. Grummon, D.S., Zhang, J.P., and Pence, T.J. Relaxation and recovery of extrinsic stress in sputtered titanium-nickel thin films on (100)-Si. *Materials Science and Engineering A: Structural Materials: Properties, Microstructure and Processing*, 1999, **275**, 722–726.
116. Suo, Z. and Hutchinson, J.W. Steady-state cracking in brittle substrates beneath adherent films. *International Journal of Solids and Structures*, 1989, **25**(11), 1337–1353.
117. Lo, Y.C., Wu, S.K., and Horng, H.E. A study of B2↔B19↔B19' 2-stage martensitic transformation in a Ti50Ni40Cu10 alloy. *Acta Metallurgica et Materialia*, 1993, **41**(3), 747–759.
118. Fu, Y.Q. and Du, H.J. Effects of film composition and annealing on residual stress evolution for shape memory TiNi film. *Materials Science and Engineering A: Structural Materials: Properties, Microstructure and Processing*, 2003, **342**(1–2), 236–244.
119. Grummon, D.S. Fabrication, microstructure and stress effects in sputtered TiNi thin films. *Materials Science Forum,* 327–328, pp. 295–302, 2000.
120. Mathews, S.A., Wuttig, M., and Su, Q.M. The effect of substrate constraint on the martensitic transformation of Ni–Ti thin films. *Metallurgical and Materials Transactions A: Physical Metallurgy and Materials Science*, 1996, **27**(9), 2859–2861.
121. Nix, W.D. Mechanical-properties of thin-films. *Metallurgical Transactions A-Physical Metallurgy and Materials Science*, 1989, **20**(11), 2217–2245.
122. Bouslykhane, K., Moine, P., Villain, J.P., and Grilhe, J. Mechanical-properties and wear-resistance of ion-beam-assisted sputter-deposited NiTi(n) coatings. *Surface & Coatings Technology*, 1991, **49**(1–3), 457–461.
123. Huang, X. and Liu, Y. Substrate-induced stress and the transformation behavior of sputter-deposited NiTi thin films. *Materials Science and Engineering A-Structural Materials Properties Microstructure and Processing*, 2003, **352**(1–2), 314–317.
124. Ishida, A., Sato, M., and Miyazaki, S. Mechanical properties of Ti-Ni shape memory thin films formed by sputtering. *Materials Science and Engineering A-Structural Materials Properties Microstructure and Processing*, 1999, **273**, 754–757.
125. Ishida, A., Sato, M., Kimura, T., and Miyazaki, S. Stress-strain curves of sputter-deposited Ti-Ni thin films. *Philosophical Magazine A-Physics of Condensed Matter Structure Defects and Mechanical Properties*, 2000, **80**(4), 967–980.
126. Miyazaki, S. and Otsuka, K. Fracture of shape memory alloys. *Science Reports of the Research Institutes Tohoku University Series A-Physics Chemistry and Metallurgy*, 1981, **29**, 111–116.
127. Hou, L. and Grummon, D.S. Transformational superelasticity in sputtered titanium-nickel thin-films. *Scripta Metallurgica et Materialia*, 1995, **33**(6), 989–995.
128. Woolman, J., Mohanchandra, K.P., and Carman, G.P. Composition and annealing effects on the mechanical properties of superelastic thin film nickel titanium. *SPIE Proceedings*, 2003, **5053**, 9.
129. Oliver, W.C. and Pharr, G.M. An improved technique for determining hardness and elastic-modulus using load and displacement sensing indentation experiments. *Journal of Materials Research*, 1992, **7**(6), 1564–1583.
130. Sergent, A., Poilane, C., Robert, L., and Delobelle, P. Nanoindentation, film growth and beam flexure; Three complementary tests for characterizing the mechanical properties of very thin materials: Application to electrodeposited nickel films. *Journal De Physique IV*, 1998, **8**(P4), 259–265.

131. Moyne, S., Poilane, C., Kitamura, K., Miyazaki, S., Delobelle, P., and Lexcellent, C. Analysis of the thermomechanical behavior of Ti-Ni shape memory alloy thin films by bulging and nanoindentation procedures. *Materials Science and Engineering A-Structural Materials Properties Microstructure and Processing*, 1999, **275**, 727–732.
132. Kumar, A.K.N., Kannan, M.D., Jayakumar, S., Rajam, K.S., and Raju, V.S. Investigations on the mechanical behaviour of rough surfaces of TiNi thin films by nano indentation studies. *Surface & Coatings Technology*, 2006, **201**(6), 3253–3259.
133. Liu, R., Li, D.Y., Xie, Y.S., Llewellyn, R., and Hawthorne, H.M. Indentation behavior of pseudoelastic TiNi alloy. *Scripta Materialia*, 1999, **41**(7), 691–696.
134. Tall, P.D., Ndiaye, S., Beye, A.C., Zong, Z., Soboyejo, W.O., Lee, H.J., Ramirez, A.G., and Rajan, K. Nanoindentation of Ni–Ti thin films. *Materials and Manufacturing Processes*, 2007, **22**(2), 175–179.
135. Small, M.K. and Nix, W.D. Analysis of the accuracy of the bulge test in determining the mechanical-properties of thin-films. *Journal of Materials Research*, 1992, **7**(6), 1553–1563.
136. Poilane, C., Delobelle, P., Lexcellent, C., Hayashi, S., and Tobushi, H. Analysis of the mechanical behavior of shape memory polymer membranes by nanoindentation, bulging and point membrane deflection tests. *Thin Solid Films*, 2000, **379**(1–2), 156–165.
137. Degen, A., Abedinov, N., Gotszalk, T., Sossna, E., Kratzenberg, M., and Rangelow, I.W. Stress analysis in Si membranes for open stencil masks and mini-reticles using double bulging and resonance methods. *Microelectronic Engineering*, 2001, **57–58**, 425–432.
138. Bonnotte, E., Delobelle, P., Bornier, L., Trolard, B. and Tribillon, G. Two interferometric methods for the mechanical characterization of thin films by bulging tests. Application to silicon single crystal. *Journal of Materials Research*, 1997, **12**(9), 2234–2248.
139. Yuan, Z.Y., Xu, D., Ye, Z.C., and Cai, B.C. Superelasticity of NiTi shape memory alloy thin films. *Journal of Materials Science & Technology*, 2005, **21**(3), 319–323.
140. Miyazaki, S. and Ishida, A. Martensitic transformation and shape memory behavior in sputter-deposited TiNi-base thin films. *Materials Science and Engineering A-Structural Materials Properties Microstructure and Processing*, 1999, **275**, 106–133.
141. Surbled, P., Clerc, C., Le Pioufle, B., Ataka, M., and Fujita, H. Effect of the composition and thermal annealing on the transformation temperatures of sputtered TiNi shape memory alloy thin films. *Thin Solid Films*, 2001, **401**(1–2), 52–59.
142. Chu, C.L., Wu, S.K., and Yen, Y.C. Oxidation behavior of equiatomic TiNi alloy in high temperature air environment. *Materials Science and Engineering A-Structural Materials Properties Microstructure and Processing*, 1996, **216**(1–2), 193–200.
143. Xu, C.H., Ma, X.Q., Shi, S.Q., and Woo, C.H. Oxidation behavior of TiNi shape memory alloy at 450–750°C. *Materials Science and Engineering A-Structural Materials Properties Microstructure and Processing*, 2004, **371**(1–2), 45–50.
144. Ko, J.H. and Lee, D.B. High temperature oxidation behavior of TiNi alloys. *Mat. Sci. Forum*, 2005, 853–856.
145. Zhang, L., Xie, C.Y., and Wu, J.S. Oxidation behavior of sputter-deposited Ti-Ni thin films at elevated temperatures. *Materials Characterization*, 2007, **58**(5), 471–478.
146. Duerig, T.W., Tolomeo, D.E. and Wholey, M. An overview of superelastic stent design. *Minimally Invasive Therapy & Allied Technologies*, 2000, **9**(3–4), 235–246.
147. Pelton, A.R., Stockel, D., and Duerig, T.W. Medical uses of nitinol. *Materials Science Fourm*, 2000, **327–328**, 63–70.
148. Mantovani, D. Shape memory alloys: Properties and biomedical applications. *JOM-Journal of the Minerals Metals & Materials Society*, 2000, **52**(10), 36–44.
149. Stepan, L.L., Levi, D.S., Gans, E., Mohanchandra, K.P., Ujihara, M., and Carman, G.P. Biocorrosion investigation of two shape memory nickel based alloys: Ni-Mn-Ga and thin film NiTi. *Journal of Biomedical Materials Research*, 2007, **82**, 768–776.
150. Stepan, L.L., Daniel, S.L., and Gregory, P.C. A thin film nitinol heart valve. *ASME Journal of Biomechanical Engineering*, 2005, **127**(6), 915–918.
151. Levi, D.S., Williams, R.J., Liu, J., Danon, S., Stepan, L.L., Panduranga, M.K., Fishbein, M.C., and Carman, G.P. Thin film nitinol covered stents: Design and animal testing. *ASAIO Journal*, 2008, **54**(3), 221–226.
152. Miyazaki, S., Hashinaga, T., and Ishida, A. Martensitic transformations in sputter-deposited Ti-Ni-Cu shape memory alloy thin films. *Thin Solid Films*, 1996, **282**(1–2), 364–367.
153. Miyazaki, S., Nomura, K., Ishida, A., and Kajiwara, S. Recent developments in sputter-deposited Ti-Ni-base shape memory alloy thin films. *Journal De Physique Iv*, 1997, **7**(C5), 275–280.

154. Mohanchandra, K.P., Shin, D., and Carman, G.P. Deposition and characterization of Ti-Ni-Pd and Ti-Ni-Pt shape memory alloy thin films. *Smart Materials & Structures*, 2005, **14**(5), S312–S316.
155. Sanjabi, S., Cao, Y.Z., Sadrnezhaad, S.K., and Barber, Z.H. Binary and ternary NiTi-based shape memory films deposited by simultaneous sputter deposition from elemental targets. *Journal of Vacuum Science & Technology A*, 2005, **23**(5), 1425–1429.
156. Mohanchandra, K.P., Shin, D.D., and Carman, G.P. Shape memory behavior of high temperature Ti-Ni-Pt thin films. *Thin Solid Films*, 2006, **515**(4), 1938–1941.
157. Sanjabi, S., Cao, Y.Z., and Barber, Z.H. Multi-target sputter deposition of Ni50Ti50-xHfx shape memory thin films for high temperature microactuator application. *Sensors and Actuators A-Physical*, 2005, **121**(2), 543–548.
158. Johnson, A.D., Martynov, V.V., and Minners, R.S. Sputter deposition of high transition temperature Ti-Ni-Hf alloy thin films. *Journal De Physique Iv*, 1995, **5**(C8), 783–787.
159. Fu, Y.Q. and Du, H.J. Magnetron sputtered Ti50Ni40Pt10 shape memory alloy thin films. *Journal of Materials Science Letters*, 2003, **22**(7), 531–533.
160. Ren, M.H., Wang, L., Xu, D., and Cai, B.C. Sputter-deposited Ti-Ni-Cu shaped memory alloy thin films. *Materials & Design*, 2000, **21**(6), 583–586.
161. Hashinaga, T., Miyazaki, S., Ueki, T., and Horikawa, H. Transformation and deformation behavior in sputter-deposited Ti-Ni-Cu thin films. *Journal De Physique Iv*, 1995, **5**(C8), 689–694.
162. Chen, J.Z. and Wu, S.K. Crystallization temperature and activation energy of rf-sputtered near-equiatomic TiNi and Ti50Ni40Cu10 thin films. *Journal of Non-Crystalline Solids*, 2001, **288**(1–3), 159–165.
163. Matsunaga, T., Kajiwara, S., Ogawa, K., Kikuchi, T., and Miyazaki, S. Internal structures and shape memory properties of sputter-deposited thin films of a Ti-Ni-Cu alloy. *Acta Materialia*, 2001, **49**(11), 1921–1928.
164. Ishida, A., Sato, M., and Ogawa, K. Microstructure of annealed Ti48.5Ni(51.5-x)Cux (x= 6.2–33.5) thin films. *Philosophical Magazine*, 2008, **88**(16), 2427–2438.
165. Sawaguchi, T., Sato, M., and Ishida, A. Microstructure and shape memory behavior of Ti-51.2(Pd27.0Ni21.8) and Ti-49.5(Pd28.5Ni22.0) thin films. *Materials Science and Engineering a-Structural Materials Properties Microstructure and Processing*, 2002, **332**(1–2), 47–55.
166. Schlossmacher, P. Microstructural investigation of a TiNiPd shape memory thin film. *Materials Letters*, 1997, **31**(1–2), 119–125.
167. Baldwin, E., Thomas, B., Lee, J.W., and Rabiei, A. Processing TiPdNi base thin-film shape memory alloys using ion beam assisted deposition. *Surface & Coatings Technology*, 2005, **200**(7), 2571–2579.
168. Wu, J. and Tian, Q. The superelasticity of TiPdNi high temperature shape memory alloy. *Intermetallics*, 2003, **11**(8), 773–778.
169. Fu, Y.Q., Du, H.J., Huang, W.M., Zhang, S., and Hu, M. TiNi-based thin films in MEMS applications: A review. *Sensors and Actuators A-Physical*, 2004, **112**(2–3), 395–408.
170. Mathews, S., Li, J., Su, Q.M., and Wuttig, M. Martensitic transformation in thin-film (TiPd)(50)(TiNi)(50). *Philosophical Magazine Letters*, 1999, **79**(5), 265–272.
171. Angst, D.R., Thoma, P.E., and Kao, M.Y. The effect of hafnium content on the transformation temperatures of Ni49TiS1-xHfx shape memory alloys. *Journal De Physique Iv*, 1995, **5**(C8), 747–752.
172. Meng, X.L., Zheng, Y.F., Wang, Z., and Zhao, L.C. Shape memory properties of the Ti36Ni49Hf15 high temperature shape memory alloy. *Materials Letters*, 2000, **45**(2), 128–132.
173. Olier, P., Brachet, J.C., Bechade, J.L., Foucher, C., and Guenin, G. Investigation of transformation temperatures, microstructure and shape memory properties of NiTi, NiTiZr and NiTiHf alloys. *Journal De Physique Iv*, 1995, **5**(C8), 741–746.
174. Tong, Y.X., Liu, Y., Miao, J.M., and Zhao, L.C. Characterization of a nanocrystalline NiTiHf high temperature shape memory alloy thin film. *Scripta Materialia*, 2005, **52**(10), 983–987.
175. Gu, H.D., Leung, K.M., and Chung, C.Y. Growth of high-temperature NiTi1-xHfx shape memory alloy thin films by laser ablation of composite targets. *Journal of Vacuum Science & Technology A-Vacuum Surfaces and Films*, 1998, **16**(6), 3420–3422.
176. Zhang, C., Zee, R.H., and Thoma, P.E. Surface evolution of NiTi and NiTiHf thin films. *Mater. Res. Soc. Symp.*, 2000, **648**, 6.
177. Tong, Y.X., Liu, Y., and Miao, J.M. Phase transformation in NiTiHf shape memory alloy thin films. *Thin Solid Films*, 2008, **516**(16), 5393–5396.
178. Wood, A.J.M., Sanjabi, S., Fu, Y.Q., Barber, Z.H., and Clyne, T.W. Nanoindentation of binary and ternary Ni–Ti-based shape memory alloy thin films. *Surface & Coatings Technology*, 2008, **202**(13), 3115–3120.

179. Ikuta, K. Micro/miniature shape memory alloy actuator. *Proceedings of the IEEE Conference on Intelligent Robotics and Automation-90*, 1990, **3**, 2156–2161.
180. Kuribayashi, K., Yoshitake, M., and Ogawa, S. Reversible SMA actuator for micron sized robot. *Proceedings of Micro Electro Mechanical Systems, (MEMS-90)*, 1990, 217–221.
181. Vanlintel, H.T.G., Vandepol, F.C.M., and Bouwstra, S. A piezoelectric micropump based on micromachining of silicon. *Sensors and Actuators*, 1988, **15**(2), 153–167.
182. Kohl, M., Just, E., Pfleging, W., and Miyazaki, S. SMA microgripper with integrated antagonism. *Sensors and Actuators A-Physical*, 2000, **83**(1–3), 208–213.
183. Huang, W.M., Tan, J.P., Gao, X.Y., and Yeo, J.H. Design, testing, and simulation of NiTi shape-memory-alloy thin-film-based microgrippers. *Journal of Microlithography Microfabrication and Microsystems*, 2003, **2**(3), 185–190.
184. Sutapun, B., Tabib-Azar, M., and Huff, M.A. Applications of shape memory alloys in optics. *Applied Optics*, 1998, **37**(28), 6811–6815.
185. Tabib-Azar, M., Sutapun, B., and Huff, M. Applications of TiNi thin film shape memory alloys in micro-opto-electro-mechanical systems. *Sensors and Actuators A-Physical*, 1999, **77**(1), 34–38.
186. Kohl, M., Dittmann, D., Quandt, E., and Winzek, B. Thin film shape memory microvalves with adjustable operation temperature. *Sensors and Actuators A-Physical*, 2000, **83**(1–3), 214–219.
187. Kohl, M., Dittmann, D., Quandt, E., Winzek, B., Miyazaki, S., and Allen, D.M. Shape memory microvalves based on thin films or rolled sheets. *Materials Science and Engineering A-Structural Materials Properties Microstructure and Processing*, 1999, **275**, 784–788.
188. Kohl, M., Skrobanek, K.D., and Miyazaki, S. Development of stress-optimised shape memory microvalves. *Sensors and Actuators A-Physical*, 1999, **72**(3), 243–250.
189. Liang, C. and Rogers, C.A. One-dimensional thermomechanical constitutive relations for shape memory materials. *Journal of Intelligent Material Systems and Structures*, 1990, **1**(2), 207–234.
190. Shin, D.D., Mohanchandra, K.R., and Carman, G.P. Development of hydraulic linear actuator using thin film SMA. *Sensors and Actuators A-Physical*, 2005, **119**(1), 151–156.
191. Shin, D.D., Mohanchandra, K.P., and Carman, G.P. High frequency actuation of thin film NiTi. *Sensors and Actuators A-Physical*, 2004, **111**(2–3), 166–171.
192. Gill, J.J., Chang, D.T., Momoda, L.A., and Carman, G.P. Manufacturing issues of thin film NiTi microwrapper. *Sensors and Actuators A-Physical*, 2001, **93**(2), 148–156.
193. Seong, M., Mohanchandra, K.P., Lin, Y., and Carman, G.P. Development of a 'bi-layer lift-off' method for high flow rate and high frequency Nitinol MEMS valve fabrication. *Journal of Micromechanics and Microengineering*, 2008, **18**(7), 10.
194. Ok, J., Chu, M., and Kim, C.-J. Pneumatically driven microcage for micro-objects in biological liquid. *Proceedings of the IEEE Annual Workshop of Microelectromechanical Systems*, Orlando, FL, 1999, pp. 459–463.
195. Hahm, G., Kahn, H., Phillips, S.M., and Heuer, A.H. *Solid State Sensor and Actuator Workshop,* Hilton Head, SC, 2000.
196. Lai, B.K., Hahm, G., You, L., Kahn, H., Phillips, S.M., and Heuer, A.H. The characterization of TiNi shape-memory actuated microvalves. *Materials Research Society*, 2000, **657**, EE8.3.1
197. Shin, D.D., Lee, D.-G., Mohanchandra, K.P., and Carman, G.P. Thin film NiTi microthermostat array. *Sensors and Actuators A: Physical*, 2006, **130–131**, 37–41.

Index

D

E

O

P

Q

R

S

T

U

V

W

X

Z